ISBN 978-1-5276-5046-6
PIBN 10877742

THE

AMERICAN
JOURNAL OF SCIENCE.

EDITOR: EDWARD S. DANA.

ASSOCIATE EDITORS

PROFESSORS GEO. L. GOODALE, JOHN TROWBRIDGE,
W. G. FARLOW AND WM. M. DAVIS, OF CAMBRIDGE,

PROFESSORS A. E. VERRILL, HENRY S. WILLIAMS, AND
L. V. PIRSSON, OF NEW HAVEN,

PROFESSOR GEORGE F. BARKER, OF PHILADELPHIA,
PROFESSOR H. A. ROWLAND, OF BALTIMORE,
MR. J. S. DILLER, OF WASHINGTON.

FOURTH SERIES.

VOL. X—[WHOLE NUMBER, CLX.]

WITH FOUR PLATES.

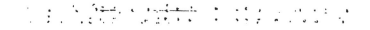

NEW HAVEN, CONNECTICUT.
1900.

253619

CONTENTS TO VOLUME X.

Number 55.

Page

ART. I.—Energy of the Cathode Rays; by W. G. CADY... 1

II.—Volcanic Rocks from Temiscouata Lake, Quebec; by H. E. GREGORY.. 14

III.—Interpretation of Mineral Analyses : a Criticism of recent Articles on the Constitution of Tourmaline; by S. L. PENFIELD... 19

IV.—Studies in the Cyperaceæ, No. XIII ; by T. HOLM.... 33

V.—Titration of Mercury by Sodium Thiosulphate; by J. T. NORTON, Jr. 48

VI.—Selenium Interference Rings ; by A. C. LONGDEN.... 55

VII.—Carboniferous Bowlders from India; by B. K. EMERSON 57

VIII.—New Bivalve from the Connecticut River Trias ; by B. K. EMERSON 58

IX.—Statement of Rock Analyses; by H. S. WASHINGTON. 59

X.—String Alternator; by K. HONDA and S. SHIMIZU 64

XI.—Action of Light on Magnetism; by J. H. HART...... 66

SCIENTIFIC INTELLIGENCE.

Chemistry and Physics—Radio-active Barium, B. v. LENGYEL, 74,—Mercuric Antimonide, and Stibonium Compounds, PARTHEIL and MANNHEIM: Atomic Weight of Iron, RICHARDS and BAXTER: Physical Properties of Cæsium, ECKARDT and GRAEFE, 75.—Compounds of Telluric Acid with Iodates, WEINLAND and PRAUSE: Simple Method of Decomposing Chromite, FIEBER: Theory of Electrolytic Dissociation and some of its Applications, H. C. JONES: Optical Activity and Chemical Composition, H. LANDOLT: Variation of the Electric Intensity and Conductivity along the Electric Discharge in Rarefied Gases, WILSON, 76.—Mechanical possibilities of flight, RAYLEIGH: Magnetic Screening, H. DU BOIS and A. P. WILLS: Sun's Corona, E. PRINGSHEIM, 77.—Some Properties of Light-struck Photographic Plates, F. E. NIPHER, 78.

Geology and Mineralogy—Monograph of Christmas Island (Indian Ocean): Physical Features and Geology, C. W. ANDREWS, 79.—Lower Silurian (Trenton) Fauna of Baffin Land, C. SCHUCHERT: Preliminary Report on the Geology of Louisiana, G. D. HARRIS and A. C. VEATCH, 81.—Geology of Panama, BERTRAND and ZURCHER: Enrichment of Mineral of Veins by later Metallic Sulphides, W. H. WEED, 82.—Mineralogical Notes, 'O C. FABRINGTON, 83.—Tabellen zur Bestimmung der Mineralien mittels äusserer Kennzeichen, A. WEISBACH: Repertorium der Mineralogischen und Krystallographischen Literatur, E. WEINSCHENK and F. GRÜNLING, 84.

Botany and Zoology—Carex de l'Asie orientale, A. FRANCHET, 84.—Systematische Anatomie der Dicotyledonen, H SOLEREDER, 85.—Eléments de Paléobotanique, R. ZEILLER: Lehrbuch der Pflanzenpalaeontologie mit besonderer Rücksicht auf die Bedürfnisse der Geologen. H. POTONIÉ, 88 —Birds of Eastern North America, C. B. CORY: Zoological Results based on material from New Britain, New Guinea, Loyalty Island and elsewhere, collected during 1895–1897, A. WILLEY: Das Tierreich, E. HARTERT, 89.

Miscellaneous Scientific Intelligence—Solar Eclipse, 89.—Lessons in Elementary Physiology, T. H. HUXLEY: Ostwald's Klassiker der Exakten Wissenschaften, 90.

Number 56.

Page

ART. XII.—Rowland's new Method for measuring Electric
Absorption, and Losses of Energy due to Hysteresis and
Foucault Currents, and on the Detection of short Circuits
in Coils; by L. M. POTTS 91

XIII.—New Jurassic Vertebrates; by W. C. KNIGHT 115

XIV.—Carnotite and Associated Vanadiferous Minerals in
Western Colorado; by W. F. HILLEBRAND and F. L.
RANSOME 120

XV.—Restoration of Stylonurus Lacoanus, a Giant Arthropod from the Upper Devonian of the United States; by
C. E. BEECHER. (With Plate I.).................... 145

XVI.—Iodometric Estimation of Arsenic Acid; by F. A.
GOOCH and J. C. MORRIS 151

XVII.—Notes on Preglacial Drainage in Michigan: by E.
H. MUDGE 158

SCIENTIFIC INTELLIGENCE.

Chemistry and Physics—Various forms of Sulphur, BRAUNS, BÜTSCHLI: Diurnal
Variation of Atmospheric Electricity, A.-B. CHAUVEAU, 161.—Lehrbuch der
Photochromie (Photographie der natürlichen Farben), W. ZENKER, 162.—Joint
Transmission of Direct and Alternating Currents, F. BEDELL, 163.

Geology and Mineralogy—Geology of the Narragansett Basin, N. S. SHALER, J. B.
WOODWORTH and A. F. FOERSTE, 163 —View of the Carboniferous Fauna of
the Narragansett Basin, A. S. PACKARD, 164.—Geological Survey of Canada.
Summary Report of the Geological Survey Department for the year 1899, 165.
—Fossil Flora of the Lower Coal Measures of Missouri, D. WHITE, 166.—
La Face de la Terre (Das Antlitz der Erde) E. SUESS, 167.—Gneisses, Gabbro-
schists and associated Rocks of Southwestern Minnesota, C. W. Hall: Notices
of some recently described Minerals, 168.

Miscellaneous Scientific Intelligence—American Association for the Advancement of
Science: British Association for the Advancement of Science, 169.—Catalogues
of the Collections in the British Museum of Natural History: Norwegian
North-Atlantic Expedition, 1876–1878: The Grammar of Science, K. PEARSON,
170.

CONTENTS.

Number 57.

Page

ART. XVIII.—Gas Thermometer at High Temperatures; by
L. HOLBORN and A. L. DAY 171

XIX.—Certain Schists of the Gold and Diamond Regions of
Eastern Minas Geraes, Brazil; by O. A. DERBY 207

XX.—Monazite; by O. A. DERBY 217

XXI.—Spectra of Hydrogen and the Spectrum of Aqueous
Vapor; by J. TROWBRIDGE 222

XXII.—New Effect produced by Stationary Sound-Waves;
by B. DAVIS 231

XXIII.—Interesting Developments of Calcite Crystals; by
S. L. PENFIELD and W. E. FORD 237

XXIV.—Method of Measuring Surface Tension; by J. S.
STEVENS 245

SCIENTIFIC INTELLIGENCE.

Geology and Mineralogy—Glacial Gravels of Maine and their associated deposits, G. H. STONE: Illinois Glacial Lobe, F. LEVERETT, 247.—Geography of the Region about Devils Lake and the Dalles of the Wisconsin with some notes on its Surface Geology, R. D. SALISBURY and W. W. ATTWOOD, 248.—Moraines of Southeastern South Dakota and their Attendant Deposits, J. E. TODD: Preliminary Report on the Copper-bearing Rocks of Douglas County, Wis., U. S. GRANT: Bulletins of the Geological Survey of Finland, J. J. SEDERHOLM and V. HACKMANN, 249.—Analyses of Rocks, Laboratory of the U. S. Geological Survey, 1880 to'1899, F. W. CLARKE, W. F. HILLEBRAND and H. N. STOKES: Bjurböle, Finland, Meteorite, 250.

Obituary—Professor G. H. F. ULRICH, F.G.S, 250.

Number 58.

Page

ART. XXV.—Notes on the Colorado Canyon District; by W. M. DAVIS 251

XXVI.—Determination of Minerals in thin Rock-sections by their maximum Birefringence; by L. V. PIRSSON and H. H. ROBINSON 260

XXVII.—Studies in Cyperaceæ; by T. HOLM 266

XXVIII.—Experiments on High Electrical Resistance; by O. N. ROOD 285

XXIX.—Occurrences of Corundum in North Carolina; by J. H. PRATT 295

XXX.—Products of the Explosion of Acetylene, and of Mixtures of Acetylene and Nitrogen; by W. G. MIXTER .. 299

XXXI.—Scapolite Rocks from Alaska; by J. E. SPURR 310

XXXII.—Qualitative Separation of Nickel from Cobalt by the action of Ammonium Hydroxide on the Ferricyanides; by P. E. BROWNING 316

SCIENTIFIC INTELLIGENCE.

Physics—Radio-activity of Uranium, W. CROOKES, 318.—Electrical conductivity of Gases traversed by Cathode Rays, J. C. McLENNAN, 319.—Kleiner Leitfaden der Praktischen Physik, FRIEDRICH KOHLRAUSCH: Photometrical Measurements and Manual for the General Practice of Photometry, W. M. STINE, 320.

Geology and Mineralogy—Status of the Mesozoic Floras of the United States, L. F. WARD, 320.—La Flore Wealdienne de Bernissart, A. C. SEWARD, 323.—Notes on Some Jurassic Plants in the Manchester Museum, A. C. SEWARD: "The Maidenhair Tree," A. C. SEWARD: New Minerals from Greenland, 323.

Obituary—JAMES EDWARD KEELER, 325.

Number 59.

ART. XXXIII.—Elaboration of the Fossil Cycads in the Yale Museum; by L. F. WARD. (With Plates II–IV). 327

XXXIV.—Chemical Composition of Turquois; by S. L. PENFIELD.... 346

XXXV.—Quartz muscovite rock from Belmont, Nevada; the Equivalent of the Russian Beresite; by J. E. SPURR... 351

XXXVI.—Volumetric Estimation of Copper as the Oxalate, with Separation from Cadmium, Arsenic, Tin and Zinc; by C. A. PETERS..... 359

XXXVII.—Synopsis of the Collections of Invertebrate Fossils made by the Princeton Expedition to Southern Patagonia; by A. E. ORTMANN..... 368

XXXVIII.—Cathode Stream and X-Light; by W. ROLLINS 382

SCIENTIFIC INTELLIGENCE.

Chemistry and Physics—Ueber die physikalisch-chemischen Beziehungen zwischen Aragonit und Calcit, H. W. FOOTE: Atomic Weight of Radio-active Barium, M. and Mdme. CURIE, 392 —Artificial Radio-active Barium, A. DEBIERNE: Relative Values of the Mitscherlich and Hydrofluoric Acid Methods for the determination of Ferrous Iron, HILLEBRAND and STOKES, 393 —Grundlinien der anorganischen Chemie, W. OSTWALD. 394 —Elements of Inorganic Chemistry for Use in Schools and Colleges, W. A. SHENSTONE: Kohlenoxydvergiftung, W. SACHS: Hall Effect in Flames, E. MARX, 395.—Wireless Telegraphy, J. VALLOT and J. and L. LECARME: Spectrum of Radium, C. RUNGE: Magnetic effect of moving electrical charges, M. CREMIEU: What electric pressure is dangerous? H. F. WEBER, 396.

Geology and Mineralogy—Recent Publications of the U. S. Geological Survey, 397. —Department of Geology and Natural Resources of Indiana, 24th Annual Report, 1899, W. S. BLATCHLEY, 398.—Geological Report on Monroe County, Michigan, W. H. SHERZER and A. C. LANE: Elements of the Geology of Tennessee, J. M. SAFFORD and J. B. KILLEBREW: Geological Survey of Canada, G. M. DAWSON, 399.—Correlation between Tertiary Mammal Horizons of Europe and America, H. F. OSBORN. 400.—Flow of Marble under Pressure, F. D. ADAMS and J. T. NICOLSON, 401.—Report of the Section of Chemistry and Mineralogy, G. C. HOFFMANN: Florencite, a new mineral, HUSSAK and PRIOR, 404.—Elements of Mineralogy, Crystallography and Blowpipe Analysis from a practical Standpoint, A. J. MOSES and C. L. PARSONS, 405.

Miscellaneous Scientific Intelligence—Latitude-variation, Earth-magnetism and Solar activity, J. HALM. 405.—Location of the South Magnetic Pole, C. E. BORCHGREVINK: Ostwald's Klassiker der Exakten Wissenschaften: Scientia, H. LAURENT and F. M. RAOULT: Introduction to Science, A. HILL, 406.

Number 60.

Page

ART. XXXIX. — Torsional Magnetostriction in Strong
Transverse Fields and Allied Phenomena; by C. BARUS 407

XL.—Notes on Tellurides from Colorado; by C. PALACHE. 419

XLI.—New Species of Merycochœrus in Montana; by E.
DOUGLASS. (Part I.) 428

XLII.—Mohawkite, Stibio-domeykite, Domeykite, Algodonite and some artificial copper-arsenides; by G. A.
KOENIG....................................... 439

XLIII.—Heat of Solution of Resorcinol in Ethyl Alcohol;
by C. L. SPEYERS and C. R. ROSELL................. 449

XLIV.—Sulphocyanides of Copper and Silver in Gravimetric
Analysis; by R. G. VAN NAME 451

SCIENTIFIC INTELLIGENCE.

Chemistry and Physics — Action of Permanganate upon Hydrogen Peroxide,
BAEYER and VILLIGER: Instance of Trivalent Carbon: Triphenylmethyl,
GOMBERG, 458.—Phosphorescence of Inorganic Chemical Preparations, E.
GOLDSTEIN: Weight of Hydrogen Desiccated by Liquid Air, LORD RAYLEIGH,
459 —Separation of Tungsten Trioxide from Molybdenum Trioxide, M. J.
RUEGENBERG and E. F. SMITH: Radio-active Barium and Polonium: Lecture
Experiments Illustrating the Electrolytic Dissociation Theory and the Laws of
the Velocity and Equilibrium of Chemical Change, A. A. NOYES and A. A.
BLANCHARD, 460 —Physikalisch-chemische Propädeutik, H. GRIESBACH: School
Chemistry, J. WADDELL: Leçons de Chimie Physique. J. H. VAN'T HOFF:
Viscosity of Gases as affected by Temperature, LORD RAYLEIGH. 461.—Absorp-
tion of gases by glass powder, P. MÜLFARTH, 462.—Normal lines of iron,
KAYSER: Arc spectra of some metals in an atmosphere of hydrogen. H. CREW:
Recognition of the Solar Corona independent of the total eclipse, H. DES-
LANDRES: Loss of charge by evaporation, W. C. HENDERSON: Action of the
coherer, T. MIZUNO, 463.—Loss of electrical charges in air which is traversed
by ultra-violet rays, P. LENARD: Handbuch der Spectroscopie, H KAYSER, 464.
—Brief Course in General Physics, G. A. HOADLEY, 465.

Geology and Mineralogy—U. S. Geological Survey, 20th Annual Report 1898–99.
Part III, Precious Metal Mining Districts, J. S. DILLER, F. H. KNOWLTON,
W. LINDGREN, W, H. WEED, and L. V. PIRSSON, 465.—Iowa Geological Survey,
Vol X. Annual Report, 1899, with accompanying papers, S. CALVIN and H.
F. BAIN, 467.—Cleopatra's Emerald Mines, D. A. MAC ALISTER. 468 —Hand-
buch der Mineralogie, C. HINTZE: Contributions to Chemistry and Mineralogy
from the Laboratory of the U. S. Geological Survey, F. W. CLARKE: Johnstono-
tite, a supposed new Garnet, W. A MACLEOD and O. E. WHITE, 469.

Miscellaneous Scientific Intelligence—Annual Report of the Board of Regents of
the Smithsonian Institution showing the operations expenditures and condition
of the Institution for the Year ending June 30, 1898, S P. LANGLEY, 469.—
Report of the U. S. National Museum under the direction of the Smithsonian
Institution for the year ending June 30, 1898, 470 —Publications of the Earthquake Investigation Committee in Foreign Languages, H. NAGAOKA, 471.—
Bulletin of the University of Nebraska: National Academy of Sciences, 472.

INDEX TO VOLUME X, 473.

THE

AMERICAN JOURNAL OF SCIENCE

[FOURTH SERIES.]

ART. I.—*On the Energy of the Cathode Rays;* by W. G. CADY.

[An investigation carried out at the Physical Institute of the University at Berlin.]

IT has long been known that the cathode rays possess a considerable amount of energy; yet the relation between this and the other discharge phenomena has hitherto received but little attention. Ebert and Wiedemann,[*] E. Wiedemann,[†] and Ewers[‡] measured the energy of cathode rays by means of calorimeters. For exact work, however, the thermo-element is to be greatly preferred. Among the recent investigations based upon the emission hypothesis, this method has been employed by J. J. Thomson [§]; O. Berg has also used a thermo-element in his work with the cathode rays.[‖]

According to the emission hypothesis there exists the following relation between the energy of the cathode rays, potential of discharge, and amount of electricity transported by the rays. Suppose that we have a bundle of rays given off from a cathode of potential V, and striking a metallic conductor which is led to earth through a galvanometer; let the conductor be struck by N particles per second. We will at first assume that the entire charge of the particles, as well as all their energy in the form of heat, is given up to the conductor. Then if v [cm. sec^{-1}] be the velocity of the particles, and m

[*] H. Ebert and E. Wiedemann, Sitzungsb. der phys. med. Soc. zu Erlangen, Dec 14, 1891.
[†] E. Wiedemann, Wied Ann., lxvi, p. 61, 1898.
[‡] P. Ewers, Wied. Ann., lxix, p. 167, 1899.
[§] J. J. Thomson, Phil. Mag, xliv, p. 293, 1897.
[‖] O. Berg. Ber. d. naturforsch. Gesellsch. zu Freiburg i. Br. XI, vol. ii, p. 73, July, 1899.

[gr.] their mass, we have for the amount of heat given up to the conductor

$$Q = N \cdot \tfrac{1}{2}mv^2 \qquad (1)$$

(ergs per second).

If the charge on each particle be ε, we shall have flowing to earth through the galvanometer a "cathode-current" of intensity

$$i = N \cdot \varepsilon \qquad (2)$$

Now the kinetic energy of each particle is

$$V \cdot \varepsilon = \tfrac{1}{2}mv^2. \qquad (3)$$

Hence

$$Q = N \cdot V \cdot \varepsilon = i \cdot V,$$

or

$$\frac{iV}{Q} = 1. \qquad (4)$$

According to Starke's investigations,* a part of the cathode rays are reflected. This can have no effect upon the above, provided that after reflection the velocity of the rays remains unchanged; and indeed the measurements made by Merritt† upon the deflection of reflected cathode rays seem to show this to be the case. We will, however, assume for the present that upon reflection the energy of each particle is diminished in the ratio $r' : 1$, while the charge remains unchanged. Then if the fraction r of the rays be reflected, we have

$$Q = N(1-r) \cdot \tfrac{1}{2}mv^2 + Nr(1-r') \cdot \tfrac{1}{2}mv^2 = \tfrac{1}{2}Nmv^2(1-rr') \qquad (1a)$$
$$i = N\varepsilon(1-r) \qquad (2a)$$
$$V\varepsilon = \tfrac{1}{2}mv^2 \qquad (3a)$$
$$\frac{iV}{Q} = \frac{1-r}{1-rr'}. \qquad (4a)$$

If $r' = 1$ (Merritt), we have still $iV/Q = 1$. If in the limiting case $r' = 0$, then $iV/Q = 1 - r$.

These relations were investigated experimentally as described in the following paragraphs.

Apparatus.

Fig. 1 shows the arrangement of apparatus at first used. For measuring the energy Q a thermo-element was here employed; it was later displaced by a bolometer (see below.) The thermo-element, S, was of the Melloni type, consisting of 49 bismuth-antimony couples each 25^{mm} long; the exposed surface of the junctions was 196^{mm^2}. It was placed inside the

---

* H. Starke, Wied. Ann., lxvi, p. 49, 1899.
† E. Merritt, Phys. Rev., vii, p. 217, 1898.

Faraday cylinder C, which was connected to earth. The rays entered through the hole o, 6ᵐᵐ in diameter. R is the glass discharge-tube, 25ᶜᵐ long, cemented to the glass plate P.

The copper bottom of the brass tube K served as cathode; in order to reduce the heating as far as possible, a stream of water was kept constantly flowing through the tube. The anode A was a zinc diaphragm.

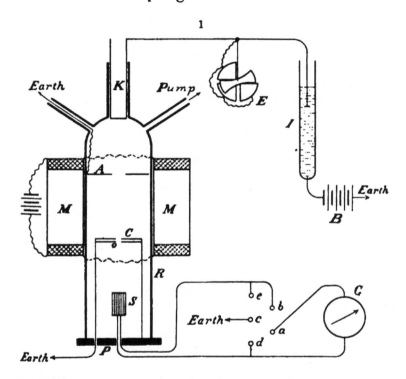

The discharge current was furnished by a secondary battery B of 2288 cells, supplemented at times by a rotary transformer, so that a maximum E. M. F. of 5400 volts could be reached. The current was regulated by the Hittorf cadmium-iodide resistance I, and the potential of the cathode measured by means of the quadrant-electrometer E. G is a Thomson galvanometer used in measuring the thermal and cathode-currents.

A serious source of error in observations lay in the evolution of gas from the cathode and other parts of the apparatus during discharge. This difficulty, so common in all gas-discharge work, was nearly always present, even after the apparatus had been in use for many hours. A partial remedy was found in running the air-pump slowly during observations, but even then the potential usually changed somewhat during the

minute or so necessary to take a set of readings. This is th
chief cause of the irregularities in the numerical values give
below.

Before the apparatus was put together, the thermo-elemen
was calibrated as follows. The radiation from the blackene
surface of a Leslie cube, filled with boiling water, passe
through a series of diaphragms and fell upon the thermo-ele
ment. The constant of radiation for lampblack at 100° wa
taken according to Kurlbaum[*] as 0·0176 gr.·cal/sec. From th
dimensions of the diaphragms could then be calculated, by us
of the Stefan-Boltzmann law, the amount of heat Q_0 radiate
from the hot surface at temperature T to the thermo-elemen
at temperature t. Like the Leslie cube, the thermo-elemen
was also covered with a layer of lampblack.

The galvanometer deflection s_0 was observed after the thermo
element had been exposed for one minute to radiation fror
the cube. The amount of heat per scale division is the
$e = Q_0/s_0$, and the amount corresponding to the deflection s i
$Q = s.e$.

During calibration the thermo-element was in air at atmo
pheric pressure; when in use with the cathode rays, it was i
a partial vacuum, and moreover the layer of lampblack ha
then been removed. This would tend to make the rise ii
temperature of the junctions in the latter case greater than ii
the former, assuming the quantity of heat supplied to be th
same: for the loss of heat by radiation and conduction to th
air must have been greater during calibration. The effect o
this error would be to make the observed energy of the cathod
rays too great; but the error is smaller, the more the loss o
heat by conduction and radiation to the air can be neglected ii
comparison to the loss by conduction through the metal of th
thermo-element itself.

In measuring the cathode-current, the contact ac (fig. 1) wa
closed, thereby connecting the thermo-element to earth througl
the galvanometer. For the thermal circuit, ab was closed, an
at the same time either ce or cd, since it was necessary to con
nect some point to earth. But when for example cd was closed
it was found that a small fraction of the cathode-current als
flowed to earth through the galvanometer, along the patl
$SbaGdc$. Closing ce caused approximately the same deflectior
in the opposite direction. Hence the mean of the two read
ings could be taken as the deflection for the thermal curren
alone.

In order to eliminate the effect of heat radiated from th
cathode, the rays were deflected before each observation b;

---

[*] F. Kurlbaum, Wied. Ann , lxv, p. 759, 1898.

means of the coils **MM** (fig. 1); the small remaining galvano-meter deflection was then subtracted from the total deflection with rays falling.

Observations with Thermo-element.

The value $i.V$ (eq. 4) in gram-calories is cathode-current in amperes $\times$ difference of potential in volts $\times$ 0·24. Pressures are always given in mm. of mercury.

The first measurements yielded values of iV/Q much greater than 1, as shown in the following table:

TABLE I.

Pressure.	$i \cdot 10^7$ Amp.	V	$Q \cdot 10^5$ gr.-cal.	iV/Q.
0·067	152	2280	90·3	9·18
0·046	138	2470	141	5·80
0·044	113	2490	123	5·50
0·042	99·4	2600	127	4·89

The junctions of the thermo-element were here, as at the calibration, covered with a layer of lampblack. It may, how-ever, be assumed that the long-wave heat radiations used in calibrating penetrate quite deeply into the interior of the lamp-black, while the cathode rays are absorbed close to its surface. If the heat conductivity of the lampblack is small compared with that of the metal of the thermo-element, it is not impossi-ble that for the same amount of heat, the rise in temperature of the junctions under the influence of the cathode rays may be smaller than at the calibration. The lampblack was there-fore removed, and in fact the value of iV/Q was thereby diminished.[*]

The following measurements were next made after removal of the lampblack:

TABLE II.

Pressure.	$i \cdot 10^7$ Amp.	V	$Q \cdot 10^5$ gr.-cal.	iV/Q.
0·046	180	3100	412	3·25
0·050	175	3100	397	3·28

TABLE III.

Pressure.	$i \cdot 10^7$ Amp.	V	$Q \cdot 10^5$ gr.-cal.	iV/Q.
0·043	122	3150	331	2·80
0·029	116	3500	416	2·34
0·026	100	3020	416	2·09
0·025	65·4	3820	309	1·94
0·022	61·7	4070	344	1·75

[*] This diminution no doubt comes in part as described above from the change in constant of the thermo-element after removal of the lampblack, also perhaps partly from the altered reflection of the cathode rays.

It is to be noted that in Table III as also in Table I, *the measured value of* iV/Q *decreases with decreasing pressure and cathode-current.*

In order to investigate this relation more closely, a series of observations was made, varying pressure and cathode current within as wide limits as possible. The cathode-current could be altered either by changing the number of accumulator cells, or by varying the cadmium-iodide resistance. Owing to the evolution of occluded gases from the cathode and its gradual discoloration, causing sudden irregularities in potential, it was impossible to get a perfectly consistent set of readings. Still in general it was evident that the ratio iV/Q diminished with decreasing i, at first rapidly, then more slowly. At high pressures ($>0.05^{mm}$), the ratio seemed to depend also upon the pressure; but when the pressure had sunk below 0.03^{mm}, the drift was hardly preceptible.

These results led to the supposition that beside the cathode rays, a fraction of the discharge current also was conducted from the cathode to the thermo-element; for as J. J. Thomson[*] has shown, the portion of gas transversed by the rays becomes a relatively good conductor, except at very low pressures. This supposition was confirmed by inserting a battery between c (fig. 1) and the earth, thereby charging thermo-element and Faraday cylinder to a positive or negative potential v.[†] This cannot affect the true value of the ratio iV/Q as long as V denotes the difference of potential between cathode and thermo-element; but the supposed disturbing current must grow stronger or weaker according as v is made positive or negative.

The following table shows how the apparent value of iV/Q was correspondingly increased or diminished according to the sign of v.

Pressure.	v	V	$i \cdot 10^7$ Amp.	$Q \cdot 10^5$ g-cal.	iV/Q.
0.050	+70	4050	3.54	34.1	1.01
0.050	0	3980	2.64	30.2	0.84
0.050	−70	3910	2.29	30.5	0.71

The potential of the cathode as measured on the electrometer is here $V-v$; as in all measurements, i denotes the observed sum of cathode- and disturbing currents.

The variations in the value of Q result from the fact that the path of the rays was visibly altered while the thermo-element was charged.

[*] J. J. Thomson, Phil. Mag., xliv, p. 293, 1897.
[†] The cylinder was here connected with the point c. Owing to incomplete insulation between thermo-element and cylinder it was not practicable to connect the latter directly to earth.

The cathode-current is here only about 1/30 as intense as in Tables II and III, hence the value of $i V/Q$ is also much smaller. The extremely small value of $i V/Q$ when the charge is negative is explained as follows: owing to the conductivity of the path of the rays, a weak current finds its way from the battery used in charging the thermo-element, along this path to the anode. This current is opposed to the cathode-current, thereby reducing the apparent value of the ratio $i V/Q$. When the thermo-element is positively charged, the ratio is on the other hand greatly increased.

Measurements similar to the above, made at a pressure of 0.012^{mm}, showed that charging the thermo-clement no longer had any influence upon the value of the ratio. These observations on the conductivity of the cathode rays agree well with those of Starke* and Thomson.†

The attempt was also made to intercept the disturbing current by means of a grating of several fine wires laid over the opening of the Faraday cylinder. This was, however, of no practical assistance, as even at high pressures the effect was very slight.

The results that have been given show that the observed value of $i V/Q$ was in general too large. The disturbing current was at times greater than the cathode-current itself, but it was impossible to measure directly its absolute value. It therefore remained to establish the lower limit of the ratio. For this purpose a series of 46 observations was selected, having a range of pressure from 0.01 to 0.05^{mm}, of potential from 2400 to 5400 volts, and of cathode-current from 40.10^{-7} to $0.23.10^{-7}$ amp. When arranged in order of decreasing cathode-current, the values of $i V/Q$ showed great irregularities, since the order was without regard to pressure and other conditions. Still a distinct drift in the ratio was evident, especially when the observations were divided into five groups and the average values taken :

Pressure.	$i.10^7$ Amp.	$i V/Q$
0·041	19·6	1·32
0·035	7·41	0·97
0·026	4·29	0·85
0·033	2·02	0·84
0·023	0·49	0·83

From this it is evident that when the cathode-current decreases to the fortieth part of its original value, the ratio $i V/Q$ is changed by only 38 per cent. This at least makes it seem

* H. Starke, Wied. Ann., lxvi, p. 52, 1899.
† J. J. Thomson, Phil. Mag., xliv, p. 293, 1897.

probable that eq. (4a) is approximately correct, although th
limiting value of the ratio is considerably smaller than unity
Still owing to the sources of error in the use of the thermo
element, no great dependence can be placed upon absolut
values. It was therefore determined to measure the energy (
in a surer manner, by Kurlbanm's bolometric method.[*]

Observations with Bolometer.

A surface-bolometer forms one arm of a Wheatstone bridge
suppose the bridge balanced, and a current of intensity I
flowing through the bolometer. Let the resistance of the bolc
meter under these conditions be w. If now an amount of hea
Q per second be supplied by the cathode rays to the bolometer
the galvanometer in the bridge will suffer a certain deflection
Suppose the same deflection caused by increasing I_1 to I_2
the cathode rays no longer acting. Then from Kurlbaum'
formula,

$$Q = w \cdot \frac{I_2}{I_1} (I_2{}^2 - I_1{}^2) \text{ watts,}$$

where I and w are expressed in amperes and ohms respectively
If the four bridge resistances are equal, then $I = \frac{1}{2}c$, wher
c is the current in the battery branch, and

$$Q = \frac{1}{4}w \cdot \frac{c_2}{c_1} (c_2{}^2 - c_1{}^2) \text{ watts.}$$

As it was not practicable to make both deflections exactl
equal, the following method was adopted:

    1. galv. defl. a_1 with current c_1, rays falling.
    2. " " a_2 " " c_2 " deflected.

We thus get the following equation:

$$Q = \frac{1}{4} w \frac{a_1}{a_2} \frac{c_2}{c_1} (c_2{}^2 - c_1{}^2)$$

The thermo-element formerly used (S, fig. 1) was replace
by a bolometer, whose terminals were led out through th
plate P. The bolometer was made after Lummer and Kur
baum's process,[†] with a few modifications. In the first obser
vations it consisted of a platinum foil about 0.001^{mm} thick
mounted with mica insulation upon a brass frame. The forr
is shown in B (fig. 2). Care was taken that the entire bolc
metric resistance should be reached by the rays; also that a
cathode rays entering the Faraday cylinder that did not strik

[*] F. Kurlbaum, Wied. Ann., lxv, p. 746, 1898.
[†] O. Lummer and F. Kurlbaum, Wied. Ann., xlvi, p. 204, 1892.

the bolometer, should give up their charge to earth through the metallic frame. Although the density of the rays is not perfectly uniform throughout the bundle, still it can be assumed that the distribution of temperature in the thin bolometric foil under the influence of the rays and of the electric current is practically identical.* The silver was not dissolved away from the platinum-silver foil until the bolometer had been mounted.

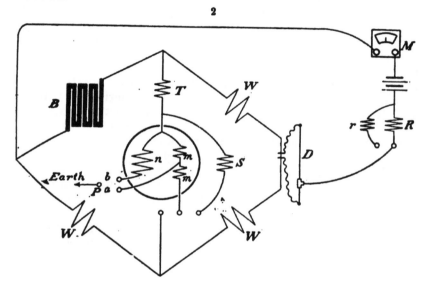

Fig. 2 shows the arrangement of the bridge. The current-intensity i in the battery branch was measured by means of the ampere-meter M. Shunting the resistance r parallel to R increased the current from i_1 to i_2 and gave an accurate means for measuring the increase. A Thomson galvanometer of du Bois-Rubens type was employed. For the thermal measurements two 20-ohm coils mm were used; in order to get rid of the cathode-current, a point between the two coils was led to earth (contact a, fig. 2) and the resistance T so chosen that with battery disconnected the cathode-current caused no deflection of the galvanometer. Two 2000-ohm coils n served to measure the cathode-current, by closing the contact b, while a remained open.

S is a shunt for reducing the sensitiveness of the galvanometer when necessary. The bridge resistances were balanced by means of the sliding contacts at D. The manganin wire resistances W were as nearly equal to the bolometric resistance

---

* F. Kurlbaum, Wied. Ann., lxv, p. 755, 1898.

as possible, but small discrepancies were always taken int
account. It was necessary to use great care in insulating th
entire circuit from earth.

The bolometer was found to be much more sensitive tha
the thermo-element, although its effective surface was smalle
The source of error from gas-evolution was also less seriou
owing to the rapidity with which observations could be mad

The process of etching and washing such small bolometers
somewhat complicated when a metallic frame is used, as th
latter has to be protected by a coating of wax, which is afte
wards dissolved away. Thus it is difficult to avoid strainin
the delicate platinum strips. This may become a very serio:
source of error, as it is impossible to know the extent of th
injury. Moreover, a brownish deposit was nearly alwa:
formed upon the strips, probably from some action of the ac
used to dissolve the silver, upon impurities in the wax. Th
may in some way have caused a further source of error.

Four different bolometers with brass frames were used. Th
results are briefly given below, though in at least two cas
great irregularities occur. It was not until a frame of slate h:
been made that reliable observations were obtained (see below

Bolometer I. Eleven strips, $11 \cdot 5 \times 0 \cdot 66^{mm}$. Resistance w
room temperature $= 26 \cdot 21$ ohms.

The observations were all made at high pressures; the rat
$i V/Q$ is therefore chiefly dependent upon the pressure and n
upon the cathode-current.

Pressure.	$i \cdot 10^7$ Amp.	V	$Q \cdot 10^5$ Watt.	$i V/Q$.
0·116	18·0	2590	213	2·21
0·113	12·3	2610	140	2·28
0·064	22·3	4090	541	1·69
0·055	13·7	4210	376	1·53

Bolometer II. Nine strips, $11 \cdot 5 \times 1 \cdot 0^{mm}$. $w = 15 \cdot 74$ ohm
An influence machine was used instead of the battery
source of discharge. As the range of values of i is not grea
only the average values are here given: $i = 1 \cdot 40 \cdot 10^{-7}$ amp
pres. $= 0 \cdot 020^{mm}$; $i V / Q = 0 \cdot 81$.

Bolometer III. Seven strips, $11 \cdot 5 \times 1 \cdot 3^{mm}$. $w = 8 \cdot 75$ ohm
A battery of 30 cells was so arranged that it could be co
nected at will between P (fig. 2) and the earth. This rais
the bolometer and of course the entire bridge to the potenti
of the battery. The Faraday cylinder was kept in contact wi
P. Readings were taken with the bolometer alternately at
and $+60$ volts.

From the table it is evident that as in the case of the therm
element, charging the bolometer to a certain potential increas
the ratio $i V/Q$ to a greater extent, the stronger the cathod
current, that is, the greater the conductivity of the gas.

Pressure.	v	$i \cdot 10^7$	V	$Q \cdot 10^5$	iV/Q.
0·033	0	5·30	3130	106	1·57
	+60	7·20	3090	119	1·91
0·028	0	2·10	4320	80·8	1·12
	+60	3·23	4200	113	1·20
0·029	0	1·96	5150	75·5	1·33
	+60	4·91	4960	154	1·58
0·030	0	1·90	4230	67·5	1·19
	+60	2·80	4130	88·5	1·31
0·030	0	0·54	3910	20·8	1·02
	+60	0·67	3880	25·3	1·02
0·031	0	0·28	4220	11·9	0·98
	+60	0·56	3900	21·9	1·01

Bolometer IV. Seven strips, $11\cdot5 \times 1\cdot3^{mm}$. $w = 9\cdot20$ ohms. Values of the ratio were irregular, but the averages from three groups of observations are here given:

Pressure.	$i \cdot 10^7$ Amp.	iV/Q.
0·070	9·43	0·81
0·040	11·48	0·76
0·023	0·18	0·72

As with Bol. I, the ratio at high pressures seems dependent upon the pressure alone. In the third group the cathode-current was diminished by increasing the cadmium iodide resistance.

These results are of only relative importance except as they indicate that the absolute value of iV/Q is somewhat less than 1.

A fifth bolometer was therefore made and mounted after the Kurlbaum process upon a slate frame. After the silver had been dissolved from the foil, the strips looked perfectly clean and seemed free from strains. In order to keep the cathode rays from the slate, a small brass diaphragm was placed close over the bolometer, with a square opening of the same size as the bolometer surface.

Bolometer V. Seven strips, $7 \times 0\cdot73^{mm}$. $w = 8\cdot72$ ohms. The following observations were taken:

Pressure.	$i \cdot 10^7$	V	$Q \cdot 10^5$	iV/Q.
0·038	6·10	2270	76·9	1·80
0·061	5·00	2690	82·3	1·58
0·035	4·80	3480	149	1·12
0·024	2·26	3770	85·5	1·00
0·024	2·09	3560	81·2	0·92
0·030	1·69	4460	92·6	0·81
0·022	1·36	3730	58·0	0·87
0·032	1·09	3740	50·4	0·81
0·032	1·06	3750	47 9	0·83
0·028	0·50	3900	23·6	0·83
0·026	0·255	4740	14·5	0·83
0·023	0·040	4800	2·29	0·84

This bolometer was apparently free from the faults of its predecessors: therefore only the usual errors of observation need be considered, especially that arising from the evolution of gas. This error can hardly have amounted to more than 3 per cent with weak cathode-current; hence it follows from the last table that the true value of the ratio iV/Q lies between 0·80 and 0·86.

Since the bolometer foil was only 0·001^{mm} thick, the question arises whether an appreciable part of the rays can have penetrated through to the other side. In this case it might easily happen that the particles on passing through would lose a part of their energy while retaining their charge. Mc Clelland[*] and Wien[†] have shown that cathode rays after traversing an aluminium window carry a negative charge; but Lenard did not succeed by means of a thermo-element in detecting any heating effect.[‡] Still if we assume according to Lenard that for solids the quotient of absorptive-power divided by density = 3200,[§] we find from his formula that not more than 1 per cent of the rays can have passed through the bolometric foil used. This can have had practically no effect upon the ratio under discussion.

Conclusion.

It only remains to compare the limiting value of the ratio a experimentally determined, with the equation (4a)

$$\frac{iV}{Q} = \frac{i-r}{1-rr'} ,$$

where r denotes the reflected part of the rays, while upon reflection the rays lose their energy in the ratio $r':1$.

Starke[‖] has found that platinum reflects about 40 per cent of the rays. Assuming, therefore, that for the bolometer, $r = 0·4$ we find that when $iV/Q = 0·83$, $r':0·7$.

This result is in contradiction with the observations of Merritt already referred to. The latter are unfortunately not above criticism, as the author himself admits. Merritt compared the deflectibility of direct and reflected rays by observing the phosphorescence on the walls of the tube. It is hardly to be supposed that the spot from the reflected ray can have been even approximately as distinct as that from the

[*] J. A. McClelland, Proc. Roy. Soc., lxi, p. 227, 1897.
[†] W. Wien, Verh. Phys. Ges. Berlin, xvi, p. 165, 1897.
[‡] Ph. Lenard, Wied. Ann., li, p. 239, 1894. This experiment was, however performed in air at ordinary pressure, which rapidly absorbs the energy of the cathode rays.
[§] Ph. Lenard, Wied. Ann., lvi, p. 274, 1895.
[‖] H. Starke, Wied., lvi, p. 58, 1899.

direct; moreover, since an induction coil was used as source of discharge, one would expect the formation of a "magnetic spectrum,"[*] which would greatly complicate the appearance of both spots. Thus it is easy to see that a systematic error may have crept into the results.

Whether such an error can have been great enough to give the value $r' = 1$ instead of $r' = 0\cdot7$ is, of course, not certain; but it may be added that the energy of the cathode rays is proportional to the square of their deflectibility, hence an error of n per cent in the measurement of the deflection would correspond to an error of $2n$ per cent in the energy. In any case it is desirable to determine both the deflectibility and the energy of reflected cathode rays by as accurate a method as possible.

My thanks are due to Prof. Warburg for his kind advice and assistance during this investigation.

[*] K. Birkeland, Comptes Rendus, cxxiii, p. 492, 1896.

Providence, R. I., May 5, 1900.

ART. II. — *Volcanic Rocks from Temiscouata Quebec;* by HERBERT E. GREGORY.

TEMISCOUATA LAKE was explored by the early geolog Canada, and Logan reported[*] the presence at Point Trembles of 'tough, green sandstone, with pebbles of morphic rock.' A survey of the region was made by and McInnes in 1886–87, and the presence of volcanic noted. Their report in reference to these rocks reads lows : "It is important to notice in connection with the aux Trembles sandstones the evidence which they ap afford of contemporaneous volcanic activity. This is, pe

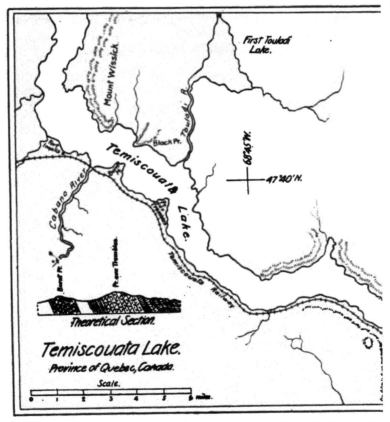

First Toulaï Lake.

Mount Wrissick

Black Pt.

68°45'W.

47°40'N.

Cabano River

Temiscouata Lake.

Burnt Pt.

Pt. aux Trembles

Theoretical Section.

Temiscouata Lake.
Province of Quebec, Canada.
Scale.

partly indicated by the color of the rock, which varie green to red and purple, but is more clearly seen in the a ance of epidote with which the rock is charged and in

* Geology of Canada, 1863, p. 423.

s somewhat amygdaloidal aspect."* During the summer of 897, the region was studied by Prof. H. S. Williams, and the pecimens then collected were turned over to the writer for tamination. The outcrops visited extend along the lake for distance of 20 miles, as shown on the accompanying map.

Mount Wissick.—The rocks examined from Mt. Wissick nd the east shore of the lake are fossiliferous limestones, andstones and shales, without admixture of volcanic materials. 'he shales and sandstones do not differ from similar rocks ound elsewhere. The arenaceous limestone in places is seprated into hexagonal prisms, probably from shrinkage, and ppears as if composed of vertical columns with quite regular utline. The thin layers of limestone in the slates at the orth of Mt. Wissick show under the microscope well-defined olitic structure in which the little spheres are broken by umerous minute faults.

West Shore.—The section on the West Shore extends from Burnt Point, below Fort Ingalls, to the outlet of the lake.

At Burnt Point the rocks are coarse conglomerates made up f materials not much water worn, and show evidence of rapid leposition. Slates and limestones form the most abundant ebbles, and occasionally attain a diameter of 1–2 feet. The otal thickness of the conglomerate at this point is about 1000', out such great thickness of the beds is reported to be quite ocal. Succeeding the Burnt Point conglomerate, to the south, ccur thin-bedded shales and sandstones, which contain fossils f Niagara age.†

At Point aux Trembles, the rocks, both along the railroad nd on the lake shore, appear at first sight to be greenish sandtones and coarse brown conglomerates. They are interstratied with the other beds of the region, and have practically he same dip and strike. A closer examination, however, hows them to be volcanic. The finer, more sandy beds, conain quantities of volcanic ash, and the coarser ones are conlomerates of typical andesitic fragments, with scarcely any oreign material. Parts of the rock contain very prominent ounded fragments of amygdaloidal andesite. As with the olcanics of northern Maine,‡ so here the gradation from the andstones of the region to pure volcanic material can be aced, and like the Maine breccias and tuffs, these rocks show iore or less rounding of their pebbles and sorting by water. is believed that the volcanic vents were near some body of ater, and that the material fell in or near the water and ceived a limited amount of wearing before final consoliition.

* Geol. Survey of Canada—Annual Report, 1887, p. 33M.
† Geol. Survey Canada, Annual Report, 1887, p. 33M.
‡ U. S. G. S. Bulletin 165.

From Point aux Trembles to the foot of the lake—a distance of about 10 miles—the rocks which outcrop are slates and impure sandstones with rare tufaceous material. In places the beds are much folded and traversed by faults of small displacement.

Petrography.

The ordinary stratified rocks of the region present no peculiarities which require detailed description. The volcanic series consists of fine tuff and coarse amygdaloidal conglomerate or breccia.

Fine Tuff.—The fine tuff appears in the hand specimen as a dense bluish-grey sandstone in beds a few feet in thickness. Under the microscope its true character is revealed, and it is found to consist of tiny fragments of andesite, broken crystals of plagioclase and olivine, and areas of devitrified glass with rare quartz grains and biotite shreds. The andesite has a hyalopilitic groundmass, which varies in amount of glass, so that some fragments appear to be practically all glass with only a few scattered laths of plagioclase. The larger crystals of andesine have albite twinning and are largely kaolinized. The olivines are represented in the slide by areas of serpentine with characteristic outlines and cleavages shown by dotted ore grains. Iron ore is sprinkled quite generally through the section.

Volcanic Conglomerate.—The volcanic conglomerate consists of subangular pebbles of andesite and amygdaloidal andesitic bombs, embedded in a finer matrix of red and green grains of the same material. The pebbles range in size from $\frac{1}{4}$ inch to 6 inches in diameter, and show in the hand specimen phenocrysts of plagioclase, and occasionally epidote and augite. The embedded bombs form a conspicuous feature of the rock. They are quite spherical and are distinctly different in appearance from the matrix. Amygdules, about the size of bird shot, make up fully one-half of the bomb and are filled with green balls of chlorite—rarely with calcite. Some of the larger amygdules are seen to be occupied by chalcedony inside the chlorite lining.

Microscopic Examination.—The microscope shows the general matrix of this coarse type to consist of crystals of feldspar, pyroxene and olivine, and fragments of andesite, devitrified glass, and jasper. Iron dust, a small amount of calcite, and a few shreds of biotite, are also present. The feldspar crystals consist of a few orthoclases, and many rather small andesines, with albite and carlsbad twins. The pyroxene crystals are represented only by their outline, in a base peppered with iron dust. The olivines are fragmentary crystals, and largely replaced by serpentine and iron. The bulk of the sections examined is formed of fragments of pyroxene-andesite of a type sparingly porphyritic with small stout phenocrysts set in a groundmass, having in some places a glassy base, in others being made up almost entirely of ragged feldspar micro-lites, arranged with flow structure. The fragments of devitri-fied glass present show occasional cusps, which represent expanded steam cavities. No close line, however, can be drawn between the fragments of glass and the fragments of andesite containing a few feldspar microlites in a glassy base. The dark-red iron dust is not generally distributed through the rock, but is present along the cracks of the olivine crystals, and it also gives color to the glassy fragments.

The Amygdaloidal Bombs.—The section cut from one of the amygdaloidal bombs reveals the composition and structure of an extrusive basic andesite. The minerals present as phe-nocrysts are plagioclase, pyroxene, olivine and iron ore, with infiltrated quartz and calcite in the amygdules. There is also a considerable development of minerals of the chlorite group, perhaps largely delessite. The plagioclases are short, rather square crystals with ragged ends and strongly-marked albite and pericline twinning. Many of them are bent and curved and irregularly broken. In composition they are between andesine and labradorite. Augite occurs as stout crystals always partly decayed and replaced by chlorite or epidote, or rarely by calcite. Olivine crystals are about as abundant as augite and show their characteristic parting along cracks now filled with iron. Serpentine has generally replaced the olivine. Iron ore occurs in a few ragged grains. The groundmass is of feldspar microlites usually untwinned and with little evidence of flowage. The structure was hyalopilitic, but the glass fill-ing the interstices is devitrified and stained brown with iron. The steam cavities now filled and converted into amygdules, though very abundant, have exerted little influence on the arrangement of the feldspars in the groundmass—a fact which suggests that these cavities formed in part before any crystal-lization of feldspars had occurred.

All the amygdaloidal cavities are filled with secondary pro-ducts which have a uniform arrangement. They are lined with a coating of a malachite-green mineral of the chlorite group, arranged in radial forms. Most of the smaller cavities are completely filled with this substance and appear on the rock surface as embedded green pellets. The larger cavities have the chlorite lining fractured, and fragments of it have floated in toward the center, leaving means of access for later infiltrations. Similar phenomena have been observed in certain

volcanic rocks in Maine.*   The inside filling of the amygdul
is silica or rarely calcite.   The material next to the chlori
lining is chalcedony, with well-developed mammillary stru
ture.   The center of the cavity is a mass of quartz grai
separated from the chalcedony by a definite boundary.
thus appears that the amygdules were filled by three or fo
successive periods of infiltration.

Summary.—While the mere description of these rocks a
the determination of their volcanic origin is, in itself, chief
of local interest, these facts have, however, important bearin
on the general geological history of the Atlantic coastal regio
That they are interbedded with Niagara sediments helps
determine the time when wide-spread volcanic activity ga
rise to the numerous small areas of tuffs and lavas in t
Maine-Quebec region, while this fact throws additional lig
on the physical conditions obtaining in that period.   A
finally, the recognition of so much distinctly contemporaneo
volcanic material in well-bedded sediments affords another pl
for the study of the sedimentaries by microscopic analysis.

* U. S. G. S. Bulletin 165, Plate xiv.

Yale University, New Haven, Conn., March, 1900.

ART. III.—*On the Interpretation of Mineral Analyses: a Criticism of recent Articles on the Constitution of Tourmaline;* by S. L. PENFIELD.

ABOUT eighteen months have elapsed since H. W. Foote and the author published in this Journal* a joint article on the chemical composition of tourmaline. Since that time two articles have appeared, presenting views differing from one another and from those of Foote and the author; one on the Constitution of Tourmaline, by Prof. F. W. Clarke† of Washington, the other *Über das Mischungsgesetz der Turmaline,* by Prof. G. Tschermak‡ of Vienna.

In order to discuss the views presented in these articles it is desirable to carefully consider some facts concerning chemical analyses. In the first place, a perfect chemical analysis cannot be made. There are, to be sure, a very few analytical processes for the determination of single constituents, which, when carefully executed, can be relied upon to give results varying less than 0·01 per cent from the theory; but when it comes to mineral analysis, necessitating the splitting up of a complex body and the determination of a number of constituents, such accuracy cannot be attained. In making a mineral analysis, one seldom feels confident that all determinations are correct, even within 0·25 per cent of the truth, although if duplicate analyses are made, it is expected that, for the majority of the constituents at least, the two determinations will agree within 0·10 or 0·20 per cent of one another. At times, of course, depending upon the difficulty of the analysis or the scarcity of available material, variations of 0·50 per cent, or even more, in duplicate determinations are not to be wondered at.

Secondly, analytical work may be of a high order, the results very accurate, and yet an analysis may not serve for the determination of a chemical formula because made on material more or less impure. The chemists of to-day have a decided advantage over those of a former generation, for the microscope enables them to study their material, select that which is best, and, if impurities cannot be avoided, to take their effect into consideration in discussing the analytical results. Then again the heavy solutions are invaluable for separating out material for analysis, and, what is considered of very great importance, for furnishing a guarantee of the purity of any given material; for if it can be stated that all of the mineral floats on a solution of a certain specific gravity and sinks when

* Vol. vii p. 97, 1899.
† This Journal, IV, viii, p. 111, 1899.
‡ Mineralog. und Petrogr. Mitth., xix, p. 155, 1899.

the specific gravity is lowered to a trifling extent, it gives one not only great confidence in the purity of the material, but, also, it enables the investigator to present data which others can make use of in judging the character of the work.

It has been the author's privilege during the past twenty-five years to make many analyses of minerals, and to superin-tend the making of many more in the Sheffield Mineralogical Laboratory ; also to discuss the analytical results and derive therefrom the chemical formulas of minerals, and this occasion will be taken to call attention to certain features which are regarded as most important in mineralogical investigations. In the first place, the utmost pains should be taken to secure pure material, and, if the results are to be published, the char-acter of the material should be described so that its degree of purity can be judged by others. Secondly, if an analysis pre-sents any especially difficult features, the method of analysis should be carefully described, and it is in almost all cases well to give at least some brief outline of the analytical methods employed. Then, too, when material is abundant, it is advis-able to make analyses in duplicate, and to give all of the deter-minations, together with the averages. Thus the investigator has from beginning to end the satisfaction of a control over all determinations, and, if agreements are close, others can form some estimate concerning the care with which the work was executed. There are those who apparently entertain the belief that closely agreeing duplicate determinations indicate great accuracy in analytical work, but that is not necessarily the case, for in some analytical methods there is a tendency for results to come too high, in others too low, and thus duplicate determinations, made under like conditions, either with faulty methods, or with good methods improperly executed, may be uniformly high or uniformly low, agreeing with one another, and yet varying considerably from the truth. Still two closely agreeing determinations carry with them a certain weight which cannot be ignored. Thirdly, with each analysis, the quotients obtained by dividing the several constituents by their molecular or atomic weights, as the case demands, should be given, and from the quotients thus obtained the ratio between the several constituents should be determined. The ratio ought not to be given simply rounded out to the nearest whole numbers, but, taking the quotient of the most characteristic or best determined constituent as unity, the ratio should be given to the second place of decimals. *It is safe to assume that the close approximation of a ratio to whole numbers constitutes the strongest argument that can be advanced in support of the excellence of an analysis and the correctness of the derived formula.* It will seldom happen that a ratio approximates to whole numbers merely as a matter of accident. Provided the compound is a simple one, instead of giving the ratio, an excellent method is to give the calculated composition, which can then be compared directly with the results of the analysis. Lastly, for determining a formula one or two good analyses are of more value than many indifferent ones, hence it will often prove best to make new analyses on material of unquestioned purity. This may be done not wholly with the idea that the new analyses are better than those made by other investigators, but, knowing all about the quality of the material and the working of the analyses, it will be possible to exercise better judgment in summing up the results of the investigation, and to present with greater force the arguments needed in support of the proposed formula.

Turning now to the consideration of tourmaline, two new analyses were made by Foote and the author, upon material of ideal purity and with the use of most carefully studied methods. The results need not be repeated here, but it will be stated that, with the exception of a single water determination, all constituents were determined in duplicate; that in twenty out of a total of twenty-three instances, the discrepancy between duplicate determinations did not exceed 0·10 per cent; and that the maximum variation in the remaining three instances was 0·18 per cent. The single water determination which was not duplicated was controlled by a closely agreeing estimation of loss on ignition. In working out the ratios from these analyses, the method was adopted of calculating for the metals their equivalent of hydrogen, including fluorine with hydrogen, since tourmaline contains hydroxyl with which fluorine is isomorphous. Thus the ratio was found between SiO_2, B_2O_3, and Total Hydrogen, from which the empirical formula of the tourmaline acid was derived. For the sake of the present discussion the ratios will be repeated in two forms: with one-fourth of the SiO_2 as unity and also with one-twentieth of the Total Hydrogen as unity. This latter method has been here adopted, because a few relations can be brought out better in the discussion by so doing. The ratios of the two analyses are as follows:

	SiO_2	:	B_2O_3	:	Total H.	SiO_2	:	B_2O_3	:	Total H.
De Kalb	4·00	:	1·01	:	19·90	4·02	:	1·01	:	20·00
Haddam Neck	4·00	:	1·02	:	19·98	4·00	:	1·02	:	20·00

These ratios approximate *very closely* to the whole numbers 4:1:20; such close approximations, in fact, are seldom obtained, and cannot in these two instances be regarded merely as matters of accident; they are the reward, rather, of careful analytical work on material of unquestionable purity. As soon as the ratios were worked out, it was seen at once that at least one important key to the solution of the tourmaline problem had at last been found: *the empirical formula of the tourmaline acid must be* $H_{20}B_4Si_4O_{21}$.

And now, for the sake of the discussion, some space will be devoted to the ratios derived from the analyses of Riggs, and Jannasch and Kalb. These have already been given by Foote and the author[*] with $\frac{1}{4}$ SiO_2 as unity, and are now repeated together with the ratios derived by taking $\frac{1}{5}$ Total Hydrogen as unity. They have moreover been arranged in series, commencing with the closest approximation to $4:20$ between SiO and Total Hydrogen, and proceeding to the maximum deviation from this ratio.

TOURMALINE RATIOS DERIVED FROM THE ANALYSES OF RIGGS

No.	No. Dana.	SiO_2 : B_2O_3 : Total H.	SiO_2 : B_2O_3 : Total H
1.	43.	4·00 : 0·94 : 20·03	3·99 : 0 94 : 20·00
2.	45.	4·00 : 0·95 : 20·03	3·99 : 0·95 : 20·00
3.	48.	4·00 : 1·01 : 20·06	3·99 : 1·00 : 20·00
4.	47.	4·00 : 0·98 : 20·08	3·98 : 0·97 : 20·00
5.	52.	4·00 : 0·94 : 20·11	3·98 : 0·93 : 20·00
6.	49.	4·00 : 1·01 : 20·12	3·97 : 1·00 : 20·00
7.	36.	4·00 : 0·90 : 20·2	3·96 : 0·89 : 20·00
8.	44.	4·00 : 0·88 : 20·2	3·96 : 0·87 : 20·00
9.	46.	4·00 : 0·96 : 20·2	3·96 : 0·95 : 20·00
10.	42.	4·00 : 0·97 : 19·8	4·04 : 0·98 : 20·00
11.	54.	4·00 : 0·98 : 19·8	4·04 : 0·99 : 20·00
12.	39.	4·00 : 0·94 : 19·7	4·06 : 0·95 : 20·00
13.	41.	4·00 : 0·92 : 19·7	4·06 : 0·93 : 20·00
14.	51.	4·00 : 0·91 : 19·6	4·08 : 0·93 : 20·00
15.	37.	4·00 : 0·93 : 20·5	3·90 : 0·91 : 20·00
16.	38.	4·00 : 0·92 : 19·5	4·10 : 0·93 : 20·00
17.	55.	4·00 : 1·01 : 20·6	3·88 : 0·98 : 20·00
18.	40.	4·00 : 0·96 : 19·3	4·14 : 1·00 : 20·00
19.	50.	4·00 : 0·98 : 19·2	4·16 : 1·02 : 20·00
20.	53.	4·00 : 0·97 : 18·9	4·23 : 1·00 : 20·00
	Average	4·00 : 0·95 : 19·88	4·02 : 0·96 : 20·00

TOURMALINE RATIOS DERIVED FROM THE ANALYSES OF JANNASCH AND KALB.

No.	No. Dana.	SiO_2 : B_2O_3 : Total H.	SiO_2 : B_2O_3 : Total H
1.	62.	4·00 : 0·80 : 20·00	4·00 : 0·80 : 20·00
2.	64.	4·00 : 0·84 : 20·01	4·00 : 0·84 : 20·00
3.	61.	4·00 : 0 95 : 20·2	3·96 : 0·94 : 20·00
4.	57.	4·00 : 0·99 : 19·8	4·04 : 1·00 : 20·00
5.	56.	4·00 : 0·96 : 19·7	4·06 : 0·97 : 20·00
6.	63.	4·00 : 0·98 : 19·7	4·06 : 0·99 : 20·00
7.	58.	4·00 : 0·95 : 20·4	3·92 : 0·93 : 20·00
8.	60.	4·00 : 0·88 : 20·4	3·92 : 0·86 : 20·00
9.	59.	4·00 : 0·92 : 18·8	4·25 : 0·98 : 20·00
	Average	4·00 : 0·92 : 19·9	4·02 : 0·93 : 20·00

* Loc. cit., pp. 114 and 115.

Before entering upon the discussion of these ratios, let it be understood that the analysis of tourmaline is one of the difficult problems of analytical chemistry, and although Riggs made duplicate and often triplicate determinations of B_2O_3 and H_2O in almost all cases, and duplicated somewhat more than half of his determinations of SiO_2 and F, only single determinations of other constituents are recorded in his paper, while Jannasch and Kalb record only single determinations. Also it is to be borne in mind that although both Riggs and Jannasch and Kalb undoubtedly used carefully selected tourmaline fragments for analysis,[*] still there is nothing to indicate that slight amounts of foreign materials might not have been present in some of the specimens analyzed. Keeping these facts then well in mind, let us examine the ratios as presented in the foregoing tables. It is granted that the ratios are not exactly 4:1:20, and to get *exact ratios* from mineral analyses is not to be expected, but the close approximation to 4:1:20 in the case of the two analyses by Foote and the author, of sixteen out of the twenty analyses by Riggs, and of eight out of the nine analyses by Jannasch and Kalb, *constitutes an overwhelming amount of evidence in support of the empirical formula of the tourmaline acid*, $H_{20}B_2Si_4O_{21}$. It is safe to state that there does not exist a series of thirty silicate analyses of any one mineral yielding ratios which approximate so closely to whole numbers as the tourmaline analyses referred to above. That some analyses fail to yield a ratio as close to rational numbers as desired, reflects discredit neither upon the analyst nor upon the character of his work, for the material for analysis might not in all cases have been pure. Take, for example, No. 17 of the series of Riggs, brown tourmaline from Hamburg, N. J., occurring in calcite. The ratio of SiO_2 : Total Hydrogen$=4:20.6$. Evidently the bases are too high, and this particular analysis is peculiar in that it shows 5·09 per cent CaO, while the next highest percentage of CaO recorded is 3·70. The material might well have contained some calcite, either as small included nodules, or as an infiltration along cracks, and if the amount of calcite be assumed as 1·78 per cent, equivalent to 1 per cent CaO, the analysis would add up to 100·82, which is not too high for such a complicated substance, and the ratio of SiO_2 : Total Hydrogen would become 4·00 : 20·3 or 3·94 : 20·00. To assume that the Hamburg material probably contained some calcite seems far more reasonable than to speculate, as Prof. Clark does, upon some complex formula especially adapted to suit this particular analysis. Again, Nos. 18, 19, and 20 of Riggs, and 9 of Jannasch and Kalb indicate either

[*] Compare foot-note, this Journal, IV, vii, p. 115, 1899.

that the amount of base is low, SiO_2 being assumed as practically correct, or, what is far more likely, that the amount of SiO_2 is too high, as seen best when one-twentieth of the Total Hydrogen is taken as unity. Does the high silica ratio indicate that for these special cases a new type of tourmaline formula is needed, or is it not simpler to assume that the material from which these analyses were made might possibly have contained a little quartz or other silicate as impurity? It would take not over two per cent of quartz as an impurity to bring about the extreme amount of variation from the ratio 4:20 recorded in the foregoing tables.

Both Clarke and Tschermak seem to place implicit confidence in *all* of the recent tourmaline analyses: they seem to regard them as perfect, and the material analyzed as necessarily pure; consequently they try to devise formulas or expressions (how complicated they are will be shown) to suit all of the analyses. Clarke states that a formula in order to be satisfactory " must adequately express the composition of the compound in question, covering all of its variations." It is evident, however, that a formula should not cover variations due to impurities in the material analyzed, nor possible inaccuracies in analytical work. Both Clarke and Tschermak have their well-known theories concerning the composition of mica, and, seeing in the lithia-, iron-, and magnesia-varieties of tourmaline certain analogies, respectively, to muscovite, biotite and phlogopite, they both endeavor to force the tourmaline formulas to conform to their ideas concerning the constitution of mica. True, as shown by the analyses of Riggs,[*] there occur at Auburn, Rumford and Hebron, Maine, pseudomorphs of muscovite after tourmaline; but the fact need not necessarily be taken to indicate that tourmaline is closely related to mica, nor that, by what may be designated as a sort of *molecular cleavage*, tourmaline is transformed into mica. The author is familiar with these Maine localities, and he does not believe that such alterations are common, nor does he believe that it is common to find similar alterations at other localities where tourmaline is found. Muscovite is evidently a very stable molecular compound; it occurs more or less pure as pseudomorphs after many minerals, and not, as it would seem, because each and every one of these minerals contains as a nucleus the muscovite molecule, but because muscovite has such a tendency to form under a variety of conditions that it develops, provided materials suitable for its formation are at hand. Thus it is supposed that an ancient mud flat becomes converted to a shale and eventually to a mica schist, as the result of indurating and metamorphic processes,

---

[*] This Journal, III, xxxv. p. 41, 1888.

nd yet mud and shale are not supposed to contain the chemi-
al nucleus of muscovite; but, rather, they contain constituents
uitable for the formation of muscovite.

It is quite common at the Maine localities to find cavities in
quartz and other minerals once occupied by tourmaline crys-
als, and often there still remain in these cavities remnants of
be fresh, unaltered tourmaline. The tourmaline fragments
ppear as if etched, and much material has evidently been dis-
olved and carried away. Hence replacement pseudomorphs
f muscovite after tourmaline might well result from the sub-
equent filling of such cavities by muscovite. Replacement
seudomorphs are well illustrated by specimens from the Maine
ocalities in the Brush Collection. The specimens referred to
xhibit cavities containing remnants of fresh unaltered tourma-
ine, and as secondary minerals we now find cookeite and quartz
eposited on the walls of these cavities, with absolutely nothing
o indicate that the material for the formation of the cookeite
nd quartz was derived from tourmaline, for it could equally
vell have been derived from lepidolite and other minerals; in
act the tourmaline originally in a cavity could scarcely have
urnished sufficient lithia for the formation of the cookeite,
rovided that not a trace of the lithia was carried away. The
act, therefore, that pseudomorphs of muscovite after tourma-
ine have been described, cannot be taken and accepted as
roof that tourmaline is closely related to the micas, and that
he formula of tourmaline must show close analogies to the
ormulas of the minerals of the mica group.

Clarke's formula for the tourmaline acid is $H_{29}B_3Si_6O_{31}$,
vhich requires the ratio of $SiO_2 : B_2O_3 :$ Total Hydrogen to be
$6:1.5:29$, or $4:1:19.33$. It is true that three of the
nalyses of Riggs yield this ratio (Nos. 18 and 19, page 22, and
lso No. 20, if it is assumed, as is by no means proved,* that
he titanium exists as Ti_2O_3). Assuming Ti_2O_3 as existing in
No. 20, the ratio becomes $4.00 : 0.97 : 19.2$.

These three analyses then, out of a total of twenty by Riggs,
lone support Clarke's formula. Two of the three varieties
No. 19, black tourmaline from Pierrepont, N. Y., and No. 20,
rown tourmaline from Gouverneur, N. Y.) have been analyzed
y Rammelsberg, a fact ignored by Clarke. It may be claimed
by some that Rammelsberg's analyses are not to be taken into
consideration because they are old. It should be said, however, that Rammelsberg undoubtedly determined the bases and
loss on ignition in his analyses with considerable accuracy, and
his results need not be wholly ignored. Assuming that SiO_2
and B_2O_3 are present in tourmaline in the ratio $4:1$, it becomes

---

*See suggestions by Foote and the author, p. 117 of their article, loc. cit.

possible to determine both SiO_2 and B_2O_3 by difference, and this treatment of Rammelsberg's analyses was fully discussed by Foote and the author. In the present instance it is only necessary to repeat here the ratios between SiO_2 and Total Hydrogen derived from the two analyses, which are to be compared with Numbers 19 and 20 (page 22) of Riggs.

	SiO_2 : Total H.	SiO_2 : Total H.
Black tourmaline, Pierrepont, analyzed in 1889,	4·00 : 20·2	3·96 : 20·00
Brown tourmaline, Gouverneur, analyzed in 1850,	4·00 : 20·7	3·86 : 20·00

Thus Rammelsberg's results for these two varieties conform to the $H_{20}B_2Si_4O_{21}$ formula, rather than to the more complicated one proposed by Clarke. The analysis of the black tourmaline from Pierrepont is one of the last, if not the very last, analysis made by Rammelsberg, and he claimed for it a high degree of accuracy.[*] The ratio derived from it is certainly very close to 4 : 20.

As an aluminium boro-silicic acid, Clarke writes his formula in linear form as follows :

$$Al_6(SiO_4)_6(BO_2)_2.BO_2H_2.H_{12}.$$

The composition of the black tourmaline from Pierrepont (I) and the brown tourmaline from Gouverneur (II), as derived from Riggs' analyses, are then expressed by him as follows :

I. $\begin{cases} 13Al_6(SiO_4)_6(BO_2)_2.BO_2Ca.Mg_4H_4 \\ 7Al_6(SiO_4)_6(BO_2)_2.BO_2Mg.Mg_4H_4 \\ 2Al_6(SiO_4)_6(BO_2)_2.BO_2Na_2.Al_4Na_4H_2. \end{cases}$

II. $\begin{cases} 5Al_6(SiO_4)_6(BO_2)_2.BO_2Ca.Mg_4H_4 \\ 3Al_6(SiO_4)_6(BO_2)_2.BO_2Mg.Mg_4H_4 \\ 2Al_6(SiO_4)_6(BO_2)_2.BO_2NaH.Al_4Na_4H_4. \end{cases}$

The above molecular mixtures when united become :

I. $Al_{114}Mg_{61}Ca_{13}Na_8H_{64}B_{66}Si_{132}O_{662}$ and
II. $Al_{54}Mg_{35}Ca_5Na_8H_{42}B_{30}Si_{60}O_{310}.$

Thus, taking multiples of a molecule containing a large number of atoms, and admitting of many variations in the replacement of the hydrogen atoms, it becomes a matter not of chemical, but rather of arithmetical skill to so choose the numbers that the calculated percentage values shall agree with the results of the analyses : and all this labor when there is good evidence for believing that owing either to slight defects in the analyses or impurities in the materials the two analyses in question fail by a little only to give ratios like those of the remaining analyses.

[*] Abhand. K. Akad., Berlin, 1890.

Another difficulty also, which Clarke evidently recognizes, is that there is no appropriate place in his molecule for the attachment of fluorine, and hence he suggests that the fluorine may sometimes replace the group BO_2, an equivalent, as stated by him, which is strongly indicated in the cappelenite group of minerals. Now from a chemical standpoint the group cited containing cappelenite, melanocerite, caryocerite and tritomite, seems poorly adapted for illustrating an important chemical principle, since the composition of all of the minerals of the group is very complicated, and only in the case of one mineral, tritomite, have direct determinations of B_2O_3 been made. Clarke's formula, it will be observed, is applicable directly to the three analyses of Riggs' (Nos. 18, 19 and 20, p. 22) which are low in bases, or high in silica: it will therefore be convenient to designate it as the *acid formula A*. In order to adapt his formula to the many analyses in which the ratio of SiO_2 : Total Hydrogen is approximately 4 : 20, Clarke presents a modification of his formula consisting in the substitution of a basic, bivalent, aluminium-hydroxide radical (AlOH) for two of the hydrogen atoms of his acid, and he always represents the (AlOH) radical as replacing the two hydrogen atoms attached to BO_2. Fluorine in this formula is considered as combined with aluminium to form a bivalent radical (AlF) isomorphous with (AlOH) instead of replacing the BO_2 group. A hypothetical molecule containing the (AlOH) radical and employed for expressing the composition of the green tourmaline from Haddam Neck is as follows :

$$Al_6(SiO_4)_6(BO_2)_2.BO_2(AlOH).Al_2Li_3H_4.$$

By substituting for the metals in the foregoing formula their equivalent of hydrogen, and simplifying, there results an expression which may be designated the *basic formula B*, so designated because it contains more hydrogen atoms than the *acid formula A*. The two are given together for comparison :

A, *Acid formula*, $H_{20}B_3Si_6O_{21}$.
B, *Basic formula*, $H_{31}B_3Si_6O_{22}$.

The empirical formula of the tourmaline acid, as derived by Foote and the author, $H_{20}B_3Si_6O_{21}$, is equivalent to $H_{20}B_3Si_6O_{21\frac{1}{2}}$, which is exactly midway between the *acid and basic formulas* of Clarke. Thus in order to find expressions that will yield calculated percentage values agreeing with the several analyses becomes again not a matter of chemical science, but rather an arithmetical problem, and one, too, which is bound to succeed; for, given the two formulas, some mixture of the molecules can be found to fit any analysis which falls within the limits

of Clarke's two types. Thus to express the composition of
the green tourmaline from Haddam Neck, where the ratio of
SiO_2: Total Hydrogen is very close to 4:20, Clarke employs an
equal number of *basic* and *acid* molecules as follows:

Basic. $10Al_2(SiO_4)_2(BO_2)_2.BO_2(AlOH).Al_2Li_2H_4.$

Acid. $\begin{cases} 3Al_2(SiO_4)_2(BO_2)_2.BO_2.Ca.Fe_4H_4 \\ 1Al_2(SiO_4)_2(BO_2)_2.BO_2.Ca.Al_2NaH_4 \\ 6Al_2(SiO_4)_2(BO_2)_2.BO_2NaH.Al_2NaH_2. \end{cases}$

Prof. Clarke is very ingenious in his use of figures, and the
agreement between the calculated percentages and the results of
the analysis is indeed most satisfactory; but what is gained?
The analysis gave a ratio of almost exactly 4 : 20 and a mixture of ten of the *basic* and ten of the *acid* types of Clarke's
molecules must yield a ratio of exactly 4 : 20 ; hence it is a fore-
gone conclusion that the theory must agree with the analysis.
In order to bring about the agreement, however, a mixture of
molecules is employed containing an aggregate of 1092 atoms.
If we are to accept the idea that formulas must in some way
be found to suit all analyses, and, therefore, it is necessary to
have pairs of formulas of acid and basic types, why not accept .
as the acid type for tourmaline the formula proposed by Rammelsberg, $H_{10}B_2Si_4O_{20}$? By substituting two bivalent (AlOH)
groups for four hydrogen atoms, and then reducing to the
simple hydrogen expression, we obtain for the basic type H_{10}
$B_2Si_4O_{20}$, and by the proper replacement of the hydrogen atoms
by metals and the combination of the two types of formulas,
theoretical compositions could be calculated to a degree with
all of the analyses. Indeed, these two formulas have some
advantage over Clarke's, in that they are simpler, and priority,
at least for one of them, can be claimed. Doubtless other
pairs of formulas could be devised which would suit all variations presented by tourmaline analyses.
One other illustration presented by Clarke in support of his
theory may here be cited, black tourmaline from Auburn,
Maine, No. 2 of Riggs (page 22). The composition is expressed as follows:

Basic. $35Al_6(SiO_4)_6(BO_2)_2.BO_2(AlOH).Fe_4H_4.$

Acid. $\begin{cases} 2Al_6(SiO_4)_6(BO_2)_2.BO_2.Ca.Fe_4H_4 \\ 27Al_6(SiO_4)_6(BO_2)_2.BO_2NaH.Al_2NaH_2. \end{cases}$

The aggregate number of atoms in the above mixture of
molecules is 3499. Let it be observed (page 22) that as far as
the ratio of SiO_2: Total Hydrogen is concerned this analysis
is one of the very best, the ratio of SiO_2: Total Hydrogen
being 4:20.03, which would require the basic and acid types of

Clarke's molecules to be present in nearly equal proportion. There are employed by Clarke, however, 35 *basic* and 29 *acid* molecules, from which the ratio of SiO_2: Total Hydrogen = 4:20·06 is readily calculated: thus the ratio as derived from Clarke's molecular mixtures is farther removed from 4:20 than that derived from the analysis itself, and yet Prof. Clarke makes the following statement: "It will be noticed that the molecule A" (with him the basic one) "is in excess of the other two; a condition which fits the analyses, but which is incompatible with the formula proposed by Penfield and Foote." It may be stated, however, that the $H_{20}B_2Si_4O_{21}$ formula of tourmaline is not based upon hypothetical molecular mixtures, but, fortunately, upon actual analyses, and good ones, like the one by Riggs last cited, where the ratio is almost exactly 4:20. This example furnishes a good illustration of the fact that ratios are more serviceable in support of a formula than comparison between percentage values.

And, finally, Prof. Clarke, in the concluding pages of this article, presents his reasons for believing that the radical ($= Al - BO_2$) may be replaced "in part by the similar groups ($= Al - OH$) and ($= Al - F$)," or, in other words, that hydroxyl and fluorine are equivalent to, and isomorphous with, BO_2. Clarke bases this conclusion upon the fact that, in many of the analyses of Riggs, and Jannasch and Kalb, the amount of B_2O_3 found is not sufficient to yield a ratio of $SiO_2 : B_2O_3 = 4:1$ (see page 22), and the deficiency seems to him to be too great to be due to experimental errors. For explaining the occasional low determinations of B_2O_3 referred to, there is a far simpler way than the one proposed by Clarke: special pains were taken by Foote and the author to ascertain the conditions for accurately determining B_2O_3 in tourmaline, and it was found that by fusing the mineral with five times its weight of sodium carbonate and extracting with water, a little boron was still retained by the residue; a fusion of the residue with another portion of sodium carbonate was therefore made, and the boron was then determined by the well known Gooch method. It was further demonstrated that two fusions with sodium carbonate were sufficient for extracting all of the B_2O_3. Professor Riggs has kindly informed the author that in making his boron determinations by means of the Gooch method, he fused only once, but used, however, ten times as much sodium carbonate as mineral. It is hence probable that a double fusion with five times the weight of sodium carbonate is better than a single fusion with ten times the weight. As may be seen on page 22 the ratio of $SiO_2 : B_2O_3$ is very close to 4:1 in the majority of Riggs' analyses. All who had occasion to determine boron prior to the description by Gooch of his admira-

ble method, unite in testifying that accurate and reliable estimations of B_2O_3 were exceedingly difficult, in fact almost impossible, to obtain. Jannasch and Kalb employed Bodewig's[*]
modification of Marignac's method for estimating B_2O_3, weighing boron as KBF_4; hence, considering the complex character
of tourmaline, their determinations are certainly as close to the
truth as could be expected (see page 22), especially when it is
taken into consideration that some boron would be lost if
a double fusion with alkali carbonate was not made.

Tschermak seeks to explain the composition of tourmaline as
a mixture of two complex silicate molecules, as follows:

$$Si_{12}B_6Al_{16}Na_4H_6O_{62} = B_6Al_4O_{15}.4(Si_3Al_3NaH_2O_{13}) = Tu.$$
$$Si_{12}B_6Al_{10}Mg_{12}H_6O_{62} = B_6Al_4O_{15}.2(Si_3Al_3H_3O_{13}.Si_3Mg_6O_{13}) = Tm.$$

The radical $(Si_3Al_3NaH_2O_{13})$, in Tu is the generally accepted
paragonite formula, and $(Si_3Al_3H_3O_{13}.Si_3Mg_6O_{13})$ in Tm is
Tschermak's typical meroxene formula, except that a part of
the hydrogen in meroxene is replaced by potash. Thus tourmaline is supposed by Tschermak to contain mica molecules in
combination with the boron radical $B_6Al_4O_{15}$.

In order to express the composition of the two varieties of
tourmaline analyzed by Foote and the author, the following
relations are employed: for Haddam Neck, Tu_{44}, Tm_7, and for
DeKalb, Tu_{13}, Tm_{37}. The foregoing relations when written
out in linear form are, respectively, $Si_{612}B_{306}Al_{714}Mg_{84}Na_{116}H_{84}$
O_{3213}, and $Si_{600}B_{300}Al_{512}Mg_{444}Na_{12}H_{22}O_{3160}$. Thus with expressions,
each containing about 5500 atoms, Tschermak shows that the
calculated percentages agree in a satisfactory manner with the
results of the analyses after the latter have been very much
simplified by making numerous substitutions and recalculating
to 100 per cent.

The two formulas of Tschermak, $Si_{12}B_6Al_{16}Na_4H_6O_{62}$ (Tu)
and $Si_{12}B_6Al_{10}Mg_{12}H_6O_{62}$ (Tm) are both derived from an acid
$H_{60}B_6Si_{12}O_{62}$, which is three times the empirical formula $H_{20}B_2$
Si_4O_{21} proposed by Foote and the author. Tschermak states
that to refer tourmaline to the simpler acid $H_{20}B_2Si_4O_{21}$ is a
decided step in the wrong direction, since it was shown ten
years ago that the three-fold formula was correct. True, by
taking suitable mixtures of Tschermak's Tu and Tm molecules,
it is possible to calculate theoretical percentages which agree
closely with the results of simplified analyses, but that fact
does not necessarily prove the correctness of the formulas
under consideration, and that the true constitution of tourmaline has been established. In his admirable *Lehrbuch der
Mineralogie* Tschermak gives the composition of both pyrite

and marcasite as FeS_2; calcite and aragonite as $CaCO_3$; rutile, octahedrite and brookite as TiO_2; and cyanite, andalusite and sillimanite as Al_2SiO_5. If, therefore, for these compounds the simplest formulas are employed, it seems certainly a great presumption to state that in the case of tourmaline the complex three-fold formula had been definitely proved.

Finally Tschermak, who, like Clarke, does not seem to take into consideration that some analyses may not be reliable, tries to explain the composition of the tourmalines from Pierrepont and Gouverneur (Nos. 19 and 20, p. 22), and Tamatawe (No. 9, page 22), by assuming the existence of a third molecule, $Tn = Si_{12}B_2Al_9Mg_{12}H_9O_{61} = B_2Al_6O_{12}.2(Si_3Al_3H_6O_{11}.Si_3Mg_6O_{12})$, in which the portion $(Si_3Al_3H_6O_{11}.Si_3Mg_6O_{12})$, is supposed to be analogous to phlogopite. It does not seem reasonable, however, to believe that the tourmalines from the localities in question are essentially different from those from other localities, and need, consequently, different formulas, especially since we have a recent analysis of the Pierrepont, and an early analysis of the Gouverneur varieties of tourmaline by Rammelsberg, both of which conform to the $H_{20}B_2Si_4O_{21}$ formula.

Summary.—As shown by the tabulation of ratios on pages 21 and 22 there exist a series of recently made and carefully executed tourmaline analyses which give ratios of $SiO_2 : B_2O_3 :$ Total Hydrogen approximating closely to $4:1:20$, from which the empirical formula of the tourmaline acid, $H_{20}B_2Si_4O_{21}$, is derived. That a few analyses do not yield ratios agreeing as closely as desired to $4:1:20$ is not to be wondered at, when the difficulties presented by the analysis are taken into consideration, together with the fact that the material analyzed might not in all cases have been perfectly pure and homogeneous. As far then as analytical evidence may be relied upon for establishing the formula of a mineral, it may be considered *as definitely proved that the empirical formula of the tourmaline acid is $H_{20}B_2Si_4O_{21}$.* The science of inorganic chemistry has not yet reached such a state of development that it can be proved, as stated by Tschermak, that the threefold formula, $H_{60}B_6Si_{12}O_{63}$, is the correct one. The empirical formula $H_{20}B_2Si_4O_{21}$, proposed by Clarke, can rest only on the analytical evidence supplied by a few analyses of Riggs and one by Jannasch and Kalb which yield ratios approximating to $4:1:19.33$ (page 22), and there are good reasons for believing that these ratios would not be obtained a second time if the analyses were repeated. Since tourmaline always yields sufficient water to form two hydroxyl radicals, it may be considered *as probably, if not absolutely, proved that the formula of the tourmaline acid should be $H_{18}(OH)_2B_2Si_4O_{19}$.* Beyond this point it seems safe only to speculate and it cannot be considered that the ideas presented are capable of being definitely proved. All of the analyses indicate that at least half of the hydrogen atoms of the tourmaline acid, are replaced by aluminium, and this fact, coupled with the idea that it seems reasonable to unite the two hydroxyl radicals with the two boron atoms, led to the suggestion by Foote and the author[*] that the characteristic feature of all varieties of tourmaline is an *aluminium-borosilicic acid*, $H_4Al_6(B.OH)_2Si_4O_{19}$. In this acid the mass effect of the $(Al_6(B.OH)_2Si_4O_{19})$ is regarded as so overwhelming that it makes no difference how the nine remaining acid hydrogen atoms are replaced, whether largely by aluminium and to a trifling extent by bivalent metals and alkalies, or largely by magnesium and to a trifling extent by aluminium and alkalies, the result in all cases is tourmaline with its characteristic crystalline structure. That trivalent, bivalent and univalent metals, playing as it were the role of isomorphous constituents, may unite in replacing the nine hydrogen atoms of the tourmaline acid, is indeed a remarkable feature of isomorphism, but it furnishes an explanation of the composition of tourmaline, and one which can be comprehended, in part at least.

Looked at from the standpoint of an instructor, what explanation of the chemical composition of tourmaline can be given to a student provided the ideas of Tschermak prevail? Only this, that the composition is exceedingly complicated; that there are two molecules *Tu* and *Tm* (page 30), exhibiting certain analogies to minerals of the mica group, which mix in varying proportions, and that by taking appropriate multiples of the two molecules theoretical compositions can be calculated to agree with the results of analyses, provided the latter are very much simplified. By taking molecular mixtures containing in the aggregate several hundred and even thousand atoms, as done by Clarke (page 28) and Tschermak (page 30), it would seem as though chemists or perhaps arithmeticians might aspire to devise formulas for expressing the chemical composition of any sort of substance of which any kind of an analysis has ever been made.

[*] Loc. cit., p. 118.

Sheffield Laboratory of Mineralogy and Petrography,
Yale University, New Haven, June, 1900.

ART. IV. — *Studies in the Cyperaceæ;* by THEO. HOLM.
XIII. *Carex Willdenowii* and its allies. (With three figures
in the text.)

TORREY, in his monograph of North American *Cyperaceæ*,
was the first author to call attention to the peculiar habit of
Carex Willdenowii Schk., which he suggested separating from
Carex proper as a distinct genus, and to which the name
"*Phyllostachys*" would be appropriate. The foliaceous bracts
and the distinctly articulated base of the style were the characters, which Torrey considered as being of sufficient importance for the establishment of this new genus. Later authors
have all agreed, however, in not adopting this genus, but
merely to accept *Phyllostachys* as a section of *Carex*, and it
seems very natural, since such articulation of the style is also
observable in several other species of very remote relationship,
besides that the foliaceous bracts are not characteristic of this
species alone. A strange coincidence is, that another botanist
made a similar suggestion, but many years later, concerning a
Carex, which in many respects is analogous with *C. Willdenowii*.
Duval-Jouve discovered in France, near Montpellier, a species
of *Carex*, which he named *C. œdipostyla* on account of its
articulated style base, and in which he noticed the rhachis to
be winged and the bracts to be foliaceous; this author suggested the establishment of a new section: "*Œdipostyla*,"
and he considered *C. moesta* Kunth and *C. phalaroides* Kunth
as its nearest allies, since, also, these possess a "stylus basi bul-
boso-incrassatus." Moreover Duval-Jouve proposed *C. phyllostachys* Mey. as a member of this same section, on account of
its small number of female flowers, its foliaceous bracts, its
winged rhachis, the shape of its utricle and finally its articulated style-basis. But, on the other hand, Duval-Jouve does
not seem to have known *C. Willdenowii* and he evidently did
not know of Torrey's monograph, since he makes no allusion
to this; he would, no doubt, have counted *C. Willdenowii*
among his *Œdipostylæ* inasmuch as the utricle of this species
differs less from that of his new species than that of *C. phalaroides* and *C. moesta*. We may thus consider *Phyllostachys*
and *Œdipostyla* as analogous forms of our genus *Carex*, but it
will be necessary to remove some of the species, which formerly have been considered as representatives of these sections.
In examining the section *Phyllostachys* as adopted by
Carey,[*] this contains, as understood by him, only *C. Willdenowii* Schk., *C. Steudelii* Kth. and *C. Backii* Boott, all of

[*] For references consult the bibliography appended to this article.

which may surely be regarded as closely related to each other, thus forming an apparently natural section or group. The same view was, also, held by Tuckermann, who placed them as his *Phyllostachyæ* next his *Careyanæ*, in which we find *C. plantaginea, C. gynomane* and *C. depauperata*. It was Professor L. H. Bailey that first enlarged the section *Phyllostachys*, and this author includes not only the three species enumerated above, *C. Willdenowii*, etc., as a subsection "*Bracteatæ*" (*Bractoideæ* Bailey), but also *C. Geyeri* Boott and *C. multicaulis* Bail. as representing another subsection "*Phyllostachyæ*," however with no very clear distinction. As a matter of fact, the section *Phyllostachys* became confused by this enlargement, which is readily noticeable from the statement by Professor Bailey that the section is not only "to be regarded, probably, as an offshoot of the *Montanæ*, but also to be connected with the *Laxifloræ* by *C. multicaulis*, being related to *C. Hitchcockiana*." It is true that Drejer considered his *Carices dactylostachyæ* as derived from his *C. sphæridiophoræ*, but this author would not have included the true *Phyllostachys* in any of these, judging from the diagnosis, while he might have included *C. Geyeri* among the *Dactylostachyæ*. In a lately published paper on the *Carices* of South America, a German author, Rev. G Kükenthal, follows Prof. Bailey's disposition of *Phyllostachys*, and includes another species, *C. Sellowiana* Schl., which, however, shows no marked affinity to either *C. Willdenowii* or to *C. Geyeri*. Rev. G. Kükenthal, moreover, considers *Phyllostachys* as a subsection of Drejer's *Sphæridiophoræ*, which does not seem to be in conformity with the original diagnosis of this section. While thus Torrey's *Phyllostachys*, even if reduced to a section, has been maintained through so many years, Duval-Jouve's *Œdipostyla* does not seem to have gained recognition at all. In the Catalogue of European species of *Carex*, Dr. H. Christ has only enumerated *C. œdipostyla* as one of the *Digitatæ* Fries and placed it between *C. Halleriana* Asso and *C. pilulifera* L., while this same author has counted *C. depauperata* Good and *C. olbiensis* Jord. among the *Phyllostachyæ* of Torrey. If we consider the *Digitatæ* as understood by Dr. Christ, these contain representatives that are so little related to each other, that we must combine the *Sphæridiophoræ*, the *Dactylostachyæ* and the *Lamprochlænæ* with each other in order to get them all together. It may be, however, that *C. œdipostyla* is in some respects related to, for instance, *C. digitata* L., as suggested by Dr. Christ, but we cannot find any characters by which we might connect this with *C. pilulifera* or with *C. supina* Wahlbg. Future research must decide the proper place for *C. œdipostyla;* in the present paper we intend only o discuss the position of the old section *Phyllostachys*, as lefined by Torrey.

We have seen from the above that the section has been conidered as an offshoot of the *Montanæ* Fr., which would mean o indicate some relationship between the *Sphæridiophoræ* on he one side and *Phyllostachys* on the other. We have, also, een that it has been regarded as a subsection of the *Sphæridiohoræ*, which would even indicate a still closer relationship to hese. The question is now whether our conception of the *Sphæridiophoræ* allows any such combination, if we take this n absolute accordance with Drejer, the author, who established the section. As long as we quote these sections as those of Drejer, and not " ex parte," we are supposed to follow the diagnosis strictly with no modifications whatsoever.

Let us read what Drejer says about the *Sphæridiophoræ*, on page 9 in his Symbolæ Caricologicæ: " Spica mascula unica. Spicæ foemineæ una vel plures, rotundatæ, sibi et masculæ approximatæ, sessiles. Bracteæ membranaceæ nervo dorsali excurrente herbaceo, vel herbaceæ ad basin membranaceæ. Perigynium membranaceum, pube hirsutie vel tomento vestitum, caryopsin arcte includens, rostro breviore longioreve apice scarioso hyalino subbilobo. Stigmata terna, caryopsis trigona. Centrum habet hic grex in *C. pilulifera, ericetorum* cet., ex quibus character sumendus est. Ortum ducit inter monostachyas, nominatim in *C. Wormskjoldiana Hornem,*" etc. The central species are all of ordinary habit; none possess foliaceous bracts replacing the *squamæ;* in none of these are there androgynous spikes on long axillary, basal peduncles; in none of these is the rhachis winged or zigzagged, nor is the male portion of the inflorescence so inconspicuous as in *Phyllostachys;* furthermore the utricle is not extended into a long, straight and rough beak in any of these. It seems difficult to understand, how any author could see such close relationship to exist between these sections, so distinct do they really appear. And even if we, agreeing with Drejer, see no objection for admitting species with almost simple or very little decompound inflorescence among such, which possess a number of separate spikes, androgynous or with separate sexes, we nevertheless cannot detect any connecting forms by which we might unite the *Sphæridiophoræ* or *Dactylostachyæ* with *Phyllostachys,* inasmuch as we cannot even consider *C. Geyeri* and *C. multicaulis* as members of this section, the *Phyllostachys* of Torrey.

It is now interesting to notice, that Drejer did not only decline to accept *Phyllostachys* as a genus, but that he did not even consider any of its species with enough significance to constitute even a special section of *Carex*, although he was well acquainted with these, at least with *C. phyllostachys, C. Willde-*

nowii and *C. Steudelii.* However, since authors of recent years have seen no difficulty in retaining this as a section, it will be necessary to analyze it again in order to decide the actual importance of its morphological characters: the structure of the inflorescence.

Carex Willdenowii possesses a terminal, androgynous spike with male flowers at the top, and a few lateral similarly androgynous, which are developed on long peduncles from the axils of the basal leaves. The plant thus shows a somewhat peculiar habit, which is not frequently met with in the *Carices genuinæ*, but which, to some extent, suggests certain species of *Vigneastra* Tuckerm. If we consider the simple androgynous spike by itself, this is just one of the fundamental characters of the old group "*Monostachyæ*" with its sections *Psyllophoræ* and *Capitatæ*; but if we, furthermore, include the basal, long-peduncled spikes as being merely a part of the main inflorescence, the plant becomes better referable to the group *Heterostachyæ*. In these the lateral peduncles are provided with a clado-prophyllon at their base, an organ which is, also, present in *C. Willdenowii.* The basal position of these lateral spikes, or to be exact "spicate inflorescences," is very characteristic in *C. Willdenowii*, while these are situated higher up on the culm in most of the *Heterostachyæ*, and are more or less remote from each other, as we remember from *C. sylvatica* Huds., *C. laxiflora* Lam., etc. This distinction, drawn from the basal position of the lateral inflorescences in *C. Willdenowii* and its nearest allies *C. Steudelii* and *C. Backii*, is not, however, applicable only to these species, but to several others, which show no affinities or immediate relationship to the members of *Phyllostachys*. It is to be observed in *C. basilaris* Jord., *C. phalaroides* Kth., *C. pedunculata* Muhl., *C. Linkii* Schk. and in *C. ædipostyla* Duv.-Jouve, which may belong to the sections *Trachychlænæ* and *Dactylostachyæ* of Drejer. While thus our species of *Phyllostachys* may be arranged very naturally among the *Heterostachyæ*, still another disposition may be made when we examine the section *Vigneastra* of Tuckermann. As described by this author, these species possess decompound, ramified inflorescences, which are always androgynous, male at the top, and may have two or three stigmata. If we, carefully, compare these species with *Phyllostachys*, we soon discover that each branch of *C. cladostachya* Wahlbg., for instance, corresponds to a complete flower-bearing stem of *C. Backii* with its terminal and axillary inflorescences.

The clado-prophylla P in *C. cladostachya* (fig. I) correspond to those same organs, P, in *C. Backii* (fig. II); the bracts L correspond to the green leaves L in *C. Backii*, of which only the scars are indicated; the androgynous spikes A, B, C and D to those of

C. Backii. Finally the terminal androgynous spike A is equivalent with the same, A, in this species, and the utriculi, painted black in the figures, correspond to those of *C. Backii.* We might even go still farther and confine ourselves to the uppermost lateral branch in *C. Backii* (B in fig. II), which showed the

EXPLANATION OF FIGURES.

Fig. I. *Carex cladostachya* Wahlbg., lateral branch of inflorescence, enlarged. A = the terminal, B, C and D = the lateral spikes, all androgynous with the male flowers at the top; P = clado-prophyllon; L = leaves, the lowest one removed. Utriculus is painted black in all the spikes

Fig. II. *Carex Backii* Boott, the complete inflorescence of a single shoot, enlarged. Letters as above; the leaves (L) have been removed. All the spikes are androgynous with the male flowers at the top; the uppermost lateral spike (B) is developed upon the rhacheola of an abnormal utriculus, of which the female flower is merely rudimentary. Utriculus is painted black.

Fig. III. *Carex œdipostyla* Duv.-Jouve; the terminal inflorescence, showing the zigzagged rhachis and the leaf-like bracts; copied from Duval-Jouve, but reduced one-half.

common case of a prolonged, flower-bearing rhacheola. In this specimen, the little branch B constitutes the rhacheola bearing four female, one-flowered spikelets and one terminal, male spike, while the utricle at the base has acquired a corresponding development as the clado-prophyllon P in *C. clado-stachya*: thus the ramified rhacheola demonstrates exactly the same structure as the branches B, C and D in *C. cladostachya*. Considered from this point of view the section *Phyllostachys* Torr. is morphologically inseparable from any of the other sections of *Eucarex*, of *Vignea* or *Vigneastra*. Utriculus shows in these sections or subgenera, as they have been considered by some authors, the same plasticity of developing either as a membranaceous clado-prophyllon at the base of peduncles or as a normal utriculus, surrounding the female flower. The extension of the rhacheola may sometimes result in the suppression of the female flower, but not always. Both Duval-Jouve, Gay, Roeper and Schulz enumerate not a few cases where the female flower had developed normally; besides that the rhacheola protruded through the orifice of the same utriculus, bearing one or two female flowers with bracts, utricles and pistils. The clado-prophyllon at the base of the lateral peduncles has, also, been observed to contain a pistil, perfect or rudimentary. But a case like that which we have observed in *C. Backii* differs from most others by the development of a complete male spike besides the female ones at the base; it is evidently an exceptional occurrence, and it illustrates exactly one of the minor inflorescences in the genus *Schœnoxiphium*. As a matter of fact, this genus, as well as *Uncinia* and *Kobresia*, show a very striking resemblance to *Carex*, when considered from a morphological viewpoint.

As regards the development of the lower bracts in *Phyllostachys* into green leaves instead of membranaceous scales, this character does not seem to be of much importance as far as concerns the systematic position of the species. Similar green bracts are, also, known from the other sections, for instance in specimens of *Carex scirpoidea* Michx. of the *Sphæridiophoræ*, where a single or sometimes even several female spikelets may be seen subtended by leaf-like bracts, a case which we noticed in the mountains of Greenland. Moreover in *Carex Hilairei* Boott, the lower bracts of the androgynous spike are developed into long leaves, as figured by Boott (Plate 468); this species may probably be referred to the *Dactylostachyæ*. Furthermore *C. phyllostachys* Mey., *C. multicaulis* Bail. and *C. ædipostyla* Duv.-Jouve show similar leafy bracts, and have, therefore, been regarded as members of *Phyllostachys*, although they do not seem to possess the most important character of this section and are, perhaps, better placed in a section parallel

ith the · *Dactylostachyæ.* The zigzagged rhachis is another
laracter, which has been claimed as important to *Phyllostachys*
orr., yet this is, also, noticeable in *C. arctata* Boott, in *C. phyl-
stachys* Mey. and *C. œdipostyla* Duv.·Jouve and several others.
he winged margins of the rhachis is, on the other hand, a rare
:ature in *Carex*, and seems only to have been observed in the
)ecies of *Phyllostachys* Torr., in *C. œdipostyla* and *C. phyllo-
'achys.* Finally to be mentioned is the utricle, which, as already
ated by Drejer, is one of the most essential organs for classify-
ig the *Carices* in sections, and he distinguishes between three
ategories within *Eucarex :* " Prima est, ubi margines tam
rcte circa caryopsin concrescunt, ut verum rostrum non
ormetur; perigynium tum aut apice perforatum est, aut bre-
·issime rostellatum ore integerrimo vel subemarginato. (*C.
:olytrichoides, baldensis, pallescens, atrata, vulgaris* cett.)
Secunda est, ubi rostrum quidem formatur, sed ab inferiore
perigynii parte non aut vix distinguitur, estque apice hyalinum,
bilobum aut irregulariter bifidum. (*C. pilulifera, ericetorum
et aff., frigida* et aff. cet.) Tertia denique et perfectissima est,
ubi rostrum verum et distinctum, apice distincte bifidum aut
bicuspidatum formatur. (*C. distans* et aff., *C. vesicaria* et aff.
cet.) These sections are again divided into "greges" : *Carices
melananthæ,* etc., which are distinguished by the consistency
of utriculus, whether membranaceous or spongy, glabrous or
hairy, and furthermore by the disposition of the sexes, though
of lesser importance.

If we now apply the classification, based upon the structure
of utriculus, it is readily to be seen that *Phyllostachys* as it
stands at present, including *Bracteatæ* and *Phyllostachyæ,* is no
very natural section: that the *Bracteatæ* are quite distinct
from the latter. We have already in the preceding given our
views about the position of Professor Bailey's *Phyllostachyæ* as
more properly to be arranged parallel with the *Dactylostachyæ,*
while *Phyllostachys* proper shows certain analogies with
Drejer's *Hymenochlœnæ.* In any case the distinction that has
hitherto been drawn between *Phyllostachys* and the other sections is by no means tenable, nor have any accordances been
proved to exist between these species and Professor Bailey's
Phyllostachyæ : C. Geyeri, etc. Moreover, when Drejer did
not exclude the distribution of the sexes as being of some
importance to the classification, he, nevertheless, was well
aware of the fact that "*formæ hebetatæ*" do exist in most,
if not in all, the "greges," which he proposed. These
"*formæ hebetatæ*" constitute such species as may naturally
be looked upon as old types of the respective sections or
"*greges,*" for instance *C. scirpoidea* Michx. as being the type
of *Sphæridiophoræ,* etc. By combining the morphological

features of the extreme forms, the simpler with the higher
developed, we cannot avoid noticing that they occasionally
unite in some characters, which may be understood as exhibiting their descent through modifications from a common, or
fundamental type.   We may in this manner take *Phyllostachys*
of Torrey to be a lesser developed form of *Hymenochlœnœ*,
and these species "sensu strictiori" do represent several analogies in habit and structure.   But we have, also, demonstrated
in the preceding, that *C. Buckii*, *C. Willdenowii* and *C.
Steudelii* show characters that are, also, common to *Vigneastra*,
which, however, does not necessarily indicate anything more
than this section possesses characters that may be compared
with those of the "*Carices genuinœ.*"   *Vigneastra*, as far as
these peculiar species are understood, represent some certain
transition between *Vignea* and *Eucarex*.   And if it were not
for the structure of utriculus and the three stigmata in *C.
Willdenowii*, we might have had just as good reason for placing this among the *Vigneœ* as a highly developed type, combining these with *Vigneastra* of Tuckermann.   And in regard
to the systematic position of *Vigneastra*, if this is to be retained
as a subgenus, it may be most naturally placed between *Vignea*
and *Eucarex*, as suggested by Rev. G. Kükenthal, instead of
as a section between *Microrhynchœ* and *Hymenochlœnœ* as
proposed by Professor Bailey.   In returning to our *Phyllostachys*, the species by which it is represented in accordance
with Torrey, may seem naturally to be referred to the *Hymenochlœnœ*, but as extreme and poorly developed forms ; they may
occupy a position almost as far from the central types: *C.
sylvatica* Huds. and *C. cherokeensis* Schw. as *C. nepalensis*
Sprgl. and *C. longipes* Don :

Hymenochlœnœ.

   I. Spicis androgynis, apice masculis, paucifloris. (*C. Willde-
      nowii*, *C. Steudelii* et *C. Backii.*)
  II. Spicis simplicibus sexu distincto plus minus densifloris. (*C.
      sylvatica* Huds., *C. cherokeensis* Schw. cet.)
 III. Spicis androgynis, apice masculis, plus minus densifloris.
     (*C. nepalensis* Sprgl., *C. longipes* Don).

In the preceding pages we have attempted to demonstrate
the systematic position of these species, formerly constituting
the genus *Phyllostachys* Torr. ; we shall, now, briefly discuss
the internal structure of these in order to demonstrate the
differences between them and *C. Geyeri*, *C. œdipostyla* and
C. multicaulis.   On the other hand, we freely admit, that
divergencies in anatomical structure may not necessarily indicate distant relationship, especially not if we adopt the classifi-

cation of Drejer, where the species are arranged not in accordance with such artificial characters as are applied for the separation of *mono-, homo-* and *heterostachyæ,* or of *di-* and *tri-stigmaticæ,* but in accordance with purely natural affinities. It becomes, thus, evident that each of Drejer's "*greges*" or sections, as we may call them, is really an aggregate of a number of apparently very diverse forms, beginning with the "*formæ hebetatæ*" and passing gradually over into "*formæ centrales,*" from which again a number of allied species, but somewhat differently developed, extend towards the limits of the section as "*formæ desciscentes.*" An anatomical investigation of some of these parallel forms, the "*hebetatæ*" for instance, will no doubt reveal certain analogies with the corresponding "*formæ hebetatæ*" of other sections, while the central types may be very distinct. In this way the anatomical characters become obscure, and it will be necessary to make many comparisons between the various forms of each section, before we shall know where to draw the distinction between the characters of the section "as its own," and those, which may be regarded as inherited from old types through its "*formæ hebetatæ.*" Our anatomical study of *Carex Willdenowii* and its allies must, therefore, necessarily be incomplete, but may, perhaps, be useful in further research, at least as a contribution to comparative studies of other species of the genus. We shall not, however, confine ourselves to these species, but we will compare their structure with that of *C. œdipostyla, C. Geyeri* and *C. multicaulis.* The structure of utriculus is, no doubt, the most important, yet, as will be shown, some interesting characters may, also, be obtained from the root, the stem and the leaves. In regard to the localities, from where our material was gathered, we might state that both *C. Willdenowii* and *C. Steudelii* were collected in the woods near Brookland, D. C.; that the former especially prefers shaded places in rich woods, while the latter inhabits drier ground in thickets or open woods; *C. Backii* was received from Mr. James M. Macoun in Ottawa, as collected on dry, grassy and rocky places in open woods and thickets near Ottawa; *C. Geyeri* from the Rocky Mountains in Montana and British Columbia, in somewhat damp ground; *C. multicaulis* from Wyoming; *C. œdipostyla* from thickets near Montpellier in southern France.

The root.

The epidermis and hypoderm show a uniform structure in all three species: *Carex Willdenowii, C. Steudelii* and *C. Backii.* The cortex is thin-walled throughout in the last species, but shows a distinctly thickened outer zone in the two other species. The inner cortex shows the usual tangential collapsing. Endodermis is exceedingly thick-walled in *C. Willdenowii* and *C. Backii*, but not so in the third species. The pericambium is more or less thick-walled in all three species and it is, also, interrupted by all the proto-hadrome vessels. In *C. Backii* we noticed that the average number of pericambium-cells between each two proto-hadrome vessels was only two or three, while there were four in the others. The conjunctive tissue was quite thick-walled in *C. Steudelii* and *C. Willdenowii*, but much less so in *C. Backii*. The root-structure is thus very uniform in these species, and agrees very much with that of *C. œdipostyla*, except that the pericambium is much more thick-walled in this species, besides that there are here mostly five pericambium-cells between each two proto-hadrome vessels. Very different is the structure exhibited by *C. Geyeri* and *C. multicaulis*. The cortex in the former consists exclusively of very thick-walled parenchyma, which is more or less collapsed though without leaving any lacunes ; in the other species, *C. multicaulis*, it is only the outermost five or six strata that are thick-walled, while the inner are quite thin-walled, but not collapsed, although the root was apparently of the same age as the others. The endodermis is exceedingly thickened, especially on the inner walls, and very porose. The pericambium, which is not very thick-walled, is in both species only interrupted by some of the proto-hadrome vessels, but in different ways : in *C. multicaulis* every other proto hadrome vessel bordered on endodermis, in almost regular alternation, while in *C. Geyeri* there was only one vessel out of thirteen which did not border on the endodermis ; in some other specimens we noticed that fourteen of twenty-three proto-hadrome vessels were situated inside the pericambium.

The rhizome

is cespitose and very short in our species with the exception of *C. Geyeri*, in which there is a creeping rhizome, but of very short internodes and sympodial as in most of the other *Curices*. The cortical parenchyma is quite broad and somewhat thickened all through in *C. Steudelii*, but not so in *C. Willdenowii*. The endodermis is thickened like an O—endodermis in both species. The stereome is only to be observed as accompanying the mestome-bundles, which it surrounds completely, while it does not occur in the cortex as independent strata or groups. The mestome-bundles are irregularly scattered in several, about four bands, and are nearly all perihadromatic ; there is, furthermore, a solid, central pith of slightly thickened cells. In *C.*

Geyeri the cortical parenchyma shows about eight thick-walled strata near the periphery, while the inner ones are very thin-walled ; otherwise the structure of the stereome, the endodermis, etc., is like that of *C. Willdenowii* and *C. Steudelii.*

The stem.

The above-ground stem is somewhat rough from prickle-like projections from the epidermis and triangular in *C. Willdenowii*, *C. Steudelii* and *C. Backii.* However, when we examine the structure closer, it is readily seen that the outline is so sharply triangular that the term "three-winged" may be well applicable to these species. The cuticle is smooth and thin in all three species ; the epidermis shows a distinct thickening of the outer walls, with the exception of the cone-cells ; stomata are quite abundant. The cortical parenchyma is very poorly developed in *C. Steudelii*, consisting only of one stratum short, roundish cells, which border on wide lacunes. In *C. Willdenowii* this same parenchyma is better developed and shows several layers with relatively narrow lacunes ; in *C. Backii* there are distinct palisades, but the lacunes are very wide in this. The stereome is thickwalled in *C. Willdenowii* and *C. Backii*, but not so in the third species ; it accompanies the mestome-bundles either as hypodermal groups or separated from the epidermis by the cortex ; it is, also, present on the hadrome side of the mestome-bundles ; it reaches its highest development in the three wings, where it, furthermore, occurs as isolated groups, not being in contact with any mestome-bundles. These, the mestome-bundles, are very regularly arranged in one band near the periphery and represent larger oval in alternation with smaller, almost orbicular bundles, when viewed in transverse sections ; the parenchyma-sheath is thinwalled, and the mestome-sheath shows its inner cell-walls to be distinctly thickened. A thinwalled pith occupies the center of the stem ; it is hollow in *C. Backii*, solid in the other species.

In *C. œdipostyla* the stem is triangular, but without being winged, and the outer cell-walls of epidermis are more heavily thickened than in the former species ; the cortex is more open, there being very distinct intercellular spaces besides lacunes of quite considerable width. The stereome is thickwalled, but does not occur separate from the mestome-bundles, which are here represented in two concentric bands, instead of but one as in the former species ; the mestome-sheath shows the cells to be thickened all around. A partly hollow pith occupies the center of the stem. In *C. Geyeri* and *C. multicaulis* the stem is longer than in the former species and shows, also, a much more solid structure; the cortex constitutes a compact palisade-tissue with relatively small lacunes, especially in the latter species. The stereome occurs in larger groups and is very thickwalled, but invariably accompanying the mestome-bundles; the sharp angles of the pentagonal stem of *C. Geyeri* have no isolated stereome, but merely cortical parenchyma; the terete stem of the other species possesses a larger number of mestome-bundles and, consequently, it contains, also, more stereome. The mestome-bundles constitute a single band and the mestome-sheath is very heavily thickened. The solid pith is thickwalled in *C. Geyeri*, but thinwalled in the other.

The leaf

is narrow, but flat in *C. Willdenowii* and *C. Steudelii*, relatively broad in *C. Backii*. The cuticle is thin and smooth in the former species, but quite thick in the latter. Epidermis exhibits several modifications; it is perfectly glabrous and smooth on the upper face in *C. Steudelii* and *C. Willdenowii*, and is differentiated into a row of bulliform cells just above the midrib; above the stereome the epidermis, viewed " en face " consists of narrower and shorter cells than those which cover the mesophyll. If we examine the lower surface of the leaf we notice a corresponding structure, but there are no bulliform cells, and stomata are observable outside the meso-phyll; these are but slightly projecting and possess a wide but shallow air-chamber. If we consider the leaf of these species in cross-sections, we notice that the outer wall of epidermis is slightly thickened and somewhat projecting in the shape of very minute papillæ on either side of the midrib, but only on the lower surface. In *Carex Backii* the epidermis-cells are very short outside the stereome, much shorter than we observed in the former two species; their outer wall is very thick and we find here small wart-like projections from the epidermis, which form several rows on each side of the bulliform-cells, besides that there are four such projections around each stoma. The mesophyll is differentiated into a distinct palisade-tissue on the upper face of the leaf of *C. Steudelii* and *C. Willde-nowii*, while it consists of roundish cells with large intercellu-lar spaces on the lower face, almost as a typical pneumatic tissue; large lacunes traverse the mesophyll in both species. In *C. Backii*, on the other hand, the mesophyll is nearly homo-geneous throughout the blade, very open and without palisade-cells. The stereome is thickwalled in *C. Willdenowii* and *C. Steudelii*, but quite thinwalled in *C. Backii*. It accom-panies the mestome-bundles as hypodermal groups on either face, at least in the larger bundles, and there is, also, an isolated group in each of the two margins. The mestome-bundles are arranged in one plane, larger alternating with smaller ones; the parenchyma-sheath is thin-walled and color-less; the mestome sheath shows distinctly thickened inner cell-walls. The bracts, which subtend the one-flowered, female spiculæ in *C. Steudelii, C. Backii* and *C. Willdenowii* are leaf-like, at least the lower ones, and show a structure corre-sponding to that of the proper leaves, only that they are rela-tively narrower and possess more stereome than these. The bracts of the·uppermost female spikelets are, on the contrary, scale-like, awned. membranaceous and with hyaline margins like the scales of the *Hymenochlœnæ* in general.

In comparing the leaf of *C. Steudelii, C. Backii* and *C. Willdenowii* with that of *C. œdipostyla* we notice almost the same structure; but the outer wall of epidermis is thicker in this species, and without any development of papillæ, as we noticed in *C. Backii* for instance; the mesophyll is very open and is not differentiated into any distinct palisade-tissue. The stereome is quite thick-walled, and the mestome-sheath has its inner cell-walls considerably more thickened than we observed in the former species. The leaves of *C. Geyeri* and *C. multi-caulis* are longer and much narrower than in any of the former species; they are slightly conduplicate with a distinct keel in the former, but in the latter the midrib is not projecting. The outer cell-wall of epidermis is quite thick, especially so in the bulliform cells, which are here relatively small and occur as a single group above the midrib, when viewed in transverse sec-tions. No papillæ are visible around the stomata, but in *C. multicaulis* there are a few rows of such projections on each side of the bulliform cells, similar to those observed in *C. Backii.* The stomata are level with epidermis and confined to the lower surface of the blade. The mesophyll is a dense palisade-tissue with a lacune between each two mestome-bundles in *C. Geyeri*, but is more open in the other species. The stereome is extremely thick-walled in both species and accom-panies the mestome-bundles, besides that it, also, occurs as isolated groups in each of the two margins in *C. multicaulis*, but not in *C. Geyeri.* The mestome-bundles themselves show · the same structure as we have found in the other species, described above, but the mestome-sheath is more thick-walled than in these.

Utriculus.

This little organ furnishes the most essential characters in distinguishing the sections and even the species of *Carex*. It is almost bottle-shaped in *C. Steudelii* with a long, straight and rough beak, and is readily distinguished from that of *C. Willde-*

nowii, where the beak is shorter and less pronounced. However the internal structure is the same for these two species. The outer wall of epidermis on the dorsal face is somewhat thickened, on the ventral face it is much less so ; there is but little mesophyll displayed in a few strata and no stereome, not even by the two mestome-bundles, in which, furthermore, the mestome-sheath is quite thin-walled. In *C. Backii* the utricle shows a similar structure with the exception of a few stereome-cells on the leptome-side of the two mestome-bundles. A much firmer structure is exhibited by *C. œdipostyla* in which the epidermis is quite thick-walled on either face, besides that the mesophyll is more compact and occurs in several layers ; moreover we find in this species very thick-walled stereome, not only on either face of the mestome-bundles, but also as several isolated groups between these. There are more than two mestome-bundles and these possess a very thick-walled mestome-sheath. The utricle of *C. Geyeri* and *C. multicaulis* has a very short beak like *C. œdipostyla ;* it is perfectly glabrous and smooth. The epidermis is heavily thickened on both faces ; the mesophyll is poorly developed, and the stereome, which is quite thick-walled, is in *C. Geyeri* confined to the mestome-bundles, while in the other species it occurs as several isolated groups between them.

It would seem from the above as if *C. Steudelii, C. Willdenowii* and *C. Backii* are closely related to each other, but not to *C. œdipostyla, C. Geyeri* or *C. multicaulis,* hence the section *Phyllostachys,* as defined by recent authors, does not constitute a natural section. If we, moreover, consider those species, which by Torrey were the fundamental ones for his genus, it appears as if they are not sufficiently characteristic among themselves to necessitate the establishment of even a section. And the same is the case with *C. œdipostyla.* In regard to *C. Geyeri* and *C. multicaulis,* these do not possess very pronounced characters either, but similarly to the others may be arranged in some of the larger sections as "formæ hebetatæ." The species of Torrey's *Phyllostachys* appear, on the other hand, as inseparable from the *Hymenochlœnæ,* of which they probably represent one of the earlier types. The earliest is unknown in this section, but was perhaps monostachyous ; but we might, also, suppose that the *Hymenochlœnæ* developed parallel with one of the other sections from one common "forma hebetata," which then branched out into two or more distinct sections.

Brookland, D. C., February, 1900.

Bibliography.

Bailey, L. H. : A preliminary synopsis of North American Carices (Proceed. American Acad. Sci., 1886.)

Bailey, L. H. : Studies of the types of various species of the genus Carex (Memoirs Torrey Bot. Club, vol. i, 1889).

Boott, Francis : Illustrations of the genus Carex, vol. i. London, 1858.

Carey, John : Carex in Gray's manual, 1st edition, 1848.

Christ, H. : Nouveau catalogue des Carex d'Europe (Bull. Soc. bot. Belgique, vol. xxiv, 1885).

Christ, H. : Appendice au nouveau catalogue des Carex d'Europe (Bull. Soc. bot. Belgique, vol. xxvii, 1888).

Clarke, C. B. : On *Hemicarex* Benth. and its Allies (Journ. Linn. Soc., vol. xx, London, 1884).

Drejer, S. : Revisio critica Caricum borealium (Naturhist Tidsskr., vol. iii, Kjœbenhavn, 1841).

Drejer, S. : Symbolæ Caricologicæ. Kjœbenhavn, 1844.

Duval-Jouve, J. : Sur un Carex nouveau (*Carex œdipostyla* J. Duv.-J.) (Bull. soc. botanique de France, vol. xvii, 1870).

Duval-Jouve, J. : Sur la présence d'un rachéole dans l'utricule du *Carex œdipostyla* J. Duv.-J. (Bull. Soc. botanique de France, vol. xxi, 1874).

Gay, I. : De caricibus quibusdam (Ann. d. sc. Botanique, Ser. 2, vol. x, Paris, 1838).

Klinge, Johannes : Vergleichend histiologische Untersuchung der Gramineen- und Cyperaceen- Wurzeln. Dorpat, 1879.

Kunth, C. S. : Enumeratio plantarum, vol. ii. Stuttgart, 1837.

Kükenthal, Georg : Die Carex vegetation des aussertropischen Südamerika (ausgenommen Paraguay und Südbrasilien) (Engler's Jahrb., vol. xxvii, 1899).

Lemcke, Alfred : Beiträge zur Kenntniss der Gattung Carex Mich. (Inaug. Diss. Königsberg in Pr. 1892).

Mazel, Antoine : Etudes d'anatomie comparée sur les organes de végétation dans le genre Carex (Thesis Genève, 1891).

Roeper, Johannes : Zur Flora Mecklenburgs. Pars 2. Rostock, 1844.

Schulz, August : Zur Morphologie der Cariceæ (Berichte d. Deutsch. bot. Ges., vol. v, 1887).

Tuckermann : Enumeratio methodica Caricum quarundam, 1843.

Torrey, John : Monograph of North American Cyperaceæ (Ann. of the Lyceum. New York, vol. iii, 1828–36).

ART. V. — *The Titration of Mercury by Sodium Thio-sulphate;* by JOHN T. NORTON, JR.

[Contributions from the Kent Chemical Laboratory of Yale University—XCV.]

ACCORDING to J. J. Scherer* mercurous nitrate, mercuric nitrate and mercuric chloride may be estimated by direct titration with sodium thiosulphate, Hg_2S, $2HgS . Hg(NO_3)_2$, and $2HgS.HgCl_2$ being the precipitates obtained in each case. I have been unable to obtain access to Scherer's original publication, but Sutton† gives the following very general directions for this process:

(*a*) "*Mercurous salts.*—The solution containing the metal as a protosalt only is diluted, gently heated and the thiosulphate delivered in from the burette at intervals, meanwhile well shaking until the last drop produces no brown color. The sulphide settles freely and allows the end of the reaction to be easily seen. One cm³ of the $\frac{1}{30}$ normal solution of thiosulphate = 0·02 grams Hg or 0·0208 HgO.

(*b*) *Mercuric nitrate.*—The solution is considerably diluted, put into a stoppered flask, nitric acid added and the thiosulphate cautiously added from the burette, vigorously shaken meanwhile, until the last drop produces no further precipitate. Scherer recommends that when the greater part of the metal is precipitated the mixture should be diluted to a definite volume, the precipitate allowed to settle and a measured quantity of the clear liquid taken for titration; the analysis may then be checked by a second titration of the clear liquid if needful. One cm³ of $\frac{1}{30}$ normal thiosulphate =·015 of Hg. or ·0162 of HgO.

(*c*) *Mercuric chloride.*—With mercuric chloride the end of the process is not so easily seen. The very dilute solution is acidified with hydrochloric acid, heated nearly to boiling, and the thiosulphate cautiously added so long as a white precipitate is seen to form; any great excess of the precipitant produces a dirty-looking color. Filtration is necessary to distinguish the exact ending of the reaction. One cm³ of $\frac{1}{30}$ normal thiosulphate =·015 Hg or ·0162 HgO."

Fresenius‡ gives practically the same directions, but omits all mention of that portion of the process dealing with mercurous nitrate.

In view, therefore, of the scant information available on the subject and of the apparent difficulty of working the process accurately according to the directions given, an attempt was

* His Lehrbuch der Chemie, i, 513.   † Volumetric Analysis, p. 230.
‡ Quantitative Analysis.

made to ascertain whether the careful regulation of temperature, dilution and amount of acid present might not produce beneficial results.

That portion of the process dealing with mercuric chloride was first taken up. The mercuric chloride used was pulverized, dried at 100° and its purity proved by several determinations as mercuric sulphide. The sodium thiosulphate was made up of approximately $\frac{1}{20}$th normal strength and standardized on decinormal iodine, which in turn was titrated against decinormal arsenious acid made from pure resublimed arsenious oxide.

For the action of sodium thiosulphate upon the mercuric chloride Scherer gives the equation,

$$3HgCl_2 + 2Na_2S_2O_3 + 3H_2O = 2HgS.HgCl_2 + 2Na_2SO_4 + 4HCl.$$

According to my experience, the action results in the formation of a dense white precipitate which refuses to settle either by shaking or standing, thus making it impossible to fix the end reaction by reading the first drop of thiosulphate which produces no further white precipitate in the solution containing the mercuric chloride. Recourse must be had therefore to filtering. By far the quickest and neatest method is to use the asbestos filter deposited on a large perforated platinum cone.[*] This cone is set in a glass funnel by means of a rubber connector and the funnel is passed through the stopper of a large side-necked Erlenmeyer connected with an exhaust pump. A little asbestos fiber shaken in the liquid to be filtered was found to be very beneficial in preventing the precipitate from running through the filter. In all the following experiments the thiosulphate was run into the solution containing the mercuric chloride in excess, the whole shaken up with asbestos fiber, filtered and the excess of thiosulphate determined by $\frac{1}{10}$th normal iodine. This method of procedure seems to be far preferable to attempting to catch the end of the reaction by running in the thiosulphate until the last drop produces no precipitate. In the experiments shown in Table I no attention was paid to the temperature of the solution and the thiosulphate was run in until the liquid turned brown. In every case the solution was allowed to stand until there was no further visible change of color.

A glance at the table shows that the results are most irregular. In Table II is seen the result of regulating the temperature and the length of standing after the addition of the sodium thiosulphate.

[*] Amer. Chem. Jour., i, 321.

TABLE I.

	HgCl$_2$ taken calc'd as Hg. grams.	Na$_2$S$_2$O$_3$ in excess. cm³.	Volume at beginning. cm³.	HgCl$_2$ found calc'd as Hg. grams.	Error. grams.
1.	0·0446	46·28	200	0·0343	0·0103 −
2.	0·0354	46·28	400	0·0326	0·0028 −
3.	0·0356	44·97	400	0·0225	0·0131 −
4.	0·0345	44·5	100	0·0308	0·0037 −
5.	0 0354	44·43	50	0·0326	0·0028 −
6.	0·0382	22·59	50	0·0354	0·0028 −
7.	0·0375	8·58	50	0·0385	0·0010 +
8.	0·0371	1·84	50	0·0304	0·0067 −
9.	0·0731	2·28	50	0·0774	0·0043 +
10.	0·1486	9·34	50	0·1489	0·0003 +

TABLE II.

	HgCl$_2$ taken as Hg. grams.	Volume at beginning. cm³.	Temperature. C.	Standing minutes.	Na$_2$S$_2$O$_3$ in excess. cm³.	HgCl$_2$ as Hg. found.	Error. grams.
1.	0·0738	50	36°	40	16·68	0·0494	0·0244
2.	0·0741	50	70	15	15·42	0·0738	0·0003
3.	0·0741	75	70	12	16·07	0·0733	0·0008
4.	0·0744	50	70	10	14·6	0·0755	0·0011
5.	0·0764	50	72	7	6·77	0·0771	0·0007
6.	0·0762	50	75	10	8·54	0·0799	0·0037
7.	0·0756	50	73	15	9·99	0·0815	0·0059
8.	0·0774	50	68	15	10·84	0·0767	0·0007
9.	0·0745	75	69	7	6·62	0·0805	0·0060
10.	0·0736	50	68	5	15·82	0·0714	0·0022

These results, although better than those of Table I, are st
very uncertain. On the supposition that the change fro
white to black, which takes place in the solution after the ad
tion of an excess of sodium thiosulphate more or less quick
according to the temperature, was due to an increased amou
of HgS in the compound 2HgS.HgCl$_2$, the next step was
ascertain whether this could be avoided by stopping the ad
tion of the thiosulphate at the first indication of a change
color in the white precipitate, diluting the solution with a lar
amount of cold water and immediately throwing it on t
filter. The following table (III) shows the result of the expe
ments.

In the case of quantities of mercuric chloride up to (
gram the results shown in Table III are very satisfactory, b
when larger amounts of mercuric chloride are used the err(
again become prominent. In Table IV, the effect of loweri:
the temperature to 60° C. and of increasing the dilution
100^{cm} is shown.

TABLE III.

	HgCl$_2$ taken as Hg. grams.	Volume at beginning. cm^3.	Temperature. C.	Na$_2$S$_2$O$_3$ in excess. cm^3.	HgCl$_2$ found as Hg. grams.	Error. grams.
1.	0·0749	50	70°	4·15	0·0751	0·0002 +
2.	0·0749	50	75	0·72	0·0728	0·0021 −
3.	0·0756	50	72	1·46	0·0759	0·0003 +
4.	0·0753	50	70	2·57	0·0750	0·0003 −
5.	0·0390	50	70	3·43	0·0395	0·0005 +
6.	0·0388	50	72	8·19	0·0390	0·0002 +
7.	0·0380	50	76	2·03	0·0393	0·0013 +
8.	0·1494	50	78	4·99	0·1498	0·0004 +
9.	0·1489	150	78	4·38	0·1512	0·0023 +
10.	0·1480	50	70	0·52	0·1438	0·0042 −
11.	0·1498	50	78	1·47	0·1540	0·0042 +
12.	0·1484	50	71	2·09	0·1517	0·0033 +
13.	0·1480	75	72	1·59	0·1509	0·0029 +

TABLE IV.

	HgCl$_2$ taken as Hg. grams.	Volume at beginning. cm^3.	Temperature. C.	Na$_2$S$_2$O$_3$ in excess.	HgCl$_2$ found as Hg. grams.	Error. grams.
1.	0·0759	100	60°	3·06	0·0766	0·0007 +
2.	0·0384	"	"	2·81	0·0387	0·0003 +
3.	0·1498	"	"	1·1	0·1500	0·0008 +
4.	0·1503			1·63	0·1506	0·0003 +
5.	0·1479			2·41	0·1480	0·0001 +
6.	0·1489			2·12	0·1503	0·0014 +
7.	0 2244			2·63	0·2259	0·0015 +
8.	0·1490			2·33	0·1484	0·0006 −
9.	0·0758			2·	0·0762	0·0004 +
10.	0·0383			2·58	0·0379	0·0004 −

From this table it is plain that Scherer's process for the estimation of mercury in the form of mercuric chloride is capable of yielding accurate results if carried out under certain fixed conditions. These conditions, which must be closely adhered to, are as follows: The solution containing the mercury in the form of mercuric chloride is placed in a liter flask, diluted to 100$^{cm^3}$ and heated to a temperature of 60° C. The sodium thiosulphate in $\frac{1}{20}$th normal solution is run in from a burette until the white precipitate formed begins to take on a brownish tinge. The solution is then diluted with cold water, some asbestos fiber added to coagulate the precipitate and the whole is quickly thrown on the filter. After careful washing of the precipitate, the filtrate is diluted to a definite volume, 3 grams of potassium iodide added and the excess thiosulphate

titrated with iodine and starch solution. The duration
process need not exceed 15 minutes. It is worthy o
that there is no necessity of using any hydrochloric a
addition to that formed in the reaction. This certainly
nates one probable source of error—the interaction of
chloric acid and sodium thiosulphate.

In dealing with the estimation of mercury in the f(
mercurous nitrate the same method of procedure was em]
as in the case of mercuric chloride. A solution of mer
nitrate was prepared by dissolving as much as possible
grams of the salt in about 200$^{cm^3}$ of water, filtering o
clear liquid and diluting to a definite volume. The st;
of the solution was determined by precipitation as m
mercury by means of the electric current. Contrary 1
statement made in Sutton, the brown precipitate of E
formed as shown in the equation,

$$Hg_2(NO_3)_2 + Na_2S_2O_3 = Hg_2S + Na_2SO_4 + N_2O_5—$$

does not settle and leave a clear supernatant liquid, b
solution remains cloudy and it is impossible to see ar
reaction. Although the conditions of dilution, tempe
and amount of acid present were carefully considere
arrangement or adjustment of these conditions was
under which satisfactory results could be obtained. T;
gives the result of experiments.

TABLE V.

	$Hg_2(NO_3)_2$ taken as Hg. grams.	HNO_3 1:3. cm³.	Volume at beginning. cm³.	Temperature. C.	$Na_2S_2O_3$ in excess. cm³.	$Hg_2(NO_3)_2$ found as Hg. grams.	E g
1.	0·0148	none	50	50°	4·28	0·0129	0·
2.	0·0148	"	75	60	4·45	0·0117	0·
3.	0·2976	"	300	. 60	12·67	0·2760	0·
4.	0·1488	"	·100	40	2·19	0·1386	0·
5.	0·1488	"	100	50	6·92	0·1378	0·
6.	0·1488	"	200	50	0·78	0·1388	0·
7.	0·0744	"	100	65	0·73	0·0636	0·
8.	0·0744	1	100	55	0·49	0·0733	0·
9.	0·0744	2	100	40	1·16	0·0660	0·
10.	0·0744	1	100	55	1·35	0·0685	0·
11.	0·0744	1	100	55	0·82	0·0686	0·
12.	0·0744	½	200	55	1·71	0·0685	0·
13.	0·0744	4	100	40	1·77	0·0649	0·
14.	0·0744	5½	100	40	1·34	0·0645	0·
15.	0·0744	1	100	45	1·71	0·0654	0·
16.	0·0744	10	100	45	1·55	0·0669	0·
17.	0·1488	1	100	40	0·73	0·1391	0·

The errors in experiments 1, 3, 4, 5, 6, 10, 11, 12 and 17 are, proportionally to the amount of material handled, practically the same and this fact caused me to make a careful revision of all standards; but no mistake could be found. The reaction upon which the process depends requires the formation of Hg_2S, but this mercurous sulphide breaks down immediately into mercuric sulphide and mercury. The latter is probably acted upon by the free nitric acid present to form mercuric nitrate, which in turn is transformed into the compound $2HgS.Hg(NO_3)_2$, by the action of the thiosulphate. At any rate, with an error so large, whatever its source may be, the process is plainly impracticable.

The third step in Scherer's process deals with the action of sodium thiosulphate on mercuric nitrate. In the following experiments a solution of mercuric nitrate was prepared either by dissolving as far as possible 20 grams of mercuric nitrate in about 200^{cm^3} of cold water, filtering off the supernatant liquid and diluting to a definite volume (exps. 1–8), or by dissolving the salt in strong nitric acid and diluting (exps. 9–19). The standard of the solution was obtained by precipitation as metallic mercury by means of the electric current. The yellow precipitate, formed according to Scherer's reaction,

$$3Hg(NO_3)_2 + 2Na_2S_2O_3 = 2HgS.Hg(NO_3)_2 + 2Na_2SO_4 + 2N_2O_5,$$

on adding the sodium thiosulphate settles much better than in the case of either mercuric chloride or mercurous nitrate; but, as the supernatant liquid takes on a permanent yellow color towards the end of the reaction, it is impossible to see when the thiosulphate produces no further precipitation. On this account, therefore, the same method of procedure was adopted as in the case of the mercuric chloride and mercurous nitrate, i. e., filtration and titration of the excess of sodium thiosulphate with iodine and starch solution. The result of the experiments is shown in the following table.

These results seem to show the impossibility of obtaining accurate results according to Scherer's method for the determination of mercuric nitrate by direct titration with sodium thiosulphate. The constant plus error cannot be accounted for on the hypothesis that the nitric acid present decomposes the sodium thiosulphate, for in that case the error would lie in the other direction. It is more probable that the constitution of the compound $2HgS, Hg(NO_3)_2$ is not definite enough to make it the basis for an analytical process.

TABLE VI.

	Hg(NO$_3$)$_2$ taken as Hg. grams.	Volume at begin-ning. cm^3.	HNO$_3$ 1:3 cm^3.	Tem-perature. C.	Na$_2$S$_2$O$_3$ in excess. cm^3.	Hg(NO$_3$)$_2$ found as Hg. grams.	Error. grams.
1.	0·1167	200	none	60°	0·46	0·1384	0·0217+
2.	0·1167	100	1	60	3·63	0·1348	0·0181+
3.	0·1167	100	2	60	0·23	0·1375	0·0208+
4.	0·1167	100	none	60	0·17	0·1360	0·0193+
5.	0·1167	300	"	21	0·12	0·1232	0·0065+
6.	0·1167	300	"	21	0·15	0·1375	0·0208+
7.	0·1167	200	"	21	11·8	0·1461	0·0294+
8.	0·1167	200	20	21	15·97	0·1403	0·0236+
9.	0·1278	200	none	21	5·23	0·1647	0·0369+
10.	0·1278	200	"	"	1·35	0·1653	0·0375+
11.	0·1278	100	"	"	1·43	0·1662	0·0384+
12.	0·0752	200	"		2·09	0·0996	0·0244+
13.	0·0255	100	"		2·01	0·0280	0·0025+
14.	0·0255	200	5		3·44	0·0264	0·0009+
15.	0·0255	200	10		0·85	0·0334	0·0079+
16.	0·0255	300	none	"	1·96	0·0264	0·0009+
17.	0·0255	200	"		1·9	0·0287	0·0032+
18.	0·0255	200	"	"	1·76	0·0290	0·0035+
19.	0·0639	200	"	60	1·53	0·0831	0·0192+

In conclusion I wish to thank Prof. F. A. Gooch for his kind advice and assistance.

ART. VI.—*Selenium Interference Rings;* by A. C. LONGDEN.

SOME months ago, while at work on thin films deposited by cathode discharge, at the request of Mr. C. C. Trowbridge of Columbia University, I attempted to deposit for him a thin conducting film of selenium. The attempt was not successful; because, although the selenium was in the conducting form in the cathode, the film was always in the amorphous form, which is not an electrical conductor. A number of films were deposited under all sorts of conditions, but they were invariably amorphous and therefore non-conducting.

During the deposition of selenium, although no conducting films were produced, a very interesting and beautiful phenomenon was observed. The selenium cathode was small in comparison with the size of the glass plates upon which the films were deposited, and therefore the distribution of selenium was not uniform, the film being considerably thicker in the center than at the edges. The deposited material is sufficiently transparent to transmit large quantities of light through thicknesses of several wave-lengths; so that when light falls upon a film of varying thickness, the beams reflected from the upper and lower surfaces present interference phenomena similar to Newton's rings. The interference bands which were at first produced were so irregular that they could hardly be called rings, because of the lack of uniformity in the upper surfaces of the films; but after arranging the cathode in the form of a small globule, films increasing uniformly in thickness toward the center were obtained.

These films, when viewed by reflected light, present concentric systems of circular interference fringes of great beauty and remarkable brilliancy. The transparent selenium film is deposited upon plane glass, and, having a convex upper surface, it constitutes a very thin plano-convex lens. The interfering beams of light, however, are not reflected from the surfaces of an air film, between the convex lens and a second plane, but from the two surfaces of the selenium lens itself. As this lens is thickest in the center, instead of being thinnest in the center as in Newton's air film, the selenium rings increase instead of decrease in width and brilliancy, counting from the center outward. For the same reason the order of the colors is red, yellow, green, blue, violet; instead of violet, blue, green, yellow, red; and, instead of having a black center when viewed by reflected light, as is the case with Newton's rings, the color of the center depends upon the order of the ring, counting from the margin inward.

It will be seen by considering the geometrical conditions which govern radiation from a point to a plane, that if our cathode is a small enough globule, the radius of curvature of the film will depend upon the distance of the cathode from the glass plane upon which the film is deposited. Accordingly, this distance determines the diameter of the rings, and the width of any particular ring. By placing the cathode at a suitable distance from the glass, it is very easy to obtain exceedingly brilliant rings as much as 8 or 10 millimeters in width . and 4 or 5 centimeters in diameter.

These rings are so brilliant that they are as easily projected upon the screen as ordinary lantern slides; and without the usual precautions in regard to size and illumination demanded in projecting Newton's rings with a glass lens and plate. They should, however, be projected by reflected light.

I am not aware that Newton's rings of this character, namely, interference rings from the reflecting surfaces of a very thin plano-convex lens, have heretofore been produced by any process whatsoever. The physical conditions connected with selenium rings are of a simpler character than those which exist in the case of ordinary Newton's rings, and they may, therefore, afford physicists new opportunities for the study of this subject.

Physical Laboratory of Columbia University,
April 20, 1900.

ART. VII.—*Carboniferous Bowlders from India ;* by B. K. EMERSON.

IN the spring of 1894, I was in the office of the Geological Survey of India and Dr. Fritz Nötling unpacked for me and explained a fine series of the bowlders of the Carboniferous Glacial period of the Salt range in northwestern India.

They had been obtained partly from the talus and partly from the conglomerate itself, and he was kind enough to give me samples of the different forms. Later, Dr. King, the Director of the Survey, observing the unconscious but silent admiration with which I examined the finest bowlder of the series, packed it up and sent it after me to my lodgings. This is the largest bowlder in the figure and is 10 inches long.

It is a dark red rock halfway between a granite and a quartz porphyry and is perfectly scratched on four sides and rough and battered on the ends.

It is in color, shape, and kind and perfection of striation, so like the great bowlder, about 9 feet long, of red sandstone which stands in front of the Geological Museum at Amherst and which was taken from one of the streets at Amherst during the lowering of the latter, that I may cite a figure of the latter for comparison from my monograph of old Hampshire Co., Mon. xxix, U. S. G. S., pl. xxxiii, p. 192.

The bowlder with the cord around it was broken in dislodging it from the ledge and shows remains of the conglomerate cemented onto its upper surface. The bowlder on the small box is scratched on several small facets separated by sharp crests and has also fragments of the conglomerate attached.

The bowlder in front was not quite in focus but a dim line runs across the middle of the front from the upper left hand corner to the lower right hand corner and the portion below that line is distinctly striated, while the part above that line is weathered several millimeters deep by long exposure to the weather as it projected on the cliff while the lower portion was protected within the solid mass of the rock. Dr. Nötling explained to me the conditions of the occurrence and there can be no doubt about the glacial character and the Carboniferous age of the deposit. I send this short article because doubt has been expressed in a recent standard text-book as to the reality of the Carboniferous Glacial period.

ART. VIII.—*A new Bivalve from the Connecticut River Trias;* by B. K. EMERSON.

A FEW days ago my former assistant, Mr. Chas. S. Merrick, of Wilbraham sent me a large slab of sandstone of a buff color, much stained by malachite, containing several indistinct casts of a unio-like bivalve.

The shell seems to be one of those fresh or brackish water forms common in the Trias allied to the Unionidæ. It may be compared with the *Anoplophora lettica* Quenstedt,[*] and may be called *Anoplophora Wilbrahamensis.* It is a distinctly unio

 shaped shell 38ᵐᵐ long, 18ᵐᵐ high 7ᵐᵐ thick. The exterior is smooth with fine lines of growth. The mantel impression is quite deep as preserved on the central portion of the length. The beak rises very slightly and the hinge line is long and straight and there is in the fig-ure what seems to be the impression of a long, posterior tooth and above this a slight groove at the place of attachment of the ligament.

The large slab contains 14 imperfect casts, all of which may well belong to one species.

I do not know of any other bivalve shell from the Trias in Massachusetts.

* Pal., plate lxiii, fig. 28.

т. IX.—*The Statement of Rock Analyses;* by Henry
S. Washington.

)ʀ late years the importance of chemical analyses in the
dy of igneous rocks is generally recognized, and their pub-
.tion and use is becoming more and more common.

ʌpart from improvements in methods and facilities, as well
increase in the number of workers, the main factor in this
rease in the number of analyses published is the growing
reciation of their vital importance for a thorough compre-
.sion of the rocks of the globe.

'his importance of rock analyses to-day lies in their disclos-
the bearing of the chemical composition of the rock masses
the deeper theoretical problems of petrology. Many of
se, such as the differentiation of magmas, the genetic rela-
iships of various rock bodies or parts of a complex to each
er, or the true character and meaning of the so-called
:trographic provinces," are only to be solved by means of
roughly good and trustworthy analyses.

'here is an increasing tendency among some petrographers
regard the magma as one of the main objects of study, of
ich the solid rocks are simply the consolidated and acces-
e portions, many of their characters being to a great extent
:nitous and dependent on extraneous conditions of solidifi-
on and the like.

'he chemical composition of the magma either persists as
h in the rocks formed from it (as in Brögger's aschistic
es), or is to be inferred if the magma has undergone changes
h as those induced by differentiation (as in Brögger's dia-
istic dikes), absorption of country rock or other causes.
is chemical composition is, as far as we can tell at present,
single original character of the magma which is open to
dy after it has undergone the physical and physico-chemical
nges which result in the formation of igneous rocks, whether
histic or diaschistic.

Hence for obtaining a knowledge of these vast and highly
portant portions of the earth's crust, of their original char-
ers, the conditions under which they exist and the changes
iich they may undergo, rock analyses are absolutely essential.
is true that physics also comes to our aid in certain direc-
ns, but this does not lessen the importance of the chemical
vestigation.

The practical use of analyses involves their collation or com-
rison one with another, either those of the rocks of one
ven region, or those of different regions with each other, or

of rocks belonging to any given group or kind. Scattered as
they are through the literature, it is often a matter of some
practical difficulty, or at least inconvenience, to bring several
together under the eye so as to observe readily the features
which it is desired to see. This inconvenience is increased by
the fact that there is no uniformly adopted sequence in which
the various constituents are stated.

In some cases, as in the publications of the U. S. Geological
Survey, the analytical order is followed. The constituents are
put roughly in the order in which they are determined in the
course of the analysis.

Again a roughly chemical order is followed, either by putting all the acid radicals at the top of the column and the basic
radicals below, as is usually done in Tschermak's Mittheil-
ungen, or by putting SiO_2 first, followed by TiO_2, then the
metallic oxides in the order R_2O_3, RO, R_2O, with H_2O, P_2O_5, Cl,
etc. last of all, as is the custom of Roth, Zirkel, Rosenbusch,
and many others, including the writer. But even here there
are variations in the order followed, as in some cases CaO is
put before MgO, and K_2O before Na_2O.

A third method is that adopted of late years by Pirsson.
This, which may be called the petrographic order, consists in
putting the nine most important oxides first, beginning with
SiO_2 and ending with H_2O, while the subordinate and more
rarely occurring constituents follow after in the same general
order.

A last method shows a lack of any system, the constituents
being put down almost at haphazard, and with little apparent
attempt at order or natural association. This method is fortunately rarely met with, and is to be dismissed at once as quite
unworthy of consideration.

In consequence of this confusion not only is the inconvenience of and time required for copying and comparing analyses
greatly increased, but there is introduced a positive danger of
error due to unavoidable slips, which, unless the results are
carefully and laboriously checked, is apt to lead to wrong
conclusions.

The benefits then of a uniform system are evident, and it is
probable that every petrographer has realized this to a greater
or less extent. After some experience in the copying and use
of rock analyses and full consideration of the subject, as well
as discussion of the matter with other petrographers, I would
propose that the third method mentioned above, which was, I
think, inaugurated by Prof. Pirsson, be generally adopted, this
presenting the greatest advantages to the petrographer, with a
minimum of disadvantage.

A rock analysis, it must be remembered, is primarily intended for, and almost exclusively used by, petrographers. Therefore for him an arrangement on analytical or strictly chemical lines is neither advantageous nor appropriate. What he needs especially is an arrangement which shall bring the essential chemical features—both the percentage figures and the molecular ratios—prominently and compactly before the eye, so that the general chemical character and the relations of the various constituents may be seen at a glance. It is also of importance that the arrangement be such as to facilitate comparison of one analysis with another.

To the petrographer the eight oxides, SiO_2, Al_2O_3, Fe_2O_3, FeO, MgO, CaO, Na_2O, and K_2O, which in practically every case are present in preponderating amount, are, and must always remain, of prime importance. They are the oxides which, by their relative amounts, determine the chemical character of the rock. H_2O and CO_2, which are also found sometimes in notable quantity, are chiefly of value as a measure of the freshness of the rock. The other constituents, while of great interest, are present in minute quantities and in general, especially as compared with the main oxides, influence the character of the rock only to a very limited extent, either entering into the composition of accessory and subordinate minerals, or replacing to a very small extent the more important oxides in the essential minerals.

Hillebrand's plea[*] for their determination is well founded, and it is of course a desideratum that all analyses should be complete as to the rarer constituents. This, however, is not always possible, and it will probably remain true as it is at present, that many otherwise good and serviceable analyses are incomplete in this respect, or only show determinations of a few of the rarer constituents, notably TiO_2, MnO, P_2O_5, and Cl.

By putting the eight main oxides together, then, the petrographer is able to see at a glance the character of the rock under study, and the molecular ratios of these oxides, which are the only ones of practical importance, may be written after them and so easily compared *inter se*. Furthermore, whether an analysis is complete or incomplete, these oxides, which are determined in every case, are always in the same relative position, so that the eye finds them without trouble, thus immensely facilitating comparison and study.

The question naturally arises as to the advisability of stating the analytical results, not in oxides, but in terms of the elements, oxygen being given as a separate constituent. This is

[*] Hillebrand, Bull. 148, U. S. G. S., p. 15, 1897.

done' by some petrographers in the reduction of analyses for certain lines of work, though in these cases oxygen is omitted. This last seems to me a somewhat unjustifiable procedure, in view of the fact that oxygen makes up by far the greatest percentage in all the known terrestrial rocks.

It is true that we do not yet know whether the oxides exist as such in the magma or not, though the theories of physical chemistry seem to favor the view that they do. But apart from such considerations, the possible benefits to be obtained are so doubtful and meager that it seems scarcely justifiable at present to make any such sweeping change. Experience has shown the convenience and practicability of the present method for the study of rocks, and as it is sanctioned by long and universal usage, it had better be retained, and the results stated in oxides.

The statement of the analysis then will be divided into two divisions, the main portion, in which are placed the principal oxides, and the subordinate in which occur all the rest.

The order in which the constituents may be put is the next consideration. That which is here proposed is as follows:

SiO_2, Al_2O_3, Fe_2O, FeO, MgO, CaO, Na_2O, K_2O, H_2O (ignit.), H_2O (110°), CO_2, TiO_2, ZrO_2, P_2O_5, SO_3, Cl, F, S (FeS_2), Cr_2O_3, NiO, CoO, MnO, BaO, SrO, Li_2O.

As regards the main portion the sequence usually adopted seems eminently proper. This is SiO_2, Al_2O_3, Fe_2O_3, FeO, MgO, CaO, Na_2O and K_2O. In this we start out with the chief acid radical and the constituent present in largest amount, and go through successively lower orders of oxides to the most positive radicals, the alkalis. At the same time they are presented in a way which brings the oxides together in their natural petrographic and mineralogic relations. The sesquioxides are together; ferrous iron follows ferric, MgO is next to FeO, as the two go hand in hand in the ferro-magnesian minerals; CaO is intermediate as is proper, since it is a constituent both of these and of the feldspars, and it is next to Na_2O, as it is associated with it in plagioclase.

H_2O should follow K_2O, as it is a highly important and regularly determined constituent. I follow Hillebrand[*] in urging the separation of ignition and hygroscopic water, putting the former first, as being an essential ingredient. Next to these should come CO_2, as this, with H_2O, is a measure of the freshness of the specimen, and this character can, therefore, be told at a glance. Together also they constitute the "loss on ignition" so often given, and in this case can be connected by a bracket.

[*] Hillebrand, Bull. 148, U. S. G. S., p. 29, 1897.

After these it seems well to put the acid radicals which have been determined, following the main principle of the other division ; TiO_2 and ZrO_2 should come first, then P_2O_5 and V_2O_3. SO_3 is put next instead of at the head of the lesser acid radicals, as it, like Cl which follows, is a constituent of the sodalite group. F comes after Cl, and then S or FeS_2 may be put.

Then come the subordinate metallic oxides, in the order R_2O_3, RO, R_2O. These would include Cr_2O_3, (Ce_2O_3), NiO, (CoO), MnO, (CuO), BaO, SrO, Li_2O. Finally in rare cases C or N may be put at the last.

While the above order of arrangement is undoubtedly open to criticism in certain respects, yet it seems to the writer, as well as to other petrographers to whom it has been submitted, to be on the whole the most logical and convenient. It may be added that practical experience of it in the collection of a large number of analyses sustains this view.

I would also urge the advisability of printing the important molecular ratios with each new analysis, wherever this is possible. The reader will be saved the trouble of calculating them for himself, and will thus immediately have a fuller comprehension of the character of the rock.

In this connection I would suggest, as a practical and convenient method for using analyses, the use of a card catalogue, the cards being printed with the oxides in their proper order. Experience has shown the utility of such a plan, and the time given to its preparation will be more than compensated for by that saved in reference and collation.

Art. X. — *A String Alternator;* by K. Honda and S. Shimizu.

The transverse vibration of a stretched iron wire was first used by M. Wien* as an interrupter for an induction coil. The vibration was produced by the mutual attraction between a magnet and the wire. Pupin,† however, replaced the driving force by the motion of a wire carrying an electric current in the magnetic field. A similar arrangement was also used by L. Arons.‡

Pupin's interrupter may be modified to serve as an alternator in the following way. As shown in the annexed figure, a

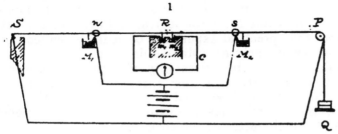

copper wire is horizontally stretched; one of the ends is fixed to a stout support S, and the other, passing over a small metallic pulley P, is attached to a weight Q. The wire is electrically insulated at the middle point R, so that no current can flow through the point but through the circuit C. Close to the middle point and just under it, two mercury cups m_1 and m_1, are placed; two short vertical wires fixed to the horizontal one are dipped into each of them. Again, at quarter distances from S and P, two short wires about 2^{cm} long are vertically fixed to the main wire; two cups M_1 and M_2 are placed under them for mercury contact. One pole of an accumulator is connected to these cups, and the other to S and P; n and s are two poles of a strong magnet (or better an electromagnet) horizontally placed close to the wire; the magnet maintains the wire in constant vibration. Evidently, there is a definite relation between the direction of the current and the poles of the magnet for continuing the vibration. This direction of the current is easily found by trials. A tap given to the wire so as to produce vibrations with a single node at the middle point sets the wire in constant vibration. By this arrangement the mercury contact is made in turn at M_1 and M_{11}, so that the current passing through the portion of the circuit C changes

* M. Wien, Wied. Ann., xliv, 683, 1891.
† M. I. Pupin. this Journal, III, xlv, 325, 1893.
‡ L. Arons, Wied. Ann., lxvi, 1177, 1898; lxvii, 682, 1899.

its direction once in a complete vibration. Thus, suitably changing the length (by sliding two wedge-shaped blocks) or the tension (by changing the suspended weight) of the string, an alternating current of desired frequency is obtained.

Now, in the above method, only a part of the current employed can be turned to the alternating current; in case where a strong alternating current is required, it is sufficient to use two equal sets of accumulators; two poles of the one set being connected to M_1 and P, and those of the other to M_2 and S, so that the currents through the portion of the circuit C are directed in opposite sense. A sensitive galvanometer, inserted in the circuit, shows whether the mean strength of the alternating current in both directions is equal or not. The small difference in the current-strength can easily be effaced by adjusting the spark gap of the mercury contact.

To cut off the electrical connection at the middle point of the wire, the following method proved to be the most satisfactory. A silk-covered wire of suitable length is cut into two parts. The ends are twisted together like a rope for the length of 1^{cm} and bent at right angles to the wire as shown in the

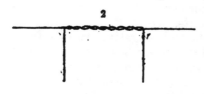

annexed cut. The twisted portion of the wire is firmly bound together by a silk thread, and the remaining portion of the cover is scraped off. Though different specimens of wire were tried, copper and aluminum wires gave the most satisfactory result.

The simplicity of the present arrangement for obtaining an alternating current of definite frequency is one of its chief merits. The actual frequency, if the tension of the string be not small, was found not to differ by more than 1 per cent from the value calculated by the formula

$$n = \frac{1}{l}\sqrt{\frac{T}{\rho}},$$

where l is the length, T the tension and ρ the linear density of the wire. It was not difficult to obtain vibrations ranging from 30 to 1000 per second.

It may not be out of place to remark that in using the present arrangement, the self-induction of the circuit is to be reduced as small as possible, inasmuch as it diminishes the strength of the current. Hence in case the self-induction of the circuit be not small, a large number of accumulators must be used; otherwise the string would not maintain its vibrations. Measuring the mean strength of the current by an ampere-balance, we found but a small fluctuation, if the current does not exceed 1·5 amperes.

Physical Laboratory of the Imperial University, Tokyo, Japan.

ART. XI.—*The Action of Light on Magnetism;* by JOSEPH
HALL HART.

THE electro-magnetic theory of light assumes the existence
of electro-magnetic waves of the required frequency in the
ether and of their identity with ordinary light-waves.   In consequence of the alternate electric displacements there should
be an alternating magnetic field and we might expect the
demagnetization of iron by light as a result, if the intensity
of the light were gradually diminished.   We have, however,
little trustworthy evidence of the action of light on magnetism.
The contradictory results obtained and the smallness of the
effect have prevented the attainment of experimental results
sufficient for definite conclusions on the subject.   In particular,
the action of light on the magnetism of iron bars presents a
field in which, if ordinary care is taken, positive results should
be obtained.

Several experimenters in the early part of the present century tried to magnetize iron and steel by the action of light.
Morichini,[*] Christie[†] and Mrs. Somerville[‡] obtained positive
results from a series of experiments, while P. Riess and L.
Moser,[§] and John W. Draper[‖] obtained negative results from
precisely similar experiments, and the thing has been generally
regarded as impossible.   Under ordinary circumstances there
can be little doubt that this is the case, but, according to Bidwell,[¶] in a paper before the Physical Society in 1889, if a certain initial condition is fulfilled, we can find evidence of the
action of radiation upon the magnetism of iron, and the ful-
fillment or non-fulfillment of this condition explains the
diverse results of previous experimenters.

The condition is that the susceptibility of the bar AB to be
operated upon shall be greater (or less) for a magnetic force in
the direction AB than for an equal one in the direction BA.
Iron bars having this property were obtained by the following
method: A piece of soft iron rod, which may conveniently be
10 or 12^{cm} long and from 0·5 to 1^{cm} in diameter, is raised to a
bright yellow heat and slowly cooled.   When cold, it is placed
inside a solenoid, through which is passed a battery current of
sufficient strength to produce a field of about 350 or 400
C. G. S. units.   The iron when removed from the coil is found
to be permanently magnetized, and its north pole is marked

---

[*] See Christie.   [†] Phil. Trans., 1826, p. 219.
[‡] Phil. Trans., 1826, p. 132.
[§] Annales de Chemie, xlii, 304.
[‖] Journ. of Franklin Institute, Feb. 1835.
[¶] Proc. Phys. Soc., London, April, 1889, p. 455.

for the sake of distinction with red sealing-wax varnish. The bar is then supported horizontally and in an east-and-west direction behind a small reflecting magnetometer, and over it is slipped a coil, which is shunted with a rheostat, the resistance of which can be gradually increased from 0 to 26 ohms. The coil can be connected by a key with a single battery cell, which is so arranged as to produce a demagnetizing force inside the coil. The resistance of the rheostat is slowly raised, so that more and more current passes through the coil, the battery key being alternately lifted and depressed until the magnetometer indicates that the iron bar as a whole is perfectly demagnetized. The strength of the demagnetizing force required varies according to circumstances: it is generally about one thirtieth or one twenty-fifth of the original magnetizing force.

A piece of iron thus treated possesses certain remarkable properties. The magnetization induced by a force acting in such a direction as to make the marked end a north pole, is greater than that caused by an equal force in an opposite direction. If such a bar be held horizontally east and west and tapped with a mallet, the marked end at once becomes a north pole. Application of heat or instantaneous application of flame causes a similar effect. If subjected to the action of a series of equal, alternating feeble magnetic forces, the marked end generally becomes a north pole, even if the last of the alternating forces may have tended to induce the opposite polarity. These were well-known effects, but in addition, Bidwell found that such a rod was remarkably sensitive to the action of light. When placed behind the magnetometer and illuminated by an oxy-hydrogen lamp about 70^{cm} distant, there occurred an immediate deflection of from 10 to 200 scale divisions, the magnitude of the effect varying in different specimens of iron. The magnetometer mirror was one meter distant from the scale and each division$=0.64^{mm}$. In these experiments, as the action of the light was continued, the deflection slowly increased. When the light was shut off, the magnetometer instantly went back over a range equal to that of the first sudden deflection, then continued to move slowly in the backward direction toward zero. With a thick rod the slow movement was barely perceptible. He believed the first quick movement to be due to the direct action of radiation, and the subsequent slow movement to the gradually rising temperature of the bar. In several cases, the magnetism was of the opposite kind, and sometimes became north when certain portions of the rod were illuminated and south when the light acted upon other portions. This was probably due to irregular annealing and a consequent local reversal of the

direction of maximum susceptibility: it indicated, probably
that the light effect was local, and confined to the illuminated
surface. The effect varied directly as the intensity of the
illumination; it was strongest for red light and independent of
direction of vibration of incident light; and blackening the
bar made the action much slower.

I first repeated and verified Bidwell's experiments. I pro-
cured eight pieces of the softest iron obtainable, of the following dimensions: each was 20^{cm} long; four were cylindrical
rods, with diameters 13^{mm}, 9·5^{mm}, 6·5^{mm}, 3·5^{mm} respectively: they
were designated rods A, B, C and D; four were of rectan-
gular cross-section, 12×3·5^{mm}, 12×2·5^{mm}, 12×1·5^{mm}, 12×0·75^{mm}
respectively: they were designated, rods E, F, G and H.
They were treated in the same manner as in Bidwell's experiments; they were carefully annealed, being heated in iron
filings almost to the point of fusion and then cooled in ashes for
about 25 hours. They were first strongly magnetized and
then carefully and totally demagnetized. As a result of numerous trials and careful manipulation, I was able to get distinct
effects from a gas jet at a distance of 6 inches. The following is a characteristic set of experiments: the light was perpendicular to the axis of the bar and only 3^{cm} of the north
end of each rod was exposed. The deflection represents
increase of magnetism as shown with magnetometer.

Rod.	Area exposed to light.	Cross-section of rod.	Deflection.
A	6·15$^{sq\ cm}$	1·33$^{sq\ cm}$	22·5
B	4·48	0·71	15·0
C	3·06	0·33	7·1
D	1·65	0·10	1·5
E	3·60	0·42	10·0
F	3·60	0.30	7·5
G	3·60	0·18	4·5
H	3·60	0·09	1·5

The effect was proportional to the area exposed, but not
directly. It depended also on the intensity of the illumination. The action of the light does not appear to be here confined to the illuminated surface, as the effect is approximately
the same, no matter which 3^{cm} of length of the bar is illuminated, and hence distance of illuminated surface from magnetometer has little or no effect. This is contrary to Bidwell's
results and is probably due to the very regular annealing. The
heating effect was also present. Light of long wave-length
had greater effect. I did not attempt Bidwell's experiments
in regards to polarized light and its action on the magnetism
of the iron, since no positive results were obtained by him.

I then extended these experiments of Bidwell in an attempt to learn more in regard to the nature of this action of light on magnetism. Every cause of molecular disturbance favors the magnetization of a bar subjected to a magnetizing force, and also favors its demagnetization after it has been withdrawn from the field. Vibration has a specially marked effect upon iron. Ewing[*] has shown that if a bar of this metal be kept from the slightest vibration one can obtain residual magnetizations much greater than those shown in steel bars, but the least vibration causes the acquired magnetism to vanish almost completely. If the action of the light is of the same nature as that of a mechanical shock on a magnet and not dependent on the special structure of the magnetic field in Bidwell's experiment, we ought to obtain effects from light in experiments like those of Ewing, if they are sufficiently delicate. I have been able to get very good results from a series of experiments with this object in view. The bars were magnetized in a magnetic field of about 400 C. G. S. units and the current was steadily and rapidly shut off by means of a continuous current rheostat. The slightest vibration caused a sudden fall in the residual magnetism. Extreme care was necessary. Only a small portion of the bar was exposed to the light, as it was found that the magnetizing coil could not be removed without causing the sudden fall in the magnetic strength of the bar. Electric disturbances due to the surface-road and electric light wires had the same effect, and it was necessary to work during the quiet portions of the day. The following is the average result of a number of experiments. Only 3^{cm} of the bar was exposed, that is, there was the same surface exposed as in previous experiments, and the light was that of an ordinary gas jet at a distance of six inches; other conditions were the same. The direction of the deflection indicated in every case a decrease of magnetism.

Rod.	Area exposed to light.	Deflection.
A	6·15 sq cm	44·0
B	4·48	26·1
C	3·06	12·5
D	1·65	6·4
E	3·60	13·1
F	3·60	8·4
G	3·60	4·3
H	3·60	0·9

These results differ from those of Bidwell in the fact that the deflections here are permanent, even after light has been

[*] Proc. Roy. Soc., 1890, p. 342.

removed. In my own repetition of Bidwell's experiment, the
magnetometer image did not always return to the same spot
after the light had been flashed upon the iron and removed.
There was in general a partial return, the law or cause of which
I was unable to discover. I then attempted a similar experiment obviously suggested by the previous one, namely, that of
increasing by light the magnetic strength of a bar which was
in a steady field. I here also had great difficulty on account
of external vibrations and electrical disturbances. I steadily
increased the current up to different strengths and then flashed
the light on the exposed portion of the bar and noted increase
with the magnetometer. The first series here given is when
the field is of about 10–15 C. G. S. units and the second, of
about 350 C. G. S. units, 3^{cm} was exposed and the light was a
gas jet at a distance of 6 inches.

Rod.	Area exposed.	Deflection—I.	Deflection—II.
A	6·15 $^{sq\ cm}$	none	none
B	4·48	none	none
C	3·06	42	none
D	1·65	6	10
E	3·60	none	none
F	3·60	36	none
G	3·60	23	15
H	3·60	7	14

It is a noteworthy fact that it is only the larger bars from
which no deflection was obtained. The conditions were precisely the same for the series and therefore for the larger bars
there is either an increased sensitiveness to vibrations coming
from without, or the action is dependent on the mass in some
manner not understood. It is probably only a small portion
of the iron which is so extremely sensitive as to be influenced
by light. A slight but sudden increase of the field up to the
maximum strength used in this experiment caused a rise in the
magnetism of one bar amounting to 125 divisions, while a
slight and sudden fall in the strength would also cause a
slight increase, when the sensitive state was obtained.

All these results can be effectively accounted for by the
assumption of a regular closed magnetic circuit (in iron bars
carefully annealed, magnetized, and immediately demagnetized
as above described) in which the lines of force return through
a thin layer at the surface. The circuit is wholly or partially
closed in every case except that of complete saturation, and
the thickness of this surface layer depends on the magnetic
state of the bar. In Bidwell's experiments, we may suppose
the larger portion of the lines of force to return through the
surface layer, which is here of considerable thickness. Here

the action of the light is to break or destroy a number of those lines of force. On removal of the acting force, or source of energy, there exists a tendency to return to the original form, owing to action of magnetic lines in the immediate vicinity. In the next two experiments, the instability is of the opposite kind and there is no tendency to return to the original form. In both cases, where the iron was demagnetized by light, and where the iron was magnetized by light, the action is similar to that of a blow under the same circumstances.

While this satisfactorily explains the mechanism of the results of the action of light on a magnetized iron bar, that of the direct action of the light is in doubt. Analogous effects can be obtained from purely mechanical actions and I have obtained them also from the action of electric waves. Whether these quantitatively remarkable results are due to the magnetic or purely mechanical action of light, is unknown. I determined therefore to make some experiment in regard to the action of polarized light on the magnetism of iron. Bidwell has shown that polarized light in the ordinary experiments has the same effect as ordinary light. In order to make a more sensitive action, I obtained from Professor A. W. Wright a number of films of iron deposited on thin glass plates by electrical discharge in exhausted tubes. The method of production and general properties of these films are given in this Journal for January and September, 1877. The films of iron had a brilliant luster and a high degree of transparency and polarized the light transmitted by them powerfully. The films were approximately 2^{cm} long and $0 \cdot 5^{cm}$ in width and the thickness was very small, about a quarter of the wave length of red light. The amount of iron in one of these magnets was extremely small; the magnetic moment of the largest one was so small that when suspended by a single silk fiber, the finest obtainable, the magnetic action of the earth's field was scarcely noticeable. But the magnet possesses relatively strong permanent magnetism for the amount of iron involved, in fact, the iron appears permanently saturated. A remarkable feature of this magnetism is its strong resistance to a reversing force. If we assume the reversal of any ordinary magnet by an opposite magnetizing force to be due to the cumulative action resulting when a few unstable molecular magnets are reversed and these produce other unstable groups which are in their turn reversed and so on, then this phenomenon can be explained by the almost total absence of unstable groupings of molecular magnets in the mass of the iron. In a magnetic field of 400 C. G. S. units it required from three to five seconds to reverse it, and in a field of 25 C. G. S. units it was impossible to reverse it without tapping. Under these circumstances it was found to be impossible to demagnetize this magnet totally by any of the ordinary methods. The magnetometer used in the previous experiments, while extremely delicate, was not of the slightest use in these, and therefore it was necessary to measure the time of vibration of the magnet itself, with and without light polarized in different directions falling on its surface. The system was almost aperiodic, the magnet coming completely to rest in five half vibrations. No difference in time amounting to 0·2 seconds was observable out of a total of 40 seconds approximately. Owing to the unexpected magnetic behavior of the iron, it was not thought that the light polarized so as to have its vibrations perpendicular to the magnetic axis of the magnet would demagnetize it completely, but that a slight fall from completely saturated state might occur, and so the experiments were continued. The magnet was kept suspended in a field of 600 to 800 C. G. S. units for some time; the field was gradually diminished to 200 C. G. S. units and then torsion for given deflection was measured with and without the application of polarized light. Five complete revolutions of the torsion head caused a deflection of 55° of the magnet without the application of light, while 4·8 revolutions caused the same deflection when light so polarized that the vibrations were perpendicular to the magnetic axis was applied to the surface. There is a possible error of 20° for each of these readings. The same experiment was repeated with light parallel to the magnetic axis of the magnet and a difference in the torsion amounting to 40° was observed. The quantitative difference in the two experiments is very small, but the results of a great number of the observations were taken and were fairly consistent. This apparently shows a difference depending on direction of the plane of polarized light analogous to the demagnetizing effect of an alternating current.

Owing to the anomalous behavior of iron films when placed in a magnetic field and their strong resistance to a reversal of the magnetization, it was thought advantageous to attempt the experiment with a nickel film, since the resistance to demagnetization would probably be much less than for iron films. Although the film which was used contained nearly fifty times the amount of metal of the heaviest of the iron films and was exposed in a field of nearly 2000 C. G. S. units, there existed no appreciable magnetic moment in the nickel and the experiment was given up.

From the results of this investigation the following conclusions may be drawn:

1. An effect of light on magnetism is, in general, easily obtained and depends only on a certain condition of sensibility artificially produced.

2. This condition is, apparently, that a peculiar and unstable distribution of the magnetic lines of force shall exist in the magnet.

3. The action is, in general, a purely mechanical one, similar to the effect of a blow.

4. Evidence of the magnetic action of polarized light has probably been obtained by the use of iron films on glass.

In conclusion, I wish to thank Prof. A. W. Wright for his assistance throughout the investigation. He suggested the subject and general plan of research, and afforded me ample facilities for performing the experiments.

Sloane Physical Laboratory, June 15, 1900.

SCIENTIFIC INTELLIGENCE.

I. CHEMISTRY AND PHYSICS.

1. *On Radio-active Barium.*—In a preliminary notice BÉLA V. LENGYEL calls attention to the fact that the bodies thus far known which emit Becquerel's rays are of five different kinds, viz., the compounds of uranium, of thorium, polonium, radium, and Debierne's body which is analogous to titanium. Uranium and thorium are well-defined chemical substances, while the three others are merely hypothetical elements. Radium is the best known of these, but from the statements made in regard to it the author believes that it can hardly be regarded as an existing element. In favor of its existence the two principal facts from a chemical point of view are the higher atomic weight found by Madam Curie for the radio-active barium, and the single foreign line found by Demarcey in the spectrum of such material. In regard to the first point, the author thinks that a body which is present in sufficient amount to raise the atomic weight eight units above barium should not fail to reveal itself in the course of various chemical decompositions. He believes, moreover, that the single spectrum-line noticed by Demarcey does not point to the presence of an element closely analogous to barium, for, as is well known, the spectra of barium, strontium, and calcium consist of many sharp lines and less clearly defined bands, similarly grouped in the three cases, and an element almost identical with barium should possess an analogous spectrum. He states also that it should be remembered that these hypothetical elements are always found adhering to other well-known chemical elements. Curie found polonium with bismuth, Giesel found it with lead; radium is attached to barium, Debierne's element goes with titanium. All these radio-active substances come from pitchblende, being separated from it by analytical processes. It is difficult to assume that elements exist which differ from others that are well known in nothing except their radio-activity.

Such considerations as the above led the author to investigate experimentally the question whether radio-active bodies contain new elements or not. He ignited uranium nitrate with two or three per cent of barium nitrate and finally fused the resulting oxides in the electric arc. The mass was dissolved in nitric acid, the solution was evaporated, whereupon a large part of the barium separated as nitrate, then from the decanted liquid a precipitate of barium sulphate was produced which showed radio-activity. The author believes that he has prepared radium synthetically, but does not consider that these preliminary experiments are sufficient to decide with certainty whether this is a definite chemical element or not.

There appears to be a weak point in this supposed synthesis of radium from the possibility that the uranium nitrate used may have contained the "element" in question, although v. Lengyel does not mention such a possibility. At all events, the result of the experiment makes it seem possible that the radio-activity of uranium may be due to the same cause as that of radium, and perhaps also of polonium and the others.—*Berichte deutsch. chem. Gesellsch.*, xxxiii, 1237.

H. L. W.

2. *Mercuric Antimonide, and Stibonium Compounds.*—PARTHEIL and MANNHEIM have prepared mercuric antimonide, Sb_2Hg_3, by passing hydrogen antimonide over finely powdered mercuric chloride which was mixed with sand and fragments of glass. The reaction took place slowly, the white mercuric chloride becoming black. This mercuric compound is interesting because like potassium antimonide it gives stibonium compounds when treated with alkyl iodides. The authors have studied its behavior with ethyl and propyl iodides, heated under pressure. They thus obtained the double salts $Sb(C_2H_5)_4I.HgI_2$ and $Sb(C_3H_7)_4I.HgI_2$. By treatment with silver oxide these bodies yielded tetraethyl stibronium hydroxide, $Sb(C_2H_5)_4OH$, and the corresponding propyl compound, from which various salts were prepared.— *Chem. Centralbl.*, 1900, 1, 1091.

H. L. W.

3. *The Atomic Weight of Iron.*—This important constant has not been revised since about fifty years ago; meanwhile analytical methods have been much improved, and the sources of error involved in atomic weight determinations have been more carefully studied. In view of these facts, RICHARDS and BAXTER have made a preliminary series of seven determinations by reducing ferric oxide to metallic iron. Their results show that the previously accepted atomic weight, 56, is somewhat too high and that in all probability the value is very near 55·88.—*Zeitschr. anorg. Chem.*, xxiii, 255.

H. L. W.

4. *The Physical Properties of Cæsium.*—ECKARDT and GRAEFE have made an elaborate examination of the physical properties of metallic cæsium which was prepared by heating pure cæsium carbonate with metallic magnesium in an iron tube in a current of dry hydrogen. The summary of their results is as follows:

Sp. gr. of liquid Cs at 40°	1·827
Sp. gr. of liquid Cs at 27°	1·836
Sp. gr. of solid Cs at 26°	1·886
Solidifying point of Cs	26·37°
Conductivity at 27° (Ag=100)	3·63
Specific heat	·04817
Atomic heat	6·406
Heat of fusion (for 1 g.)	3·73 cal.
Contraction upon solidification	2·627 vol. per cent.
Coefficient of expansion	·0003948

The specific gravity found by the authors corresponds with the value 1·88 previously determined by Setterberg, but it does not agree with the value 2·40003 recently given by Mencke. The latter is probably incorrect, not only in view of the agreement of the other investigators, but because it gives an atomic volume for cæsium, which is not in harmony with the periodic system.— *Zeitschr. anorg. Chem.*, xxiii, 378.   H. L. W.

5. *On Compounds of Telluric Acid with Iodates.*—WEINLAND and PRAUSE have prepared the salt $I_2O_5.2TeO_3.K_2O.6H_2O$, as well as an exactly corresponding rubidium compound, and an ammonium salt which varies from the foregoing in containing $8H_2O$. The potassium salt of a different acid, $I_2O_5.TeO_3.K_2O.$ $3H_2O$ was also obtained. These results are interesting inasmuch as they show that TeO_3 forms complex acids with iodates similarly to the oxides SO_3, CrO_3, MoO_3 and WO_3.—*Berichte deutsch. chem. Gesellsch.*, xxxiii, 1015.   H. L. W.

6. *A Simple Method of Decomposing Chromite.*—For getting this very refractory mineral into a soluble condition FIEBER recommends heating ·5 grams of the finest possible material with six parts of sodium-potassium carbonate for ten minutes in a platinum crucible, then after cooling adding six parts of borax and fusing at a gradually increasing heat, finally over the blast-lamp for three-quarters of an hour. This generally effects the decomposition. If it fails, it is only necessary to add another portion of sodium-potassium carbonate and fuse again.—*Chem. Zeitung*, xxiv, 338.   H. L. W.

7. *The Theory of Electrolytic Dissociation and Some of its Applications;* by HARRY C. JONES, 8vo, pp. xii–289, New York, 1900 (The Macmillan Company).—The title of this valuable addition to chemical literature hardly gives a proper idea of the scope of the work. The first chapter, 68 pages, is devoted to an exposition of "The Earlier Physical Chemistry," which serves as an excellent introduction to the main subject of the book. The four chapters are designed to answer the questions: What was physical chemistry before the theory of electrolytic dissociation arose? How did the theory arise? Is it true? What is its scientific use? The author has given very satisfactory answers to these questions, and the book will undoubtedly be useful to those who wish to gain a clear notion of some of the chief points of modern physical chemistry.   H. L. W.

8. *Optical Activity and Chemical Composition;* by Dr. H. LANDOLT, translated by JOHN McCRAE, Ph.D., 12mo, pp. xi–158, London, 1899 (Whittaker & Co., from The Macmillan Company, New York).—This work forms a chapter in Graham-Otto's "Lehrbuch der Chemie." The topics dealt with are the general principles of optical activity, the connection between the rotary power and the chemical composition of carbon compounds, and the connection between the degree of rotation and chemical constitution.   H. L. W.

9. *Variation of the Electric Intensity and Conductivity along the Electric Discharge in Rarefied Gases.*—Skinner has shown that very near the anode the electric intensity is very small or zero when the positive column is not striated. He also mentions that with a striated positive column the potential-difference

between the anode and an exploring wire near it, was a minimum when the exploring wire was at a short distance from the anode, so that the apparent electric intensity near the anode was negative. These results are confirmed by HAROLD A. WILSON, who measured the differences of potential between two exploring wires kept at a fixed distance apart in the discharge, which could be brought into any desired portion of the discharge by moving the electrodes between which the discharge took place. This discharge was produced by a storage battery of 600 cells. The gases employed were air, nitrogen, and hydrogen. The positive drop close to the anode was very apparent from the experiments. Very close to the positive electrode there is apparently a very intense ionization. In the negative glow and Faraday dark space there is also ionization in excess. In every kind of electric discharge ionization appears to occur on the metallic electrode when it is red hot and most easily there. The negative ions formed at the surface of the cathode constitute the cathode rays, and produce the ionization which the intensity curves show occurs in the negative glow.—*Phil. Mag.*, June, 1900, pp. 505–516.   J. T.

10. *Mechanical possibilities of flight.*—Lord RAYLEIGH in the Wilde lecture published in the Manchester Memoirs 1899, No. 5, discusses this proposition and shows that in order for a man to support himself by a vertical screw by working at the power an average man can maintain for eight hours a day, he would require a screw ninety meters in diameter; and in this estimate no account is taken of the weight of the mechanism or of frictional losses.—*Nature*, May 31, 1900.   J. T.

11. *Magnetic Screening.*—The great extension of electric circuits for lighting and for power makes it impossible in most cases to use the ordinary form of mirror galvanometer. H. DU BOIS and A. P. WILLS discuss the subject of the protection of galvanometers from outside changes in the magnetic field produced by the commercial employments of strong currents. The cases of iron cylindrical shells and of spherical shells is treated and the results are used by H. Du Bois and H. Rubens in the construction of a protected or armored galvanometer (Panzergalvanometer). Cast-steel cylinders having a radius ratio of air space 1·5 to 1·6 and spherical cast-steel shells 1·3 to 1·4 give in general this order of protection : one shell, 10 ; two shells, 100 ; three shells, 1000.— *Ann. der Physik*, pp. 78–95.   J. T.

12. *The Sun's Corona.*— Mathias Cantor in the Ann. der Physik for March, 1900, believes from his experiments that a rarefied gas through which an electrified discharge is passing shows no perceptible absorption corresponding to its emission spectrum, and Professor Fitzgerald in Nature, May 3, 1900, remarks that this fact confirms the suggestion that the sun's corona is an electrical discharge around the sun since the bright spectrum line of the corona is not represented by a dark line in the solar spectrum. E. PRINGSHEIM criticises severely the results of Cantor and believes that he has not taken suitable precautions in his experiments.

He points out that the discharge through the gases was an inter mittent one and the gases were only a portion of the time in the condition when, according to Kirchhoff's laws, an absorption was to be expected. For the remaining portion of the time the light of the arc lamp employed could go through the gas completely unabsorbed. Furthermore since the spectral lines in a Geissler tube represent homogeneous light, an absorption could only be expected, according to Kirchhoff's laws, in a very narrow spectral region. In order to show it, a spectrum of a large dispersion must be employed and a narrow slit, in order that no light of the neighboring portion of the arc spectrum can intrude upon the homogeneous portion under examination. These precautions were not taken by Cantor.—*Ann. der Physik.*, No. 5, 1800, p. 199–200.

J. T.

13. *On some Properties of Light-struck Photographic Plates* by FRANCIS E. NIPHER. (Abstract by the author.)—A paper with the above title recently published* by the Academy of Science of St. Louis, deals among other things with some features which seem to be new in photography.

In a general way it is probably true that any plate, upon which a camera impression has been made, may be developed either as a positive, or as a negative. Suppose the object to be a white design on a black ground. Let the camera be replaced by printing frame, containing the plate and an opaque stencil with some design punched through it. Place this plate at any distance from a known source of light. Give any exposure to the plate If the plate be now developed at a considerable distance from the lamp, a negative image will appear. The lamp should be vertically over the developing bath. If the bath in a similar exposure is taken nearer to the lamp, the negative will be poorer one. At a still nearer distance, nothing will develop In the parlance of the photographer, the plate will fog. This called a zero plate. If the plate be developed at a point still nearer the lamp, the picture will appear as a positive. The conditions of zero plate as a function of exposure is now being studied. A given exposure is a time integral of the effect due to any given illumination, at the point of exposure.

The paper referred to gives half-tone reproductions of photographs thus developed in the light as positives. In one attempt was made to "fog" a plate into a zero condition before it was put into the camera for exposure to a street scene. The plate was fogged by X-rays for two hours, and was exposed in the camera for ten minutes. It was developed in a hydrochinon bath, within one foot of a sixteen-candle lamp. The result is superb positive of the street scene in which the moving object leave no trace. The plate was an instantaneous isochromatic plate by Cramer.

Another picture reproduced was a picture on the same kind plate, obtained by an exposure of one minute in the camera. The

* No. 6, vol. x, Transactions.

street was strongly lighted, and the plate was converted into a zero condition, and the positive was then produced, during the single minute and while in the camera.

It is also found that the camera exposure may be shortened, by fogging the plate in open lamp-light before it is put into the camera. If the plate is to be developed within 20 cm. from a sixteen-candle lamp, these fast plates will be put into a zero condition by holding them for 90 seconds at a distance of one meter from the lamp. When exposed in the camera such a plate yields a perfect positive picture. If traces of fogging appear, the plate should be moved nearer the lamp. The precise conditions which yield the best results have not yet been determined, nor is it yet known how short the exposure may be made. Some of the plates turn out badly until experience has been gained.

It is evident that if a similar change can be produced in the operation of printing, so that a positive print may be obtained from a positive plate, the dark-room may perhaps be dispensed with for purposes of developing.

In some cases the most sensitive plates have been exposed for from three to four hours to brilliantly lighted street scenes, with the diaphragm fully open. There is not the least trouble in developing such plates, as positives, in a strongly lighted room. In general, the greater the exposure, the darker the developing room must be in order to get a zero or a negative result.

Large pin-hole images of the sun may be developed in this way, and there appear to be many ways in which the process may yield good results. One of the incidental features of value is that a little of what photographers most fear does not interfere with the results.

II. GEOLOGY AND MINERALOGY.

1. *A Monograph of Christmas Island (Indian Ocean): Physical Features and Geology ;* by CHARLES W. ANDREWS. With descriptions of the Fauna and Flora by numerous contributors. Pages x, 337 ; plates xxii. London, 1900 (British Museum of Natural History).—The unique character of Christmas Island, in its position, history, and life, gives peculiar interest to this account of the results obtained from the ten months' vigorous explorations made by Mr. Andrews, of the Geological Department of the British Museum.

Christmas Island has an area of 43 square miles, and rises in places to a height of 1000 feet; it is covered with a dense tropical vegetation. It is situated in the eastern part of the Indian Ocean, 190 miles to the south of Java, 900 miles northwest of the coast of Australia, and 550 miles east of the atolls of Cocos and North Keeling. The submarine slopes about it are so steep that a depth of 1000 fathoms is found within two or three miles of the coast, while to the north, a depth of 3200 fathoms was found (Maclear Deep), and to the south and southwest, of 3000 fathoms

(Wharton Deep). The island is described as forming the summit of a submarine peak, the base of which rises from a low saddle which separates the two abysses named, and on the western end of which the Cocos-Keeling Islands are situated. Its peculiarly isolated position, hence, is most striking. Its history is also unique, since, although known to navigators since the middle of the 17th century, no one seems to have penetrated into the interior until 1887, and, as remarked by Dr. Murray, down to a few years ago it was probably the only existing tropical island of any large extent that had never been inhabited by man, savage or civilized. Its animal and vegetable life, therefore, are thus far almost unchanged by the conditions introduced by human life.

Geologically, the island consists largely of elevated Tertiary limestones with extensive series of eruptives; briefly, it may be considered as an ancient atoll raised to a considerable height above the level of the sea. The "central nucleus" is made up of compact yellow limestone in places very hard and showing no traces of bedding or jointing. This is referred to the Eocene (or Oligocene) and is accompanied by basalts and trachytes both beneath and between the beds. The total thickness of these older Tertiary and accompanying volcanic rocks is estimated to be 600 feet. Forming the mass of the island is the Miocene Orbitoidal limestone, separated from the older rocks by basalts and basic tuffs. The higher elevations are dolomitic limestones containing 34 to 41 p. c. of magnesium carbonate; these show traces of coral structure and imperfect remains of Foraminifera. Thick beds of phosphate of lime, in part limestone beds altered by overlying guano, in part phosphatized volcanic tuffs, occur on some of the elevated points and have proved to be of economic value. The Tertiary limestone, especially the Miocene, forms abrupt vertical cliffs, sometimes 250 feet in height, along a large part of the coast line. A series of terraces is also noted around the shore, and outlying the whole is the fringing coral reef. The author remarks upon the remarkable development of elevated Tertiary rocks and the difficulty in explaining their deposit over an area so isolated. He adds that the great thickness of reef limestone, required by the Darwinian theory of atoll formation, is not found, and although there may be some evidence that subsidence did occur in the history of the island, it is clear that it was not for any long period nor of any great extent. It is interesting to recall in this connection the similar observations recently made by Agassiz on the elevated Tertiary limestones of the Fijis and other islands of the Pacific.

The life of the island is fully described from the collections made and it is shown to be to a remarkable extent endemic. Thus of the 319 species of animals recorded, about 45 per cent are described as peculiar to it, although this percentage may perhaps be reduced when the fauna of Java and other neighboring islands is more minutely known. For example, of the mammals, all are peculiar species except one; of six reptiles, four are peculiar; of fourteen species of land shells, six are peculiar, etc. The author gives an interesting summary of the conditions, as to winds and ocean currents, which have been instrumental in the introduction of the fauna and flora.

2. *On the Lower Silurian* (*Trenton*) *Fauna of Baffin Land;* by CHARLES SCHUCHERT. Proc. U. S. Nat. Mus., vol. xxii, pp. 143–177 (with plates xii–xiv), 1900.—The author reports upon several collections placed in his hands for study. The following excellent summary is quoted from page 175 of the report:

"The only Lower Silurian horizons known in northeastern Arctic America are of Trenton and Utica age. The latter zone appears only on the north shore of Frobisher Bay, but the Trenton is found in various places from the north shore of Hudson Strait to latitude 81° north. The Lower Silurian is thickest on Akpatok Island, where it is from 400 to 500 feet in depth. Dr. Bell, however, estimates the entire thickness of these strata in this region to be not less than 900 feet.

" In Baffin Land, and apparently elsewhere in Arctic America, the Lower Silurian strata rests unconformably on old crystalline rocks. To the north of Baffin Land, the former are overlain by beds of Niagara or Wenlock age.

" The Trenton faunas, occurring in various places around the insular Archæan nucleus of North America, have much in common, and this indicates that the conditions at that time were very similar, while the sea was in communication throughout. As yet, however, the distribution of the strata, together with their faunas, are well known only to the south and southeast of the Archæan nucleus, yet that of the west (Manitoba) and of the northeast (Baffin Land) show direct communication.

" The Baffin Land fauna had an early introduction of Upper Silurian genera in the corals *Halysites*, *Lyellia*, and *Plasmopora*. In Manitoba similar conditions occur in the presence of *Halysites*, *Favosites*, and *Diphyphyllum*. Other Upper Silurian types do not appear to be present.

" The Trenton fauna of Silliman's Fossil Mount, at the head of Frobisher Bay, has seventy-two species, of which twenty-eight are restricted to it. This fauna shows an intimate relationship with that of the Galena of Minnesota, Iowa and Wisconsin. Fifty-seven per cent of the species of Baffin Land also occur in the Galena of the regions just mentioned.

" The Trenton fauna of Baffin Land shows that the corals, brachiopods, gastropods, and trilobites have wide distribution, and are therefore less sensitive to differing habitats apt to occur in widely separated regions. On the other hand, the cephalopods, and particularly the pelecypods, indicate a shorter geographical range. The almost complete absence of Bryozoa in the Baffin Land Trenton contrasts strongly with the great development of these animals in Minnesota and elsewhere in the United States."

3. *A Preliminary Report on the Geology of Louisiana;* by GILBERT D. HARRIS and A. C. VEATCH, Part V, Geology and

Agriculture, plates 1–62, figures 1–7, map, pp. 4–354, 1899. (Geological Survey of Louisiana, Wm. C. Stubbs, Director.) The author recognizes in Section II the following formations :

Cretaceous series—Ripley stage.

Eocene series—Midway, Lignitic, Lower Clarborne, and Jackson stages.

Oligocene—Vicksburg and Grand Gulf.

Lafayette—

Quaternary—(including the Columbia of McGee and Biloxi sands of Johnson) Basal gravel, Port Hudson, Loess and Yellow Loam, Alluvium and recent coastal formations.

In Section III—special reports—nine miscellaneous reports are given by various authors, the last on " Wood-destroying Fungi," by Professor Atkinson.

4. *Geology of Panama.*—In a study of the geology of the isthmus of Panama recently published by Messrs. BERTRAND and ZURCHER,* Bertrand draws the following important conclusions regarding the probability of earthquake shocks in the Panama region :

"There are no volcanoes near Panama, all eruptions having ceased since the Miocene; this is the first and most important point of all.

Since the earthquake of 1621, which is in reality disputed, there have been in the region only very feeble shocks, a part of which were due to the echo of far distant earthquakes.

The depression made use of by the Panama canal project is not a transverse fracture.

The sinking of the Pacific coast, and especially the subsidence of the bay of Panama, of which there are numerous indications, are not phenomena now going on but finished, so far at least as regards the present geological epoch. There is there no special reason for crustal movement.

Finally, the plan of the lines of folding and the distribution of volcanic and seismic activity following these lines, shows that Panama is situated in a sort of dead angle, in a tranquil zone, at an equal distance north and south from the lines of disturbance.

Thus all considerations, whether statistic, volcanic or seismic, lead to the same conclusion, that Panama is the most stable and least menaced region of Central America."   L. V. P.

5. *Enrichment of Mineral of Veins by later Metallic Sulphides;* by WALTER HARVEY WEED. Bulletin of the Geological Society of America, vol. II, pp. 179–206.—Secondary sulphide enrichment is certainly one of the most interesting subjects of economic geology, and, as noted by Mr. Weed in the opening pages of his article, is one upon which surprisingly little has been written, considering its scientific and practical importance. Among prominent writers on ore deposits, De Launay, Posepny, Emmons and

* *Études geologiques sur l'isthme Panama* and *Phénomènes volcaniques et les tremblements de terre de l'Amerique Centrale.* Paris, 1900.

Kemp have mentioned the subject, but the present paper is the rst to treat it separately and comprehensively.

After some introductory remarks and definitions a discussion ollows of the chemical reactions which take place during the eaching of the mineral contents of the gossan zone; then the hemical processes concerned with mineral deposition in the nrichment zone and of the alterations of various vein minerals, re considered. A more extended description and discussion of he manner of occurrence of secondary enrichments is then entered pon, dealing with the deposits of copper, silver and zinc in promnent localities, here and abroad, which have been carefully tudied.

To the material taken from various outside sources Mr. Weed as added much valuable information, gained as a result of his wn extensive observations and studies of the ore deposits of the West.

The conclusions reached are worthy of careful notice and are quoted here in full. Conclusions : "From what has been shown it is concluded that later enrichment of mineral veins is as important as the formation of the veins themselves, particularly from the economic standpoint. The enrichment is usually due to downward-moving surface waters, leaching the upper part of the vein and precipitating copper, silver, et cetera, by reaction with the unaltered ore below. In many cases the enrichment proceeds along barren fractures and makes bonanzas. In others it forms films, pay streaks, or ore shoots in the body of leaner original ore. In still other cases the leaching, transportation, and redeposition are performed by deep-seated uprising waters acting upon the vein.

As a consequence of this, veins do not increase in richness in depths below the zone of enrichment.

The practical bearing of the phenomena described and the deduction drawn from them will, I think, be apparent to every mining engineer and geologist. If my views be correct, the future of many ore deposits is to be judged in the light of these facts, and the value of the mine must not be based on the presumption that the ore will continue in unabated richness in depth."

A forthcoming paper, which is promised on the secondary sulphide enrichments of Butte, Mont., will be awaited with no little interest.
                                                                    C. H. W.

6. *Mineralogical Notes.*—Number 7 (vol. i) of the Geological Series of Publications by the Field Columbian Museum (pp. 221–240), by Dr. O. C. FARRINGTON, contains an account of several new mineral occurrences and also a description of the fine calcite crystals from Joplin, Missouri. The rare species *inesite* is noted from a mine near Villa Corona, Durango, Mexico. It occurs in tufts of radiating crystals of flesh-red color. These correspond n angle with the results of Scheibe and show the new forms : (11·0·12) and *s* (946). The following analysis was made :

	SiO$_2$	MnO	FeO	CaO	MgO	H$_2$O (cryst.)	H$_2$O (const.)
G.=2·965	44 89	36·53	2·18	8·24	tr.	5·99	2·21=100·34

From this, the composition H$_2$(Mn, Ca)$_4$Si$_4$O$_{12}$+3H$_2$O is deduced, which varies somewhat widely from the results hitherto obtained.

Caledonite in distinct crystals is noted from the Stevenson-Bennett mine, Organ Mountains, New Mexico ; also *gay-lussite* from Sweet Water Valley, Wyoming. The use of *dolomite* as money by the Indians in Lake County, California, is noted. The tokens are shaped in cylindrical forms and burned, which brings out reddish streaks in the oxidation of the iron, and then polished and perforated. In this form they are highly valued by the natives.

The various types of the beautiful Joplin crystals of calcite from Joplin are described and well figured. The observations form an important addition to the literature of the species.

7. *Tabellen zur Bestimmung der Mineralien mittels äusserer Kennzeichen.* Herausgegeben von Dr. ALBIN WEISBACH. Fünfte Auflage. Pp. 106. Leipzig, 1900 (Arthur Felix).—The Determinative Tables of Prof. Weisbach, first issued in 1866, are so well known by those interested in mineralogy that their value hardly needs to be remarked upon here. Based upon external characters alone, they give the student a simple means of determining species, applicable in ordinary cases, and teach him to use his powers of observation with accuracy and discrimination.

8. *Repertorium der Mineralogischen und Krystallographischen Literatur vom Anfang d. J. 1891 bis Anfang d. J. 1897, und Generalregister der Zeitschrift für Krystallographie und Mineralogie,* Band xxi–xxx. Herausgegeben und bearbeitet von E WEINSCHENK und F. GRÜNLING. II. Theil (Generalregister von F. Grünling). Pp. 394. Leipzig, 1900 (Wilhelm Engelmann).— The First Part of this Index has already been noticed in the present volume of this Journal (p. 229). Part II, now issued, completes the work, which is invaluable for every one interested in the progress of Mineralogy. The fact that this general index extends to nearly four hundred pages shows strikingly the immense amount of material in the Zeitschrift to which references are made.

III. BOTANY AND ZOOLOGY.

1. *Les Carex de l'Asie orientale ;* by A. FRANCHET, Nouv. archiv. du muséum d'hist. nat., Ser. 3, vols. 8-10, Paris, 1896-1898.—The present comprehensive work deals with the vegetation of *Carex* in eastern Asia, of which 274 species are described and a number of new ones are figured ; the geographical distribution is given to each species. Inasmuch as the number of species of *Carex* is in the neighborhood of some 800, the author has rendered excellent service to the study of the genus by presenting figures of the new with their general habit and other details. Eastern Asia, and especially Japan, appears to be ex-

lingly rich in representatives of this genus, in spite of the
; that many large areas of, for instance, China and even Japan
yet unexplored, and the extreme north-east is almost un-
wn, save the collections brought home by Eschscholtz and
llman. In looking over the vast material treated in the pres-
work, one notices readily that a number of very peculiar
es are characteristic of that part of the world ; there are,
ed, several which are utterly unlike the ordinary types of
-ex : *C. podogyna*, for instance, possesses a utricle borne on
lose stipe, 4–8^{mm} in length ; in *C. Nambuensis* there are sev-
lateral androgynous spikes on long, filiform peduncles;
achygyna has inflated, bladeless sheaths subtending globose,
illate inflorescences ; the leaves of *C. capilliformis* are numer-
, long and capillary; in *C. hakkodensis* and *C. rhizopoda* we
t with types of very much the same habit and structure as
singular *C. lejocarpa* and *C. circinata ; C. gentilis* shows the
it of the *Indicæ* Tuckm., while *C. moupinensis* reminds one
a *Scirpus* or *Rhyncospora* rather than a *Carex*. Only two
cious species are known from eastern Asia: *C. Redowskiana*
A. Mey. (*C. gynocrates* Wormskj.) and *C. grallatoria*. Among
monœcious we meet with the circumpolar *C. rupestris*, and
h *C. pyrenaica*, of which the latter shows a most singular
graphical distribution : Mountains of Middle- and South-
:ope, western Asia, New Zealand, Rocky Mountains of Col-
do and Alaska. *C. pauciflora* and *C. microglochin* are, also,
resented in this Flora ; *C. incurva* is reported from West
na, and is known besides from so remote localities as the
tic region, the Magellan strait, Himalaya, the coast and
her mountains of Europe. The cosmopolitan *C. vulgaris* is
course, included, besides some species which are very abundant
Europe, as for instance : *C. vesicaria, C. filiformis, C. Pseudo-*
erus, C. Buxbaumii, etc.

t appears, altogether, as if the genus in eastern Asia possesses
umber of species in common with other parts of the globe,
many that are not known from elsewhere. But in these special
es, special to eastern Asia, are several which actually repre-
t forms analogous to those which inhabit other parts of the
rld. There is a series of *Microrhynchœ*, of *Melananthœ* and
n · of *Dactylostachyœ*, which correspond very well with such
es as exist in Europe, Asia and America. It is this part
the work, the geographical distribution in connection with the
racterization of the various types, so excellently done by
nchet, which makes his treatise of the East Asiatic *Carices*
most important and instructive in this line of studies. T. H.

. *Systematische Anatomie der Dicotyledonen ;* by H. Sole-
der. Stuttgart, 1899—"Ergo species tot sunt, quot diversæ
mæ seu *structuræ* plantarum, rejectis istis, quas locus vel casus
rum differentes (Varietates) exhibuit, hodienum occurrunt"—
se words of Linnæus may be well understood as an indication
the importance of structural characters to systematic work in

Botany, but fully a century elapsed before this thought ripened
and became realized as "the anatomical method" nowadays
almost universally adopted in the scientific world.   French
botanists, prominent among whom were Mirbel and Chatin, were
the first to take up the idea, that affinities and divergences in the
vegetable kingdom might be sought in the internal structure, and
the rapid increase in the number of genera and species made it
almost necessary to invent other characters than those supplied
by the study of morphology alone.   In large genera it had,
already, become difficult to distinguish the species of closely allied
types satisfactorily, and although the very beginning of the
anatomical method was purely for the sake of applying the internal characters in the service of systematic work, the study of
plant-anatomy soon broadened into other lines, where the anatomy
became a branch of its own, making the first and principal found-
dation for development of physiological research.   It is the most
rational development of science, when we look back at the work
of the earlier botanists, when they began to discriminate organs
of plants and compose the systems, and then gradually became
aware of natural groups of plants, until genera and species
became adopted as a means of expressing in brief the mutual
affinities.   Then followed the doctrine of morphology, first as a
mere guide in systematic research, the terminology; then it gradually developed into the study of analogies and homologies
amongst the plant-organs themselves, while contemporarily anatomical research was found to be useful for controlling the validity
of morphological identities.   Soon anatomy was applied as an
aid in systematic work, and it is this branch of Botany which
Dr. Solereder has illustrated in his present book.

   In looking through the pages of this elaborate work, one gets
an idea of the history of the anatomical method from its beginning to its present stage, and it is, really, an enormous quantity
of labor that has already been bestowed upon this line of
Botany.   It seems as if the first, or at least the most effective,
impulse was given in the works of Radlkofer and Vesque, and
since then a number of other botanists have taken the subject up
with strenuous efforts to prove the validity of anatomical characters.   And so far have we reached now in the last decennia that
no systematic work seems complete unless this method is considered.   Engler's and Prantl's systematic treatise of the natural
orders is an excellent example of what great importance is
attributed to anatomical characters, and the results of such
studies are, really, twofold: they bring to light a number of
structural details, more or less applicable to demonstrate the
affinities between many genera and species, and, moreover, we
obtain by these same results an accumulation of data useful to
the understanding of the internal life of the plants, the structure
of the tissues, etc.   From these may again be drawn conclusions as to the functions of the structural elements, their
necessity to plant-life.   The study of anatomy became thus

reatly encouraged when botanists discovered that such observa-
ons might, also, be useful to classification, and it is no exagger-
tion to say, that the anatomical method has brought out a keener
iterest in the study of anatomy, than if plant-structures were
orked up for the mere sake of learning how plants are built.

Never before have botanists cared to give so many detailed
ccounts of the internal structure of the reproductive and vege-
ative plant-organs as they do now, and moreover, this same
iethod has given rise to another still more modern branch of
otany, generally known as "ecology," which has attracted so
iuch attention lately. From the knowledge of the structure,
onsidered by itself, investigations have been broadened into the
tudy of its application to systematic botany, while at the same
ime these same anatomical characters have thrown light upon
he connection between these and the conditions under which
lants live. But investigations of that kind require long time
nd steady attention, and it is very complimentary to European
otanists that so much work has already been accomplished in
his line: "the anatomical method." The large number of
apers published upon the subject are, however, very scattered,
nd there has been great need for a book in which all the facts
itherto known, were brought together in systematic form. It
iust be said that the author has been very successful in his com-
ilation of this enormous material to which he has himself con-
ributed extensively. Besides giving a skillful and thoroughly
cientific representation of the anatomical characters of dicotyle-
lonous plants, the author has rendered excellent service to fur-
her studies by appending a bibliography to each of the orders
reated. It would be impossible to review all the results which
re laid down in this book, inasmuch as it is not a book to be
imply read, but to be studied. A very detailed account is given
f each order, including a large number of genera and species,
nd much can be learned about the anatomy of the root, the
tem and the leaves, as these have been treated in their various
modifications. The anatomical characters are represented as
generic or specific, and it is very interesting to see the great variation that exists in a number of species, even of the same genus,
when we consider, for instance, the mere structure of epidermis
with its stomata and hairs, which furnish so many and such prominent distinctions. Very important characters are, also, derived
from the modifications that are frequently met with in the
mesophyll of the leaf or the bark of the stem with its reservoirs,
as cells or ducts, the contents : crystals or liquids ; moreover,
the very varied development of the mestome in root, stem and
leaf, the structure of the pith, etc.

But when we look at the material, upon which these investigations are based, it is at once noticed, that only a relatively small
number of North American plants have, so far, been studied
from this viewpoint. This ought to give an impulse to similar
studies in this country, inasmuch as the North American vegeta-

tion contains so many types which no doubt are of great interest anatomically, and might even serve to solve various difficult problems in systematic botany. It is sad to see, however, that anatomical work is so little appreciated in this country, while physiological research seems to be "fashionable." And it is very astonishing that American investigators seem constantly to overlook the importance of morphological and anatomical research as the principal foundation of physiological work.

Systematic Botany in this country had prominent leaders in Nuttall, Elliott, Torrey and Gray, but we see no reason why this branch of Botany should not be extended still further, in the same scope as abroad. · Europe has the advantage, however, of having had a systematic epoch, a morphological and an anatomical, followed by a physiological, while in this country physiological research was taken up before the systematic had been more than justly commenced. While recommending Dr. Solereder's book to students in this country, we hope that American investigators will take so much interest in the work that the second edition may, also, contain results gained by anatomical studies of North American genera.   T. H.

3. *Éléments de Paléobotanique,* par R. ZEILLER, Ingénieur en Chef des Mines, Professeur à l'École Nationale Supérieure des Mines. Pp. 421, 8vo. Paris, 1900. (Georges Carré et C. Naud Éditeurs, 3, Rue Racine.)—The present handsome volume has been prepared from the botanical standpoint, and forms an important addition to the small but growing group of text-books on the subject which Lesquereux twenty-five years since considered yet in its "infancy." The work of an accomplished systematist, the various groups of fossil plants and their principal forms and relationships are treated with charming clearness and precision. This text is not only a most timely one to the special student, but will be indispensable to the general reader since it displays so clearly the progress which has been made in Paleobotany, and the light which fossil plants shed upon problems of development and descent. The more important structural details are succinctly treated, and a chapter on Floral succession is added. The 210 figures inserted in the text add much to convenience of use, and the volume is a highly satisfactory one in every respect.
         G. R. W.

4. *Lehrbuch der Pflanzenpalaeontologie mit besonderer Rücksicht auf die Bedürfnisse der Geologen ;* von Dr. H. POTONIÉ. Pp. 402, 8vo; with 3 plates and 355 text figures. Berlin, 1899. (Ferd. Dümmlers.)—Differing essentially from Zeiller's text-book, that of Potonié approaches the subject of fossil plants with reference first to the needs of the stratigrapher; thus these two works are in a large measure supplementary to each other. The treatment is in fact more distinctly geological than in the case of any of the text-books on Paleobotany which have yet appeared.

The opening chapter on Vermeintliche und Zweifelhafte Fossilien is very interesting, and the closing one—Charakterisirung der Fossilen Floren—is a useful résumé. The profuse introduction of illustrations throughout the body of the work is a most commendable and labor-saving feature in which it excels. The ferns are given extended treatment in consonance with their stratigraphical importance, while the generalized group of ancient plants now known as the *Cycadofilices* receive a careful structural description. Though giving, in accordance with the general plan of the work as outlined, the greater prominence to earlier forms, the author has yet found space to include many of the more fundamental details of structure characterizing the representative groups of fossil plants. The colored frontispiece showing a Carboniferous landscape is new and effective. Both this and the preceding text-book must prove widely useful. G. R. W.

5. *The Birds of Eastern North America. Part II. Land Birds.* Key to the Families and Species; by CHARLES B. CORY, Curator of Department of Ornithology in the Field Columbian Museum. Pp. i–ix, 131–387. Chicago, 1899 (Special edition printed for the Field Columbian Museum).—The second part (this Journal, viii, 398) of Mr. Cory's classified catalogue of North American Birds has recently been issued. It is liberally illustrated and will be found very useful by the many who are interested in this subject.

6. *Zoological Results based on material from New Britain, New Guinea, Loyalty Island and elsewhere, collected during the years 1895–1897 ;* by ARTHUR WILLEY. Part iv, pp. 357–530, plates xxxiv–liii; May, 1900. Cambridge (University Press).—The earlier parts of this important series of papers have already been mentioned in this Journal (vii, 79, 322 ; viii, 398). The present part contains ten papers (Nos. 18 to 27) by different authors. Among these may be mentioned as of especial interest the description by J. J. Lister of *Astrosclera willeyana,* the type of a new family of Sponges.

7. *Das Tierreich. Eine Zusammenstellung und Kennzeichnung der rezenten Tierformen.* Herausgegeben von der Deutschen Zoologischen Gesellschaft. Generalredakteur, FRANZ EILHARD SCHULZE. 9 Lieferung. *Aves,* Redakteur, A. REICHENOW. *Trochilidæ* bearbeitet von ERNST HARTERT. Pp. i–ix, 1–254, with 34 text figures and alphabetical Index. Berlin, 1900. (R. Friedländer u. Sohn.) Earlier numbers of this great work have been noticed in these pages. The present part includes the family *Trochilidæ* of the Birds and has been prepared by Mr. Ernst Hartert of the Zoological Museum at Tring, England.

IV. MISCELLANEOUS SCIENTIFIC INTELLIGENCE.

1. *The Solar Eclipse.*—The total solar eclipse of May 28th was observed under exceptionally favorable weather conditions both in this country and Europe. Nowhere on the line of totality did clouds interfere with the observations. In consequence a great quantity of material was obtained, but as yet no results have been announced.

In addition to photographic and spectroscopic work, the bolometer was used by at least two parties with special reference to variations of heat radiation in the rifts and streamers of the corona near the sun's limb. The corona showed a conformity to the types of the last three periods of minimum sun spots so close as to be very striking even to the naked eye. The equatorial streamers were of moderate extent, and the curved polar rays sharply defined to very near the photosphere.　　　　W. B.

2. *Lessons in Elementary Physiology;* by THOMAS H. HUXLEY, LL.D., F.R.S., edited for the use of American Schools and Colleges by FREDERIC S. LEE, Ph.D. Pp. 577, 8vo. New York, 1900 (The Macmillan Company).—Huxley's Physiology is so well and favorably known to teachers and students of biology that an extended review of the aim and methods of the book is scarcely necessary in this place. Despite the marked advances in physiology since the preceding revision of the Lessons in 1885, many teachers have felt reluctant to abandon the book, owing to the many points of excellence which it retained. Indeed, it can fairly be stated that few text-books of science equal Huxley's Lessons in the clearness of exposition and a type of analytic treatment which has contributed largely to the development of a proper scientific attitude. The American publishers are therefore to be complimented in having secured the assistance of Professor Lee in a new revision which is extremely satisfactory. The histological portions have appropriately been transferred from their former separate chapter to the descriptions of the various tissues to which they apply. The descriptions of the chemistry of the blood and lymph have been revised and extended the chapter on digestion has been corrected and greatly improved, statistics of nutrition have been added and the metric system has been introduced. The parts dealing with the nervous system and innervation have been rewritten almost entirely and have been adapted to the recent progress in this department of physiology. Many new illustrations and diagrams have been inserted and though the book has grown by nearly 200 pages, the spirit of the earlier work is retained. We cannot refrain from quoting from Dr. Lee's Preface the concluding sentence, which describes the attitude of more than one physiologist towards Huxley's Lessons: "The present writer has performed his task with a long-standing feeling of affection for the pages which introduced him to the study of Physiology, and first gave him a clear insight into the nature of scientific conceptions and scientific reasoning."

L. B. M.

3. *Ostwald's Klassiker der Exakten Wissenschaften.* Leipzig, 1899 (Wilhelm Engelmann).—Number 109 of this valuable series has recently been issued, it contains a memoir entitled "Ueber die Mathematische Theorie der elektrodynamischen Induction von Riccardo Felici." Translated by Dr. B. DESSAU (Bologna) and edited by E. WIEDEMANN (Erlangen).

THE

AMERICAN JOURNAL OF SCIENCE

[FOURTH SERIES.]

- - -

ART. XII. — *On Rowland's new Method for measuring Electric Absorption, and Losses of Energy due to Hysteresis and Foucault Currents, and on the Detection of short Circuits in Coils;* by LOUIS M. POTTS.

THE following investigation has had as its object the testing of methods devised by Professor Rowland* for the measurement, in the first place, of electric absorption; further, of the energy losses due to hysteresis and Foucault currents; and, finally, for the detection of short circuits in coils.

I. ELECTRIC ABSORPTION.

Historical.—It has long been known that a Leyden jar, which has been charged and then discharged, will show another charge after standing a short time. If this is discharged, after a short time the jar will show another charge; this may be repeated indefinitely. These "after-charges" are known as residual charges and are due to the phenomenon now called electric absorption. Faraday† made some experiments on this phenomenon in Leyden jars, and seems to have attributed it to a conduction of the charge into the interior of the dielectric, and after discharge creeping back again to the coatings and manifesting itself as the residual charge. Kohlrausch‡ was the first to make any elaborate investigation of the subject. He charged the condenser and then measured the potential at certain intervals with an electrometer. In this way he obtained the relation between the potential and time. He advanced the idea that the phenomenon was due to an electric polarity of

* See this Journal for July, 1899, pp. 35-57.
† Faraday, Experimental Researches.
‡ Pogg. Annalen, vol. xci, pp. 59-82, 179-214.

the particles of the dielectric, produced by the electric for‹
between the plates of the condenser. Rowland and Nichol‹
have shown that certain homogeneous crystals show no electr
absorption. H. Hertz[†] has shown that pure benzine possess‹
no electric absorption, while impure does.[‡] The great sen‹
tiveness of this phenomenon to change of temperature has be‹
noted. The energy loss in condensers due to it has also be‹
studied.[§]

Theory of Electric Absorption.

The theory of electric absorption has been developed b
Clausius,[‖] Riemann,[¶] Maxwell[**] and Rowland.[††] The fo
lowing development is that of Maxwell applied by Prof. Rov
land to the case of a dielectric acted upon by an e.m.f. var‑
ing harmonically.

A dielectric such as paraffin paper is made of a substance ‹
a certain dielectric capacity and specific resistance havir
imbedded in it particles of a different dielectric capacity ac
different specific resistance. Now we can very closely appro
imate to this case by considering a plane plate condenser, ‑
which the dielectric is made up of a number of layers of di
ferent substances. An ordinary condenser is merely a gre
number of very small condensers like this, joined in multip‹

The theory of electric absorption as extended by Prof. Ro‑
land shows that a condenser possessing electric absorptic
should act as a capacity in series with a certain resistanc‹
The value of each depends on the period of the current.
b_1, b_2, etc., are constants and T the period of the current, t‹
resistance is of the form

$$R = b_1 T^2 - b_2 T^4 + b_3 T^6, \text{ etc.}$$

and if a_1, a_2, etc., are constants the capacity is of the form

$$\frac{1}{c} = a_1 - a_2 T^2 + d_3 T^4, \text{ etc.}$$

General Theory.—The arrangement adopted is essentiall‑
Wheatstone bridge, in which the fixed coils of an electro-dyr‹
mometer were placed in one arm of the bridge and the ha‹

[*] Phil. Mag., p. 414, 1881.
[†] Wied. Annalen, p. 281, 1883.
[‡] Phil. Trans., p. 599, 167; Proc. Roy. Soc., p. 468, 1875.
[§] Physical Review, 1899, p. 79.
[‖] Théorie Méchanique de la Chaleur; deuxième partie.
[¶] Riemann, Mathematische Werke, p. 48.
[**] Electr. and Mag., vol. i, p. 452.　[††] This Journal, Dec. 1897, p. 429.

ing coil in the cross-connection, in place of the galvanometer in the direct-current method of use. The adjustment of the bridge thus used depends upon the fact that there will be no deflection of the electro-dynamometer if the phase difference of the current in the fixed coils and those in the hanging coil is 90°.

Fig. 1 is the arrangement used. Let R_1, R_2, R_3 and R_4 be the resistances of the different arms and r that of the cross-connection. Let Cn be the current in the arm n of the bridge, and C_s in the cross-connection.

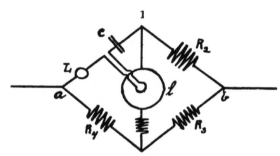

If we apply at a and b a direct electromotive force E, we shall have the following expressions for the currents:

$$C'_s = E\frac{R_2 R_4 - R_3 R_1}{\Delta} \qquad (1)$$

$$C'_1 = E\frac{R_4(R_2 + R_3) + r(R_2 + R_4)}{\Delta} \qquad (2)$$

where

$$\Delta = r(R_3 + R_4)(R_1 + R_2) + R_1 R_2(R_3 + R_4) + R_3 R_4(R_1 + R_2)$$

If in place of using a direct current we apply to the terminals a, b, a simple alternating e.m.f. Ee^{ibt}, we shall get the corresponding quantities by the following substitutions:

$$C'_1 = C_1 e^{i(bt + \phi_1)} \qquad C'_s = C_s e^{i(bt + \phi_s)}$$

and if we place in the arm 1 a coil of self-induction L and a capacity c, we must substitute for R_1, R'_1, where

$$R'_1 = R_1 + ibL - \frac{i}{bc}$$

and, if l is the self-induction of the hanging coils of the electro-dynamometer, we must replace r by r', where

$$r' = r + ibl$$

Making these substitutions and dividing (1) by (2), we have

$$\frac{C_2}{C_1}\epsilon^{i(\phi_2-\phi_1)} = \frac{R_3R_4 - R_1R_2 - i\left(bL - \frac{1}{bc}\right)R_2}{R_4(R_2+R_3) + r(R_3+R_4) + ibl(R_3+R_4)} \qquad (3)$$

Now taking only the real part of the above quantities we have

$$\frac{C_2}{C_1}\cos\phi =$$

$$\frac{(R_3R_4 - R_1R_2)\{R_4(R_2+R_3) + r(R_3+R_4)\} + \left(\frac{l}{c} - b^2lL\right)R_2(R_3+R_4)}{\{R_4(R_2+R_3) + r(R_3+R_4)\}^2 + b^2l^2(R_3+R_4)^2} \qquad (4)$$

If $\phi = 90°$, $\cos\phi = 0$ or

$$(R_3R_4 - R_1R_2)\{R_4(R_2+R_3) + r(R_3+R_4)\} + \left(\frac{l}{c} - b^2lL\right)R_2(R_3+R_4) = 0$$

or

$$R_1 = \frac{R_3R_4}{R_2} + \left(\frac{l}{c} - b^2lL\right)\frac{R_3+R_4}{R_4(R_2+R_3) + r(R_3+R_4)} \qquad (5)$$

This then is the condition satisfied when there is no deflection of the electro-dynamometer.

The first term of the above is the same as the expression for R_1, when the adjustment is conditioned by no deflection of a galvanometer in the cross-connection, and a direct current is used. The second is a correction term, always very small, at most one per cent and often entirely negligible. In a circuit carrying an alternating current the effective values of R_1, etc., are not usually the same as their actual ohmic values, but are larger. They include that part of the impedance against which work must be done to maintain the current. Let R_2, R_3 and R_4 be as nearly as possible pure ohmic resistances, i. e. let their actual values be equal to their effective ones. And let the arm 1 contain iron, a condenser possessing electric absorption, or any piece of apparatus using energy which is not expended in heating the conductors of that arm. The value of R_1 calculated by the above formula will be the effective resistance, and this, less the actual value as measured by a direct current, will be the increased resistance due to the hysteresis of the iron, the energy-loss by electric absorption, etc.

In the measurement of certain quantities (*e. g.* hysteresis loss in iron) by this method, it is necessary to insert in arm 1 of the bridge a large coil. Such a coil acts not as a pure self-induction, but on account of the numerous turns of the wire so close to one another as a self-induction in *multiple* with a capacity. On this account formula (5) is not strictly accurate but should include a term involving the capacity of the coil. Let the capacity which in parallel with the self-induction, L, will have the same effect as the given coil be k and let R'_1 be the resistance of the coil and R''_1 the resistance of the remainder of arm 1. Substitute in (3) in place of

$$R_1 + ibL - \frac{i}{bc}$$

$$R''_1 + \frac{R'_1 + ibL}{(1 - b^2kL) + ibkR'_1}$$

and also put

$$R_4(R_2 + R_3) + r(R_4 + R_2) = A$$

Then

$$\frac{C_2}{C_1} \varepsilon^{i(\phi_2 - \phi_1)} =$$

$$\frac{(R_2R''_1 - R_3R_4)(1 - b^2kL) + R'_1R_3 + i\{bkR'_1 R''_1 R_3 + bLR_3 -}{A(1 - b^2kL) - b^2klR'_1(R_2 + R_4) + i\{AbkR'_1 + (1 - b^2kL)}$$
$$\frac{bkR_1R_3R_4\}}{bl(R_3 + R_4\}} \quad (6)$$

As before, the condition for no deflection is that the real part of (6) vanish, or

$$[(R_2R_4 - R_3R''_1)(1 - b^2kL) - R'_1R_3][A(1 - b^2kL) - b^2klR'_1(R_2 + R_4)]$$
$$= [AbkR'_1 + (1 - b^2kL)bl(R_3 + R_4)][bLR_3 - bkR'_1(R_2R_4 - R_3R''_1)]$$

Expanding this it becomes,

$$A(1 - b^2kL)^2(R_2R_4 - R_3R''_1) - AR'_1R_3(1 - b^2kL)$$
$$+ b^2klR'^2_1(R_2 + R_4)R_3 - (R_2R_4 - R_3R''_1)(1 - b^2kL)b^2klR'_1$$
$$(R_2 + R_4) = Ab^2kLR'_1R_3 - Ab^2k^2R'^2_1(R_2R_4 - R''_1R_3)$$
$$(1 - b^2kL)(R_2 + R_4)R_3b^2lL - (1 - b^2kL)(R_2 + R_4)(R_2R_4 - R''_1R_3)$$
$$b^2klR'_1$$

Now since k is in all cases small and l is also small, the terms above which involve k^2 and kl may be dropped, whence we get,

$$R_1 = \frac{A(1 - b^2kL)^2(R_2R_4 - R_3R''_1) - b^2lL(R_3 + R_4)R_3(1 - b^2kL)}{AR_3}$$

$$= (1 - b^2cL)^2 \frac{R_2R_4 - R_3R''_1}{R_3} - b^2lL \frac{R_3 + R_4}{A}(1 - b^2kL)$$

Now since the last term is very small and $1 - b^2kL$ is nearly 1, it may be dropped; and we have for the final formula,

$$R'_1 = \frac{R_2R_4 - R_3R''_1}{R_3} + \left(\frac{l}{c} - b^2lL\right)R_4 \frac{R_3 + R_4}{(R_3 + R_2) + r(R_3 + R_4)}$$
$$- 2b^2kL \frac{R_2R_4 - R_3R''_1}{R_3} \quad (7)$$

In the measurement of electric absorption and hysteresis loss, it is necessary to know the period of the current used. In this investigation a speed counter and chronograph were used ; the speed counter was placed at the end of the dynamo shaft, and directly connected to it, a contact was so arranged that for every one hundred revolutions of the dynamo arma- ture a circuit was closed and a record made on a chronograph sheet. On a table beside the electro-dynamometer was a key, which also could be used to make a record on the chronograph sheet. As soon as the bridge had been balanced, this key was pressed. And by the measurement of this sheet the period of the current at the time of the observations was quite accu- rately determined. The error from this source was usually not more than one part in 1000, never more than 1 in 100.

After each adjustment, the resistance of each arm was deter- mined by the use of a "Post-office Box" (when the current through any part of the apparatus was not large, it was not necessary to measure its resistance after each adjustment). The resistance of that part which was affected the most by heating was measured first, and in this way the actual value at the time of adjustment was quite closely determined. In cases where the heating was large, the error from this cause might reach several parts in 1000 in $R'_{,,}$ and consequently a con- siderably greater amount in the value of the electric absorption resistance.

Errors due to induction and electrostatic action of the dif- ferent portions of the apparatus were carefully guarded against by the arrangement. And induction was tested for by reversal of the relative directions of the currents in different portions of the apparatus. Usually no effect was noticed or at most it was very small. The errors introduced by the self- induction of the electro-dynamometer coils and also that caused by the electrostatic action of the turns of a large coil on one another, were determined and corrected for when suf- ficiently large in amount.

Apparatus.

Electro-dynamometer.*—The self-induction of the fixed coils was ·0165 henry, and of the hanging coil ·0007 henry.

Dynamos.—The current used in this investigation was fur- nished by one of three dynamos. The Westinghouse alternator in power house of the University furnished a current of period ·0075, i. e. 133 complete periods per sec. This was used for only a few observations. In most of the work two small dynamos constructed in the University workshop were

*See this Journal, July, 1897, p. 35.

Both were directly connected to small electric
Both had armatures of the pancake type. The one
coils in the armature and four poles and thus pro-
rrent of two complete periods for each revolution of
ire. The other was larger but of similar construc-
ng six coils in the armature and six poles, and gave
plete periods for one revolution. If the load on the
as not changed, these dynamos would run at a very
stant speed. With the second dynamo, the number
te periods per second could be varied from 6 to 70.
at the lower speed the electro-dynamometer was
balance, since the hanging coil would vibrate with
it and blurr the image of the scale. The voltage
could be controlled very well by changing the
f the field. In any one series of observations the
mo was used, as the results using different dynamos
be comparable, on account of the different har-
troduced. The small dynamos which were almost
used gave, however, very good sine curves.
ces.—The high resistances and those which were
o carry very small currents were made of fine ger-
wire wound on thin sheets of fiber. The self-induc-
lectrostatic action of these was practically zero. The
stances, and those required to carry larger currents
of a special resistance wire, which had a very slight
emperature coefficient, and would bear considerable
th a very small change of resistance. This wire was
slates. Each slate contained sufficient number 30
ave nearly 2000 ohms resistance. These were con-
subdivided for adjustment. For the final adjust-
dinary resistance box was used, but never more than
were used in this box, and then the total resistance of
vas at least 1000 ohms.
uctance.—Two coils were used.
ernal diameter 35·46^{cm}; internal diameter 23·8^{cm},
No. 20 B. and S. Self-inductance 5·30 henrys.
188 ohms.
e dimensions as A except depth. Self-inductance
ys, 1747 turns No. 22 B. and S., single cotton cov-
r wire. Resistance about 78 ohms.
ers.—2 and 3. Paper condensers made by Marshall
micro-farads capacity.
ng.—8 micro-farad wax condenser, made by Will-
Jo. and divided into sections of one micro-farad each.
ndensers.—½ M. F. standard condenser made by

⅛ M. F. standard condenser made by the Troy Electric Co.

The Bridge.—The bridge was set up permanently on a table; all connections were soldered; and the wires used in connecting different parts of the bridge and instruments were No. 25 cotton-covered copper wire. All wires were made as short as possible and no wires were twisted.

Investigation.—In fig. 2, is shown the arrangement used in the measurement of electric absorption; s_1, s_2 and s_3 are current

2

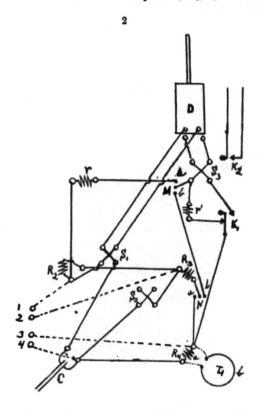

reversing switches, which were used in testing for induction of one part of the bridge on another. In most cases, however, there was no appreciable error due to this cause, so that it was not usually necessary to take a set of readings for all positions of these switches. M and N are two switches which were used in a later experiment described below. In the measurement of electric absorption M and N were in the position *a*. In the fourth arm of the bridge there was a resistance R_4. K_1 is a key to close the circuit through the hanging coil of the electro dynamometer. When K_1 was not pressed down, it closed the cross-connection of the bridge through a resistance

equal to the resistance of the hanging coil of the electro-
ynamometer. This avoided any sudden change of current
id hence a change of speed of dynamo. The key R_2 was the
ironograph key described above and was closed just at the
me the bridge was balanced.

TABLE I.

$T = \cdot0075 \qquad r = 6535$

Condenser.	R_1	R_2	R_3	R_4	Calculated R'_1	A $R'_1 - R_1$
3	34·04	2022	4600	99·7	43·86	9·82
3	34·45	2020	9090	203·4	45·22	10·77
2	34·45	2020	2475	99·7	81·46	47·01
2, 3[S] ...	34·45	2020	8920	407·0	92·23	57·78
2, 3[P] ...	34·45	2020	2205	51·7	47·37	12·92
2, 3[P] ...	34·48	1991	4130	99·7	48·09	13·61
2, 3[S] ...	34·19	1992	2031	99·6	97·83	63·64
3	34·45	1992	3325	99·6	59·72	25·27
2	34·49	1992	2414	99·6	82·28	47·79

A few measurements were first made using the paper con-
lensers 2 and 3. Table I shows the results. In this and the
following tables, R'_1 denotes the effective value of R_1 calculated by formula (5) and $A(= R_1' - R_1)$ the resistance due to
the electric absorption. T is the period of the current. The

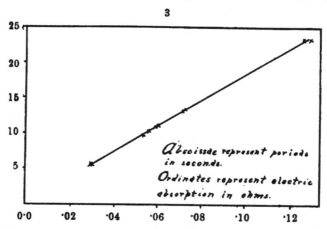

results with these condensers was very unsatisfactory, as the
heating was so great that it was difficult to make accurate
determinations.

The wax condenser made by Willyoung was next used.
This condenser had been made in a vacuum under pressure,
and showed quite small heating by the current. By taking a
series of measurements during an afternoon, the results were
not affected to any great extent by the changes in temperature

of the room, as these were comparatively slow. Of course there was still some slight heating by the current; but this in extreme cases did not amount to enough to occasion an error of more than one or two per cent.

Table II gives results for the absorption of all eight sections of this condenser in parallel, for different periods of the current. In Fig. 3 the above results are plotted; the ordinates

TABLE II.

Date, etc.		$r=6020$		$R_2=1060$ Calculated.		$R_4=303·1$	
		R_3	R_1	R_1'	$\begin{array}{c}A\\(R_1'-R_1\end{array}$	T.	$\dfrac{A}{T}$
2-23-99	1	7192	34·68	44·81	10·13	·0557	181·7
-22-	2	7251	34·70	44·45	9·75	·0544	179·2
	3	5564	34·67	57·91	23·24	·1269	183·2
	4	5556	34·66	58·00	23·34	·1291	180·8
	5	6694	34·73	48·15	13·42	·0735	182·6
	6	6718	34·77	47·97	13·20	·0717	183·6
	7	7016	34·85	45·93	11·08	·0602	184·1
	8	7044	34·88	45·75	10·87	·0592	183·6
	9	7921	35·11	40·68	5·57	·0301	185·2
	10	7910	35·10	40·74	5·64	·0301	187·4

represent A, the resistance, which in series with the condenser, would be equivalent to the electric absorption; and the abscissæ represent the period of the current in seconds. The curve proves to be a straight line, or A/T is a constant within the limits of error of the experiment. On account of this very simple relation connecting A and T, this condenser was exceedingly convenient for a test of the method.

The first test applied was to change the relations of the resistances in the different arms of the bridge. Table III

TABLE III.

Date, etc. 3-16-99	$r=6536$ R_2	R_3	R_4	R_I	R_1' Calculated.	$\begin{array}{c}A\\(R_1'-R_1)\end{array}$	T	$\dfrac{A}{T}$
1	509·2	5561	507·2	34·61	46·57	11·96	·0640	186·7
29								
2	509·2	5581	507·2	34·66	46·40	11·74	·0628	187·0
3	2071·	4598	99·56	34·65	44·97	10·32	·0558	184·9
4	2071·	4622	99·56	34·65	44·74	10·09	·0541	186·5
5	1009·5	6750	302·9	34·71	45·42	10·71	·0572	187·2
6	1009·5	6785	302·9	34·68	45·19	10·51	·0563	186·7
7	302·1	6732	1009·2	34·83	45·41	10·58	·0563	187·5
8	302·1	6728	1009·2	34·84	45·38	10·54	·0563	186·8
9	99·46	4511	2044·5	35·00	45·19	10·19	·0542	188·0
10	99·46	4511	2044·5	35·04	45·19	10·16	·0536	189·7

gives the results for R_2R_4 nearly constant and R_3/R_4 varied. The variation in R_3/R_4 was about 200 per cent and A/T is practically constant. The slight increase of A/T in the last two measurements is due to the larger current in arm 1 and a consequent heating of the condenser and also to the fact that after the coil of the electro-dynamometer had been heated slightly, it would be cooled a small amount before its resistance could be measured. Table IV shows the results when R_4

TABLE IV.

$R_4 = 1505\cdot8$ $r = 4810$

Date, etc.		R_2	R_3	R_1	Calculated. R'_1	T	$\frac{A}{(R'_1 - R_1)}$	$\frac{A}{T}$	Mean.
4-19	1	2070·	76,600	34·68	40·70	·0312	6·02	192·9	} 194·8
118	5	2070·	74,200	34·80	42·02	·0367	7·22	196·7	
	2	1009·	36,120	34·76	42·07	·0375	7·31	194·9	} 196·0
	4	1009·	35,970	34·80	42·25	·0378	7·45	197·1	
	3	99·48	3,542	34·75	42·30	·0386	7·55	195·6	195·6

is kept constant and R_2 and R_3 are varied. These also show A/T constant. It appears then from the above facts that A is a quantity independent of the relative values of the resistances in the different branches of the bridge.

In the next test the period was kept constant and the electromotive force acting on the condenser was varied about 300 per cent. The results of this test are given in Table V. The values for the higher electromotive forces are slightly greater, owing to the two heating effects mentioned above. Aside from

TABLE V.

$R_2 = 1008\cdot9$ $R_4 = 303\cdot8$

Date, etc.	R_3	r	R_1	Calculated. R'_1	$\frac{A}{R'_1 - R_1}$	T	$\frac{A}{T}$	Volts e.m.f. acting on condenser $= V$
3-10-99								
1	6649	24,410	35·45	46·14	10·69	·0523	204·3	240·
27								
2	6593	24,410	35·34	46·53	11·19	·0562	199·3	240·
3	6797	12,490	35·13	45·17	10·04	·0508	197·6	152·
4	6777	12,490	35·15	45.31	10·16	·0511	198·9	152·
5	6606	6,348	35·03	46·56	11·53	·0584	197·3	77·
6	6473	6,348	34·96	47·57	12·55	·06418	195·5	77·
3-14								
6	6591	4,807	34·73	46·60	11·87	·0590	201·2	76·
28								
7	6803	6,371	34·79	45·14	10·35	·0515	201·1	142·
8	6732	6,371	34·78	45·61	10·83	·0540	200·6	142·
9	6555	18,830	35·01	46·74	11·73	·0567	203·5	243·

this, A/T is constant, i. e., A is independent of the current flowing through the condenser.

The variation in A/T due to changes of temperature was obtained as follows: in an opening made in the side of the box containing the condenser, a mercury thermometer was placed. The temperature indicated by the thermometer was of course not that of the inside of the condenser but that of the outer edge. The condenser, however, was kept within a degree or two at least of the desired temperature for some six or more hours before being used.

This method gave sufficiently accurate results, as there was no occasion for an accurate determination of the temperature.

TABLE VI.

Date, etc.	Temp. cent. t	$r=4810$ R_3	R_2	R_4	R_1	Calculated. R'_1	$\dfrac{A}{R'_1-R_1}$	T	$\dfrac{A}{T}$
11–16–99	19·3	709·0	2557	203·62	35·73	56·48	20·75	·0920	226·0
102									
1	22·0	707·8	2195	202·6	34·90	65·35	30·45	·1167	258·3
11–2–99									
7	22·2	707·8	2614	202·6	34·97	54·88	19·91	·0816	242·7
1	16·5	709·9	3271	203·0	34·40	44·08	9·68	·0543	178·6
12–8									
2	16·7	709·6	2862	203·0	34·44	50·04	15·60	·0816	191·3
106									
	20·2	2985·	2985	203·3	34·93	48·27	13·34	·0569	234·5
1	4·5	302·4	2696	407·3	34·55	45·87	11·32	·0752	150·9

Table VI gives the results of this investigation, and in fig. 4 they are plotted with temperatures as abscissas and A/T as

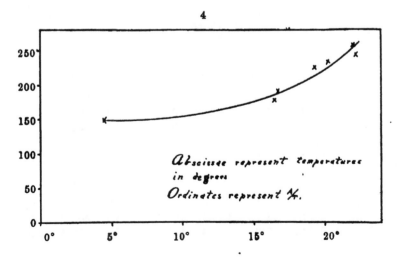

4

ordinates. It appears that at ordinary laboratory temperatures a variation of ·1° C. will cause a change of about 1 per cent in the value of A/T and consequently in A for a given T. From this it appears that the slight variation noted above in the constancy of A/T would be easily accounted for by the changes of temperature due to the current or the gradual changes due to changes in the temperature of the room, since a series of observations usually occupied three hours or longer.

Table VII gives the results for the two ⅓ microfarad condensers described above connected in parallel; and in fig. 5 the results are plotted as before.

TABLE VII.

$R_4 = 303·3$ $r = 4811$

Date, etc.		R_2	R_3	R_1	Calculated. R'_1	A $(R'_1 - R_1)$	T
11–29–99	1	409·7	2984·	34·07	41·66	7·59	·0271
105	2	409·7	2723·	34·07	45·59	11·52	·0345
	3	409·7	1807·	34·07	68·79	34·72	·0752
	4	409·7	1604·	34·07	77·49	43·42	·0893
	5	409·7	2454·	34·07	50·66	16·59	·0428
	6	409·7	3010·	34·07	41·30	7·23	·0239
	7	410·6	3157·	34·04	39·38	5·34	·0187

It was necessary to use the two in parallel in order to get sufficient current through the fixed coils of the electro-dynamometer. This was especially true for long periods, as then the impedance of the condenser increased and at the same time the available electromotive force from the dynamo decreased.

Capacity of a Condenser which shows Electric Absorption.

From the theory of electric absorption as based on the heterogeneous nature of the dielectric it appears that there should be a variable value of the capacity of such a condenser depend-

5

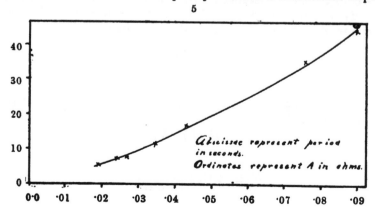

Abscissæ represent period in seconds.
Ordinates represent A in ohms.

ing upon the period of the current flowing through the condenser. The method chosen for the measurement of capacity
was one described in this Journal, December, 1897.

Method.—This consisted in the use of a divided circuit.
One branch, *a*, contained a resistance R_a, either the fixed or
hanging coils of the electro-dynamometer, and a condenser
whose capacity *c* was to be studied. The other arm, *b*, contained a resistance R_b, a coil with which the capacity *c* is com-

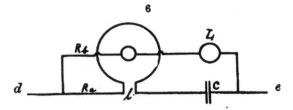

pared, and either the hanging or fixed coils of the electro-
dynamometer. Let L be the coefficient of self-induction of
the coil plus that of the coils of the electro-dynamometer in
that arm, and let *l* be the self-induction of the coils of the
electro-dynamometer in the arm *b*. Let an electromotive force

$$E = E_\circ e^{(ibt)}$$

be applied to the terminals *de*. Now representing the maximum values of the current by C_a and C_b and the phases by ϕ_a
and ϕ_b we have for the branch *a*

$$C_a e^{i(bt+\phi_b)} = E_\circ \left(R_a + ibl - \frac{i}{bc} \right)$$

and for branch *b*

$$C_b e^{i(bt+\phi b)} = E_\circ (R_b + ibL)$$

If we adjust the resistances until there is no deflection of the
electro-dynamometer, we shall have the difference of phase
$(\phi_b - \phi_a) = 90°$ or $\cos (\phi_b - \phi_a) = 0$. Further,

$$\frac{C_b}{C_a} e^{i(\phi_b - \phi_a)} = \frac{R_b + ibL}{R_a + ibl - \dfrac{i}{bc}} \qquad (12)$$

Hence since $\cos (\phi_b - \phi_a) = 0$ we must have the real part of this
equal to zero, or

$$(1 - b^2 cl)(b^2 cL) = b^2 c^2 R_a R_b$$

or

$$\frac{L}{c} = R_a R_b \left(\frac{1}{1 - b^2 cl} \right) \qquad (13)$$

As in the case considered above, a coil does not act as a self-induction alone but as a self-induction in *parallel* with a capacity due to the electrostatic action of the turns of the coil on one another. For this reason the above formula is not exact for any actual case; but there must be substituted above in place of $R_b + ibL$

$$R'_b + \frac{\dfrac{R'_b + ibL}{ibc'}}{R'_b + ibL - \dfrac{i}{bc'}} \qquad (14)$$

Substituting this in (12) we have

$$\frac{C_b}{C_a} e^{i(\phi_b - \phi_a)} = \frac{R'_b + \dfrac{R'_b + ibL}{ibc'R'_b + 1 - b^2c'L}}{R_a + ibl - \dfrac{i}{bc}}$$

$$= \frac{R'_b(1 - b^2c'L)bc + R''_b bc + i[(b^2cc'R'_b R''_b + b^2cL)]}{R_a(1 - b^2c'L)bc + (1 - b^2cl)bc'R''_b - i[(1 - b^2c'L)(1 - b^2cl) - b^2cc'}$$
$$[R''_b R_a$$

As before, the condition for no deflection is that the real part of this equals zero. Hence

$$b^2c[R'_b(1 - b^2c'L) + R''_b][R_a(1 - b^2c'L)c + R''_b(1 - b^2cl)c']$$
$$= b^2c[R'_b R''_b c' + L][(1 - b^2c'L)(1 - b^2cl) - b^2cc'.R''_b.R_a]$$

Expanding,

$$R'_b R_a(1 - b^2c'L)^2 c + R''_b(1 - b^2cl)(1 - b^2c'L)R'_b c'$$
$$+ R''_b R_a(1 - b^2c'L)c + R''_b{}^2(1 - b^2cl)c' =$$
$$R'_b R''_b c'(1 - b^2c'L)(1 - b^2cl) + L(1 - b^2c'L)(1 - b^2cl)$$
$$- b^2cc'R''_b R_a(L + R'_b R''_b c')$$

Since c' is small, we can drop the term in c'^2. We have, on dividing by $(1 - b^2c'L)(1 - b^2cl)c$ and rearranging terms,

$$\frac{L}{C} = \frac{R'_b R_a(1 - b^2c'L)}{1 - b^2cl} + \frac{R'_b R_a}{(1 - b^2cl)} + R''_b{}^2 \frac{c'}{c}(b^2 - b^2c'L)$$
$$+ \frac{b^2c'R''_b R_a L}{(1 - b^2cl)(1 - b^2c'L)}$$

or

$$\frac{L}{C} = \frac{(R'_b + R''_b)R_a}{(1 - b^2cl)} + R''_b{}^2 \frac{c'}{c} \frac{1}{(1 - b^2c'L)}$$
$$- \frac{b^2c'LR_a R'_b}{(1 - b^2cl)} + \frac{b^2c'LR_a R'_b}{(1 - b^2cl)(1 - b^2c'L)}$$

Now, since in any case the last three terms are small, and $(1-b^2cl)$ and $(1-b^2c'L)$ are nearly one, they may be dropped from the last terms and we have

$$\frac{L}{C} = \frac{(R'_b + R'_b)R_a}{(1-b^2cl)} + \frac{c'}{c}R'_b{}^2 + (R'_b - R'_b)b^2c'L \qquad (15)$$

Investigation—The arrangement is the same as in fig. 2. In position a and with a resistance in arm 4, electric absorption can be found. In position b and with a coil L in arm 4 in place of the resistance, the capacity of the condenser can be compared with the standard coil L. In formula (13) R_a includes not only the ohmic resistance of branch a, but also the added resistance due to electric absorption.

A preliminary investigation was carried out to find whether the correction due to the capacity of the coil C' were appreciable and, if so, to ascertain its amount.

The method chosen was as follows: An arrangement was made as in fig. 2, except that in place of the condenser C the coil L with which the capacity of the condenser is to be compared is placed in arm 1 of the bridge. The arrangement was first balanced with a direct current; and the value of R_1 as measured and as calculated from R_2, R_3, R_4 were the same. In Table VIII are given the results for R'_1 as calculated for three

TABLE VIII.

Date		R_3	R_1	R'_1 By formula 13	T	$D = R'_1 - R_1$
		$R_2 = 1010$		$R_4 = 912\cdot5$	$r = 4810$	Coil A = 5·3 henry
4·20	1	4080	224·6	225·8	·0188	1·0
	2	4092	224·8	225·2	·0488	·4
	3	4074	225·0	226·2	·0182	1·2
	4	4034	225·0	228·4	·0182	3·4 with ·01 microfarad in parallel with coil.

If formula (15) is used in calculating R'_1 it is 226 2. Hence $D = 1\cdot2$ as without condenser.

periods of the current. In all three cases R'_1 is greater than R_1 by an amount D. A third observation was made with a condenser of ·01 microfarad capacity shunted across the terminals of the coil. The corrected formula (7) was tested in this way. The result gives the same value of D, with or without the condenser, thus verifying the formula. By assuming the quantity D as entirely due to the electrostatic action of the coil, which if not absolutely true, the formula will at least give a value of c' the equivalent capacity of the coil, which may be used as the limiting value. The value of c' is ·006 microfarad. In formula (15) the last term will be very small as compared to the first for any values of b used in this work. The second term will in the most unfavorable circumstances amount to only 1 part in 10,000, so that the corrections due to the electrostatic action of the coil may be entirely neglected.

The next point investigated was the correction due to the electric absorption. In these observations the absorption was determined, the capacity then measured, and the electric absorption again determined. In Table IX are given the

TABLE IX.

		$r=4811$		$R_3=707.6$			$R_4=202.6$	$L=5.302$ in 2 and 5.318 in rest.
, etc.		R_3	R_1	Calculated. R'_1	$\frac{A}{(R'_1-R_1)}$	T	$\frac{A}{T}$	Mean. $\frac{A}{T}$
6	1	2195.	34.90	65.35	30.45	.117	261.	
								253.
	7	2614.	34.97	54.88	19.91	.0816	244.	

		R_a	R_b	T	A	Corrected for electric absorption. R_a	R_b	$10^6 \times \dfrac{L}{R_a R_b}=C$
	2	238.2	2604	.121	30.6	268.8	2604	7.574 M.F.
	3	228.8	3053	.103	26.1	——	3079.	7.549
	4	328.2	2119	.103	26.1	——	2145.	7.554
	5	431.5	1602	.102	25.8	——	1628.	7.569
	6	531.5	1302	.0908	23.0	——	1325.	7.552

results with the condenser in one arm and then changing to the other, and also changing the resistance in series with the condenser. The results are corrected for electric absorption and the change in L due to the change of c and the coil from one arm to the other, caused by the coils of the electro-dynamometer having different coefficients of induction. The greatest difference between two determinations under these different conditions is 3 parts in 1,000.

The change of capacity with the period of the current was now tried. Table X shows the results of the investigation. The error of each observation has a limit of about 1 part in 1,000, if the observations are compared among themselves, while the actual error, as compared with the true ratio $\dfrac{L}{C}$ may be in error two or three times this; but we are not particularly concerned here with the actual value, but merely the change with change of period. In fig. 7 the results are plotted on two scales. The results for the capacity seem to agree very well

TABLE X.

L=5·31

Date, etc.	R_2	R_3	R_4	R_1	r	R'_1	A	T
709·0	2556·8	203·62	35·73	4811	56·46	20·73	·0919	2

R_a	R_b	T	A	Corrected for electric absorption. R_a	R_b	$10^6 \times \dfrac{L}{R_a R_b} = C_1$	$C = \dfrac{}{1-}$
429·2	1589·1	·160	36·2	429·2	1605·3	7·719	7·7
429·3	1628·1	·0654	14·8	429·3	1642·9	7·540	7·5
429·3	1640·0	·0389	8·8	429·3	1648·8	7·513	7·5
429·4	1651·0	·0151	3·4	429·4	1654·4	7·487	7·4
429·4	1628·0	·0601	13·8	429·4	1641·8	7·543	7·5

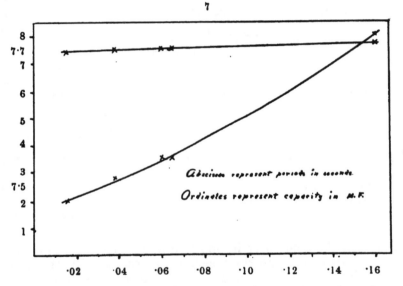

with the theory. The condenser has a capacity which shows a slight variation with the period of the current. The capacity increases with increase in the period of the current, as shown by the theory; and in amount it is somewhat less than as the square of the period.

II. Detection of Short Circuits.

A very useful application of the Wheatstone bridge is its use for the detection of short circuits in coils of wire. If a mass of metal or a closed coil of wire is held near a coil of wire carrying an alternating current, there will be induced in it certain currents. Hence there must be more energy supplied to the primary circuit to keep up the current. This extra consumption of energy will manifest itself by an apparent increase in the resistance and consequently a greater i^2R loss. This fact is made use of in the following method for the detection of short circuits in coils and was suggested by Professor Rowland.[*]

The method is as follows: the connections are as in fig. 2, with the exception that a coil of wire is introduced in arm 1 in place of the condenser. If the resistances are now balanced until there is no deflection of the electro-dynamometer, and a mass of metal approaches the coil, there will be a deflection of the electro-dynamometer; owing to the increase in the effective resistance of the arm 1. If a coil of wire whose ends are not connected be laid on top of the coil in arm 4 there will be no deflection, while if the ends are connected or if there is a short circuit in the coil there will be a deflection.

TABLE XI.

No. of turns.		$R_2 = 1010$ Dia. of wire.	$R_4 = 1507$ R_3	$R_1 = 225·7$ Uncorrected. R'_1	$T = ·015$ D	R_c
0		—	6711	226·8	0	
2·	14	1·59	5882	258·8	9·15 cm.	32·0
3·	22	·62	6497	234·3	2·21	7·5
5·	25	·44	6520	233·4	1·90	6·6

Table XI shows the sensitiveness of the method. The coil A of 5·3 henrys was used. The bridge was balanced and then small coils of wire the same size as the inner diameter of the coil were placed on the coil and the deflection noted; and the apparent increase of resistance was determined in the same manner as the electric absorption in the case when the coil L was replaced by a condenser. Column D gives the deflection after the coils were placed on the large coil in arm 1 and R_c is the apparent increase in resistance of arm 1. It appears from this table that with a coil of the size used, a short circuit in another coil of same size could be detected, even though the coil were of quite fine wire and only one turn was crossed. Other conditions being the same, the sensitiveness varies directly as the cross section of the wire in the coil to be tested, if the resistance of the contact between the two ends of wire is neglected. In cases where small coils are to be tested, the sensitiveness may be increased by filling the center of the coil with iron. And of course as short a period of current as available should be used.

[*] This Journal, December, 1897.

III. HYSTERESIS.

The arrangement used for the determination of losses of energy due to hysteresis and Foucault currents is the same as for the determination of electric absorption, with several additional elements.

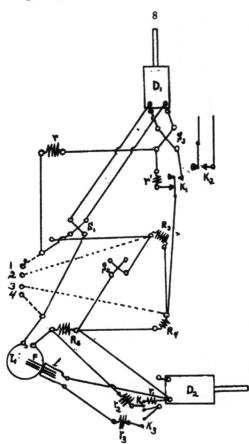

Fig. 8 shows the arrangement used. All the arms are the same as in fig. 2 except arm 1. Arm 1 contains a coil L. In this coil is placed the iron to be tested, F. l is a small coil of wire surrounding the iron, and used to determine the induction through the iron. The two coils of the electro-dynamometer D, are connected in series. The key K_2 serves to put the coil l, the resistance r_2 and the coils of the electro-dynamometer in series. By noting the deflection produced by the current induced in l, the induction may be calculated, the electro-dynamometer having been previously calibrated.

R_s is a small resistance in arm 1. A small current is shunted off from the terminals of R_s and can be sent through the resistance r_2 and the electro-dynamometer by the key K_1; and thus the total current in arm 1 may be determined.

Measurement of the current.—The calibration of the electro-dynamometer showed the current to be quite accurately proportional to the square root of the deflection ($\sqrt{D}$). A deflection of 1 cm. corresponds to a current of ·00213 ampere. We shall then have the total current in arm 1

$$C_1 = ·00213 \sqrt{D} \frac{r_2 + R_s}{R_s}$$

In place of measuring the current by an electro-dynamometer, in some cases a small ammeter might be used, but as the current enters as a square in the energy loss, it must be accurately determined.

Determination of the induction: For the determination of the induction several different forms of the secondary coil were tried. Coils of the same size as the internal diameter of the coils A and C were first used. These were found unsuitable, however, as a considerable current was induced in them, even when they contained no iron, and their coefficient of self-induction was not negligible. The best form was found to be a coil of fifty turns, just large enough to contain the iron used. No measurable current was induced in this, when it contained no iron, and its self-induction was negligible. The e.m.f. around the circuit when K_4 is closed will be E_1 where

$$E_1 = \sqrt{D} \times \cdot 00213 \times R \text{ volts}$$

Now let N be the number of turns in the coil l and S the average cross section of the iron surrounding the coil L and we have the induction per sq. cm. B

$$B = \frac{\sqrt{D} \times \cdot 00213 \times R}{N \times 4 \cdot 44 \times r} \times 10^8 \text{ c. g. s. units}$$

where N is the number of complete periods of the current.

Energy loss.—The energy loss due to hysteresis is ordinarily expressed as a certain loss per c.c. of iron per cycle. Energy loss $= i^2 R t$.

If C_1 is the current in arm 1, H the total added resistance due to the hysteresis and Foucault currents, v the volume of the iron used, we have the energy loss per cycle,

$$C^2_1 \frac{H}{V} T$$

since t will equal T, the period of the current.

Experiment.—The iron used was ordinary transformer iron. The plates were L-shape, and could be fitted about the coil C very nicely. In the first place the uniformity of B was tested for different quantities of iron. A slot was cut in the center of one side of an L plate, thus dividing the side into two parts which were made as nearly equal as possible. A coil of wire of one hundred turns was wound on each of these parts of the iron. By noting the deflection produced by the two coils in succession when the iron was placed in coil A, the relative value of B close to the coil A and farther away could be tested. By placing in the coil different numbers of the plates and placing the test plate at different points, the uniformity of B was tested.

Table XII shows the results of this test. It appears from this table that with either 8 or 10 pieces of iron B is practi-

TABLE XII.

No. of pieces of iron.	Test plate on outside of bundle.		Per cent var. B.	Test plate next to outside plate.			Two plates outside of test plate.		
	Lower coil.	Upper coil.		Lower coil.	Upper coil.		Lower coil.	Upper coil.	
2	6·51	6·03	3·	—	—	—	—	—	—
4	4·14	3·82	3·	—	—	—	—	—	—
6	3·60	3·48	1·9	3·60	3·62	·7	—	—	—
8	2·65	2·55	2·	2·71	2·71	0·	—	—	—
10	—	—	—	—	—	—	2·75	2·85	2·

cally uniform. With fewer pieces the induction was greatest next the coil, and when more were used the magnetism was not as great in the central pieces.

Table XIII shows the results of a series of measurements, and in fig. 9 a curve is plotted showing the relation between B and the energy loss per cycle.

TABLE XIII.

Weight of iron=340 g.   Volume=43·3 cc.   Area section= 981 sq. cm. D_1=deflection d; mometer by current C. D_2=same for secondary current.

3-21-00
-115-

	R_2	R_3	R_4	R_1	R'_1	H	T	D_1	D_2	r_1	
1	913·2	2945·	·1011·	148·3	313·5	165·2	·0277	3·62	3·18	166·7	5ξ
2	505·9	2140·	1508·	148·3	356·5	207·8	·0280	5·85	7·40	166·7	59
3	208·6	1106·	2016·	148·3	380·2	231·9	·0275	8·78	14·70	166·7	59
5	1769·	2408·	507·5	148·3	372·8	224·5	·0276	9·00	3·00	166·7	129·
6	1563·	2100·	507·5	148·3	377·7	229·4	·0273	10·61	3·86	166·7	129·
8	1060·	1431·	507·5	153·0	375·9	222·9	·0267	4·00	6·95	377·0	129·7
9	503·1	1671·	1010·5	153·0	304·2	151·2	·0254	9·60	14·10	377·0	129·7
11	206·9	1639·	2014·	153·1	254·1	101·1	·0255	18·10	14·36	377·0	149·8

	B	C_1	H per c.c.	C_1^2HT
1	$2·88 \times 10^3$	·0262	3·815	$7·10 \times 10^3$
2	$4·32 \times 10^3$	·0337	4·799	$1·524 \times 10^4$
3	$6·17 \times 10^3$	·0412	5·356	$2·51 \times 10^4$
5	$6·07 \times 10^3$	·0418	5·185	$2·50 \times 10^4$
6	$6·86 \times 10^3$	·0454	5·298	$2·98 \times 10^4$
8	$8·92 \times 10^3$	·0576	5·148	$4·56 \times 10^4$
9	$1·207 \times 10^4$	·0891	3·482	$7·00 \times 10^4$
11	$1·318 \times 10^4$	·1226	2·335	$8·93 \times 10^4$

SUMMARY.

From the above it appears that the method described is a perfectly good method for the measurement of electric absorption. In all cases tried the electric absorption has acted as a resistance in series with a capacity. This resistance is independent of the current. The temperature has a decided effect. The value of this absorption increases very rapidly with rising temperature. The theory as given above appears to be verified

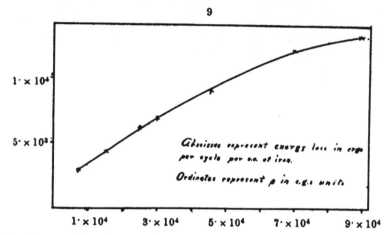

by the results in as far as a condenser possessing electric absorption may be considered as a capacity in series with a resistance, both of which depend upon the period of the current. While the variation is in the proper direction in each case, its rate of change with the period does not agree with the theoretical formula, especially in the case of the wax and paper condenser.

It also appears that the method given is a good one for the determination of the capacity of a condenser, which shows electric absorption. If electric absorption is corrected for, the capacity of such a condenser is a quantity which can readily be determined, and may be compared with a self-induction standard to 1 part in 1,000. It also appears that no correction is necessary for the electrostatic action of the turns of the standard coil on one another, at least if the relative size of coil and condenser is properly chosen, and if the resistance of the coil is not too large as compared with the total resistance in the branches of the circuit.

The method given for the detection of short circuits in coils proves to be an exceedingly sensitive one.

The method described for the measurement of losses due to hysteresis and Foucault currents gives very good results. Its chief advantage lies in the fact that a very small quantity of the material is necessary. In the experiment only 360 grams were used. Much less than this amount could be used, with nearly as great accuracy as in the case given. For the value of the resistance H could be increased by decreasing the period of the current used.

In conclusion I wish to express my sincere appreciation of much assistance received from Professors Rowland and Ames during three years spent in study at the Johns Hopkins University. The investigation was suggested by Professor Rowland, and the methods used were those devised by him and described in the articles noted above.

Johns Hopkins University,
    May 1, 1900.

ART. XIII.—*Some New Jurassic Vertebrates;* by WILBUR
C. KNIGHT. Paper No. III.

FOR several years the Geological Department of the University of Wyoming has been directing its entire energy and
what little money it could procure toward building up a collection of Jurassic vertebrates. On account of splendid success in the field the collection has already assumed considerable
proportions and from time to time new animals are being discovered. The bulk of the material at the present time is
Dinosaurus, but the marine beds of the Rocky mountain
Jurassic have not been neglected and a few marine reptiles
have been discovered. Among these are two new species of
the order of Sauropterygia, which belong to two genera,
neither of which have been reported from the Rocky mountain
Jurassic; one of them being a Plesiosaur and the other a Cimo-
liosaur. With this addition to the fauna of the Eastern Rocky
mountain region, there are now known two species of Ichthyo-
saurs[*] and three species of the Plesiosaur[†] type. The two
skeletons which form the basis for this paper are not as complete as one would like to have them for generic and specific
determinations, but there are sufficient remains in each case to
give a very good idea of the species.

Plesiosaurus shirleyensis sp. nov.

The remains of this species consists of vertebra from all
parts of the column; there being many from the cervical
region; numerous teeth and fragments of teeth; a distorted
portion of the lower jaw and many phalanges. Teeth large
and numerous, incurved, elliptical in cross-section, interior surface of the teeth covered with numerous very fine angular
striæ, exterior surface nearly smooth and showing faint marks
of striæ. The greater portion of one side of the lower jaw
measures $\cdot280^m$ in length and has a depth of $\cdot035^m$. The vertebra are slightly biconcave, and all wider than long; but in
the dorsals and posterior cervicals the length and breadth are
nearly equal. The neural spines were found attached to the
centra of the cervical vertebra, but the suture was obliterated.
On the anterior caudles the neural spines are of considerable
height.

Nothing of importance is known of the dorsal vertebræ,
excepting that they are slightly biconcave and circular in

[*] Baptanodon discus Marsh; Baptanodon nateus Marsh.
[†] Megalneusaurus rex Knight. See this Journal, vol. v, p. 378.

transverse sections. Anterior caudles are flattened beneath and have two large circular facets for the articulation of the chevrons; neural arches firmly attached to centra. The basal phalanges have compressed, angular shafts, their shape depending largely upon their position in the paddle. Those in the central region approach a quadrangular section; while those on the exterior are V-shaped in cross-section with the opening of the V to the interior of the paddle. Terminations of phalanges rugose and strongly biconvex.

Measurements.

	M.
Length of fragment of lower jaw	·280
Depth of fragment of lower jaw	·035
Length of teeth about	·049
Transverse diameter of teeth at base of striations	·006
Length of striated portion	·014

Dorsal vertebra.

Length	·048
Width	·055
Height to neural platform	·045

Posterior cervical vertebra.

Length	·050
Width	·055
Height to neural platform	·045
Height of process from neural platform	·090

Caudle vertebra (anterior).

Length	·029
Width	·046
Height of neural platform	·034

Phalanges.
Basal.

	M. large.	medium.
Length	·049	·045
Width, distal end	·022	·020
Width, proximal end	·023	·020
Width of shaft	·015	·012

Intermediate.

Length	·035	·030
Width, distal end	·020	·018
Width, proximal end	·022	·019
Width of shaft	·013	·013

Terminal (nearly).

Length	·025	·021
Width, distal end	·012	·012
Width, proximal end	·013	·013
Width of shaft	·008	·007

This animal had a long neck and large paddles; but did not attain a length of over fourteen to sixteen feet. It was discovered in the Shirley* stage of the Jurassic rocks of Albany County, Wyoming, and was associated with *Baptanodon discus, Ostrea strigilecula, Camptonectes bellistriatus* and *Astarta Packardi.* Type specimen marked letter H in the collection of the University of Wyoming.

Cimoliosaurus laramiensis sp. nov.

This specimen consists of numerous vertebra and nearly a complete front limb. The humerus conforms to the general shape of Cimoliosaurs, but whether it belongs to the division where ulna and radius articulate with the humerus or to the other group where the pisiform is present, cannot be satisfactorily determined. There are two very distinct facets at the distal end of the humerus; the remainder of the margin is rounded and shows no sign of a pisiform. The facet for the radius occupies over one-half of the width of the expanded end, and is nearly flat. The humerus is a short, heavy bone, broadly expanded distally, and the post-axial border slightly recurved. Trochanteric ridge very prominent; shaft transversely elliptical with a rugose prominence on the anterior border, ·120^m below the head; apparently for the attachment of a large muscle; head spherical. Radius V-shape with the angle truncated and toward the ulna. The articulations of the ulna slightly convex; the proximal end very wide, distal narrow, and the ulna and radius evidently uniting without a central opening; exterior margin curved and very thin. Ulna not known. There are six carpal bones, all angular, but their relative position has not been determined.

Metacarpals and basal phalanges only slightly compressed and with flattened shafts, the upper surface slightly rounded, with a depression on either side, making a transverse section as figure in plate B, No. 1; they are also biconvex with pitted terminations. The terminal phalanges are much more compressed and without the depressions on the surface. All the vertebra wider than long and moderately biconcave, with neural arches firmly attached. Dorsal vertebra with a forward overhanging of the centra as is usually found in Cimoliosaurs, and circular in transverse section. Cervical vertebra numerous and anterior ones very small.

Caudle vertebra large, elliptical in transverse section. Anterior ones with large transverse processes, and large angular chevron facets.

*This is a new name applied to the Rocky Mountain marine Jurassic. See paper presented by me to the Geological Society of America, December 30, 1899.

Measurements.

	M.
Length of humerus	·312
Width of humerus, distal end	·136
Thickness, maximum, of distal end	·041
Height of head	·100
Width of head	·068
Width of shaft humerus midway	·066
Thickness of shaft of humerus midway	·051

Radius.

Length	·095
Width of { proximal articulation	·085
distal articulation	·064
Thickness on interior edge	·033

A. Phalanx.
  Basal.

Length	·053
Width of base	·030
Width of shaft	·023
Width at distal end	·027

Vertebra.
  Dorsal.

Length	·037
Width	·045
Height of centrum	·040
Concavity of ends of centrum	·006

Anterior caudle.

Length	·032
Width	·050
Height of centrum	·035
Concavity of ends of centrum	·006

Cervicals.
  Anterior.

Length	·022
Width	·030
Height	·032

  Posterior.

Length	·026
Width	·041
Height	·032

This was a small animal, probably not more than twelve fe
in length. It was found associated with Baptanodon remai
in Shirley stage of the Freezeout Hills, Carbon Count
Wyoming. This specimen is marked letter T in the collecti
of the University of Wyoming.

Geological Department,
    University of Wyoming, April 6, 1900.

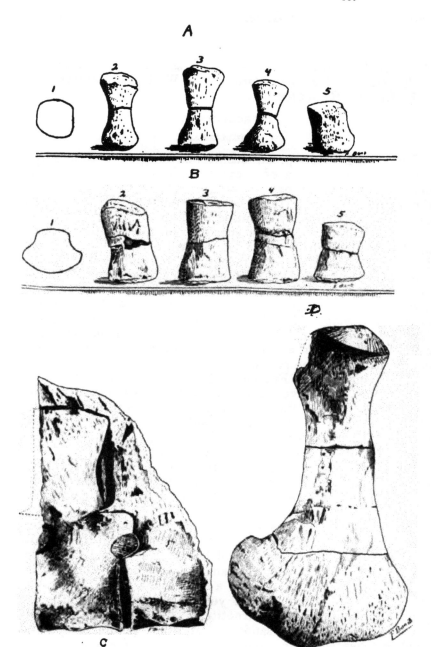

EXPLANATION OF FIGURES.

A and C.—*Plesiosaurus shirleyensis.*
A, No. 1.—Transverse section of a basal phalanx. (× ⅔)
Nos. 2, 3, 4, 5, basal phalanges. (× ₇⁄₁₆)
C.—Side view of cervical vertebra. (× ⅔)
B and D — *Cimoliosaurus laramiensis.*
B, No. 1, transverse section of a basal phalanx. (× ⅔)
Nos. 2, 3, 4, 5, basal phalanges. (× ₇⁄₁₆)
D.—Humerus. (× ½)

ART. XIV. — *On Carnotite and Associated Vanadiferou.*
Minerals in Western Colorado; by W. F. HILLEBRAND an(
F. LESLIE RANSOME.

INTRODUCTION.

W. F. HILLEBRAND.

THE rather wide-spread occurrence in western Colorado o
considerable quantities of a canary-yellow ore of uranium ha
been known for a few years past. It has been marketed t
the extent of a few tons, but its mineralogical nature wa
unknown or incorrectly surmised until about a year ago.

In the spring of 1899 a specimen of this ore first reache(
my hands through Dr. Geo. P. Merrill of the U. S. Nation
Museum, and was speedily recognized to contain a mineral o
minerals new to science. Within a week appeared the announce
ment by MM. C. Friedel and E. Cumenge* of a new minera.
carnotite, a hydrous vanadate of uranium and potassium
obtained through Mr. Poulot of Denver, from Roc Creek
Montrose Co., Colo. Mr. Poulot had already identified vana
dium in it. It was at once seen, despite certain differences i
composition, that the two were identical.

According to the French authors the mineral is of simpl
composition, as above expressed, with only a little iron an(
mere traces of Al, Ba, Cu, Pb, and also according to M. an(
Mme. Curie, of the radio-active substances radium and polo
nium. The empirical formula $2U_2O_3, V_2O_5, K_2O, 3H_2O$, wa
assigned to it, some doubt attaching to the water.

Since then I have been able to examine carnotite from sev
eral localities in western Colorado, finding in each case th(
same lack of agreement with the analyses of Friedel an(
Cumenge.

While engaged in this work there were brought to my atten
tion certain more or less greenish sandstones from the vicinit;
of Placerville on the San Miguel River, San Miguel Co., Colo.
which were said to be highly vanadiferous and of considerabl(
extent, and in which a zone a few inches thick was rathe1
strongly impregnated with a yellow mineral resembling an(
probably identical with carnotite. This latter appeared als(
scattered through the sandstone at other points in sporadi(
small patches, sometimes only visible by aid of a lens.

This occurrence led to the thought that the carnotite bodie:
farther west might also be associated with existing or depend-

---

* Bull. Soc Chim. de Paris (3), xxi, 328, 1899; Bull. Soc. Franc. Min., xxii, 26
1899; Comptes Rend.. cxxviii, 532, 1899; Chemical News, lxxx, 16, 1899. Th(
papers as published in French differ slightly.

ent on preëxistent vanadiferous sandstones. For the carnotite of Montrose and Mesa Counties, as mentioned by the French authors, occurs mixed in all proportions with quartz-sand grains, the remnants beyond doubt of former sandstone bodies, and it was soon recognized that in the carnotite bodies the vanadium existed in two conditions, the larger part by far as pentavalent vanadium in the easily soluble carnotite, and a smaller and sometimes hardly distinguishable portion as trivalent vanadium in a much less soluble silicate which was free from uranium.

It was hoped that by a field reconnaissance, observations bearing on these points would be obtained and material assembled which would on analysis help to solve the nature and explain the association of these two entirely different classes of mineral substances.

The field and microscopical observations of Messrs. Ransome and Spencer are embodied in the pages immediately following these introductory remarks. Unfortunately their collections, except from Placerville, came to hand so late that the chemical work on the more western occurrences has been confined to the carnotite bodies alone, of which material was already in my hands. Greenish sandstones have been observed by Messrs. Ransome and Spencer in those regions, but whether any of them are highly vanadiferous, or what their connection with the carnotite may be, remains yet unknown. The only two examined did not owe their color to vanadium. Meanwhile the chemical results thus far obtained, which it is advisable to put on record at once, will be found in the concluding section of this paper.

In this place it is my pleasure to record my appreciation of the readiness with which the following gentlemen have supplied me with material for study : Messrs. Poulot and Voillequé of Denver, Mr. A. B. Frenzel of Placerville, and Mr. J. R. Duling of Paradox. These gentlemen have, also, not hesitated to furnish me with all information at their disposal as to occurrence, etc., of these interesting ore bodies.

OCCURRENCE OF THE URANIUM AND VANADIUM ORES.

F. L. RANSOME.

General.—In the autumn of 1899, a brief reconnaissance trip was undertaken into the western portions of San Miguel, Montrose and Mesa Counties, near the Utah-Colorado line. I was accompanied by Dr. A. C. Spencer, whose knowledge of the stratigraphy of the region was of great assistance. The primary object of the expedition in accordance with a suggestion from Mr. S. F. Emmons, was to investigate the

copper deposits of La Sal Creek, Paradox and Sinbad Valleys,
and the vicinity, which had been responsible for some mining
excitement a few months previously. A memorandum from
Dr. W. F. Hillebrand, received just before starting, indicated that it would be well also to examine certain prospects
on which some preliminary work had been done, looking
toward the extraction of ores of uranium and vanadium. It
is to the latter that these present notes are confined.

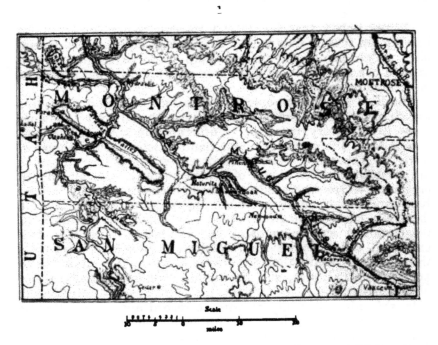

The most convenient way of reaching Paradox Valley was
found to be the stage road, which, starting from Placerville, a
settlement and station on the Rio Grande Southern R.R., runs
by way of Norwood, Shenandoah and Naturita to Paradox, a
distance of 60 or 70 miles.

The region can also be reached from the west by way of
Moab, in Utah. The Placerville route and the general geographical relations are roughly indicated in the accompanying
sketch map, fig. 1.

The topography of the region west of Placerville is that
characteristic of the "mesa country" of western Colorado and
southeastern Utah. Broad stretches of plateau are intersected
by steep-walled cañons and cliff-encircled valleys. The underlying rocks comprise the "red beds" of the Dolores forma-

tion* (Triassic), the La Plata formation (Jurassic), the McElmo formation (Jurassic), the Dakota sandstone (Cretaceous), and the Mancos shales (Cretaceous). Carboniferous rocks occur in Sinbad Valley, and a series of gypsum-bearing shales of unknown age in the bottoms of Sinbad and Paradox Valleys, but as the ore-deposits to be described all occur in the beds of the La Plata and McElmo formations, these older rocks need not be again referred to. The sediments making up these various formations lie usually nearly horizontal, but they are sometimes flexed and frequently faulted.

In all of the prospects examined, the ore of uranium occurs in the form of the recently described bright yellow carnotite. In one case this is intimately associated with a dull olive-green mineral which according to Dr. Hillebrand is either identical with, or very closely allied to, the vanadium-mica roscoelite. Deposits of one or both of these minerals occur widely scattered over San Miguel and Montrose Counties, Colorado, and in the Blue Mountain (Sierra Abajo) district of southeastern Utah; but a portion only of the known deposits were personally examined.

The Placerville Deposits.—These are essentially vanadium deposits, and occur 4500 feet nearly northeast of the railway station near Placerville, and about 1000 feet above the San Miguel River. The lower 900 feet of the San Miguel cañon is cut in the typical "red beds" of the Dolores formation. Above the Dolores comes the La Plata sandstone in typical development as described in the Telluride folios, viz., two heavy beds of light-colored sandstone, separated by a much thinner bed of dark limestone. The roscoelite occurs as an impregnation in the lower bed of the La Plata sandstone, about 100 feet above the base, and just beneath the bed of dark limestone. All the beds at this point are practically horizontal. The dark green vanadiferous sandstone occurs in a nearly continuous band, approximately parallel to the bedding planes, and varying in thickness from a few inches up to five or six feet. This band extends along the sandstone cliffs for an estimated distance of about 2000 feet. The roscoelite occurs more or less thoroughly impregnating portions of the fine-grained sandstone which constitutes the mass of the bed. It sometimes makes up more than 20 per cent of the vanadiferous facies. Normally the sandstone of this lower bed of the La Plata is light buff in color, with patches, mottlings, and stripes of pale pink, the latter color being apparently due to small amounts of ferric oxide. But when richly impregnated with roscoelite, this light-

---

* The formation names used in these notes are those adopted by Cross in the text of the Telluride Folio, of the U. S. Geol. Survey. The reader is referred to this folio for fuller descriptions.

colored sandstone becomes dark olive-green, the contrast bei
most marked when the cliffs are wet.   Although fairly co
tinuous for the distance stated, this zone of vanadiferous san
stone is by no means regular.   It varies much in thickness, a
in one place splits into two or more branches.   At some poin
two or more distinct streaks of roscoelite-bearing sand we
found at different horizons in the main sandstone bed.   T
carnotite is not nearly so abundant as the roscoelite at t
locality.   It occurs as minute yellow specks in the sandstor
and particularly as thin horizontal seams or streaks near t
bottom of the vanadiferous zone.

2

The work thus far done on the several claims which ha
been located along these deposits is of the most superfic
character.   On the Canary claim a tunnel of about 18 feet h
been run by Mr. A. B. Frenzel which exposes a typical s
tion of the impregnated zone (fig. 2).   The roof of the tunr
is formed by the underside of the bed of dark limestone pre
ously referred to.   This limestone is underlain by a few inch
of sandy limestone which passes into the light-buff La Pl
sandstone without break.   The latter contains abundant calci

as a cement or matrix for the sand grains. At a distance of a foot below the limestone, the sandstone shows, on close inspection, numerous specks of carnotite and gives a qualitative reaction for vanadium (Hillebrand). Examined in thin section under the microscope, this portion of the bed shows a fine-grained homogeneous sandstone, in which well rounded grains of quartz are held together by a rather abundant matrix of calcite. The latter is crystallized as a fine granular aggregate and includes small indeterminable particles of various kinds. Many of the quartz grains as seen in section are entirely surrounded by calcite, and so isolated from adjacent grains. A pale lemon-yellow substance occurs sporadically in the section as an interstitial material between the quartz grains. This is probably the carnotite which was seen as small yellow specks in the hand specimen. It is very minutely crystalline, recalling the habit of some obscure chloritic aggregates in decomposed igneous rocks, and is too indistinct for successful optical study. Small crystal grains of zircon are scattered through the sandstone, and are readily concentrated in the residue after treating with hydrofluoric acid.

In the next two feet below where it exhibits the foregoing facies, the sandstone becomes pinkish in color, due to the presence of ferric oxide, and the lens fails to reveal any roscoelite or carnotite. But just below the pink coloration, the sandstone begins to show yellow and green specks. The latter especially become more numerous and larger, until at from 3 to 4 feet below the limestone the sandstone has a decided green tint. This deepens, going downward, until near the floor of the tunnel the sandstone is a deep uniform olive-green, rich in roscoelite (over 20 per cent) and showing many small yellow specks of carnotite. This is regarded as the first-class ore, and the chief value of the deposit is supposed to be in its vanadium, the uranium being very subordinate. In this respect the Placerville deposit differs from those on La Sal and Roc Creeks, presently to be described, where the mineral sought for and occurring most abundantly is carnotite.

Thin sections of the green-spotted sandstone, when examined microscopically, resemble those of the light-buff sandstone just above it, as far as character of quartz grains and relative abundance of matrix are concerned. But the character of the matrix or cement is different. Calcite is much less abundant and its place is largely taken by roscoelite. This is grass-green in thin section, and might readily be mistaken for indistinct wisps and areas of chlorite. It sometimes forms a distinct envelope around the quartz grains, showing an indistinct foliated or fibrous structure normal to the surface of the latter.

The uniformly dark green sandstone richest in roscoelite does not effervesce with acids and shows no calcite cement in thin section. The usual quartz grains are held together by roscoelite. This is crystalline, but the highest available powers show only an indistinct and minute foliation, such as may be observed in some very finely crystallized chlorites.

Close to the floor of the tunnel is a fairly regular, nearly horizontal streak of carnotite, varying in width, but usually less than an inch, and showing noticeable diminution in thickness in the face of the tunnel. This small seam is not solid carnotite, but is merely a zone in the sandstone impregnated with the bright-yellow uranium mineral. It is not nearly so continuous as the main vanadiferous belt, and was seen only at three or four places along the 2000 feet or so of outcrop of the latter. When this seam is closely examined, it is seen that narrow bands rich in carnotite alternate with green bands carrying mostly roscoelite. There is also usually present a seam generally about an eighth of an inch in thickness (though often thicker) which is almost wholly quartz. The microscope shows it to be a true quartzite, in which the original rounded detrital grains of quartz have been cemented by fresh quartz in optical continuity with the older granules. A similar quartzite occurs in the green vanadiferous sandstone above the carnotite, where it forms concretionary knots and nodules. It was not noted in the buff sandstone however, where the cementing material is calcite.

Immediately below the carnotite seam there is a parting or "floor" in the sandstone, probably originally a very thin layer of shale, which forms the working floor of the tunnel. The sandstone below this floor is plainly impregnated for a short distance with roscoelite, but the thin shale seam is regarded as the practical bottom of the deposit.

The questions of the origin and actual extent of this deposit are closely related, and of much interest. Their discussion will be deferred until the other deposits visited have been described.

Some distance below the Placerville vanadium deposit, sandstone, presumably belonging to the Dolores formation, was observed to be colored green, of a somewhat brighter hue than the vanadiferous sandstone higher up the slope. As there are some copper prospects near by, from which ore has been taken, this was supposed to be a copper stain. Qualitative tests by Dr. Hillebrand show however that the color is due to a compound of chromium.

Similar green sandstones occur on the western side of Sinbad Valley in what is apparently the La Plata formation, and were originally supposed to be impregnated with roscoelite. Dr. Hillebrand's investigations, however, show that they too owe their color to some chromium mineral. These occurrences are interesting as showing that a green color in sandstones may result from various causes, and that even a bright green tint cannot be taken as an infallible indication of copper.

The La Sal Creek deposits.—These occur in the extreme western portion of Montrose County, southwest of Paradox, and about six miles up La Sal Creek from Cashin P. O. They are reached by trails from Paradox Valley and from Cashin. The deposits are on the south side of La Sal Creek and about 700 feet above the stream. They occur for a distance estimated at more than a quarter of a mile, along the sandstone cliffs which descend from the mesa into the cañon of La Sal Creek, and only a few feet below the level of the mesa surface.

In the absence of continuous stratigraphic work, it is impossible to correlate certainly and finally the rocks on La Sal Creek with the divisions established by Cross and Spencer in the Telluride quadrangle to the eastward. It seems probable, however, that the La Plata sandstone attains a much greater thickness in this portion of western Colorado than it does between Telluride and Placerville. The limestone bed, so characteristic a feature of the formation near Placerville and further east, is not uniformly present in this western region, and the La Plata sandstone (Variegated Beds, in part, of the Hayden Survey) is not always readily differentiated from the underlying Dolores formation (Red Beds). For a vertical distance of about 400 feet above the bed of La Sal Creek the rock is a heavy-bedded, rather fine-grained, light-colored sandstone, which is considered by Dr. Spencer to be the La Plata. Above this come thinner-bedded sandstones, with some conglomerates and shales, which are included in the McElmo formation. It is in this upper series that the uraniferous deposits occur. All the beds are here approximately horizontal.

As revealed by numerous small openings near the crest of the bluff, the carnotite, which is the material here sought, is found chiefly in a massive bed of nearly white sandstone. Some of the ore, however, lies between the sandstone and a lower bed of light-gray shale. Although the prospecting openings all lie at about the same level along the cliffs, the deposit is not nearly so regular as the vanadiferous band near Placerville.

The carnotite of La Sal Creek occurs as irregular, bunchy "pockets" in the sandstone, or along the contact of the sandstone with the underlying shale. These have all the appearance of being impregnation deposits, the solutions carrying the uranium compounds having deposited the ore wherever they found ready passage through the rock—usually along bedding planes. No roscoelite was detected with the carnotite.

The most remarkable and interesting fact in regard to the La Sal Creek deposits is their very superficial character. The ore bodies are usually flat-lying streaks, a few inches thick, which grade above and below into the common light-buff sandstone, and which die out and disappear when followed into the hillside. In tunnels run but a few feet underground the yellow impregnation of carnotite can be seen to gradually die out, to be succeeded by light-colored sandstone, showing no apparent trace of the mineral. It is doubtful whether any appreciable quantity of carnotite occurs as much as 20 feet from the surface, on any of the locations, although this distance is given from memory and not from measurements on the ground. As before stated, the impregnation has usually taken place along bedding planes ; it has also proceeded along surfaces of minor and superficial movement in the rocks. In one case it was observed that a portion of the overlying sandstone had moved upon the underlying shales, the disturbance being apparently a superficial one, of a kind commonly enough observed where massive beds rest on yielding shales on a steep hillside. In other words, the movement appeared to be directly related to the present topography. The deposition of carnotite was here plainly subsequent to the movement and had taken advantage of the small openings and dislocations in the shale afforded by this very recent disturbance. It was reported that some of the best nests of ore had been found in "slide rock," i. e., rock which had slipped to same extent down the slope ; but I was unable to verify this statement further than is indicated in the preceding description. There can be little doubt but that the deposits of carnotite on La Sal Creek are not only very superficial in character but very recent in age.

The Roc Creek Deposit.—The principal claim is the Copper Prince, owned by J. R. Duling, on the north side of Roc Creek, 3 or 4 miles above its mouth, and near the foot of the Miller trail to Paradox. It is reached by this trail from Paradox Valley, or by trail from Hydraulic on the Dolores River. This deposit is in the La Plata sandstone according to Dr. Spencer's observations, and has been more extensively exploited than any other seen. The sandstone, which at this point is nearly horizontal, is cut by an east and west fault, the fault plane dipping north at about 70°. The amount and character of the throw could not be determined. It is probably less than 50 feet. The carnotite occurs in the hanging wall of the fissure as small irregular branches in a loose mass of crushed sandstone and also as an impregnation of some of the firmer portions of the bed. No roscoelite was seen. Several small tunnels have been run in on the hanging wall side of the fissure (which itself carries no vein or ore-body), but they had been securely closed by planking, and no examination of their inner ends could be made. I am thus unable to tell what becomes of the deposit as it is followed in from the surface. A few hundred pounds of picked carnotite ore has been shipped from this claim and is reported to have sold for $1.25 a pound in Denver. The character of this deposit is similar to others examined, except that in this case a well-defined fault has provided a zone of crushed and porous rock in the hanging wall, along which impregnation could take place. A few hundred feet farther west the crushed sandstone adjoining the fault has been impregnated with cupriferous solutions and is spotted with stains of the blue and green carbonates of copper.

Other Deposits.—It is known that several carnotite claims have been taken up in Gypsum Valley in what is known as the Disappointment District. The impregnated sandstone is said to cap a hill and to constitute an extensive deposit, but it was not visited. Carnotite is also reported from the Blue Mountain District, but I have no personal knowledge of these occurrences. It seems highly probable that the material will be found widely distributed in the Mesozoic sandstones of western Colorado and eastern Utah, although perhaps nowhere in very extensive bodies.

Origin of Deposits.—That the deposits of carnotite and roscoelite were formed subsequently to the deposition of the sandstones is evident from the facts presented in the preceding pages. It is equally plain that the minerals could not have resulted from the alteration, in place, of other compounds of vanadium and uranium originally deposited with the sands. The shape and position of the deposits indicate clearly that the ores have been deposited in their present position only after transportation from a greater or less distance. Moreover, the recency of the deposits and the fact that they are sometimes directly connected with faults and dislocations in the sandstones shows that the vanadium and uranium compounds could not have been the original cementing material of the quartz grains, but have in all probability locally replaced the calcite which acts as matrix to the ordinary light-colored sandstone in which the ore-bodies occur. The deposits of roscoelite appear to be comparable to the impregnations of the sandstones with cupriferous solutions observed in many places in this region, particularly on La Sal Creek near Cashin and in Sinbad Valley, whereby the sandstone becomes colored bright green with the carbonate of copper. In these cases, however, the copper appears to have been previously deposited in part as chalcocite. An analogy might also be drawn with the green chromiferous sandstone near Placerville already referred to. In all these cases the actual sources of the materials which have been deposited by solutions in their present position are not known.   Dr. Hillebrand has shown,[*] however, that vanadium in small amounts is widely distributed in sandstones, limestones, and igneous rocks.   It is perhaps present in very small amounts in the bulk of the sandstone, and the deposits described may simply represent a concentration of this material under certain favorable conditions of solution and re-deposition.

In the absence of exploitation it is manifestly impossible to predict the probable shape and size of ore-bodies formed in this manner.   The roscoelite as seen near Placerville appears, however, to be much more persistent than the carnotite. There is no apparent reason why a mass of sandstone, impregnated with roscoelite, which is continuously exposed for several hundred feet along a cliff, should not extend for a considerable distance inward from the cliff face.   The carnotite, on the other hand, appears to be a much more superficial occurrence and, in fact, to have a not yet fully understood connection with the present surface of the ground.   This would indicate that the carnotite results from a local concentration of material already existent in the sandstone, and the deposition of this material in the form of carnotite under conditions determined by proximity to the surface, and probably partly dependent upon a semi-arid climate.

CHEMICAL ANALYSES AND DISCUSSION.

W. F. HILLEBRAND.

1. *The Green Coloring Constituent of the Placerville Sandstone.*

As noted by Mr. Ransome, the greener the sandstone, the greater has been the replacement of the calcareous cementing material by the crystalline vanadiferous mineral.   Fortunately the greenest available specimen was entirely free from carbonate, and this was used for analysis, 10 grams being taken for the main portion.

The green substance was but slightly attacked by cold acids and not rapidly by hot ones; nevertheless after 12–24 hours digestion on the water-bath with diluted nitric acid, the greenish color had been transferred to the liquid, the residue of sand was nearly white, and a rather voluminous separation of non-gelatinous silica had taken place.   The filtration and washing of the insoluble matter was accomplished without difficulty. The free silica in it was extracted by two or three digestions of 15 minutes each on the water-bath with sodium carbonate solution of 5 per cent strength.   The flocculent matter had entirely disappeared and the sandy residue was collected in a

* This Journal, IV, vol. vi, pp. 209–216, 1898.

Gooch crucible and washed with hot sodium carbonate solution, followed by very dilute nitric acid and finally by alcohol to prevent turbid filtration. Drying of the sand was effected by allowing the pump to draw air through the crucible until no further loss in weight resulted.

The water in the sand was then determined at 105°, at 300° and above 300°. Similar water determinations having been made on the unattacked substance, the differences gave the water expelled at those temperatures from the soluble part.

The sodium carbonate solution was acidified, evaporated, and filtered, and the filtrate again evaporated, to recover the last of the silica. To this was added a very small portion which the nitric acid extract of the sandstone held in solution. The sum represents the total silica of the decomposed portion of the sandstone.

The further analysis of the nitric solution presented difficulty because of the vanadium it contained. It was carried out in a variety of ways, of which the following seemed to be most satisfactory.

Barium was first precipitated by sulphuric acid and subsequently separated from traces of lead, calcium and vanadium. Hydrogen sulphide then threw out further traces of lead and copper with much sulphur from reduction of V_2O_5 to V_2O_4. After evaporation of the filtrate to reoxidize vanadium and iron, the alumina, iron and trace of uranium with much vanadium were separated from calcium, magnesium and alkalies by three precipitations by ammonia solution. The combined filtrates were evaporated, ignited, the residue transferred to a porcelain boat with nitric acid, evaporated therein to complete dryness, and exposed in a glass tube to a current of dry hydrochloric acid gas, as recommended by Smith and Hibbs[*] for the expulsion of vanadium from alkaline vanadates. The removal of vanadium being incomplete, the contents of the boat were again evaporated with nitric acid and again distilled, and these operations were repeated till no further evidence of a brown distillate appeared. The boat now contained only magnesium and alkalies with a little calcium. The distillates, containing all the vanadium that was in the boat, had been collected in U-tubes charged with water, and were eventually obtained in sulphuric solution for further treatment by evaporating with sulphuric acid in porcelain.

The precipitate of Al_2O_3, etc., was dissolved in nitric acid, evaporated almost to dryness in platinum and boiled with sodium hydroxide solution. These operations were repeated on the precipitate after filtration. The iron and titanium thus

---

[*]Jour. Am. Chem. Soc., xvi, 578, 1894.

purified were redissolved, precipitated by ammonia, weighed redissolved by acid potassium sulphate, the ferric iron wa reduced by hydrogen sulphide and titrated with permanganate after boiling out the reducing agent. The titanium was ther found colorimetrically by hydrogen peroxide.

The combined alkaline filtrates were heated with excess o ammonium nitrate to separate alumina from most of the vanadium. The precipitate was redissolved in nitric acid reprecipitated by ammonia, then ignited and weighed and redis solved by acid potassium sulphate. The sulphuric solution wa reduced by hydrogen sulphide, boiled and filtered from plati num sulphide, again boiled in a current of carbon dioxide and titrated hot with permanganate. Sulphur dioxide gas wa then introduced, boiled out in a current of carbon dioxide, and the titration repeated, this second result being taken as repre senting the V_2O_4 equivalent to the V_2O_5 that still contaminated the Al_2O_3. Deducting the V_2O_5 thus found from the combined weight of Al_2O_3 and V_2O_5 gave the Al_2O_3 of the soluble con stituent of the sandstone.

The combined filtrates from the alumina were evaporated ignited to remove ammoniacal salts, the residue converted to sulphates, and united with the earlier solution of vanadium sul phate. The vanadium in it was then determined precisely in the manner already described, and when added to that found with the alumina and calculated to V_2O_5, gave the total in the sandstone.

A check was made by dissolving the sandstone in hydro fluoric and sulphuric acids in a current of carbon dioxide and titrating the V_2O_4 with permanganate. The figure thus obtained was a trifle below that found as above detailed, and this is to be attributed to the presence of a very little vanadium as V_2O_5 in carnotite, and probably as a lead vanadate. All the uranium and most of the trifling amount of lead could be extracted by cold dilute nitric acid, thus showing them to be foreign to the green substance.

It is preferable to regard iron as in the ferric state rather than as ferrous, otherwise an equivalent amount of vanadium would have to be considered as V_2O_4, for which there is no occasion. It is quite possible that the iron is in part, if not wholly, foreign to the green matter, for it would be surprising to find an exposed sandstone free from ferric oxide.

If the iron is disregarded as probably extraneous, the ratio given below afford the following empirical formula :

$$H_{396} \ R'_{194} \ R''_{39} \ R'''_{610} \ Si_{763} \ O_{2775}.$$
$$\text{or } H_{396} \ R'_{194} \ R''_{39} \ R'''_{610} \ (SiO_4)_{555} \ (Si_2O_5)_{69}.$$

Simplified this is $H_{40} \ R'_{20} \ R''_4 \ R'''_{61} \ X_{62\cdot4}.$

Considering that the mineral is probably not perfectly pure and the uncertainties affecting the amount of water to be considered, etc., the above affords a close approach to the typical phengitic muscovite formula $H_2KR'''_3X_3$ in which X comprises the groups SiO_4 and Si_2O_6.

Analysis of the vanadiferous sandstone.

		Sandstone. Per cent.	Ratios.	Silicate. Per cent.	Roscoelite.
	SiO_2	12·56	·763	46·06	45·17
	TiO_2	·02			·78
	V_2O_3	3·50*	·0846 ⎫ ·305	12·84	24·01
	Al_2O_3	6·15	·2206 ⎭	22·55	11·54
	Fe_2O_3	·20	·0046	·73 FeO 1·60	
	CaO	·12	·0078 ⎫	·44	
Soluble	BaO	·37	·0088 ⎬ ·039	1·35	
in	MgO	·25	·0228 ⎭	·92	1·64
Nitric acid.	K_2O	2·41	·0938 ⎫ ·097	8·84	10·37
	Na_2O	·06	·0035 ⎭	·22	·06
	H_2O at 105° ..	·54	·1100	1·98	·40
	H_2O 105°–300°	·14	·0283	·51	·17
	H_2O above 300°	·97	·1978 ·198	3·56	4·12
	UO_2	·05			
	PbO	·06		100·00	99·86
	V_2O_5	·05			
Insoluble in Nitric acid.	Quartz, etc. ..	72·24†			
	H_2O at 105° ..	·04			
	H_2O above 105°	·20			

99·93

Traces of Li, Cu, Mo, Bi. No Cl, SO_3, or P_2O_5.

An attempt to satisfactorily distribute the constituents among different chemical molecules would be futile, since, for instance, we do not know whether to ascribe the barium to a biotite or muscovite molecule, but the close approach to the muscovite ratio leaves little room to doubt the character of the body. Notwithstanding the chloritic aspect of the mineral under the microscope, the analysis shows that it cannot be a chlorite, and we have undoubtedly to do with a body closely related to the mica roscoelite, wherein, however, the percentage proportions of Al_2O_3 and V_2O_3 are reversed, thus affording a further striking example of the mutual replaceability of these two oxides.

* Oxidation by permanganate of the HFl and H_2SO_4 solution of the sandstone indicated 3·43 per cent V_2O_5 (mean of 3·50, 3·32, 3·48, 3·43).

† Containing about 65 per cent SiO_2, 7 per cent Al_2O_3 (Fe_2O_3, TiO_2) and about 0·3 per cent of zircon and other minerals not decomposed by repeated evaporations with HFl and H_2SO_4.

For comparison, the latest analysis of roscoelite from Placer-ville, Calif., is given a place in the above table. Peculiar, though unimportant, is the coincidence that the only two known localities for this mineral should bear the name Placer-ville.

II. *Green Sandstone Colored by Chromium.*

Other sandstones much brighter green than the vanadiferous one were observed and collected by Dr. Ransome, both at Placerville and 60 miles distant in Sinbad Valley. The color suggested a salt of copper as its cause, but analysis showed it to be due to chromium. Time has failed for an examination as to the nature of this coloring body. It is very difficultly soluble, thus presenting greater hindrance to analysis than was the case with the vanadium compound. It would be interest-ing to find it to be a micaceous mineral analogous to the one just described. Under the microscope it presents a chloritic appearance (Ransome). If opportunity offers, the problem of its nature may yet be attacked.

In still another greenish sandstone from the west bank of the Dolores River, near the mouth of La Sal Creek, analysis failed to show either vanadium or chromium.

III. *The Carnotite Ores.*

Optical Examination.—Dr. Geo. P. Merrill has kindly sub-mitted the following notes.

"The carnotite powder appears under the microscope in the form of exceedingly minute dust-like particles, without crysta outlines and acting so faintly on polarized light as to at first seem almost amorphous. Much of the matter appears merely as a fine brownish clay, stained yellow by an amorphous pig-ment, but occasionally a well-defined fragment of a light yel-low, translucent mineral is met with which doubtless represents the vanadium compound in its condition of ideal purity.

Working over a considerable amount of the powder, I have found occasional clusters of this yellow mineral in the form of flattened radiating crystals with pyramidal terminations which are without evident pleochroism, polarize only in dull colors, and give extinctions always parallel to the axis of elongation. These are so minute (not over 0.25^{mm} in length) and so thin that I have never been able to find a crystal so oriented as to give an opportunity of determining its exact character, and I can only say that the general shape is such as to suggest a hexagonal mineral, though this is by no means certain."

Chemical Examination of the Carnotite Ores.—The chemical problems involved in the analysis of the carnotite ores were peculiarly intricate. Ideal material was quite unobtainable and mechanical separation impossible. The first specimen received happened to be of a higher grade than any of the subsequent ones, containing about 5 per cent of sand grains and showing only very faint delicate reddish tracings indicative of some foreign iron mineral. In mass it was of a beautiful canary-yellow color and easily broken down by pressure. Other specimens were more coherent, the degree depending altogether on the extent to which the sandstone had been impregnated and altered, and they sometimes showed more of the peculiar reddish admixture.

A fact only suspected in making the first analysis was confirmed by subsequent work on lower grade ores, namely, that the vanadium existed in two conditions, in entirely distinct minerals; the greater part by far as pentavalent vanadium in the easily soluble carnotite, and a smaller, much less soluble portion, almost vanishing in the purest ores, in the trivalent state as a constituent of a silicate free from uranium. This observation explained the statement of Messrs. Poulot and Voillequé that they had found the low grade ores to be relatively richer in vanadium as compared with uranium than the high grade ores.

While the carnotite dissolves at once in cold dilute nitric acid, unfortunately the vanadiferous silicate is not quite insoluble, hence arose an important difficulty in the way of arriving at the true composition of the carnotite. It is true that in one case the silicate has been analyzed (p. 143), but it would be unsafe to correct the carnotite analysis on that basis. Moreover, the analyses were not all made on the same plan; some are less complete than others; they cannot in some respects be rigidly compared with each other.

As to the carrying out of the analysis, various procedures were tried and no one found which gave altogether satisfactory results as to each constituent, though many could be determined with the usual degree of accuracy. The difficulty was mainly due to vanadium and the small amount of phosphoric acid usually present. These two constituents were likely to be found in different precipitates and could not be separated from them at one stroke. Their complete removal from other bodies was at times impossible and the weight of the latter had then to be corrected for these residual amounts.

Two entirely different lines of attack were open. One, that of Friedel and Cumenge, described in their paper on carnotite, had already been used with apparent success. It depends on rendering the vanadium insoluble in water by evaporating the nitric acid solution to dryness. Water extracts the alkalies and uranium without dissolving the vanadium, iron, or aluminum. Repetition of this process yielded Friedel and Cumenge satisfactory results.   They recommended washing with water containing ammonium nitrate.

The other method contemplated the quantitative removal of vanadium by dry hydrochloric acid gas as already described (p. 131).   This had been shown to be perfect for alkaline pyrovanadates by Smith and Hibbs and it was hoped might succeed even with such complex mixtures as the present.   The distillations were made on the crude ore and also on the nitric acid solution after filtration from the insoluble matter and evaporation to thorough dryness.   The action is immediate in the cold, copious red-brown vapors coming off and condensing in part as a dark red liquid in the tube.   But repeated distillations and the application of heat were required to effect complete removal of the vanadium accompanied by the arsenic and molybdenum.   After each distillation the contents of the boat had to be evaporated with nitric acid.   The blue and green colorations apparent on adding this acid showed that the hydrochloric acid gas had reduced a good deal of the vanadium to a lower state of valence, and in this condition it was incapable of forming the volatile body.

A somewhat more satisfactory separation sometimes resulted when this distillation method was combined with that of Friedel and Cumenge by subjecting both the evaporated uranium nitrate solution and the residue insoluble in water to the action of hydrochloric acid gas.

The distillates obtained by either way were evaporated with sulphuric acid, the arsenic and molybdenum were separated by hydrogen sulphide, and the vanadium was then titrated by permanganate at a temperature near boiling and again after reduction by sulphur dioxide gas.   In one or two cases when the temperature of distillation had been high and it was feared some iron had passed over, the distillates were evaporated in porcelain with nitric acid, transferred to a platinum crucible evaporated therein with sulphuric acid and fused with sodium carbonate.   The aqueous extract was then treated as above for arsenic, molybdenum and vanadium.

The separation of Friedel and Cumenge, while perhaps adequate for technical purposes in ores free from phosphorus,[*] does not in my hands give perfect satisfaction even then.   It is impossible to prevent a little of the vanadium, also of the iron and aluminum, from going with the uranium, and on the other hand a little uranium may stay with the vanadium

---

* With even only half a per cent of P_2O_5 many times that amount of UO_3 is rendered insoluble in water after evaporation to dryness with nitric acid. The compound formed is of a lemon-yellow color, which is masked by the separate vanadic acid until this has been removed by ammonia. This insoluble body was treated as follows in order to arrive at the P_2O_5 and UO_3 it contained. After solution in nitric acid the phosphorus was precipitated by ammonium molybdate and from the filtrate the uranium by three precipitations by ammonia.

Again when the vanadic acid is extracted from the residue by ammonia, as prescribed by Friedel and Cumenge, a little of the other constituents of the residue accompany the vanadium into solution.

If the combination of the two methods is used, it is better after removal of the vanadium by distillation to further treat the two residues separately at first instead of to unite them at once. Full details of this treatment are unnecessary and would unduly extend this paper. Suffice it to say that from the uranium portion the little iron and aluminum present are separated by ammonium sulphide and carbonate, and after removal of the latter, and acidification, the uranium can be thrown down by ammonium sulphide, and then by at least two precipitations by ammonia, or far better, by precipitating the neutral hydrochloric solution by freshly precipitated and alkali-free mercuric oxide at boiling heat, as prescribed by Alibegoff.[*] As pointed out by von Foullon and also by Alibegoff, contrary to certain still widely disseminated statements, ammonium sulphide does not afford a good separation of uranium from calcium. This is perhaps especially true if the solution contains any phosphorus. Again, contrary to another statement, it seems perfectly possible to separate uranium completely from alkalies by a few ammonia precipitations.

The finally ignited and weighed U_3O_8 was redissolved in nitric acid, filtered if necessary (SiO_2, Al_2O_3, Fe_2O_3), and divided into two parts of which the one was tested for the very little P_2O_5 usually present, and the other for vanadium by conversion into sulphate, reduction by sulphur dioxide gas, and titration by very dilute permanganate solution.

Table of Analyses of Carnotite Ores.

I. Copper Prince Claim, Roc Creek, Montrose Co., Colo.

According to the donor of the specimen, Mr. J. R. Duling, I-*a* is from the same lot as that from which Mr. Poulot obtained the specimens afterwards analyzed by Friedel and Cumenge.

II. Yellow Boy Claim, La Sal Creek, Montrose Co., Colo.
III. Yellow Bird Claim, La Sal Creek, Montrose Co., Colo.

The last two claims belong to the same group and the ore is from the same "blanket" (Voillequé).

I-*b* and II-*b*, the first analyses made, are not strictly comparable with the other analyses since they represent the effect of warm dilute hydrochloric acid, whose greater action is shown by the nearly complete solution of the ferruginous admixture. Cold, dilute nitric acid was used for the other analyses.

[*] Ann. Chem. u. Phar., vol. ccxxxiii, 133, 1886; Zeit. für anal. Chemie, xxvi, 632, 1887.

	I.			II.		III.
	a	b	c	a	b	
Insol. ...	7·10*	8·34†	19·00‡	10·33§	‖	¶
UO_2	54·89	52·25	47·42	54·00	52·28	20·51
V_2O_5	18·49	18·35	15·76	18·05	17·50	7·20
P_2O_5	·80	·35	·40	·05	tr.	none
As_2O_5 ...	tr.	·25	none	none	none	none
Al_2O_3 ...	·09	?	·08	·29	?	·08 ?
Fe_2O_3 ...	·21	1·77	·72	·42	3·36	·25
CaO	3·34	2·85	2·57	1·86	1·85	1·64
SrO	·02	?	?	tr.	tr.	?
BaO	·90	·72	·65	2·83	3·21	·29
MgO ...	·22	·20	·24	·14	·17	·07
K_2O	6·52	6·73	6·57	5·46	5·11	1·51
Na_2O ...	·14	·09	·07	·13	·02 ?	·01
Li_2O	tr.	?	?	tr.	?	?
H_2O 105°	2·43	2·59	1·85	3·16	4·52	1·85
H_2O 350°	2·11	3·06	2·79	2·21	3·49	1·64 (300°)
$H_2O + 350°$	none	none	none	none		·19(+300°)
PbO	·13	·25	·18	·07		·09
CuO	·15	·20	·22	tr.		tr.
SO_3	none	·12	·18	none		none
MoO_3 ...	·18	·23	·18	·05		·04
SiO_2	·15	·06	·13	·20		·07
TiO_2	·03	·10	?	?		
CO_2	·56	·33	none	none		·06 ?
	98·46	98·84	99·01	99·25		

Note for H_2O rows II: total H_2O in ore (braced).

* Containing ·54H_2O, ·09V_2O_5.

† The insoluble matter Had the following composition: SiO_2 5·18, V_2O_5 ·21, P_2O_5 ·09, K_2O ·26, Na_2O ·04, H_2O (105°) ·56, (350°) ·32 (+350°) ·48, Al_2O_3, TiO_2, Cr_2O_3, etc, by diff. 1·20. The SO_3 of this analysis is not combined with BaO, for it is wholly extracted by dilute acids. Once in solution the acid used is sufficient to prevent its immediate precipitation as $BaSO_4$ by union with some of the barium present. The same holds true for the SO_3 of I-c.

‡ This material was obtained by floating off the finer matter, allowing it to settle, collecting on a Gooch filter, and drying it in a current of air drawn through the crucible. The insoluble matter held in addition to 16·41 quartz and silicates including ·39 V_2O_5 and a little UO_3), H_2O (105°) ·83, (300°) ·73, (+300°) 1·03; total H_2O 2·59.

§ Containing ·16 V_2O_5 and 1·90 H_2O.

‖ The insoluble matter contained besides quartz and silicates ·25 V_2O_5, ·21K_2O, ·05 Na_2O.

¶ This analysis was made purposely on a relatively poor ore, furnished by Messrs Poulot and Voillequé, with the object of determining, if possible, the composition of the vanadiferous silicate which it contained (see p. 143). The data for calculating the H_2O values of both analyses are as follows:

	Ore. (a)	After extraction of carnotite by cold. dilute nitric acid. (b)	After extraction of residue by hot nitric acid, sod. carb., etc. (c)
H_2O 105°	3·53	1·68	·02
H_2O 300°	2·11	·47	·03
$H_2O + 300°$	·83	·64	·02
	6·47	2·79	·07

Discussion of the Carnotite Analyses.—It will be noted that a somewhat marked deficiency appears in most of the analyses, the cause of which is quite unknown. Great care was exercised in most cases and especially in those which show the greatest loss. It seems hardly possible that any serious constant loss of a known constituent should have occurred, but the only alternative demands the presence of an element or elements unnoticed and which cannot have been weighed with the known constituents. The researches of M. and Mme. Curie have shown that these ores contain traces of radio-active elements, precipitated the one by hydrogen sulphide, the other by sulphuric acid. Their presence, however, in quantity sufficient to account for the observed losses in the above analyses, specially when 10 grams of ore were operated on, could not possibly have escaped observation. To whatever cause it may be due, this loss alone suffices to render somewhat uncertain any calculations based on the analytical figures, though if the loss is to be ascribed to uranium or vanadium the ratios would not be sufficiently affected to obscure any simple relations that might exist.*

Another difficulty is the impossibility of knowing what constituents to exclude and what to include in deducing molecular ratios. It is certain that most if not all of the iron is foreign to the yellow body. It is probable that phosphorus is likewise so, since its extraction by dilute acids does not keep pace with that of the uranium and vanadium. It may possibly be in combination with the iron, in part at least. The alumina doubtless is derived from the vanadiferous silicate which seems to exist in all the ores and which is not quite insoluble in cold dilute acids. If so, a small portion of the vanadium, potassium, magnesium, and water are to be attributed to this mineral, but a general correction based on the analysis of this compound (see p. 143) would not be justifiable. Its application leads to nothing definite, even in the case of the particular ore No. III, from which the silicate analyzed was derived.

In the following tables are given first the recalculated analyses and then the molecular ratios. All those constituents have been excluded which pretty certainly do not belong to the carnotite, but small portions of some of those retained are unquestionably extraneous. In two cases (I-*a* and I-*b*) a certain proportion of lime has been arbitrarily excluded equivalent to

---

*According to Dr. Harry C. Jones of Johns Hopkins University, who very kindly undertook to examine a specimen of the Copper Prince ore for rare gaseous elements, helium is not present. Faint hydrogen lines were observed, the source of which was ascribed to water vapor. Other lines, due probably to hydrocarbons, were fairly strong, but the specimen had been long enough exposed in our laboratory and elsewhere to have accumulated enough dust to account for them.

the acid anhydrides CO_2, MoO_3, and SO_3, less what is needed to offset PbO and CuO.

Carnotite Analyses Recalculated.

	I-a	I-b	I-c	II-a	II-b	III
V_2O_5	20·72	21·09	20·12	20·54	19·85	20·62
P_2O_5	·90	·40	·51	·06		
As_2O_5....		·29				
UO_3	61·53	60·06	60·55	61·44	59·31	58·75
CaO	3·03	2·77	3·28	2·11	2·10	4·70
BaO(SrO) _	1·03	·83	·83	3·22	3·64	·83
MgO	·25	·23	·31	·16	·19	·20
K_2O	7·31	7·73	8·39	6·21	5·80	4·33
Na_2O	·15	·10	·09	·15	·02	·03
$H_2O-105°$	2·72	2·98	2·36	3·59	5·13	5·30
$H_2O+105°$	2·36	3·52	3·56	2·52	3·96	5·24
	100·00	100·00	100·00	100·00	100·00	100·00

Ratios.

	I-a		I-b		I-c		II-a		II-b	III	
V_2O_5	1133		1208		1101		1124		1081	1128	
P_2O_5	63	1196	28	1248	36	1137	4	1128			
As_2O_5....			12								
UO_3	2140		2088		2105		2136		2062	2043	
CaO	541		495		586		377		375	839	
BaO(SrO) _	68	671	54	606	54	718	210	627	238	660	54	943
MgO	62		57		78		40		47	50	
K_2O	776		826		891		659		616	460	
Na_2O ...	33	809	22	848	20	911	33	692	4	620	6	466
$H_2O-105°$	1511		1656		1311		1994		2850	2944	
$H_2O+105°$	1311		1956		1978		1400		2200	2911	

These ratios lead to the following empirical formulas, ic which only the water given off above 105° is considered. Since the water is wholly removable below 350° it is regarded as water of crystallization and not of constitution.

	R′	R″	R^v	U	O	H_2O
I-*a* ..	1618	671	2392	2140	13880	1311
I-*b* ..	1696	606	2496	2088	13958	1956
I-*c* ..	1822	718	2274	2105	13629	1978
II-*a* .	1384	627	2256	2136	13367	1400
II-*b*	1240	660	2162	2062	12871	2200
III ..	932	943	2256	2043	13178	2911

If it be assumed that the bivalent elements offer the most accurate determinations, the above ratios may be reduced to the following simpler terms on that basis :

	R′	R″	R^v	U	O	H_2O
I-*a*	2·41	1	3·56	3·19	20·68	1·95
I-*b*	2·80	1	4·12	3·45	23·03	3·23
I-*c*	2·54	1	3·17	2·93	18·98	2·75
II-*a*	2·21	1	3·60	3·41	21·32	2·23
II-*b*	1·88	1	3·27	3·12	19·50	3·33
III	·99	1	2·39	2·16	13·98	3·09

The results, however, show a great lack of agreement and wide variation. It is plain that no probable formula can be calculated for the yellow body. The variations are of such a nature as to indicate in the plainest manner that it is a mixture of several substances.

Such a detailed discussion as the foregoing would hardly have been justified in view of the negative conclusions arrived at, but for the fact that Messrs. Friedel and Cumenge in their paper announced a simple formula for the body examined by them and gave it the specific name carnotite. Their published analyses are as given below, from which they have excluded considerable sand and traces of barium, aluminum, lead, copper, and radio-active bodies as present in excessively small quantities. They make no mention of calcium, and admit that their values for water are open to doubt. The formula deduced by them is $2U_2O_3,^* V_2O_5, K_2O, 3H_2O$.

Analyses of Carnotite by Friedel and Cumenge.

	Found.				Calculated.
$U_2O_3^*$ --------	64·70	62·46			63·54
V_2O_5 --------	20·31	19·95			20·12
K_2O ---------	10·97	11·09			10·37
H_2O -------			5·29	·481	5·95
Fe_2O_3 -------	·96	·65			
					99·98

It appears from these analyses that Messrs. F. and C. by great good fortune obtained a variety of samples of the pure potassium compound, free from calcium and without appreciable admixture of barium. This is very remarkable in view of the fact that all the ores from different localities examined by me show large admixture of calcium or barium salts or both, even that which is certified to have come from the same lot as that from which their material was taken. The French authors give a brief outline of their methods of analysis. That one which afforded them the best results would involve the weighing of any calcium present as sulphate along with the potassium, on the assumption that the presence of that element had been overlooked. It is much to be desired that a re-analysis of their material should be made, if there is any of it still available, in order to clear up the doubt connected with the first analysis.

In the light of the evidence herein set forth, the existence of a distinct mineral species having the composition claimed for carnotite can by no means be considered as established.

---

* Old notation equivalent to the modern UO_3.

Average quality of marketed ore.—As these carnotite ore
bodies are being exploited for the market, it is of some interest
to know the average quality of each commercial lot.   A care
fully prepared sample representing several tons of ore was
received from one of the commercial houses of Denver and was
found to carry 11·49 per cent of uranium counted as U_3O_8, and
6·40 per cent of vanadium counted as V_2O_5.   Over one-sixth
of the vanadium existed, however, in the trivalent state, not as
a constituent of the yellow body, but doubtless of a silicate like
the one whose composition is given on p. 143.

Commercial assay.—The commercial assay of these ores has
presented difficulties to the technical chemist, the results being
sometimes very discordant.

As to uranium this is not surprising.   The methods that have
probably been commonly employed will give varying results
according to the contents of the ore in phosphorus and alkaline
earths.   Possibly the old Patera process, described in most
text-books on analytical chemistry, might be made to serve
with modifications called for by the large amount of vanadium
present.

The assay for vanadium presents little difficulty and does
not require much time.   The ore is fused with sodium carbo
nate, leached with water and the fusion repeated on the residue
The combined filtrates are acidified by sulphuric acid, arsenic
and molybdenum are precipitated in the hot solution by hydro
gen sulphide, whereby the V_2O_5 is reduced to V_2O_4.   After
filtration and expulsion of hydrogen sulphide by boiling, the
vanadium is titrated in hot solution by permanganate.   It i
then reduced by sulphur dioxide gas, and after boiling this out
the titration is repeated.   The results are exact, and they are
not affected by the uranium that may be present.

IV. *Composition of the Vanadiferous Silicate in the   Carnoti
Ore.*

As already mentioned the ores contain a vanadiferous silica
free from uranium.   To the end of ascertaining its compositio
if possible, ore No. III (p. 128), from the Yellow Bird claim
low in carnotite but relatively rich in vanadium, was treate
as follows :

The carnotite from 10 grams was extracted by cold, dilu
nitric acid, and the well washed residue, consisting of coar
sand and an utterly amorphous mud, by 4 per cent sodium
carbonate solution to get rid of the small amount of silica pre
sumably set free but not dissolved by the acid.   This amounte
to 0·35 per cent in duplicate determinations, and together wit
0·06 per cent in the acid solution or 0·41 per cent in all, may

serve as a maximum figure by which to gauge the action of the cold acid on the silicate or silicates in the ore. The residue was then digested for several hours with warm nitric acid of about 1·2 sp. gr. until as shown by a companion test its action had ceased. It was then filtered, washed and digested with 5 per cent sodium carbonate solution to dissolve the copious deposit of silica. The final residue was collected in a Gooch crucible, washed with sodium carbonate, followed by dilute nitric acid to remove all alkali, then by alcohol to prevent turbidity in the filtrate, and dried by suction of the pump. In it the water was determined at different temperatures, also its general composition.* From the sodium carbonate filtrate the silica was obtained by two evaporations and filtrations, also the trifling amount held by the acid solution. This latter was then fully analyzed and the complete results follow.

Composition of vanadiferous silicate in carnotite ore.

	Per cent in ore.	Per cent calculated to 100.	Ratios.	
SiO_2	6·48	43·94	·7275	
Al_2O_3	2·445 (2·45 and 2·44)	16·58	·1622	
V_2O_3	·965 (·93 and 1·00)	6·54	·0434	·2426
Fe_2O_3	·875 (·87 and ·88)	5·93	·0370	
CaO	·035	·24	·0043	
MgO	·654 (·652 and ·657)	4·43	·1099	·1142
K_2O	·546	3·70	·0393	
Na_2O	·03	·20	·0037	·0430
H_2O 105° C.	1·60†	11·26	·9222	
H_2O 300° C.	·44†	2·98	·1655	
H_2O above 300° C.	·62†	4·20	·2333	
	14·75	100·00		

Also traces of titanium, manganese, and lithium.

On the improbable assumption that the iron is to be wholly included, and regarding only the water given off above 300°, the following ratios result:

$$H_{467} R'_{86} R''_{114} R'''_{468} Si_{727} O_{2873} ;$$

which become, if the iron is excluded,

$$H_{467} R'_{86} R''_{114} R'''_{411} Si_{727} O_{2462}.$$

These figures, while strongly suggesting definite ratios between certain of the constituents, do not under the circumstances warrant the deduction of a formula, nor do they lead

* SiO_2 48·89, Al_2O_3, Fe_2O_3, zircon etc., ·44, TiO_2 ·08, MgO ·01, K_2O ·11, Na_2O ·03, H_2O ·07; total 49·63.

† Determined on a separate portion of the same powdered sample. See p. 138 footnote for data.

to the same conclusion as in the case of the green cementing material of the Placerville sandstone. The phengite-muscovite ratio of that is not apparent here. Yet it is not at all unlikely that a mineral like the one from Placerville is present, but contaminated with some other. In fact it would be surprising to find in such thoroughly altered sandstones anything but a mixture. The fact of the existence of such vanadiferous transition products is itself highly interesting, and these tedious analyses cannot therefore be considered as made in vain. The mud-like amorphous character of this material precludes any hope of aid from the microscope in solving the question of its homogeneity.

Summary.

The body called carnotite is probably a mixture of minerals of which analysis fails to reveal the exact nature. Instead of being the pure uranyl-potassium vanadate, it is to a large extent made up of calcium and barium compounds. Intimately mixed with and entirely obscured by it is an amorphous substance—a silicate or mixture of silicates—containing vanadium in the trivalent state, probably replacing aluminum.

The deposits of carnotite, though distributed over a wide area of country, are, for the most part, if not altogether, very superficial in character and of recent origin.

The green coloring and cementing material of certain sandstones near Placerville, Col., is a crypto-crystalline alumino-vanadio-potassium silicate resembling roscoelite, but with the percentage proportions of Al_2O_3 and V_2O_3 reversed. It constitutes over 25 per cent of the sandstone at times, and contains nearly 13 per cent of V_2O_3, the latter amounting in the maximum case observed to 3·5 per cent of the sandstone.

As yet these highly vanadiferous sandstones have been found only at Placerville, where it is intended to work them for vanadium. Carnotite is associated with them in only trifling amount.

Other sandstones noticed owe their bright green color to chromium. In yet another case where the color was dull green, this was not due to either chromium or vanadium.

U. S. Geological Survey, April, 1900.

ART. XV.—*Restoration of Stylonurus Lacoanus, a Giant Arthropod from the Upper Devonian of the United States;* by CHARLES E. BEECHER. (With Plate I.)

In the animal kingdom, the attribute of bigness has come to be regarded as one of the prerogatives of the vertebrates. On this account, invertebrates seldom receive credit for having a size of more than a fraction of a cubit, and are looked upon as objects to be held in the hand or viewed under a lens. As a matter of common experience, and probably also of congratulation, large invertebrates *are* rare, and some whole classes cannot furnish a single individual measuring more than a few inches in greatest diameter.

In a list of arthropod giants, the subject of the present note must be included, and will take equal rank with the Giant Spider-Crab of Japan (*Macrocheira Kaempferi*) and the great "Seraphim" of the Scotch quarrymen (*Pterygotus anglicus*). The former can safely claim to be the largest representative of the Brachyurans that has ever existed, and to the latter may be accorded the same distinction among the Merostomes.

The living species of the Merostomata comprise only the American and Moluccan Horse-Shoe Crabs, *Limulus polyphemus* and *L. moluccanus.* The latter sometimes attains a length of three feet, and measures eighteen inches across the carapace. To find other species in this order worthy of comparison with the huge Brachyuran of Japan, it is necessary to go back to the Paleozoic forms, and among these the larger species of *Pterygotus*, and the *Stylonurus* here noticed, fill all the requirements. It should be borne in mind, however, that these statements are based upon comparative lengths and breadths. If bulk alone were considered, the common lobster (*Homarus americanus* and *H. vulgaris*) should be mentioned, though in length and extent of limbs it would be considerably smaller.

Concerning the size of the Scotch "Seraphim," Dr. Henry Woodward[*] states that "From our present knowledge of the almost perfect remains of *Pterygotus anglicus*, and on the evidence of the numerous detached portions of this extinct genus, we are justified in concluding that it attained a length of six feet, and a breadth of nearly two feet at the widest part of its body." This huge Merostome has been found in the Lower Old Red Sandstone of Scotland, at a horizon nearly equivalent to the one furnishing the remains of *Stylonurus* in America. Thus what seem to be the two largest species of this class were contemporaries, though not associates.

Historical.

The first specimen found in America that can be referred
the genus *Stylonurus* was collected by the writer about 187
and loaned to Professor James Hall.   It remained in his han
unnoticed until 1884, when he described it as *Eurypter
Beecheri.*   The specimen preserves the abdomen and portio
of two of the large posterior limbs.   No species of *Eurypter*
known possessed such greatly elongated limb joints, and the
seems to be no good reason for not referring it to *Stylonuru*
in which this is a normal character.   The specimen of *Styl
nurus Beecheri* is uncompressed, and apparently retains t
proportions of form and convexity as in life.   On this accoun
it was of considerable importance in the restoration of t
larger species.

In 1882, Hall was furnished with a plaster cast of the ca
pace of a large arthropod, by Dr. Cook, then State Geologi
of New Jersey.   The original specimen was from the Catsk
group at Andes, Delaware County, New York, and had be
sent to the museum at Rutgers College, New Brunswick, Ne
Jersey.   Prof. D. S. Martin' made the first reference to th
species in some remarks on "A New Eurypterid from t
Catskill Group," before the New York Academy of Science
October 16, 1882, an abstract of this note appearing in t
transactions of the same society some time after June, 188
In this abstract, the species is neither described nor figure
and Hall is not mentioned in any connection.   Martin stat
that he saw the specimen (=cast sent to Hall) in the Sta
Museum at Albany, and it bore the name *Stylomurus excelsi*
(evidently a misprint for *Stylonurus*).

The next reference to this form in point of time, and t
first publication of a generic and specific name accompani
with a description and accurate illustration, was given by
W. Claypole,' in a paper read before the American Ph
osophical Society, September 21, 1883, under the title "No
on a large Crustacean from the Catskill Group of Penns
vania."   It is stated on the signature containing this pap
that it was printed November 2, 1883.   Claypole's descripti
was based upon a second specimen found in Wyoming Count
Pennsylvania, which preserves about three-fourths of the cep
alothorax, and belonged to the collection of R. D. Lacoe
Pittston.   This was given the name *Dolichocephala Lacoan*
and rightly classified with the Merostomata.   It therefo
appears that, up to this time, the name *Stylonurus excelsi*
was simply *nomen nudum*, and as such cannot be recogniz
as valid.

In 1884; Hall' published his description and figure of t

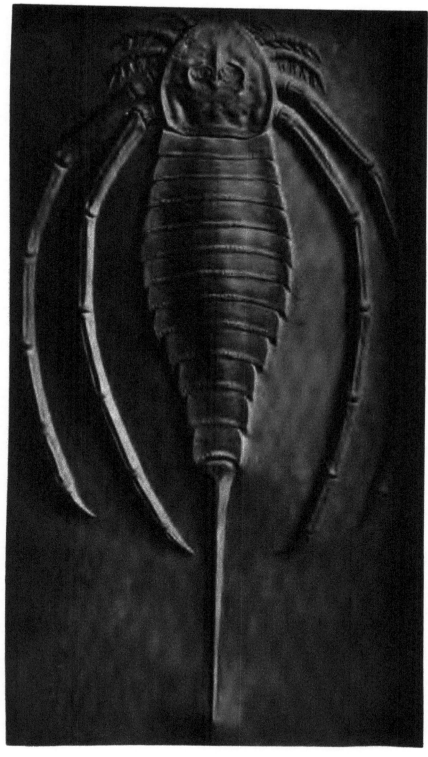

STYLONURUS (length nearly five feet).

New York specimen, in the thirty-sixth Annual Report of the New York State Museum, in a paper entitled "Description of a New Species of Stylonurus from the Catskill Group." It is here that the New York specimen was first figured and a description given, and the date of publication of this paper is the one to be considered in deciding the claims of *Stylonurus excelsior* as Hall's species.

At the Philadelphia meeting of the American Association for the Advancement of Science, September, 1884 (Proc. A. A. A. S., vol. xxxiii, published 1885), Hall[*] presented a note on *Stylonurus excelsior*, merely referring to its occurrence, and citing Martin's abstract with page and month of publication. This citation is repeated by Hall in each of his notices of this species, for only by thus establishing the species could he have any claim to priority. As already mentioned, Martin's paper does not attempt any description of this form, and Hall is not mentioned. Hall further says: "The carapace is described and figured in the 36th Report of the N. Y. State Museum of Natural History," without reference to plate, page, or year, and it is therefore quite possible that this description was not published until after the meeting of the Association. In any case, it appeared some months later than Claypole's paper, and the name *Dolichocephala Lacoana* has priority over *Stylonurus excelsior*, and must be recognized.

Claypole failed to point out the affinities of this form with *Stylonurus*, and proposed a new generic term for his species. Although there are differences that may prove of generic value when more complete specimens of the American species have been studied, yet at the present time there seem to be no strong reasons why the specimen in question should not be considered as belonging to *Stylonurus*, and it is upon this ground that the present restoration is attempted.

Material available for a Restoration.

Restorations of extinct organisms are largely exhibits of mental architecture, based upon the personal interpretation of a certain number of real things. Some statement, therefore, should be given of the character and amount of the material that has been collated to furnish a restoration of *Stylonurus Lacoanus*.

(1) The specimen of the cephalothorax described by Hall shows the complete outline and upper surface of this part, and a cast from the original was taken to represent this portion in the restoration. (2) The type of *S. Lacoanus* Claypole includes a large part of the cephalothorax of an individual nearly the same in size as the preceding. (3) Dr. J. M. Clarke

discovered that this specimen also preserved considerable evidence as to the nature of the appendages, and he succeeded in developing what appear to be the chelate antennæ, the first pair of gnathopods, and the mandibular bases of at least three others. (4) The length and number of joints in the limbs are taken from the English species *S. Logani* and *S. Powriei*, of which quite complete, though smaller, individuals have been described by Woodward.[10] (5) The outline and proportions of the abdomen follow closely those of the English forms and of *S. Beecheri*, the latter giving the natural convexity. (6) A portion of a large abdominal segment found by the writer in the Chemung group at Warren, Pennsylvania, and apparently belonging to a nearly related species, has an ornamentation closely approaching that on the cephalothorax of the type, and was used to elaborate the sculpture over the abdomen of the restoration. (7) The form and character of the telson spine correspond to *S. Logani* and also to some large fragments found by F. A. Randall at Warren and Ackley, Pennsylvania, in the Chemung group, and probably belonging with the abdominal somite already mentioned.

With the date furnished by the foregoing material, the restoration was undertaken. The first model in relief was constructed in clay, and from it a plaster mold was taken. A number of casts have been made since, and a photograph of one of them is represented in the accompanying plate (Plate I).

In this connection, it may be suggested that the type specimen of *Stylonurus* (?) (*Echinocaris?*) *Wrightianus* (Dawson sp.) represents two proximal joints of one of the large crawling feet of a form related to *Stylonurus*, and not two somites of the abdomen as indicated by Hall.[9] Any reference is at present somewhat uncertain, owing to lack of positive knowledge, and the fact that the specimen in question was first described as a plant (*Equisetides*), then referred to the Phyllocarida (*Echinocaris*), and lastly appeared as a possible Merostome, shows how this form may be interpreted by different observers. No one can doubt its arthropod nature, on account of the characteristic surface markings. Its elliptical or ovoid section without any flattening of the epimera, the very considerable overlapping of the joints, and the configuration of the suture, are more strongly indicative of the nature and requirements of a limb than of abdominal segments.

The Relief Model.

In this restoration, the animal is represented as lying on a slab, with the entire dorsal surface exposed. The cephalothorax has an axial length of 25^{cm} and a width of 22^{cm}.

The chelate antennæ were doubtless carried in a folded position, as in most related genera, and seldom were visible from the dorsal side. They are, therefore, not shown. The three pairs of short gnathopods, serving partly as swimming organs, are seen extending outward from the antero-lateral margins of the cephalothorax. Several of their distal joints are each provided with a pair of flat, ridged, spinous processes, and a similar spine at the termination of the limb.

The two pairs of great crawling feet extend outward and backward from the postero-lateral margins of the cephalothorax. The anterior pair expose 109^{cm} of length, and the posterior pair about 108^{cm}. The elements of the limbs are represented as grooved, as this character seems necessary to give the needed strength to long slender joints, and also because a similar conformation is present in *S. Beecheri*.

The abdomen measures 30^{cm} in greatest width at the fifth segment, and 66^{cm} in length exclusive of the telson. The posterior abdominal segments are represented without detachable epimera, as this feature is not as yet known to be constant for the genus, although present in some species.

The telson spine agrees proportionally in length with the same member in *S. Logani* and *S. Powriei* as described by Woodward,[10] and was given a slight upward curvature as in *Limulus*. It measures 54^{cm} in length and 7·5^{cm} across at the proximal end.

Altogether the animal as restored has a length of nearly five feet (147^{cm}), and with the legs extended it would measure about eight feet (242^{cm}) across.

It is not intended to claim any high degree of accuracy for this restoration, but merely to represent in some graphic form an animal approximating in size and character an individual of the species *Stylonurus Lacoanus*. Its size alone was the chief incentive for attempting a reconstruction, and some sacrifice of exact detail may well be allowed, in order to make any presentation of this magnificent arthropod.

Yale University Museum, New Haven, Conn., September 5, 1899.

References.

1. Claypole, E. W.—Note on a large Crustacean from the Catskill Group of Pennsylvania. Proc. Amer. Phil. Soc., vol. xxi, April, 1883–January, 1884.

2. Dawson, J. W.—Notes on New Erian (Devonian) Plants. Quar. Jour. Geol. Soc., London, vol. xxxvii, 1881.

3. Hall, J.—Description of a New Species of Stylonurus from the Catskill Group. Thirty-sixth Ann. Rept. N. Y. State Mus. Nat. Hist., 1884.

4. Hall, J.—Note on the Eurypteridæ of the Devonian and Carboniferous Formations of Pennsylvania. Second Geol. Surv. Penn., PPP, 1884.

5. Hall, J.—Note to Explanation of Plate 15 of paper on Geology of Yates County, N. Y., by B. H. Wright. Thirty-fifth Ann. Rept. N. Y. State Mus. Nat. Hist., published 1884.

6. Hall, J.—Note on the Eurypteridæ of the Devonian and Carboniferous formations of Pennsylvania; with a supplementary note on Stylonurus excelsior. Proc. Am. Assoc. Adv. Sci., vol. xxxiii, Philadelphia meeting (held September, 1884), 1885.

7. Hall, J., and Clarke, J. M.—Trilobites and other Crustacea of the Oriskany, Upper Helderberg, Hamilton, Portage, Chemung, and Catskill Groups. Geol. Surv. of the State of New York, Palæontology, vol. vii, 1888.

8. Jones, T. R., and Woodward, H.—Notes on Phyllopodiform Crustaceans, referable to the Genus Echinocaris, from the Palæozoic Rocks. Geological Magazine, n. s., dec. iii, vol. i, September, 1884.

9. Martin, D. S.—A New Eurypterid from the Catskill Group. Trans. N. Y. Acad. Sci., vol. xi, 1882–1883.

10. Woodward, H.—A Monograph of the British Fossil Crustacean belonging to the Order Merostomata. Part I., 1866; Part IV, 1872. Palæontographical Society, vol. xix.

ART. XVI.—*The Iodometric Estimation of Arsenic Acid;* by F. A. GOOCH and JULIA C. MORRIS.

[Contributions from the Kent Laboratory of Yale University—XCVI.]

THE interaction of a soluble arseniate and a soluble iodide in a suitably acidulated solution results, as is well known, in the reduction of the arsenic acid (more or less completely according to conditions of temperature and proportions of reagents and solvents) with the corresponding liberation of two atoms of iodine for every molecule of arsenic acid (H_2O_2AsO) reduced. Inasmuch, however, as the reaction of this process is reversible, it is necessary, in order that the reduction may be complete, to nullify the oxidizing action of the iodine liberated. Theoretically this end may be accomplished in either of two ways, by volatilizing the free iodine bodily or by destroying the oxidizing power of the iodine by converting it to hydriodic acid. The former method was followed in a process devised for the estimation of arsenic acid and elaborated in this laboratory.* This method, as originally put forward, consisted in adding to the solution of the arseniate potassium iodide in excess of the amount theoretically indicated, with 10^{cm^3} of sulphuric acid of half strength, and so arranging the dilution that the total volume of the liquid should be about 100^{cm^3}, boiling until the volume decreased to 40^{cm^3}, bleaching by the cautious addition of sulphurous acid the trace of free iodine still held by the hydriodic acids, diluting, cooling, neutralizing with acid potassium carbonate, and titrating with iodine, after adding the starch indicator. This process, depending upon the removal by volatilization of all but the last traces of liberated iodine and the conversion of this minute residue by sulphurous acid, involves no secondary reactions of a sort likely to influence the main effect. It is exact and fairly rapid.

The method of Williamson,† brought forward more recently, depends upon the conversion of the liberated iodine to hydriodic acid. The interaction at ordinary temperatures of a suitably strong acid, hydrochloric or sulphuric acid, upon the mixture of the arseniate and iodide sets free iodine, and the liberated iodine is converted to hydriodic acid by the action of sodium thiosulphate, the end point being the disappearance of the iodine color.

According to Williamson's directions, 25^{cm^3} portions of the solution of the arseniate are treated with potassium iodide and mixed with an equal volume of hydrochloric acid of sp. gr.

* Gooch and Browning, this Journal, xl (1890), 66.
† Jour. Soc. Dyers and Colorists, 1896, 86–89.

1·16. The precaution is recommended that the strength of the solution of the arseniate shall not exceed the decinormal value, in order that the dilution consequent upon titration by the thiosulphate may not be too great—the reducing action brought about by the action of the strong acid upon the arseniate and iodide being reversible upon the dilution of liquid with water. This procedure thus limits the process to the determination of about 0·18 gram of arsenic acid in 25^{cm^3} of the solution to be treated with an equal volume of hydrochloric acid of sp. gr. 1·16. Obviously, however, the process should, so far as the reduction is concerned, be applicable to larger amounts of arsenic provided the strength of the acid is kept up proportionately. It is essential that the liquid at the end of the titration should contain approximately ten per cent of its mass of absolute hydrochloric acid or about one-third of its volume of the aqueous acid of sp. gr. 1·16.

The arsenic acid is measured either by the amount of standard thiosulphate required to bleach the iodine or by the amount of iodine required to reoxidize the reduced arsenious acid, after neutralizing with acid potassium carbonate. If the former alternative is followed, the end-reaction must be the disappearance of the yellow color of the iodine, since in solutions so strongly acid it is impossible to place dependence upon the starch indicator; in using the latter alternative, the starch indicator is, of course, permissible and preferable.

In the direct titration of the iodine by thiosulphate two sources of error present themselves as possibilities; first, the excessive liberation of iodine by the action of air upon the strongly acidulated iodide ; and second, the liability of the thiosulphate,* if present even in momentary or local excess during the process of titration, to break down under the action of strong acid, thus changing its capacity to convert iodine to hydriodic acid. The latter contingency should be remote in proportion to the caution used in adding the thiosulphate and in keeping the liquid well stirred ; the former must of necessity vary with the acidity of the solution containing the iodide, the time of exposure to atmospheric action, and the degree of contact with the air incidental to stirring. We have thought it desirable, therefore, to see how far each of these possibilities is likely to interfere in the practical conduct of an ordinary analysis.

The effects likely to result simply from the strong acidification of the solution containing potassium iodide and their variation for conditions of dilution representing the beginning and the end of a titration on the lines laid down are shown in

* Norton, this Journal, vii, 287.

the following table. The solution of potassium iodide was diluted as indicated before the addition of the acid and the iodine set free was titrated by thiosulphate.

HCl Sp. gr. 1·16 taken. cm³.	KI taken. gm.	Total volume. cm³.	$Na_2S_2O_3$ added at once. In terms of H_2O_2AsO. gm.	$Na_2S_2O_3$ added after 5 minutes. In terms of H_2O_2AsO. gm.	$Na_2S_2O_3$ added after stirring 5 minutes. In terms of H_2O_2AsO. gm.
25	2	50	0·0013		
25	2	75	0·0004		
25	2	50		0·0035	
25	2	75		0·0019	
25	2	50			0·0042
25	2	75			0·0021
50	2	100	0·0017		
50	2	150	0·0004		
50	2	100		0·0035	
50	2	150		0·0019	
50	2	100			0·0035
50	2	150			0·0014

The proportionate strength of acid and the time before titration are, obviously, the essential factors. The absolute amount of acid present and the stirring seem to make little difference.

As to the action of the hydrochloric acid on small amounts of the thiosulphate, we have the evidence of the experiments detailed in the following statements, in which 1, 2, and 5^{cm³} of nearly $\frac{n}{10}$ thiosulphate are exposed to the action of 25^{cm³} hydrochloric acid, sp. gr. 1·16, without dilution or diluted with

HCl Sp. gr. 1·16. cm³.	Volume before titration. cm³.	$Na_2S_2O_3$ nearly $\frac{n}{10}$ in terms of H_2O_2AsO. cm³. gram.		Iodine to color without dilution in terms of H_2O_2AsO. gram.	Error of titration without dilution in terms of H_2O_2AsO. gram.	Iodine to color after diluting to 75^{cm³} in terms of H_2O_2AsO. gram.	Error of titration after dilution in terms of H_2O_2AsO. gram.
25	26	1	0·0071	0·0062	—0·0009	0·0071	0·0000
25	50	1	0·0071	0·0071	0·0000	0·0071	0·0000
25*	50	1	0·0071	0·0079	+0·0008	0·0079	+0·0008
25	50	2	0·0141	0·0146	+0·0005	0·0146	+0·0005
25*	50	2	0·0141	0·0157	+0·0016	0·0157	+0·0016
25	30	5	0·0353	0·0336	—0·0017	0·0374	+0·0024
25	50	5	0·0353	0·0359	+0·0006	0·0359	+0·0006
25*	50	5	0·0353	0·0411	+0·0058	0·0411	+0·0058

* In these experiments the acid stood in contact with the thiosulphate 5 minutes before titration.

an equal volume of water, were titrated with nearly $\frac{n}{10}$ iodine.

The condition of acidity when the volume of 50^{cm^3} contains 25^{cm^3} of hydrochloric acid, sp. gr. $1\cdot16$, is that of the beginning of titration of Williamson's process. In order that the effect of error due to such action upon the determination of arsenic acid may appear immediately, the thiosulphate and iodine used are expressed in terms of that acid.

These two sources of error, the one due to a liberation of iodine and the other due to decomposition of the thiosulphate, would naturally tend to overcome one another, but the completeness of such neutralization would naturally be largely a matter of chance in the varying conditions of actual analysis. The experiments of the following table, however, in which $\frac{n}{10}$ thiosulphate, to the amount of 1, 2, and 5^{cm^3}, was added to the liquid, 50^{cm^3} and 75^{cm^3}, containing 25^{cm^3} acid, and titrated with iodine at once, and after five minutes, were made to test the matter for the conditions of dilution at the beginning and at the end of a titration.

HCl Sp. gr. $1\cdot16$. cm^3.	KI. gm.	Volume. cm^3.	$Na_2S_2O_3$ nearly $\frac{n}{10}$ in terms of H_3O_3AsO. cm^3.	Iodine in terms of H_3O_3AsO, at once. gram.	Iodine in terms of H_3O_3AsO, after 5 min. gram.	Error in terms of H_3O_3AsO. gram.
25	2	50	1	$0\cdot0071$	$0\cdot0057$		$-0\cdot0014$
25	2	75	1	$0\cdot0071$	$0\cdot0071$		$0\cdot0000$
25	2	50	2	$0\cdot0141$	$0\cdot0131$		$-0\cdot0010$
25	2	75	2	$0\cdot0141$	$0\cdot0143$		$+0\cdot0002$
25	2	50	5	$0\cdot0353$	$0\cdot0322$		$-0\cdot0021$
25	2	75	5	$0\cdot0353$	$0\cdot0357$		$+0\cdot0004$
25	2	50	1	$0\cdot0071$		$0\cdot0028$	$-0\cdot0043$
25	2	75	1	$0\cdot0071$		$0\cdot0067$	$-0\cdot0004$
25	2	50	2	$0\cdot0141$		$0\cdot0116$	$-0\cdot0025$
25	2	75	2	$0\cdot0141$		$0\cdot0139$	$-0\cdot0002$
25	2	50	5	$0\cdot0353$		$0\cdot0314$	$-0\cdot0041$
25	2	75	5	$0\cdot0353$		$0\cdot0361$	$+0\cdot0008$

It is clear that under the conditions covered by the experiments of the two preceding tables the decomposition of the thiosulphate is likely to occur in greater or less degree, and that when the acid of sp. gr. $1\cdot16$ is not much diluted, the products of decomposition are not oxidized by the iodine completely. The latter observation is quite in harmony with the fact that sulphur dioxide bleaches iodine in strong hydrochloric acid only slowly and incompletely. In such cases dilution favors further action of the iodine, but results obtained by titration with iodine in the acid solution diluted with an equal amount of water are unmodified by further dilution.

In the following tables are recorded actual determinations of arsenic according to Williamson's process. To each 25$^{cm^3}$ of the arseniate were added 1, 2, or 3 grams of potassium iodide and 25$^{cm^3}$ hydrochloric acid, sp. gr. 1·16. The iodine was bleached by nearly decinormal thiosulphate without addition of the starch indicator, which loses all delicacy in the presence of strong acid. The time occupied by each titration was about five minutes. The standards of the arseniate were determined by the vaporization process,[*] the pur ty of reagents employed in that process having been proved by trying the process in the estimation of a solution of arsenic acid made by oxidizing pure decinormal arsenious acid by iodine.

HCl cm³.	KI. gram.	Volume at beginning of titration. cm³.	Volume at end of titration. cm³.	H₂KAsO₄ in terms of H₃O₂AsO. gram.	H₃O₂AsO found. gram.	Error. gram.
25	2	50	51	0·0062	0·0085	+0·0023
25	2	50	52	0·0125	0·0156	+0·0031
25	2	50	55	0·0312	0·0350	+0·0038
25	2	50	55	0·0624	0·0666	+0·0042
25	2	50	73	0·1559	0·1588	+0·0029
25	2	50	73	0·1559	0·1587	+0·0028
25	2	50	73	0·1559	0·1591	+0·0032
25	2	50	73	0·1559	0·1595	+0·0036
25	3	50	73	0·1559	0·1595	+0·0036
25	1	50	73	0·1559	0·1581	+0·0022
25	2	50	73	0·1559	0·1581	+0·0022
25	2	50	73	0·1559	0·1588	+0·0029

The range of error in these results is from +0·0023 gram to +0·0042 gram with a mean of +0·0031 gram—not very different from what might be expected from the effect of the interaction of the strong hydrochloric acid and the iodide alone. The counter-effect due to the decomposition of the thiosulphate is not large, yet it is probably real, as will appear in the sequel. In the following series of determinations, made with new solutions and new standards throughout, the arsenic acid was determined, first, by titrating the iodine set free by 25$^{cm^3}$ of hydrochloric acid, sp. gr. 1·16 and 3 grams potassium iodide, the solution having a total volume of 50$^{cm^3}$ at beginning and of 75$^{cm^3}$ at the end of titration and, secondly, the arsenious acid produced in the first reaction was titrated, after being neutralized with acid potassium carbonate by iodine in the presence of the starch indicator.

[*] Gooch and Browning, loc. cit.

H_2KAsO_4 taken. in terms of H_2O_3AsO. gram.	H_2O_3AsO found by the thiosulphate. gram.	Error. gram.	H_2O_3AsO found by titration of H_2O_3As with iodine. gram.	Error. gram.
0·1767	0·1798	+0·0031	0·1776	+0·0009
0·1767	0·1798	+0·0031	0·1777	+0·0010
0·1767	0·1795	+0·0028	0·1785	+0·0018
0·1767	0·1793	+0·0026	0·1785	+0·0018
0·1767	0·1794	+0·0027	0·1780	+0·0013
0·1767	0·1798	+0·0031	0·1785	+0·0018

The average error of the first operation is 0·0029 gram, not far from that of the previous series; the error of the second operation, the titration of the arsenious acid, amounts on the average to 0·0014 gram. In the second operation the error due to over-use of the thiosulphate by iodine set free outside the main reaction is obviously eliminated. The tetrathionate present after neutralization with acid potassium carbonate is unaffected by iodine, as we have found by titrating $25^{cm^3} \frac{n}{10}$ iodine mixed with 25^{cm^3} hydrochloric acid, sp. gr. 1·16, by the thiosulphate, neutralizing with acid potassium carbonate,* adding starch and getting the starch blue with a single drop of $\frac{n}{10}$ iodine. The average error of this process, therefore, 0·0014, is probably due to the products of decomposition of the thiosulphate in the first operation.

From the foregoing experiments it is clear that an arbitrary correction of about 0·0030 gram must be deducted from the indications of Williamson's process of direct titration by thiosulphate, made with the greatest care under the conditions mentioned; and that a correction varying from one-half that amount (0·0015 gram) to nothing (according to the amount of arsenious acid present) when the determination is made by iodine after neutralization with acid potassium carbonate. After making these arbitrary corrections in the results of the preceding table, the individual variations fall within reasonable limits.

On the other hand, the vaporization process, in which the arseniate is reduced by boiling with sulphuric acid and potassium iodide in the manner described,† gives indications reason-

---

* It is worthy of note, that, as we have found by experience, it is not possible to substitute an alkaline hydroxide for the carbonate in the early stages of the process of neutralization, on account of the decomposing effect of the former reagent upon the tetrathionate. This effect is in proportion to the heating of the solution, but is never wholly absent even when ice is intermixed with the liquid and the greatest care taken to prevent a rise of temperature.

† Loc. cit.

ably regular and accurate without the application of an arbitrary correction. This process, moreover, may be shortened by restricting the volume at which heating begins so that the boiling need not be extended beyond five or six minutes. According to this slight modification, the solution of the arseniate is heated in an Erlenmeyer flask with potassium iodide to an amount about 0·5 gram in excess of the amount theoretically required and 10^{cm^3} of sulphuric acid of half strength in a total volume of between 50^{cm^3} and 75^{cm^3}. The liquid is boiled till the iodine vapors are no longer visible in the flask above the liquid, the iodine color in the still hot liquid is bleached by the cautious addition of sulphurous acid, the whole is diluted with cold water, and cooled quickly. The solution is nearly neutralized with potassium hydroxide and the neutralization is completed with acid potassium carbonate. The reduced acid is titrated with iodine after adding the starch indicator. By this procedure the results of the following table were obtained.

Volume. cm^3.	H_2O_3AsO taken. gram.	H_2O_2AsO found. gram.	Error. gram.
35	0·1559	0·1559	0·0000
35	0·1559	0·1560	+0·0001
40	0·1559	0·1559	0·0000
65	0·1559	0·1559	0·0000
50	0·2495	0·2499	+0·0004
50	0·2557	0·2449	−0·0008
60	0·3119	0·3117	−0·0002
60	0·3119	0·3120	+0·0001
75	0·3119	0·3124	+0·0005
75	0·3119	0·3132	+0·0013
75	0·3119	0·3121	+0·0002
75	0·3119	0·3115	−0·0004
75	0·3119	0·3124	+0·0005

ART. XVII.—*Further Notes on Preglacial Drainage in Michigan;* by E. H. MUDGE.

THE most complete study and discussion of the surface geology of Michigan that has recently been published is contained in No. 30 of the Water Supply and Irrigation Papers of the U. S. Geological Survey, under the title, "Water Resources of the Lower Peninsula of Michigan," by Alfred C. Lane, State geologist. Besides much information of economical importance, this pamphlet contains a series of very useful maps, an attempt to work out more fully than has heretofore been done the system of Preglacial drainage, and other matters of more special scientific interest. Prof. Lane has probably given more study to the records of deep wells and other similar data than any other man in the state, and concludes that the theory of Dr. Spencer, heretofore adopted by the writer, that in Preglacial times a large stream flowed across the state from west to east at about the locality of the Grand-Saginaw valley, is incorrect. He believes that the principal Preglacial stream had its rise in the vicinity of Saginaw bay, flowing west and northwest across the peninsula to the vicinity of Manistee, on the Lake Michigan shore, a route far to the north of the Grand-Saginaw valley.

This conclusion is based, first, on the fact that a limestone ridge is believed to extend clear across Saginaw bay at a level so high as to render a large drainage channel in that direction extremely improbable; second, on the condition of the rock surface in the vicinity of Manistee, which so far as known is extremely broken and uneven, characteristic of a locality well down the course of a stream rather than near its head; and third, deep borings at scattered intervals along the line find the rock-surface at lower levels than is known elsewhere in the state.

The writer has no preconceived theories which he is bound to maintain, and concedes that Prof. Lane's theory rests on coherent even if somewhat meager evidences. However, the element of speculation is necessarily large, and the problem can scarcely yet be considered definitely settled. I desire to present some new facts bearing on the question, which render it necessary to modify quite materially Prof. Lane's contour map of the rock surface (Plate VI of his pamphlet). As a matter of course, in the present state of knowledge, large areas of the rock surface being totally unknown, such a map must contain a large element of guesswork, and Prof. Lane, I am sure, will welcome any corrections of the same. His map is constructed with contour intervals of 100 feet, the contour lines being arranged in harmony with the theory of a deep central valley in the location above indicated. The valley corresponds to the 300-foot contour (above sea level), the lines rising to the north and south. There is but a brief space to the south between the 300-foot and 400-foot contours; thence to the 500-foot contour the distance is from five to ten miles, and fifteen or twenty miles farther to the 600-foot line. The 600-foot line enters Ionia County near the northwest corner and traverses the county in a southeasterly direction, crossing Grand river a short distance east of Ionia City, at the locality of the outcrop of Carboniferous sandstone, known locally as the Ionia sandstone quarry. This outcrop in its bearing upon another problem has been described more fully in a previous paper.* It is the only outcrop in this part of the state, is located on the flat bottom of the river valley, and has an altitude of about 650 feet, or fifty feet higher than the contour line, which presumably is intended to represent the average altitude of the rock surface in this vicinity.

A comparison of this outcrop with some neighboring localities reveals some surprising irregularities. To the east the rock surface is quite unknown, but at Lyons, three miles east, a well has been sunk to a point a little below the 600-foot level, and another at Pewamo, six miles farther east, nearly to the 500-foot level, without penetrating rock. To the west, however, more definite results have been attained. In the city of Ionia, three miles down the river from the sandstone outcrop, two deep wells were put down two years ago, penetrating rock at about the 550-foot level. As Ionia is located between the 600-foot and 700-foot contour lines on the map, it is seen that the 600-foot line must be removed several miles to the southwest at this point.

But this is not all. Nine miles still farther down the valley the village of Saranac now has in process of construction a well which at last reports was down to a depth of more than 250 feet, the rock surface having been found at a depth of 248 feet. The location of the well has an elevation of about 650 feet, hence the rock surface is at about the 400-foot level. It follows that not only must the 600-foot contour line be shifted so as to include this locality, but also the 500-foot line, which appears on the map 25 miles to the north, while the 400-foot line would pass through Saranac instead of bordering the central valley 30 to 40 miles away. That is, the rock surface at Saranac is only about 100 feet above the bottom of the central valley.

* This Journal, November, 1895.

Probably the most natural conclusion to be derived from the above facts is that a branch of the main valley has its rise somewhere to the south of the Saranac. Prof. Lane's contours indicate a branch valley coming in from the south along the east line of Ionia and Montcalm Counties. As no deep borings are known along this line, it is presumed that such branch valley is simply a probability, based upon little known conditions. It would not require a very great change in the direction of this supposed branch valley to bring it in line with the Saranac boring.

It is not likely that the Saranac well indicates an east and west valley, because the rocks to the east rise rapidly, as shown by the Ionia wells and the sandstone outcrop above mentioned, while to the west in the vicinity of Grand Rapids the rock surface is above the 600-foot level. Neither is it probable that the valley opened to the south. The only other theory is that the main valley had become so extensive in Preglacial times as to include the Saranac location, but this presumes an amount of base-leveling out of harmony with the general situation. It may, therefore, be taken as at least fairly well settled that an important branch of the drainage system, probably not inferior to the eastern division, exists beneath the drift not far from the line indicated. Future borings along this line in Ionia and Montcalm Counties will be watched with interest.

Grand Rapids, Mich.

SCIENTIFIC INTELLIGENCE.

I. CHEMISTRY AND PHYSICS.

The various forms of Sulphur.—A paper by R. Brauns in a
t number of the *Jahrbuch für Mineralogie** discusses with
fullness the various forms assumed by sulphur on crystalliz-
om a state of fusion. Various new points of interest and
·tance are brought out, and the author concludes that there
ight modifications of sulphur to be recognized, as follows :
rthorhombic; (2) monoclinic-prismatic, Mitscherlich; (3)
o-fibrous-radiated, with concentric structure, monoclinic ?
3 of Muthmann); (4) fibrous-radiated, monoclinic (soufre
of Gernez); (5) fibrous-radiated, orthorhombic; (6) tri-
; (7) monoclinic, in six-sided plates (Muthmann); (8) rhom-
lral (Engel).
addition to these he mentions, also, two forms of fluid fused
ar and two or three forms of sulphur vapor, differing in
vapor density. The conclusion is, therefore, that sulphur can
ie twelve states differing in the degree of energy character-
them.
other investigation on a similar subject, published just
e this, is by O. Bütschli, Professor of Zoology at Heidel-
┤ The author has been led, through his investigation of the
as forms of microscopic structure belonging to different
iisms, to extend his study also to inorganic substances. He
discusses the behavior of drops of sulphur produced from
nation : he·shows that a vaporization of the solid sulphur
be noted at 58°, and that at this temperature, as well as at
oint of fusion, minute superfused drops are yielded which
n fluid for a long time, and solidify spontaneously or through
are in the form of doubly refracting spherulites. The micro-
: structure of these is given with minuteness, and a series of
:—one of them colored—shows the various forms observed.
The Diurnal Variation of Atmospheric Electricity.—For
iod of seven years, from 1892 on, observations have been
icted on the summit of the Eiffel Tower in Paris, during the
hs from May to October, having as their object the deter-
:ion of the electrical condition of the atmosphere; continu-
bservations have been also made at the *Bureau central* in
for a slightly longer period. A discussion of these by A.-B.
VEAU shows that they lead to the conclusion that there
in our temperate regions two very different types of diurnal
tion near the surface of the earth. One corresponds to the

lage-Band, xiii, pp. 39–89.
tersuchungen über Mikrostrukturen des erstarrten Schwefels nebst Bemerk-
über Sublimation, Überschmelzung und Übersattigung des Schwefels
iiger anderer Körper. Pages 96, with 6 text figures and 4 plates. Leip-
)0 (Wm. Engelmann).

warm, the other to the cold season. During the summer there is a pronounced minimum in the hot hours of the day, which is more pronounced the nearer the point of observation is to the surface of the earth, and the more it is free from the influence of trees or buildings. There is another minimum in the night, the importance of which varies in an inverse direction; the existence however, of a double daily oscillation is well marked. During the winter the afternoon minimum disappears more or less completely, while that of the night increases; there is thus a single oscillation with a maximum by day and a minimum at night; the hour of the latter corresponds to that of the second minimum of summer, about 4 P. M. Observations at other localities show very much the same phenomena.

It is concluded, in the second place, that the diurnal variation at the summit of the Eiffel Tower during the summer differs entirely from that observed simultaneously at the *Bureau centra* a few hundred meters distant, but offers a close analogy in its single oscillation to the variation observed in winter. Similar results are obtained at the summit of a wooden pylon used to support anemometers at the observatory at Trappes.

The final conclusions reached are to this effect: that the influence of the soil and the maximum in summer (due, according to the suggestion of Peltier, to the water vapor emanating from the surface, and negative like that) is the cause of the perturbation in the diurnal variation; that the general law of this variation is very simple, having a maximum in the day and a minimum between 3:30 and 4:30 A. M.—*Séances Soc. Franç. de Physique*, 1899, p. 91.

3. *Lehrbuch der Photochromie (Photographie der natürlichen Farben)* von WILHELM ZENKER. Neu herausgegeben von Prof. Dr. B. SCHWALBE. Mit einem Bildniss des Verfassers und einer Spectraltafel. Pages 157. Braunschweig, 1900 (Fr. Vieweg u. Sohn.—This volume is for the most part a re-publication of a work by Zenker (1829–1899) issued by the author, in 1868, but thus far little known. The sketch of the author's life by G. Krech, with which the volume opens, shows that he was a remarkable man of original talents, whose contributions to science covered a wide range of subjects. Early interested in the work of E. Becquerel (1848) on color photography, he devoted much time and study to the subject, and his results are given in the volume before us. The chief point which he regarded as being established was this: that silver chloride, particularly the violet silver sub-chloride, takes the same colors as the light rays which act upon it, and shows them both by reflected and transmitted light. This result was explained physically as due to the formation of stationary light-waves which caused the separation of particles of metallic silver in the silver chloride. It will be recognized at once that this is the line in which such brilliant work has since been done by Lippmann, and the relation of the experiments and theoretical discussion of Zenker to the labors of those who have followed him is given in the closing pages of the volume (133–157) by E. Tonn.

4. *On the Joint Transmission of Direct and Alternating Currents.*—In a paper presented before the recent meeting of the American Association, Dr. FREDERICK BEDELL discussed the simultaneous transmission of direct and alternating currents by the same conductor. He shows that each current acts as though it had the whole conductor to itself and the other current were absent. Further there results a saving in weight of copper and copper losses, which saving, as shown by calculation, may amount to as much as fifty per cent.

II. GEOLOGY AND MINERALOGY.

1. *Geology of the Narragansett Basin;* by N. S. SHALER, J. B. WOODWORTH and A. F. FOERSTE. Monograph XXXIII, United States Geological Survey, pls. i–xxxi, figs. 1–30, pp. v–xx, 1–402, Washington, 1899.—The investigations reported in this Monograph were begun as early as 1865 by Professor Shaler, at first in connection with his university classes in geology. Ten years ago the completion of the work for publication as a monograph was undertaken in connection with the United States Geological Survey, and has been completed with the assistance of Mr. Foerste, working in the southern section, and Mr. Woodworth, in the northern portion of the field. The study of the construction of the basin has led to one important conclusion, which requires special notice, which may be best expressed by quoting the words of Professor Shaler: "The judgment as to the nature of the mountain-building work rests in part upon observations—in the main unpublished—which I have made in other somewhat similar basins that lie along the Atlantic coast from Newfoundland to North Carolina. The general proposition that the basins are characteristically old river-valleys which have been depressed below the sea level, filled with sediments —the sedimentation increasing the depth of the depression—and afterwards corrugated by the mountain-building forces, will derive its verification in part, if at all, from the study of other troughs of the Atlantic coast. It may, however, fairly be claimed that the facts set forth in this memoir show that this succession of actions has taken place in the Narragansett field."

Mr. J. B. Woodworth has written a report upon the northern and eastern portion of the basin (pp. 99–214). In his report the pre-Carboniferous rocks recognized are classified as follows:

Algonkian Period
    Blackstone series
        Cumberland quartzites
        Ashton schists
        Smithfield limestones.

Rocks of Cambrian age are recognized in drift pebbles, but not in place in the northern basin.

The rocks of the Carboniferous system are classified under the following names: from below upward:

Pondville group	100 ft.
Millers River conglomerates	
Wamsutta group	1,000 ft.
Wamsutta conglomerates	
Attleboro sandstone	
Wamsutta slates and shales.	

These rocks are regarded as below the Coal Measures which begin with the following :

Rhode Island Coal Measures	10,000 ft.
Cranston beds	
Pawtucket shales	
Sockanasset sandstones	
Mansfield beds	
Tenmile River beds	
Seekonk sandstones	
Westville shales.	
Dighton group	1,000–1,500 ft.
Seekonk conglomerate	
Rocky Woods conglomerate.	

The classification of the rocks of the southern field is made and reported on by Mr. Foerste (pp. 223–393). His basal conglomerate and arkose are the equivalent of Mr. Woodworth's Pondville group. The Wamsutta beds are not traceable south of Providence. The Kingstown series of Dr. Foerste corresponds in part to the Mansfield, Cranston, Sockanasset and Pawtucket beds of Woodworth, and Foerste's Aquidneck shales correspond to the Westville, Seekonk and Tenmile River beds of Woodworth. The Dighton formation of the north is called Purgatory conglomerate in the southern field. It is unfortunate that the two parts of a field so small as the Narragansett Basin could not receive the same nomenclature. An insect fauna was found in the Pawtucket shales, which has already been described by Dr. S. H. Scudder. Special reference is made to this paper in the following article. The plants are also chiefly from the Pawtucket shales, and the list of species already reported in this Journal (3d series, vol. xxxvii, p. 229) were identified as equivalent to the flora of the Upper Carboniferous of Pennsylvania, by Lesquereux.　　w.

2. *View of the Carboniferous Fauna of the Narragansett Basin;* by Alpheus S. Packard, Proc. Amer. Acad. Arts and Sci., vol. xxxv, No. 20, fig. 1, pp. 399–405, April, 1900.—The author having recently studied new material, collected by Mr. J. H. Clark, of Providence, from this coal field, enumerates the following animal remains at present known from the Narragansett Coal Basin :

Spirorbis carbonarius.　Pawtucket plant beds.
Impression of an Annelid?　Pawtucket plant bed.
Impression of a plant or worm?　South Attleboro, Mass.
Sections of worm holes.　One mile south of East Attleboro.
Anthracomya arenacea (Dawson) Hind.

Track of a gastropod mollusc? Pawtucket plant beds.

Protichnites narragansettensis, n. sp. Pebble, dark arenaceous shales, North Providence.

Remains of a Crustacean? Black shales, Valley Falls.

Ostrakichnites carbonarius (*Protichnites carbonarius*) Dawson. Boulder, red shale, South Attleboro.

Anthracomartus woodruffi Scudd. Pawtucket plant beds.

Mylacris packardii Scudd. Bristol plant beds.

Etoblattina, 9 species, Scudd. Pawtucket plant beds, etc.

Gerablattina scapularis Scudd. Pawtucket plant beds.

Gerablattina fraterna Scudd. Silver Spring, East Providence.

Rhapidiopsis diversipenna Scudd. Cranston plant beds.

Paralogus œschnoides Scudd. Silver Spring, East Providence.

"The presence of the Spirorbis and of the tracks of two marine Arthropods suggest that the Rhode Island plant beds, even if in general of fresh-water origin, were deposited where the sea had access to them. The presence of these marine fossils, with the fresh-water naiad, *Anthracomya arenacea,* strongly suggests that the horizon of the black shales of Providence and also of the red and greenish beds of Attleboro, Mass., belong to the same horizon as those of the South Joggins of Nova Scotia, which is Upper Carboniferous, the rocks there consisting of sandstones and dark carbonaceous shales, frequently becoming reddish. The South Joggins shales also contain the remains of Anthrapalæmon, which should be looked for in the Narragansett Coal Measures. Thus far, then, the animal remains confirm Lesquereux's reference of the dark plant-beds to the Upper Coal Measures.

These beds also appear to be higher in the series than the Middle Carboniferous Mazon Creek beds of Illinois, which contain a larger number of marine animals, viz: Belinuridæ (Euproöps, Prestwichia and Belinurus), besides Anthrapalæmon and Acanthotelson, together with the impressions of marine annelid worms."
w.

3. *Geological Survey of Canada. Summary Report of the Geological Survey Department for the year* 1899; by GEORGE M. DAWSON, Deputy Head and Director, pp. 1–224, Ottawa, 1900. —The Canadian Survey was conducted with its characteristic energy during the year 1899. The finding of gold in the Yukon District, and the search for petroleum on the Saskatchewan has given these regions special importance, but activity has been shown in all parts of the country. Sixteen field parties were engaged during the summer, viz: in British Columbia, 3; Yukon District, 1; Great Slave Lake, 1; Alberta (boring operations), 1; Saskatchewan, 1; Ontario, 3; Ontario and Quebec, 1; New Brunswick, 2; Nova Scotia, 2; Ungava (East coast of Hudson Bay), 1; while several of the staff were engaged on special investigation upon collections, or materials, requiring laboratory study.

The investigations in the Yukon District, under direction of

Mr. McConnell, have been carried to considerable detail on account of the economic importance of the field. Gold is found in rich quantities in the stream-gravels, and "the product of a few of the 500-foot claims on Eldorado and Bonanza Creeks will exceed a million dollars each; while a considerable number on the same two creeks (in fact, the majority of the lower Eldorado claims and a few on Hunker Creek) will yield over half a million each, and claims running from a quarter to half a million are common on all these creeks and also on Dominion and Sulphur Creeks. Assuming a quarter of a million as the average, and that three-quarters of the claims in the district given above are rich enough to work, the total value approaches $95,000,000, a figure which is well within the mark."

The borings in Northern Alberta, in attempts to reach the petroleum-bearing strata at the base of the Cretaceous, have still failed to reach the "tar-sands," which, it is estimated, lie at a depth of about 2000 feet in the Victoria region. Only 1840 feet have been penetrated, owing to the extreme difficulties of holding the bore-hole open at great depth. The soft and incoherent character of the great mass of the overlying rocks requires casing to be carried on *pari passu* with the drilling. It is estimated that the Victoria bore-hole is down to within 250 feet of the "Tar-sands." At Athabasca Landing, the bore-hole reached to very near the top of the Tar-sands. At the Pelican River locality the Tar-sands were reached, and penetrated 87 feet before the gas and tar closed the hole and the working. It is believed that the underlying Devonian limestones are the natural source of the petroleum or maltha accumulated in the "Tar-sands."

Gold was discovered in the alluviums of Serpentine River, New Brunswick, but in small quantities. In York County, New Brunswick, Silurian fossils have been discovered, indicating the probable inclusion of Silurian rocks in folds of older Cambro-Silurian rocks which are more or less altered by metamorphism. Gold-bearing rocks were exploited in several localities in Nova Scotia. The Plant beds of Harrington River, Nova Scotia, were shown to be of Devonian age. Upper Cambrian fossils were discovered in the older Paleozoic rocks of Cape Breton Island by G. F. Matthew. Trenton fossils from Akpatok Island, Ungava Bay, already reported by Dr. Bell in this Journal (June, 1899), are referred to. Marine fossils of the Windsor series from Cumberland County, Nova Scotia, were brought to light by Dr. Ami. Discussion of the age of the Riversdale, Horton and Plant beds of St. John of New Brunswick and Nova Scotia, tend to confirm their Carboniferous affinities. The fish of McArra's Brook, Antigonish County, N. S., are correlated with the Old Red hornstones of Hereford district of England, above the passage beds, by A. Smith Woodward. The appropriation used by the Survey was a little over one hundred and eighteen thousand dollars.　　　　　　　　　　　　　　　　　　　　**w.**

9. *Fossil Flora of the Lower Coal Measures of Missouri;* by

DAVID WHITE, Monograph, vol. xxxvii, U. S. Geological Survey, pp. 1–467, plates i–lxxiii, 1899.—The material upon which this monograph is based was collected from the Coal Measures of Henry County, Missouri, by Dr. J. H. Britts of Clinton, Mo., and by Mr. Gilbert VanIngen and Dr. W. P. Jenney of the U. S. Geological Survey. The whole flora has been exhaustively studied, many new species described, and the old species subjected to careful criticism and revision.

Comparison of the flora with those of other regions in the United States and other countries has led to the following determinations of correlation of horizon for the Missouri flora, viz: "The Lower Coal Measures of Missouri, as represented by the coals of Henry County, were laid down soon after the Morris coal of Illinois, though probably earlier than the Upper Kittanning of western Pennsylvania, or very likely about the time of the formation of the D coal in the Northern Anthracite field."

In the European sections, close affinity is found between this flora and that of the zone of Bully-Grenay in the Valenciennes Basin, but as the author remarks, the fauna is perhaps in a measure transitional, "while it is probably contemporaneous with a portion at least of the upper zone of the Valenciennes Basin, as presented in the basins of Commentry or the Saar, . . . it may represent a slight paleontological transgression on the Stephanian ('Houiller supérieur')." It is represented in Silesia and Bohemia by the Schatzlar and Radnitzer Schichten, and corresponds with the "Transition Series" of Great Britain.   w.

5. *La Face de la Terre* (*Das Antlitz der Erde*) par ED. SUESS, translated by EMMANUEL DE MARGERIE, vol. ii, plts. 1 and 2, figs. 1–128, pp. 1–878. Paris, 1900 (Armand Colin & Cie).— Those familiar with the original edition of Professor Suess' Antlitz der Erde will not need to be reminded of the extreme value of the French edition, the second volume of which has recently appeared. It furnishes a store-house of information regarding the structure of the land surfaces and the geological structure of the earth not met with elsewhere, except as distributed in a very large number of publications. The first volume has already been noticed in this Journal (see vol. v, p. 152). In the translation, M. de Margerie has had the assistance of MM. Bernard, Depéret, Kilian, Poirault, Six and Zimmerman. The translations are admirably made, and the editors have added 85 new cuts to the French edition. These are many of them charts, and add much to the value of the volume. The translators have also added, in foot-notes, a large number of new references to literature, bringing the bibliography up to 1899, which will increase the value of the book for investigators. The present volume discusses the oceans and the geological conditions of the land and ocean surfaces during the Paleozoic, the Mesozoic, the Tertiary; and in the latter part of the volume, a number of chapters are devoted to special subjects, such as the Temple of Sérapis, Baltic and North Seas, the Mediterranean during historic period, etc.   w.

6. *The Gneisses, Gabbro-schists and associated Rocks of South-western Minnesota ;* by C. W. HALL. (Bull. U. S. Geol. Surv. No. 157, 1899, pp. 160, pl. xxvii.)—The territory from which the material described in this paper was gathered embraces the valley of the Minnesota River and the southwest corner of the State of that name wherever gneisses are found. In the 120 miles of the river valley there project through its flood plain many exposures of gneisses, gabbro-schists, diorites and diabasic effusives. They often fill the entire valley one to two miles wide, rising in a profusion of knobs and hills 50 to 100 feet above the river. These exposures of crystalline rocks are described by districts, with maps of each district, on which they are located. Their geological relations are discussed and an account of their petrography, which appears to be that of well known types, is given with the addition of numerous colored plates.   L. V. P.

7. *Brief Notices of some recently described Minerals.*—MÜLLERITE is a hydrosilicate of ferric iron described by Zambonini from Nontron, France, which is the original locality of the related nontronite (chloropal). It occurs in opaque incrusting masses of a yellowish green color; it is quite soft and has a specific gravity of 1·97. The mean of three analyses gave:

SiO$_2$	Fe$_2$O$_3$	Al$_2$O$_3$	MnO	MgO	H$_2$O
48·82	35·88	4·30	0·63	0·35	9·66=99·64

This corresponds to the formula $Fe_2O_3.3SiO_2 + 2H_2O$, which differs from the accepted composition of nontronite only in having three molecules less water.—*Zeitschr. Kryst.* xxxii, 157.

MELITE is a name given by the same author (l. c., p. 161) to partially investigated hydrosilicate of alumina and ferric iron known from a single specimen, labeled "Allophane, Saalfeld, Thuringia." It occurs in imperfect crystals (?) and stalactitic forms of bluish-brown color and opaque; the hardness is 3 and the specific gravity 2·18. An analysis gave: SiO$_2$ 14·97, Al$_2$O 35·24, Fe$_2$O 14·70, CaO 0·78, H$_2$O 33·75=99·64.

ROBELLAZITE. Briefly mentioned by M. Cumenge as occurring with the carnotite in Colorado. A partial chemical examination by M. Debierne, it is stated, shows that it contains considerable vanadium and also niobium, tantalum and tungsten with alumina, iron and manganese as bases.—*Bull. Soc. Min.,* xxiii, 17, 1900.

CUPRO-GOSLARITE. A cupriferous variety of goslarite described by A. F. Rogers as forming a light greenish blue incrustation in an abandoned zinc mine at Galena, Kansas. An analysis gave: SO$_2$ [27·02], ZnO 23·83, CuO 6·68, FeO 0·13, H$_2$O 41·76, insol. 0·58=100.—*Kansas Univ. Q.,* viii, No. 2.

CUBOSILICITE is a name given by Bombicci to the bright blue form of silica in cubic crystals occurring at Tresztyan, Transylvania, and ordinarily regarded as pseudomorphous chalcedony, perhaps after fluorite. This he classifies as a definite form of silica (pseudo-isometric, mimetic) related to melanophlogite, sulfuricine and cristobalite.—*Mem. Accad. Bologna,* viii.

III. Miscellaneous Scientific Intelligence.

1. *American Association for the Advancement of Science—New York Meeting.*—The forty-ninth meeting of the American Association was held in New York City, during the week from June 25 to 30. The President of the meeting was Professor R. S. Woodward. The attendance was satisfactorily large, though perhaps not so numerous as the time and place of meeting might have seemed to promise; the total registration was about 450. In scientific spirit, however, the occasion was highly successful, the number of the papers presented being large and their character and interest above the average. The sessions of the Association were held in the buildings of Columbia University and the efforts of the officers of this institution and of those of the American Museum of Natural History contributed largely to the general success of the meeting.

The retiring President, Mr. G. K. Gilbert, delivered on Tuesday evening an admirable address upon the subject " Rhythms and Geologic Time." Other addresses were delivered by the Vice Presidents of several of the sections, as follows: Section A, on the teaching of Astronomy in the United States, by Asaph Hall, Jr. ; Section B, on the Cathode Rays and some related phenomena, by Ernest Merritt; Section C, on the eighth group of the Periodic System and some of its problems, by Jas. Lewis Howe ; Section D, on Some Twentieth Century problems, by William Trelease ; Section E, on Precambrian sediments in the Adirondacks, by J. F. Kemp. These addresses are published in full in Science (issues of June 29 and following). In addition to the work before the different sections, fifteen affiliated societies had meetings in connection with the Association, and this fact added much to the scientific interest of the occasion.

The place selected for the meeting of 1901 is Denver, Colorado ; the meeting will begin on Monday, August 26th. Pittsburg was recommended for the meeting of 1902. The officers elected for next year are as follows :

President : Charles S. Minot, Harvard Medical School.

Vice-Presidents : Section A, James McMahon, Cornell University; Section B, D. D. Brace, University of Nebraska; Section C, John H. Long, Northwestern University; Section D, H. S. Jacoby, Cornell University; Section E, C. R. Van Hise, University of Wisconsin; Section F, D. S. Jordan, Leland Stanford University; Section G, B. T. Galloway, U. S. Department of Agriculture, Washington ; Section H, J. W. Fewkes, Bureau of Ethnology, Washington; Section I, John Hyde, Department of Agriculture, Washington.

Permanent Secretary, L. O. Howard of Washington; General Secretary, William Hallock, Columbia University; Treasurer, R. S. Woodward, Columbia University.

2. *British Association for the Advancement of Science.*—The coming meeting of the British Association will be held at Brad-

ford, England, commencing Sept. 5. The President is Professor Sir William Turner of Edinburgh.

3. *Catalogues of the Collections in the British Museum of Natural History.*—Recent issues of this series include the following :

The Cretaceous Bryozoa, volume i, pp. xiv, 457 with 17 plates; by J. W. GREGORY, London, 1899. This follows an earlier work (1896) on the Jurassic Bryozoa, in which there was given an Introduction upon the structure and affinities of the group. The second volume upon the Cretaceous Bryozoa is promised for the present year.

Catalogue of the Arctiadæ (Nolinæ, Lithosianæ) ; by Sir GEORGE F. HAMPSON. Pp. xx, 589; plates xviii–xxxv. London, 1900. This is the second of the volumes devoted to the Lepidoptera Phalenæ.

4. *The Norwegian North-Atlantic Expedition, 1876–1878.*— The following publications have recently been issued, containing the results of the further study of the zoological collections made by the Norwegian North Atlantic expedition.

XXV, Thalamophora by Hans Kiær, with 1 plate and 1 map.

XXVI, Hydroida by Kristine Bonnevie, with 3 figures, 8 plates and 1 map.

XXVII, Polyzoa by O. Nordgaard, with 1 plate and 1 map.

5. *The Grammar of Science;* by KARL PEARSON, 2d ed. revised and enlarged, pp. 1–548, figs. 1–33 (Adam and Charles Black, London; Macmillan Company, New York), 1900.—This book illustrates the undoubtedly strong trend of opinion among leaders of scientific thought toward some form of idealism. The inadequacy of the cruder forms of materialism to satisfy the questioning of the thinking mind has led men to look inward to the form of their conceptions of phenomena, in order to ascertain the relation these phenomena bear to each other.

The following passage expresses tersely the author's purpose : "The object of the present work is to insist . . . that science is in reality a classification and analysis of the contents of the mind; and the scientific method consists in drawing just comparisons and inferences from the stored impresses of past sense-impressions, and from the conceptions based upon them. Not till the immediate sense-impression has reached the level of a conception, or at least a perception, does it become material for science. In truth, the field of science is much more consciousness than an external world. In thus vindicating for science its mission as interpreter of conceptions rather than as investigator of a 'natural law' ruling an 'external world of material,' I must remind the reader that science still considers the whole contents of the mind to be ultimately based on sense-impressions " (p. 52). The volume is full of keen observations and suggestions. The chief additions made in this edition are the chapters discussing the biological conceptions of science.   w.

THE

AMERICAN JOURNAL OF SCIENCE

[FOURTH SERIES.]

————◆◆◆————

ART. XVIII.—*On the Gas Thermometer at High Temperatures ;* by LUDWIG HOLBORN and ARTHUR L. DAY.

Second Paper.

[Communication from the Physikalisch-Technische Reichsanstalt, Charlottenburg, Germany.]

IN a former paper[*] we described our measurements with the gas thermometer in some detail. The investigation was concerned principally with establishing the conditions under which this instrument may be used with certainty as a standard at high temperatures. We have shown that this is possible using pure nitrogen as the expanding gas in a bulb of platin-iridium, this material being much superior to the porcelain so long in use for the purpose, both in the accuracy of the results obtainable and the convenience in handling.

Since then the work has been continued in the direction indicated at the close of the previous paper. In order to obtain the full advantages of the exactness which the use of the platin-iridium bulb has rendered possible, the first step was to measure the undetermined coefficient of expansion of this metal at high temperatures. This was very necessary because the correction which the observations with the gas thermometer require on account of the expansion of the bulb increases more rapidly than the temperature to be measured. With the platin-iridium bulb, for instance, it amounts to 10° at 500°, 30° at 1000° and 40° at 1150°, while the increase in the coefficient of expansion with the temperature, which of course is not taken into account in the earlier paper, affects the measurement 1°, 5°, and 7° at the above temperatures respectively.

Secondly, the influence of the pressure which the expanding gas exerts on the glowing walls of the bulb was investigated

[*] Ludwig Holborn and Arthur L. Day, this Journal (IV), viii, 165, 1899.

by a simple method, and finally a second platin-iridium bulb with walls of double the thickness of the first was used in a somewhat different oven in verification of the earlier results. The use of the same thermo-element throughout enabled the observations to be compared with those of the previous series. The thermo-electric force of this element (platinum—platin-rhodium), like the others involving the use of combinations of metals of the platinum group, may be represented by an equation of the second degree within the limits of temperature covered by the investigation (250–1150°). The form of the curve is therefore known when such elements are calibrated at three known temperatures.

To render such calibration independent of the gas-thermometer, a series of melting points of pure metals was determined in which we took advantage of the opportunity to investigate rather carefully the conditions which may affect the melting temperature in certain cases.

To this is further added a chapter on the measurement with thermo-elements, the accuracy obtainable and their lasting qualities.

1. *Determination of the Expansion.*

To determine the expansion of the platin-iridium bulb, a bar of the same alloy (80 Pt, 20 Ir) was obtained, 500^{mm} in length and 5^{mm} in diameter and its linear expansion studied up to 1000°. In connection with this we also undertook the measurement of the expansion of several other substances at high tem-

1

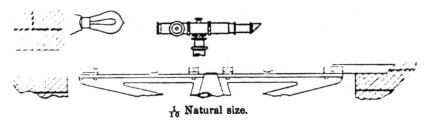

$\frac{1}{10}$ Natural size.

peratures and hope to communicate in detail the full results of the investigation at another time, restricting ourselves for the moment to such as are necessary to the discussion of the gas thermometer. Slight changes which the figures here communicated may undergo in the further investigation of the problem cannot sensibly affect the matter in hand.

The method may be described briefly as follows : The bar is laid for the measurement of its expansion symmetrically in an electrically heated fire-clay tube, 80^{cm} long, in which two lateral holes are bored 10^{mm} in diameter and 485^{mm} apart.

This tube is protected against loss of heat by radiation by two concentric fire-clay jackets with corresponding lateral openings, and the whole placed in a horizontal position upon two brick pillars in such a way that the openings are directed downward. Underneath these, two microscopes are mounted with the aid of which readings can be made of the position of certain marks cut upon a flat polished surface near the ends of the expanding bar.

The heating coil which is wound upon the innermost tube extends for a good distance beyond the ends of the bar so that the temperature opposite the opening differs in maximum only 10 per cent from that at the middle. On account of the increase in the coefficient of expansion with the temperature an average temperature for the bar is not sufficient if considerable local differences exist, hence it is eminently desirable that the differences of temperature along the bar be not too great. We are accordingly making an effort to improve upon our present coil in this direction in the hope that an even greater uniformity in the temperature distribution may be secured.

The microscopes were mounted upon sliding bases upon a long horizontal support and could be fixed in the proper positions with clamping screws. To protect them from the radiation of the oven a diaphragm through which water circulated constantly was introduced between.

With this apparatus the expansion was then measured with the eyepiece micrometers of the microscopes, the reading before heating being verified after the oven had cooled down again. No measurable displacement of the microscopes was observed throughout the series but displacements of the bar occurred now and then at the higher temperatures; symmetrical observations upon both ends, however, eliminate any error from this cause in the results.

The observations were usually made at 250°, 500°, 750° and 1000°, and for the same temperature upon different days agree to within $\pm 0\cdot01^{mm}$, whereby it should be remarked that one whole turn of the micrometer screw corresponds to about 1^{mm}.

The divisions, which were situated in the axis of the bar to diminish the effect of a possible slight bending or warping, were cut with the dividing engine, rather deep but with sharp edges, and rubbed with Parisian red (iron oxide). They were illuminated from below with incandescent lamps. After the temperature has reached a bright red heat where the bar itself gives out light, the divisions still show very sharply by contrast with the polished metal surface as bright marks upon a dark ground.

But with the metals which oxidise and even with platin-iridium which tarnishes before reaching 1000°, it is better to illuminate the glowing rod from outside also. The divisions then appear as dark lines upon a bright ground exactly as at the lower temperatures.

The temperature of the bar thus heated was measured at nine different points with a thermo-element, both wires being insulated with thin porcelain tubes so that the junction could be brought to any desired point along its length.

Platin-iridium (80 Pt, 20 Ir).—The following values were obtained for the expansion λ, of the platin-iridium bar, which together with a similar bar of pure platinum mentioned below, was kindly loaned by the firm of Heräus in Hanau:

| t | λ, (mm) | |
	Observed.	Calculated.
250°	1·033	1·033
500	2·150	2·151
750	3·357	3·355
1000	4·645	4 645

The mean length included between the two groups of five divisions each at the two ends of the bar was 483·07ᵐᵐ at 0°. From these figures the expansion λ of a bar of unit length at 0° may be represented by the formula

$$\lambda = [8198\,t + 1·418\,t^2]\,10^{-8}.$$

Platinum.—We had no bar of the alloy 90 Pt, 10 Ir, of which the second bulb was made, but one of pure platinum was measured under the same conditions and yielded the following results, which we combined with the foregoing and used the mean for correcting the gas thermometer measurements made with the bulb of the lower alloy:

| t | λ, (mm) | |
	Observed.	Calculated.
250°	1·114	1·113 .
500	2·309	2·304
750	3 570	3·571
1000	4·909	4·914

The length at 0° was 483·52ᵐᵐ and the corresponding formula for a unit bar takes the form

$$\lambda = [8889\,t + 1·274\,t^2]\,10^{-8}$$

The difference between the observed expansions and those calculated according to the formula amounts in both cases to less than 0·01^{mm}.

The largest measured difference between the observations upon the cold bars before and after heating to 1000° was 0·025^{mm}; it was generally much smaller.

Porcelain.—The expansion of Berlin porcelain was also measured upon a bar of similar dimensions—unglazed in order that it might afterward be used at still higher temperatures. Plane surfaces were ground upon each end in the same positions as for the metallic bars, and upon these, divisions about 0·15^{mm} wide but very sharply defined were cut upon a lathe with a thin copper disc, five at each end. The readings at the highest temperature were at first somewhat difficult, for the divisions are but little brighter than the background. With illumination from outside, however, excellent readings could still be obtained after the candle power had been increased.

This bar had a mean length at 0° of 483·47^{mm} and yielded the following results for the expansion:

	λ_t (mm)		λ(mm)
t	Observed.	Calculated.	
250°	0·382	0·391	0·790
375	0·628	0·612	1·299
500	0·845	0·850	1·749
625	1·093	1·105	2·261
750	1·307	(1·377)	2·703
875	1·615	(1·666)	3·340
1000	1·977	(1·972)	4·089

Observations were made with the porcelain bar at temperatures between the original four, when it was found that the observed expansions could not be represented by a curve of the second degree. The following formula is approximately true between 250° and 625°,

$$\lambda = [2954\,t + 1\cdot125\,t^2]\,10^{-9}$$

and yields the observed value again at 1000°, while for 750° and 875° the differences are considerably larger than the errors of observation. Below 250° also the values obtained from the formula are too large. The coefficient of expansion clearly increases more rapidly after red heat is reached than below 500°, and for the interval between remains nearly constant. The observed value at 1000° coincides with the calculated value again, which shows this temperature to be the point of intersection of two curves.

It may be further remarked that the various series of observations, after the preliminary heating to remove any tension remaining from the grinding, agree quite as well for porcelain as for the metals. These observations were all made with a rising temperature, however, and it is not impossible that different values would be obtained for the expansion of porcelain if the temperatures observed were approached from the other direction, i. e. in the order 1000°, 875°, 750°, etc. It also remains to be investigated what differences exist between different bars of the same manufacture.

The expansion of Berlin porcelain formed the subject of an earlier investigation by Holborn and Wien,[*] who heated porcelain plates 9^{cm} in length in a gas furnace and measured them at the temperatures 550° and 1050° nearly. The mean value then arrived at, $44 \cdot 10^{-7}$ for the linear expansion coefficient, agrees with the present value at the higher temperature within the limits of accuracy then obtainable; at the lower temperature the earlier value is too large.

More recently Bedford[†] determined the expansion of French porcelain (Bayeux) between 0° and 800°. His formula is

$$\lambda = [3425 t + 1 \cdot 07 t^2] \, 10^{-9}$$

A comparison with our results is not instructive, however, for the expansions of porcelains from entirely different sources are not properly comparable.

Our measurements with the gas thermometer and porcelain bulbs, for the calculation of which the Holborn and Wien value was used, require only a slight correction for the new determination of the expansion, partly because the absolute expansion is very small and partly because the use of too large a value for the bulb expansion between 0° and 100° makes it also necessary to correct (diminish) the coefficient of expansion of the gas about $\frac{1}{700}$. The required correction, therefore, remains less than 1° for the whole range from 500° to 1000°.

Jena Glass 59[III].—For the measurement with the gas thermometer below 500° a bulb of Jena glass 59[III] and a nitre bath were used. We have, therefore, undertaken to measure the expansion of this material up to 500°, but the results which have so far been obtained only serve to emphasize how much depends upon the way the glass is treated. We were able, however, to establish the fact that the mean error of the gas thermometer observations in this region, $\pm 0 \cdot 5°$, is not exceeded in assuming for all temperatures up to 500° the constant value $18 \cdot 10^{-6}$. For more accurate measurements it is certainly advisable to employ the platin-iridium bulb for these temperatures also, the expansion of which is much better defined. In the nitre bath it will be necessary to protect it with a thin glass tube, however, as the bath becomes alkaline with use.

---

* L. Holborn and W. Wien. Wied. Ann., xlvii, 121, 1892.
† T. G. Bedford, Phil. Mag. V, xlix, 90, 1900.

2. *Influence of Pressure on the Gas Thermometer Bulb.*

At the highest temperatures measured the inside pressure in the platin-iridium bulb exceeded that outside by nearly an atmosphere. Although no permanent change in the volume of the bulb after heating could be detected, it seemed, nevertheless, not impossible that a temporary yielding of the glowing walls under the comparatively high pressure might take place. We have investigated this question by redetermining the higher temperatures with a much diminished pressure.

This was accomplished by allowing a portion of the expanding gas to pass over into the closed tube of the manometer, thus varying pressure and volume simultaneously. The expansion-coefficient of the gas suffers a small change thereby which we estimated to be about 0·1 per cent and therefore neglected it.

The closed manometer tube had already been provided with two points (see first paper—loc. cit.) at which readings could be made for volumetric purposes, so that by bringing the mercury tangent to the lower point a definite volume could be added to that of the bulb. To enable a more accurate measurement of the temperature of this supplementary volume to be obtained than was possible with the small thermometer in the cap, a second and more sensitive thermometer was fastened to the tube and the whole carefully packed in a thick layer of cotton batting containing only the two small openings necessary to make the readings above and below.

As a rule the observations were made in sets of three, the first and last with the mercury adjusted to the lower, the second to the upper point. Since it was necessary to wait some time between observations until the temperature conditions in the manometer became uniform, the temperature in the oven had opportunity to change a few degrees. These changes (δ) were measured with the thermo-elements. The supplementary volume between the two points was measured by weighing the mercury required to fill it, the results being as follows:

$$50·630^{cm^3}$$
$$50·609$$
$$50·641$$
$$50·658$$

Mean $50·635^{cm^3}$

The differences between the determinations are largely due to an imperfectly formed meniscus at the lower point, the adjustment being necessarily made with a falling column as the mercury was drawn off through a fine capillary tube. The adjustment with a rising column as used for the temperature measurements gives much more exact results. Also the determination of the distance between the two points, 161·60^{mm} contained variations of only 0·02^{mm} in the individual readings. An error of 0·1 per cent in the determination of the supplementary volume would cause an error of 1° in the temperature at 1000° as determined at the lower point.

For the verification of the method several measurements of the volume of the bulb at 0° were made. For the first platinum bulb the determination by filling with water and weighing yielded 208·49$^{cm^3}$, the volumetric determination 208·45 and 208·51$^{cm^3}$; for the second bulb the results obtained by the two methods were 195·87 and 195·91$^{cm^3}$ respectively.

Table XII contains the temperature measurements at the two different pressures. H and t represent the pressure meas-

TABLE XII.

1899	H (mm.)	t	H' (mm.)	$t'-\delta$	δ	$t-t'$
July 22	871·71	553·0°	524·13 526·38	550·9° 556·6	+2·0° −3·4	+0·1° −0·2
	877·37	558·7	526·80 528·35	557·8 562·0	+1·5 −2·9	−0·6 −0·4
	1339·50	1022·7	663·06 666·71	1015·4 1030·1	+6·3 −7·6	+1·0 +0·2
Aug. 2	848·68	531·1	514·76 517·07	527·9 533·8	+2·5 −2·9	+0·7 +0·2
	855·11	537·4	517·76 519·05	535·5 539·1	+1·6 −1·6	+0·3 −0·1
	1047·86	728·2	581·57 585·81	721·9 734·9	+5·3 −7·3	+1·0 +0·6
	1055·69	736·1	585·80 586·89	734·8 737·9	+1·1 −1·9	+0·2 +0·1
Aug. 3	952·97	633·9	551·09 558·78	629·4 651·4	+3·5 −18·4	+1·0 +0·9
	974·06	654·8	559·11 560·58	652·5 656·7	+1·6 −2·4	+0·7 +0·5
	1242·43	924·5	638·41 639·87	921·4 926·7	+1·5 −3·1	+1·6 +0·9
	1243·62	925·8	639·62 639·63	925·5 925·6	−0·6 −0·6	+0·9 +0·8

ured upon barometer and manometer, and the temperature calculated from it for the readings made at the uppermost point; H' and $t'-\delta$, the corresponding magnitudes for the readings below. If to $t' - \delta$ the change in the oven temperature δ be added, the results contained in the column $t - t'$ or the difference between the results obtained at the two pressures is obtained. The mean value of this difference is $\pm 0.7°$ and exceeds $1.0°$ only in a single instance.

It follows from this that within the given limits of accuracy, the measurements with the gas thermometer equipped with the platin-iridium bulb were not disturbed by the pressure upon the walls of the hot bulb. The initial pressure of the gas at $0°$ was 294.40^{mm}, nearly the same as in the earlier measurements, and its coefficient of expansion 0.003666.

It has been suggested that this method be employed for obtaining the volume of the bulb at high known temperatures. It would then be possible to arrive at its coefficient of expansion in this way. In addition to the usual equation for the calculation of the temperature with constant volume,

$$\frac{HV}{1 + at} = H_0 V_0$$

a second relation is, to be sure, obtainable

$$H'\left(\frac{V}{1 + at} + \frac{V_{\prime}}{1 + at_{\prime}}\right) = H_0 V_0$$

if $t_{\prime}$ be used to represent the temperature of the supplementary volume $V_{\prime}$. The equations are nevertheless not independent of each other and are therefore not sufficient for the determination of the two unknown quantities V and t.

3. *Comparison of the Thermo-element with the Gas Thermometer.*

Second Platin-iridium Bulb.—The second platin-iridium bulb had nearly the same form and size as the first. It was, however, made from the alloy 90 Pt. 10 Ir., and its walls (1^{mm}) were twice as thick as those of the earlier bulb. Its volume at $0°$ was 195.87^{cm^3}.

After being boiled with concentrated nitric acid and several times rinsed with distilled water the bulb was carefully dried and attached to the manometer. Then it was evacuated with a mercury pump and maintained for several hours at a temperature of $1300°$, being "rinsed" from time to time with fresh nitrogen and again evacuated. The gas for the final filling was also admitted at the high temperature and the exact pressure regulated after the bulb had cooled.

The new bulb was used for several series of measurements and proved to be quite as satisfactory as the old. The gas pressure at $0°$ remained constant to within 0.1^{mm} throughout.

The comparisons of the gas thermometer and the thermo-element T_2 which are contained in Table XIII agree among themselves exceedingly well and with the earlier comparisons also. The differences, which amount to about 1·0°, are of the same order of magnitude as in all the other measurements made under the given conditions, the bulb with its extended volume and the minute thermo-electric junction being contained in an air bath. The highest temperatures are most favorable in this respect, as the radiation acts most strongly there to equalize the temperature distribution.

Below 500° where the air bath gave place to nitre, the uniformity is also greater of course.

The conditions under which the second bulb was heated were also considerably varied in order that the substantiation of the earlier results might be as complete as possible. The observations under " October 12 " in the table, for example, were made with the bulb in the same oven which had served for the measurements with the first bulb, the others in a newer and larger one, but with two different heating coils on the different days—both wound logarithmically as described in the former paper. The tubes carrying these new coils were 42^{cm} long and 6^{cm} inside diameter while the older one was only 35^{cm} long and 4·8^{cm} in diameter. The fire-clay jackets were also correspondingly larger.

TABLE XIII.

Platin-iridium Bulb No. II (90Pt, 10Ir) Gas-Nitrogen.

$V_0 = 195\cdot87$cm^3 $V_t = 0\,904$cm^3 $H_0 = 276\cdot35$mm $a = 0\cdot003666$

1899		t	e_2(MV)	Obs.—Calcul.	1899		t	e_2(MV)	Obs.—Calcul.
Oct.	9	562·1°	4759	0·0°	Oct.	12	822·9°	7497	—1·5
		571·7	4861	—0·5			822·5	7496	—1·9
		573·2	4877	—0·6			1093·5	10558	—0·7
		829·5	7568	—1·5			1096·1	10598	—1·5
		1080·9	10423	—1·8			1097·4	10616	—1·7
"	12	552·0	4668	—1·0	"	21	669·1	5836	+1·1
		552·2	4670	—1·0			917·4	8526	—0·2
		552 6	4674	—1·0			1063·3	10189	+0·5
		821·2	7478	—1·5					

In Table XIII, t represents the temperature observed with the gas thermometer, e_2 the thermo-electric force of the element T_2 in microvolts. The last column (Obs.—Calcul.) contains the difference between t and the temperature in degrees calculated from the curve for e_2. This curve will be referred to again further on.

The Observations with the first Platin-iridium Bulb.—The earlier observations (loc. cit. Tab. IX, p. 189) with the first bulb (80 Pt, 20 Ir), which were calculated with a constant coefficient of expansion, can now be corrected for the increase in the expansion with the temperature. The corrected values are contained in Table XIV, the numbers printed in italics in Table IX being here omitted. They were made with a uniformly wound oven coil, and a comparatively large (former paper, Table VIII, p. 189) fall in temperature along the bulb, roughly corrected with the help of the thermo-elements.

TABLE XIV.

Platin-iridium Bulb I (80Pt, 20Ir).   Gas-Nitrogen.

After filling.	t	$e_2(MV)$	Obs.–Calcul.	After filling.	t	$e_2(MV)$	Obs.–Calcul.
2d day	529·1°	442^9	+0·3°	22d day	1052·8°	10073	−0·1°
	614·8	5289	+0·1		1102·5	10653	+0·4
	1008·8	9555	+0·6				
				24th "	541·2	4552	+0·1
20th "	625·6	5401	−0·1		542·8	4569	−0·2
	812·1	7356	+0·6		615·5	5297	0·0
	907·3	8396	+1·2		717·3	6343	+0·7
	1014·8	9622	+0·9		814·3	7385	−0·2
	1113·9	10770	+1·7		919·7	8550	+0·2
					1017·6	9661	+0·1
22d "	511·8	4252	+1·1		1106·6	10700	+0·4
	512·7	4268	+0·2				
	551·5	4655	−0·2	25th "	516·4	4304	+0·2
	659·3	5754	−0·8		616·3	5303	−0·1
	703·8	6217	−0·9		721·3	6382	+0·7
	757·0	6775	−0·5		821·9	7462	+1·6
	859·4	7881	−0·3		913·4	8458	+1·8
	949·2	8890	−0·6		1026·0	9756	+0·5
	1001·5	9496	−1·6		1134·5	11031	+0·7
	1050·4	10044	−0·1				

The temperatures t are raised by the new coefficient of expansion 0·7° at 500°, 4·9° at 1000° and 7·5° at 1150°.

In the recalculation a small correction has also been added for the slight over-compensation of the heating coil (Table VIII)—13 microvolts at 500°, and 5 at 1150°. This corresponds to the measured temperature distribution according to which the junction of T_2 opposite the middle of the bulb reads smaller than the integral temperature—at 620° 1·2° smaller, at 820° 1·0°, at 1010° 0·5° and at 1150° 0·4°.

Observations below 500°.—After the observations in the electric oven between 500° and 1150° were completed, the thermo-element T_2 was compared with the gas thermometer equipped with a bulb of Jena glass 59^{III} between 250° and 500° in a nitre bath. These observations were designed partly to verify the earlier observations below 500°, with which the element T_2 was connected only indirectly by comparison with T_1; and partly to enable us to compare the absolute values obtained from observations in the nitre bath and in the electrically heated air bath.

This nitre bath was considerably smaller than the one used for the earlier observations and was heated electrically, whereby we were enabled to secure constant temperatures more quickly and more exactly. A wrought iron cylinder 12^{cm} in diameter, such as is used for the transportation of mercury, was cut off at a height of 27^{cm} and fitted with a cover 2^{cm} thick, which carried a turbine driven by an electric motor for stirring. The heating coil was of bare constantan wire 1·5^{mm} in diameter wound in two parallel coils upon the cylinder, the insulation being provided by a layer of asbestos board, and held in place by smearing with clay. The whole was enclosed in a larger fire-clay jacket and the space between closed in with asbestos wool, giving a layer of air a centimeter or more thick about the coil. This bath can be used up to 700°.

2.

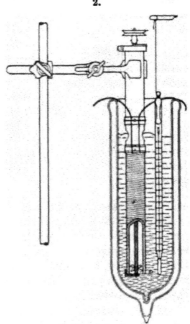

¼ Natural size.

For temperatures below 300° where insulating liquids may be used for the baths, an arrangement which did us exceedingly good service certainly deserves mention in passing. A cylindrical glass vessel, with double walls enclosing a vacuum, such as is commonly used for liquid air (Prof. Dewar's design), was filled with olive oil. Into the oil dipped a porcelain tube with two spiral furrows upon the outer surface, carrying coils of bare constantan wire. Inside the tube a turbine was inserted and held in place above and below with brass clamping rings. The stirrer and heating coil were thus in one piece and could be introduced into any insulating bath at pleasure. A vessel with a vacuum jacket is preferable however, on account of the very small quantity of electrical energy required to heat it, even when the walls are left unsilvered for reading a completely immersed thermometer. With 1·2 l. of oil only 100 watts were necessary to maintain a temperature of 250°. Although in using this apparatus we did not have an accident from explosion of the vessel, still, as this now and then occurs with such vessels at low temperatures without any apparent cause, it is well to make provision against possible personal injury in such an event. Table XV contains the comparison of thermo-element and gas thermometer in the nitre bath. The bulb was filled with nitrogen under a pressure of 471·84^{mm} at 0°.

TABLE XV.

Observations in the Nitre Bath.

	t	e_2 (MV)	Obs.—Calcul.		t	e_2 (MV)	Obs.—Calcul.
Feb. 6	340·7°	2639	−0·8°	Feb. 8	395·8°;	3153	−0·8°
	340·5	2635	−0·5		395·8	3151	−0·6
	387·5	3076	−0·9		492·6	4082	−0·7
	386·9	3071	−1·0		492·8	4084	−0·7
	439·8	3573	−1·0				
	439·8	3572	−0·9	March 29	359·6	2803	+0·4
	501·3	4171	−1·3		359·5	2802	+0·4
	500·9	4168	−1·4		400·5	3195	−0·6
					402·4	3217	−1·0
Feb. 8	267·4	1970	−0·6		498·5	4140	−1·0
	291·6	2191	−0·8		498·3	4139	−1·1
	291·7	2193	−0·9				

4. *Formulae for Thermo-elements.*

In a previous paper[*] we were able to show that the relation between the thermo-electric force and the temperature in metals of the platinum group together with gold and silver could be represented within wide limits with an accuracy of ± 1° by a function of the second degree.

[*] Ludwig Holborn and Arthur L. Day, this Journal (IV), viii, 46, 1899.

The measurements made at that time were upon wires of pure rhodium, iridium, palladium, gold and silver together with alloys of platinum with palladium and ruthenium, whose thermo-electric force referred to pure platinum was measured up to 1300°. The elements with palladium or palladium alloys differed from the others in that the range within which the parabolic formula applies is smaller.

The formula for the normal element T_2 derived from the comparisons with the gas thermometer follows, together with certain of the others; those metals being chosen which are best suited to the measurement of high temperatures and which offer a basis for exterpolation.

These have been compared with the normal element from 50° to 50° approximately and the thermo-electric forces corresponding to the exact temperature intervals as contained in Table XVI, obtained from the observed data by graphical interpolation.

Since the publication of the earlier paper some observations above 1300° have been added with the aid of an oven equipped with a coil of platin-iridium wire.

Normal Element Pt—90 Pt, 10 Rh.—The thermo-electric force e_2 in microvolts of the platinum—platin-rhodium standard element T_2 is given by the following equation from 250° upwards

$$e_2 = -310 + 8{\cdot}048\,t + 0{\cdot}00172\,t^2$$

when the hot junction is maintained at the temperature $t°$ and the cold junction at 0°. The formula is based upon the observations with the platin-iridium bulb above 500° and with the glass bulb and nitre bath below, as contained in Tables XIII, XIV and XV. In the columns "Obs.—Calcul." of these tables will be found the differences between the observed values and those obtained from the formula.

It will be recalled that the element T_1 was compared (first paper, Tables I to VII) above 500° with the gas thermometer equipped with porcelain bulbs. These observations differ among themselves somewhat, due in part to the behavior of the porcelain bulbs and in part to the less perfect conditions of measurement then obtaining. In the normal curve these observations are therefore not included, the curve being based entirely upon the temperatures measured with T_2 and the platin-iridium and glass bulbs. It is nevertheless not without interest to compare the temperatures obtained with T_1 and the porcelain bulbs, with the later ones.

This may be done by referring the normal values for T_1 (Table II former paper) to those of T_2, by adding the differences $T_2 - T_1$ (Table XI, former paper) which were obtained by direct comparison of the two elements, or in other words if T_3 be referred to the observations made with the gas thermometer as used with porcelain bulbs.

In Table XVI in the first column the values of the electromotive force for T_3 calculated from the formula are contained. The column Δ contains the differences between these values and those obtained from $T_1 + (T_3 - T_1)$ as suggested above. If now these values of Δ be compared with the earlier ones (Table XI, former paper) it will be seen that the differences have become very small now that the proper expansion coefficients of both porcelain and platin-iridium have been applied.

From this the conclusion follows that the porcelain bulbs lead to substantially the same results as the p.at.n-iridium as long as the former are not used above 1100°.[1] Under these conditions of course only bulbs glazed inside and out need be used, as the glazing does not melt up to that point.

Regarding the precautions necessary to be observed in the use of porcelain bulbs and the degree of accuracy obtained in temperature measurements with them, reference is made to the former communication.

Element Pt—90 Pt, 10 Ru.—The element platinum—platin-ruthenium follows a curve very similar to that of the normal element. Its equation is

$$e_3 = -359 + 9{\cdot}260\,t + 0{\cdot}00150\,t^2$$

Element Pt—Ir.—For the element platin-iridium a formula has also been calculated which like the others holds from about 300° on. Like the following curves it is based on the thermo-electric force of two pure metals and possesses a much greater curvature then those of the alloys. It is—

$$e_4 = -248 + 7{\cdot}282\,t + 0{\cdot}00554\,t^2$$

Element Pt—Rh.—For the combination platinum-rhodium two specimens of pure rhodium were at our disposal. The first was from Heräus, the second was especially prepared in the *Reichsanstalt.* The two formulae are

$$e_5 = -228 + 7{\cdot}230\,t + 0{\cdot}00660\,t^2$$
$$e_6 = -235 + 7{\cdot}410\,t + 0{\cdot}00660\,t^2$$

The second equation is derived under the assumption that the differences between the two curves e_5 and e_6 can be represented by a linear relation. The observed results between 300° and 1150°—the range of the gas thermometer—are made the basis for the derivation of this as well as of the other formulae.

Table XVI contains the observations upon these elements as well as the differences between them and the values calculated from the above formulae in microvolts and degrees.

TABLE XVI.

t	Pt — 90Pt, 10Rh. Calcul. MV	Δ MV	Δ Degrees	Pt — 90Pt, 10Ru. e₃ Obs. MV	Obs.—Calcul. MV	Obs.—Calcul. Degrees	Pt — Ir e₄ Obs. MV	Obs.—Calcul. MV	Obs.—Calcul. Degrees	Pt — Rh₁ e₅ Obs. MV	Obs.—Calcul. MV	Obs.—Calcul. Degrees	Pt — Rh₂ e₆ Obs. MV	Obs.—Calcul. MV	Obs.—Calcul. Degrees	t
250°	1810	−10	−1.0°	2062	+12	+1.2°	1966	+47	+4.7°	2004	+12	+1.1°	2042	+12	+1.1°	250°
300	2260	+3	+0.3	2550	−4	−0.4	2474	+39	+8.7	2536	+1	+0.1	2580	−2	−0.2	300
350	2718	+9	+1.0	3056	−10	−1.0	3002	+23	+2.1	3112	+1	+0.1	3160	−7	−0.6	350
400	3185	+9	+1.0	3580	−5	−0.5	3550	−1	−0.1	3714	−6	−0.5	3780	−5	−0.4	400
450	3661	+11	+1.1	4112	+1	+0.1	4156	+5	+0.4	4362		0.0	4432	−4	−0.3	450
500	4145	+5	+0.5	4646	0	0.0	4778	+9	0.0	5032	−5	−0.4	5118	−2	−0.2	500
550	4638	+4	+0.4	5184	−4	−0.4	5424	−9	−0.7	5744	−1	−0.1	5834	−3	−0.2	550
600	5139	+3	+0.3	5738	+1	+0.1	6102	−14	−1.0	6484	−2	−0.2	6582	−5	−0.3	600
650	5649	−1	−0.1	6290	−4	−0.4	6810	−16	−1.1	7260	0	0.0	7372	+2	+0.1	650
700	6168	−6	−0.6	6860	−2	−0.2	7564	0	0.0	8086	+9	+0.6	8192	+6	+0.4	700
750	6695	−9	−0.8	7430	0	0.0	8320	−10	−0.7	8906	−1	−0.1	9038	+3	+0.2	750
800	7231	−10	−0.9	8012	+3	+0.3	9116	−7	−0.4	9790	+10	+0.6	9928	+11	+0.6	800
850	7775	−11	−1.0	8596	0	0.0	9948	+4	+0.2	10688	+2	+0.1	10840	+8	+0.4	850
900	8328	−11	−1.0	9196	+6	+0.5	10798	+5	+0.3	11644	+19	+1.0	11802	+22	+1.1	900
950	8890	−9	−0.8	9804	+8	+0.7	11670	0	0.0	12614	+17	+0.9	12782	+21	+1.1	950
1000	9460	−8	−0.7	10410	+9	+0.7	12588	+14	+0.8	13600	−2	−0.1	13774	−1	−0.1	1000
1050	10039	0	0.0	11018	0	0.0	13518	+12	+0.6	14616	−24	−1.1	14800	−22	−1.0	1050
1100	10626	+13	+1.1	11646	+4	+0.4	14482	+16	+0.8	15692	−19	−0.9	15884	−18	−0.8	1100
1150	11222			12274	0	0.0	15462	+9	+0.5	16800	−15	−0.7	16976	−39	−1.7	1150
1200	11827			12918	+5	+0.4	16482	+21	+1.0	17946	−6	−0.3	18130	−31	−1.3	1200
1250	12440			13556	−4	−0.3	17498	−5	−0.2	19106	−17	0.7	19280	−60	−2.5	1250
1300	13063			14208	−6	−0.5	18580	+7	+0.3	20314	−11	−0.5	20492	−60	−2.5	1300
1350	13692			14864	−12	−0.9	19662	−9	−0.4	21556	5	0.2	21728	−69	−2.7	1350
1400	14331			15528	−17	−1.3	20766	−20	−0.9	22800	−30	−1.2	22988	−87	−3.4	1400
1450	14979			16220	−2	−0.2	21904	−35	−1.5	24136	+4	+0.2	24276	−110	−4.1	1450
1500	15635			16914	+8	+0.6	23056	−74	−3.0	25486	+19	+0.7	25684	−46	−1.7	1500

Attention should perhaps be called to the fact that in the
mparisons between these elements and the normal element,
fferent sections of the same curve are separated in particular
ses by considerable intervals of time so that the distribution
temperature from junction to junction was not always the
me even through the observations of a single series. Small
nsistent differences between " Obs." and " Calcul." may be
irly attributed to this cause, though for the temperature
nge 300° to 1150° when the gas thermometer was used, the
fferences rarely exceed 1°; above 1150° where the figures
e based upon an exterpolation of the formula for T_s they are
metimes larger, though nowhere exceeding 5°.

5. *Melting Points of Metals.*

To render the calibration of thermo elements independent of
e gas thermometer we have determined the melting points
various metals lying between 300° and 1100°. Two
ethods were used: According to the first, a short wire
out 1^{cm} long) of the metal to be melted was introduced into
e hot junction of the thermo-element itself, and the thermo-
ctric force at the moment of the melting and consequent
erruption of the circuit, observed. This method is simple
t leads easily to doubtful results if the melting point is at
affected by the surrounding atmosphere, as is the case with
st of the metals which oxidize in air. The element also
comes considerably shortened by the continual cutting off
the drop of melted metal after the observations and renew-
g the junction.

By the second method a larger quantity of metal is melted
a crucible and the thermo-element, protected by light porce-
n tubes, inserted. When the heating is properly regulated
e beginning of the melting or freezing point is readily recog-
able as the temperature remains stationary for a considerable
gth of time. In the following, the words " wire method "
d " crucible method " may serve to distinguish the two proc-
es.

For the latter method the apparatus was arranged as indi-
ed in fig. 3. A porcelain or graphite crucible is contained
hin a short fire-clay tube carrying a coil of bare nickel wire
ose separate turns were insulated by smearing with clay.
e whole is protected against loss of heat by a layer of loose
estos A and a thick protecting tube also of fire-clay. The
es of the thermo element are insulated from each other by
ans of thin porcelain tubes and from the molten metal by a
newhat thicker tube of the same material, 5^{mm} in inside
meter and 1·5^{mm} in thickness, which dipped at least 4^{cm} into

the melting metal (1ᶜᵐ from the bottom of the crucible) and was clamped in position there.

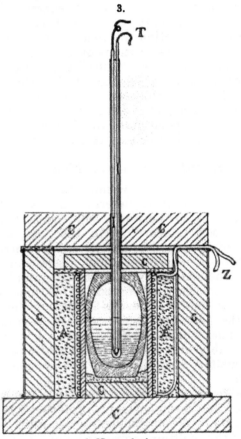

¼ Natural size.

The figure shows the oven with a graphite crucible in position and a second slightly smaller one with a hole bored in the bottom inverted over it as a cover.

The porcelain crucibles were similar in size and form, 1ᵐᵐ thick and without cover.

Inasmuch as the electric heating furnished a much more uniform distribution of temperature than the gas oven and as no systematic differences between the melting and freezing temperatures were observed even with the most varied conditions of current and quantity of metal, we did not attempt to stir the metals in general. Only in one section of the observations with silver an effort was made to expose the melted metal more thoroughly to the air by stirring.

In the preliminary experiments the coil was wound directly upon a crucible of fire clay, but that was soon given up—first because the clay was too porous to hold the melted metal in all cases (molten Cu for instance penetrated the wall and short-circuited a section of the coil), and second, the change from one crucible to another which happened nearly every day was much more convenient with the coil wound upon a separate tube. This oven consumed about 600 watts at 1000°

Some caution was necessary in heating up to the highest melting point (copper), otherwise the oven coil (1^{mm} in diameter) burned out—its renewal is a simple matter however.

For the temperature measurement thermo-elements were used which were cut from the same wire as T, and which have shown themselves to be perfectly identical with it in their results. Temperatures above 500° are derived from the normal curve (Table XVI) and below from the measurements in the nitre bath.

In each determination the thermo-electric force was observed from minute to minute and afterward plotted as a function of the time (mentioned hereafter as the "time curve"). A change of 10 microvolts could be compensated directly on the compensation apparatus by inserting 0·1 ohm. Smaller differences were interpolated from the galvanometer deflection, one scale division corresponding to from 0·5 to 1·4 microvolts according to the resistance in the branch circuit.

The accuracy of the crucible method is only limited by the accuracy with which it is possible to measure with the thermo-element, in regard to which more will be said further on. The wire method is less exact because the interruption of the circuit which indicates the melting point is not independent of any slight tension which may exist in the wires of the thermo-element, particularly the platin-rhodium wire—a factor which varies from one observation to another.

The highest temperatures observed for a particular metal by this method may therefore be expected to come nearest the truth. The inserted wire may even at times be seen to break just before the real melting takes place.

Gold.—The melting point of gold was determined by the wire method only. Two different specimens of gold had been placed at our disposal, one (I) from the *Gold und Silber Scheideanstalt* in Frankfurt-am-Main, the other (II) specially prepared and purified in the Reichsanstalt. No difference could be detected in their melting temperature.

The observations on the first two days (following table) were made in the gas thermometer oven, the others in the small oven used for the determinations by the crucible method (fig. 3) in which an empty crucible had been placed.

The arrangement of the thermo-element in the latter case was the same as above described for the crucible method with the omission of the porcelain protecting tube. The heating was so regulated that the slowly rising temperature could be easily observed at any time. It was even possible to observe the stationary temperature at the melting point just before the interruption of the circuit. The observed melting temperatures in microvolts and in degrees follow:

		Microvolts.	Degrees.
Oct. 17	(I)	10209	1064·5°
		10209	1064·5
		10206	1064·2
		10208	1064·4
Oct. 18	(II)	10202	1063·9
		10201	1063·8
Nov. 15	(II)	10211	1064·6
		10211	1064·6
		10211	1064·6
Jan. 2	(II)	10190	1062·9
		10190	1062·9
		10200	1063·7
		10189	1062·8
Jan. 3	(II)	10203	1064·0
		10197	1063·5
		10213	1064·8

The mean is 1064·0° ± 0·6°.

Silver— Wire Method.—The melting point of silver was also first determined by the wire method and two specimens were used, both from the Frankfurt *Scheideanstalt.* One of these was in the form of wire 0·5^{mm} and 0·25^{mm} in diameter and the other a piece of sheet silver 0·25^{mm} in thickness, from which thin strips were cut and inserted in the junction of the element. Here again no difference in the melting temperature of the two specimens could be established. The same ovens were used as for the gold determination. The results follow:

	Microvolts.	Degrees.
Sept. 25	8928	953·3°
	8936	954·0
Oct. 16	8937	954·1
	8922	952·8
	8922	952·8
Oct. 18	8914	952·1
	8927	953·3
	8930	953·5
	8935	953·9
Nov. 15	8939	954·3
	8958	956·0
	8925	953·1
	8941	954·5
	8924	953·0

The mean is 953·6° with a mean error of ± 0·9°.

Callendar as well as Heycock and Neville* have called attention to the fact that silver in an oxidizing atmosphere melts
and solidifies at a lower temperature than in a reducing atmosphere, the difference being caused by the fact that silver
absorbs oxygen in melting which is given off again below the
melting temperature with considerable violence ("spitting").

Silver—Crucible Method.—In order to ascertain whether
the melting point as determined by the wire method differs
from the normal melting temperature we made determinations
of it by the crucible method also.

(1) *In an Oxidizing Atmosphere.*—At first the silver was
melted in an open porcelain crucible without stirring. Under
these conditions the results differed from those with other
metals in that the silver showed no definite melting temperature. In melting as well as in solidifying the "time curve"
takes the form A (fig. 4), i. e. throughout its length it contains

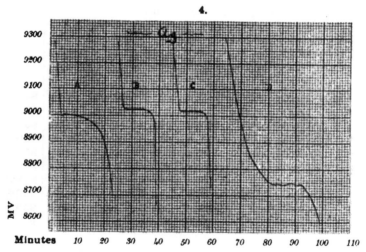

no horizontal section but falls or rises slowly over an interval
of 6° to 8°, between which limits one point may be assumed as
the melting point nearly as well as another. After substituting mica sheets for the clay oven-covers so as to be able to see
the metal during the process, it was observed that it remained
liquid through the greater part of this interval. The melting
and solidifying occurred at about 954·5° (8940 microvolts).
It may be added that the substitution of three mica covers
separated by air layers, for the two clay covers, produced no
essential change in the conditions within the oven.

If the thermo-element with its protecting tube were moved
about or the liquid metal otherwise stirred, the form of the

*C. T. Heycock and F. H. Neville, Journ. Chem. Soc., 1895, 160 and 1024.

time curve remained in general the same except perhaps that
the limits of the slow rise or fall of the temperature were
moved closer together, and if the stirring was energetic, could
be lowered some degrees.

Table XVII contains a number of results partly with stirring and partly without. G indicates the weight of the melted
metal in grams, t_1 and t_2 the upper and lower limits of the slow
change of temperature about the melting point, and t their
arithmetical mean. The letters F and M distinguish freezing
and melting points. The strength of the oven current i may
serve to give an idea of the comparative expenditure of energy
in the different determinations though only a rough one, as the
resistance of the coil on different days was not always the
same. The mean value of t for the observations where the
metal was stirred is 954·9° and differs only 1·3° from the result

TABLE XVII.

Melting Point of Silver in Air.

1899. Novem'r.	G (gr.)	*i* (Amp.)		t_1		t_2		t
				MV	Degrees	MV	Degrees.	
Without stirring.								
16	350	6·0	F	8940	954·4°	8983	958·2°	956 3°
		8·3	M	8960	956·1	9016	961·0	958·6
		0·0	F	8973	957·3	9000	959·6	958·5
		8·0	M	8970	957·0	9014	960·9	959·0
		6·5	F	8926	953·2	9006	960·2	956·7
		9·0	M	8970	957·0	9014	960·9	959·0
17	500	5·0	F	8968	956·8	9001	959·7	958·3
		8·0	M	8963	956·4	9018	961·2	958·8
		5·5	F	8942	954·6	8995	959·2	956·9
		8·0	M	8950	955·3	9011	960·6	958·0
With stirring.								
18	500	5·0	F	8931	953·6	8947	955·0	954·3
20	380	6·0	F	8950	955·3	8967	956·7	956·0
		10·7	M	8931	953 6	9004	960·0	956·8
		8·0	F	8890	950·0	8943	954·6	952·3

obtained by the wire method. We assume therefore 955° as
the mean melting point of silver in air, but at the same time
we wish to emphasize the fact that the point is not well defined
and, especially with the wire method, may suffer considerable
variations with the nature of the surrounding atmosphere.

The "spitting" was also frequently observed through the
mica covers. It took place between 933° and 940°, in the
mean at about 936°, and was much more violent when the melted
metal had been previously stirred.

The attempt to saturate the melted silver by allowing oxygen to bubble through it and thereby to reach a definite point of solidification for this condition led to no satisfactory result. The thermo-element showed considerable temperature oscillations throughout the solidification, but the mean, as was to be expected, lay much below the other determinations. Four observations gave the temperatures 939°, 938°, 942° and 938° — approximately the same temperature at which the absorbed oxygen is given off as noted above. Curve D (fig. 4) shows one of the time curves taken under these conditions, though it is by no means characteristic of all. In one case for instance with the temperature falling gradually, oxygen began to be given off at about 940°, whereupon the temperature began to rise, and reached 963° in the next five minutes, a violent spitting going on throughout the interval. At the latter temperature the metal became suddenly rigid and closed the porcelain tube through which the oxygen was entering.

(2) *In a Reducing Atmosphere.*—Higher results and better defined are obtained by preventing the access of oxygen. We sought to accomplish this first by melting the silver under a layer of common salt and afterward in a plumbago crucible over which a second crucible was inverted as a cover.

The time curves in both cases, for both melting and solidifying points, showed well defined horizontal segments, as may be seen from fig. 4, curves B (under NaCl) and C (in graphite).

Table XVIII contains a series of such observations from which 961·5° is obtained as the melting or solidifying tempera-

TABLE XVIII.

Melting Point of Silver in Absence of Oxygen.

Date.	G. (gr.)	t. (Amp.)		t.	
				MV.	Degrees.
In Porcelain Crucible under NaCl.					
Nov. 23	400	5·3	F	9023	961·7°
		10·7	M	9026	961·9
		6·7	F	9023	961·7
		11·3	M	9026	961·9
In Graphite Crucible.					
Dec. 8	352	6·7	F	9018	961·2
		9·5	M	9019	961·3
		6·0	F	9018	961·2
		9·3	M	9018	961·2

ture of pure silver. No spitting was observed under these conditions.

For the wire method neither of these courses is open for making determinations in the absence of oxygen. The bare junction cannot be placed in melted salt on account of its conductivity nor exposed to the reducing atmosphere of a plumbago crucible without danger of affecting the thermo-element.

Copper.—The melting point of copper was not determined by the wire method on account of the oxidation of the metal. In the crucible, the material used was pure copper in 100 gr. blocks from the *Haddernheim Kupferwerk* and determinations were made in both oxidizing and reducing atmospheres.

TABLE XIX.

Melting Point of Copper.

Date.	G. (gr.)	i. (Amp.)		*t.* MV.	Degrees
In Oxidizing Atmosphere.					
Nov. 7	900	——	F	10216	1065·1°
29	330	7·5	F	10213	1064·8
		11·0	M	10215	1065·0
		7·5	F	10213	1064·8
		10·8	M	10214	1064·9
Dec. 5	350	10·3	M	10219	1065·3
		8·0	F	10210	1064·6
		10·5	M	10216	1065·1
6	350	7·5	F	10211	1064·7
		10·3	M	10220	1065·4
		7·0	F	10211	1064·7
		9·9	M	10219	1065·3
Mar. 19	370	6·8	F	10213	1064·8
		10·3	M	10213	1064·8
		6·8	F	10211	1064·7
		10·3	M	10211	1064·7
21	370	6·5	F	10212	1064·7
		8·5	M	10214	1064·9
		8·5	M	10216	1065·1
23	370	8·2	M	10216	1065·1
		8·2	M	10219	1065·3
In Reducing Atmosphere.					
Dec. 7	355	7·5	F ·	10442	1084·3
		10·2	M	10440	1084·2
		7·2	F	10440	1084·2
		10·4	M	10437	1083·9

(1) *In an Oxidizing Atmosphere.*—Copper differs distinctly from silver in that when melted in an open porcelain crucible it possesses perfectly definite and coincident melting and solid-

ifying points. This seems to indicate that the copper under these conditions becomes saturated with a definite quantity of oxygen or rather with a copper oxide.

To assure ourselves regarding the constancy of this point we have made a great many observations of it without being able to notice any essential variation.

It was also quite indifferent whether the copper was fresh or had been several times used for the same purpose before.

After many determinations with a particular mass of metal it seemed to melt with increasing difficulty, a phenomenon which points to a decrease in its conductivity for heat, but should receive rather more investigation. Table XIX contains the results of the observations, giving 1064·9° as the melting point of copper in air.

(2) *In a Reducing Atmosphere.*—If the same metal which has been repeatedly melted in air be heated in a plumbago crucible it gradually reduces and the melting point rises, ultimately reaching 1084° after the metal has remained for a long time in a molten condition and become thoroughly reduced.

Melting points obtained between 1065° and 1084° where the reduction is still incomplete give poorly defined curves.

At the close of Table XIX some observations with the pure metal are given. Fig. 5 shows time curves for observations in

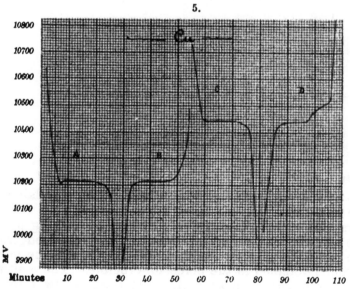

5.

porcelain crucibles (curves A and B) as well as in graphite (curves C and D).

If the melted copper contained oxygen, the unglazed porcelain tube which served to protect the thermo-element was always colored black while pure copper had no noticeable effect upon it.

Antimony.—The antimony with which our determinations were made, as well as the following metals, were obtained from C. A. F. Kahlbaum, chemist, in Berlin; with the exception of aluminium they may be regarded as pure.

For melting the antimony only graphite crucibles were employed, and even then a slight superficial oxidation was observed after cooling. Reducing gases were not introduced on account of the thermo-elements.

In solidifying, the temperature first sank far below (as much as 20°) the melting point, to which it then rose suddenly (see fig. 6, curve A). The time curves are otherwise normal. The mean value for the melting temperature was 630·6°.

6.

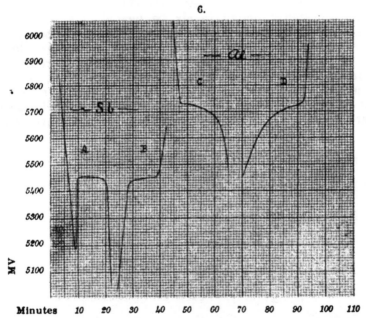

Aluminium.—Aluminium was melted in both porcelain and graphite crucibles. As has been noticed already by former observers, this metal shows no sharply defined melting point. Its time curves are similar in form to those obtained from silver in air. We give the melting point as 657·3°—the mean of the results obtained in the porcelain crucible. The melting temperatures observed in graphite were slightly lower to be sure, but the difference is more probably due to the smaller quantity of metal used.

The melting point of aluminium is not a sufficiently well-defined point for use in calibrating thermometrical apparatus;

should a temperature in this region be desired, the melting point of antimony is preferable.

Zinc.—The melting temperature of zinc can also be determined in an open porcelain crucible provided the same zinc be not too often used—with too much oxide present, the curves were less well-defined. In graphite the form of the time curve and the temperature are the same as in porcelain. The mean melting temperature may be given as 419·0°.

Lead.—The only noticeable feature in the determination of the melting point of lead is its small melting or "latent" heat, which makes some attention to the regulation of the oven temperature necessary. The mean melting point is 326·9°.

Cadmium.—The melting heat of cadmium is much greater and sharp curves are obtained with little trouble. Its mean value is 321·7°.

In Table XX are contained the determinations of the melting and freezing temperature of antimony, aluminium, zinc, lead and cadmium.

Tables XXI and XXII contain examples of the time curves as they were observed, the numbers representing the thermoelectric force in microvolts as observed from minute to minute.

Alloys.—In conclusion certain alloys remain to be mentioned whose melting points we investigated by the wire method only.

The metals were in thin strips cut from sheets of perhaps 0·2 to 0·3^{mm} thickness, prepared especially for the purpose by the Frankfurt *Scheideanstalt.*

As the investigation could only be undertaken in an oxidizing atmosphere, alloys of silver and copper gave no definite results whatever. The silver melts and separates out of the partly oxidized copper, leaving it to interrupt the circuit by breaking at a considerably higher temperature.

The two alloys 95Ag5Au and 90Ag10Au were especially prepared to ascertain whether the addition of a small quantity of gold would prevent the action of oxygen on the melting point of silver. The observed results seem to negative this assumption.

The temperature scale for the highest melting points is based upon exterpolation from the curve of the normal element T$_2$ (Table XVI).

6. *Measurement with Thermo-elements.*

Some ten years ago when the first platinum—platin-rhodium thermo-elements were obtained from the firm of Heräus for the measurement of high temperatures, both the platinum and the rhodium were so imperfectly purified chemically that the duplication of an element with approximately the same thermo-

TABLE XX.

Date.	G (gr.)	i (Amp.)		t MV	t Degrees	
			Antimony.			
Dec. 14	260	0 0	F	5453	630·8°	In graphit
		8·5	M	5449	630·4	
		5·0	F	5452	630·7	
		7·5	M	5448	630·3	
			Aluminium.			
Dec. 4	180	0 0	F	5729	657·7°	In porcela
		8·5	M	5723	657·1	
		4·5	F	5727	657·5	
		7·5	M	5723	657·1	
		4·3	F	5725	657·3	
Dec. 9	110	0·0	F	5717	656·6	In graphit
		8·0	M	5707	655·6	
		5·0	F	5718	656·6	
		7·5	M	5705	655·4	
			Zinc.			
Dec. 2	389	4·2	F	3377	419·2°	In porcela
		7·9	M	3375	419·0	
		0·0	F	3376	419·1	
		7·0	M	3374	418·9	
Mar. 30	293	0·0	F	3376	419·1	In graphit
		7·3	M	3375	419·0	
		0·0	F	3376	419·1	
		7·3	M	3373	418·8	
			Lead.			
Dec. 15	500	0·0	F	2507	326·9°	In porcela
		7·0	M	2508	327·0	
		0·0	F	2505	326·7	
		6·0	M	2505	326·7	
Mar. 26	493	0·0	F	2505	326·7	In graphit
		5·8	M	2508	327·0	
		4·2	M	2510	327·2	
		0·0	F	2510	327·2	
		4·2	M	2510	327·2	
			Cadmium.			
Dec. 2	430	0·0	F	2461	321·9°	In porcela
		6·0	M	2460	321·8	
		0·0	F	2458	321·6	
		6·7	M	2459	321·7	
		0·0	F	2457	321·4	
		6·5	M	2459	321·7	
Mar. 27	430	0·0	F	2459	321·7	In graphit
		4·3	M	2458	321·6	
		0·0	F	2461	321·9	
		4·3	M	2460	321·8	

electric force was impossible. The platinum was the first to be successfully purified, then, sometime after, the rhodium. Thermo-elements from different meltings now show variations of only 1 per cent, when formerly 10 per cent, and more was observed. The wires are also much more homogeneous now than formerly.

TABLE XXL.

"Time Curves" (MV).

Ag under NaCl	Ag under NaCl	Ag in Graphite	Ag in Graphite	Cu in Porcelain	Cu in Porcelain	Cu in Graphite	Cu in Graphite
9423	8513	9439	8766	10590	10003	10680	10081
9330	8599	9354	8861	10475	10106	10609	10179
9247	8684	9255	8956	10370	10183	10530	10268
9130	8775	9156	9016	10310	10201	10455	10353
9089	8857	9056	9018	10240	10207	10444	10409
9024	8937	9018	9018	10210	10210	10442	10426
9024	9017	9017	9019	10210	10210	10442	10431
9023	9023	9018	9019	10212	10211	10442	10433
9022	9024	9018	9019	10214	10213	10442	10436
9022	9024	9018	9019	10213	10213	10442	10437
9022	9025	9017	9022	10213	10213	10443	10438
9022	9026	9017	9033	10214	10213	10443	10439
9021	9026	9016	9043	10213	10214	10443	10440
9021	9026	9013	9055	10213	10214	10442	10440
9020	9026	9006	9064	10212	10215	10442	10441
9018	9026	8976	9074	10212	10215	10441	10444
9014	9027	8763	9121	10211	10216	10439	10456
9007	9028		9290	10210	10219	10437	10472
8990	9027			10210	10227	10433	10484
8800	9029			10208	10245	10427	10491
8600	9029			10203	10268	10410	10497
	9029			10200	10300	10342	10503
	9031			10193	10340	10195	10509
	9034			10177	10460	10070	10516
	9037			10097		9968	10564
	9050			9900			10720
	9069			9740			
	9232						
	9350						

Since the platinum wire at temperatures above 1000° is considerably less stable than the platin-rhodium, much would be contributed in the direction of the durability of the elements if the platinum could be replaced by some related metal or alloy of greater stability.

Up to the present, however, no satisfactory substitute has been found. The alloys of platinium with iridium and ruthenium, in part also with palladium, differ little in their thermo-electric properties from platin-rhodium.

Pure palladium does possess a higher thermo-electric force when combined with platin-rhodium than platinum, but can-

TABLE XXII.

"Time Curves" (MV).

Min.	Sb in Graphite	Sb in Graphite	Zn in Porcelain	Zn in Porcelain	Pb in Porcelain	Pb in Porcelain	Cd in Graphite	Cd in Grap
1	6216	5223	3551	3265	2718	2309	2852	239:
2	6080	5294	3500	3316	2671	2353	2809	242:
3	5960	5360	3449	3344	2621	2401	2767	244:
4	5840	5408	3400	3351	2574	2450	2726	244:
5	5720	5430	3376	3355	2529	2496	2685	2454
6	5600	5437	3376	3359	2506	2503	2646	2456
7	5495	5441	3376	3362	2506	2504	2607	245:
8	5390	5443	3376	3365	2505	2505	2570	245:
9	5280	5445	3376	3368	2505	2505	2533	245:
10	5190	5445	3376	3369	2505	2505	2496	245:
11	5450	5446	3376	3369	2505	2506	2462	246(
12	5454	5447	3375	3370	2504	2507	2460	246(
13	5454	5448	3375	3372	2503	2510	2459	246(
14	5453	5448	3375	3372	2502	2514	2459	2461
15	5454	5448	3375	3373	2501	2521	2459	2461
16	5452	5449	3375	3373	2497	2529	2459	2463
17	5452	5449	3374	3374	2488	2572	2459	2469
18	5451	5450	3373	3374	2473	2676	2459	2485
19	5451	5456	3374	3374	2444		2459	2499
20	5448	5468	3373	3374	2365		2458	2525
21	5438	5488	3373	3375			2458	
22	5310	5512	3372	3375			2457	
23	5060	5555	3370	3376			2456	
24		5655	3366	3389			2454	
25			3357	3404			2452	
26			3342	3433			2449	
27			3300				2422	
28							2367	
29								
30								

not be considered in this connection, for it is even less stable at high temperatures than platinum.

Thermo-elements made from different alloys of platin-rhodium such as were formerly used by Holborn and Wien* for temperatures above 1000° are also unsuited for general use on account of their low sensitiveness.

* L. Holborn and W. Wien, loc. cit.

TABLE XXIII.

Melting Points of Alloys.

Alloy.	t		Mean.	Alloy.	t		Mean.
	MV	Degrees.			MV	Degrees.	
Ag, 5 Au	8989	958·7		95 Au, 5 Pt	10805	1115·0	
	8990	958·8			10799	1114·5	
	8980	957·9	958·5		10826	1116·8	
					10827	1116·8	1116·0
Ag, 10 Au	9067	965.5					
	9070	965·8		90 Au, 10 Pt	11551	1177·2	
	9080	966·7			11522	1174·9	
	9060	964·9			11533	1175·7	
	9070	965·8	965·7		11495	1173·6	
					11578	1179·7	
Ag, 60 Au	9753	1025·3			11482	1171·5	
	9756	1025·5			11599	1181·2	1176·3
	9730	1023·3					
	9760	1025·9		85 Au, 15 Pt	12285	1237·4	
	9757	1025·6	1025·1		12282	1237·1	
					12273	1236·4	
Au, 5 Pt	10821	1116·4			12317	1240·0	1237·7

Thermo-elements were first intended only, for the measurement of temperatures from red heat upwards; afterwards it became evident that lower temperatures could be measured by this means as well. There is no advantage gained by going below 250°, however, as the errors due to the decrease in the sensitiveness at the lower temperatures begin to be more noticeable.

The accuracy of thermo-electric measurements in the earlier investigation of Holborn and Wien (loc. cit.) was stated to be 5°. The present results carry it further. It could fairly be said that, for temperatures up to 1150°, the present error would be about 1°; that is to say, for any distribution of temperature between the junctions, the temperature of the hot junction is defined by the thermo-electric force to within ±1°.

This does not in the least mean however, that with constant conditions considerably more accurate relative measurements cannot be made. The determinations of the melting points of the metals contain many examples of this.

The chief reason for this progress lies in the facilities for electric heating whereby the whole investigation of high temperatures gains in accuracy and simplicity.

For the thermo-elements the change is especially important since the electric oven allows a definite temperature distribution to be exactly reproduced at any time, whence it becomes possible even with inhomogeneous wires to carry out accurate temperature measurements by always working under the same conditions. It should be borne in mind however that this accuracy is only relative and that the same thermo-elements with a different temperature distribution between the junctions may show considerable variations.

To obtain the limits of absolute accuracy the fall in temperature along the wire must be varied as much as possible.

In the investigation here presented this was accomplished in two ways, either by measuring the temperature at a given point with two elements having a common hot junction and so arranged that the fall in temperature along the wire from the hot to the cold junctions could be varied; or secondly, by using as many different points of an element as possible successively, as junctions, and comparing them in a very short oven under exactly similar conditions of temperature distribution.

For example, by the first method the temperature near the end of a long electrically heated tube would be measured with two elements joined at their hot junctions, and the one passing out at the end near by, while the other extended through the middle of the tube which is hotter than near the ends, and out at the far end. It is clear that the distribution of temperature from hot to cold junction along the wires of the two elements would be very different and could be varied at will by simply moving the common junction to and fro.

Any lack of homogeneity in the wires would then become apparent at once in differences between the readings of the two elements.

We rather preferred the second method, the small melting-point oven (fig. 3) being well adapted to the purpose.

The wires of the elements are best simply bound together with a short platinum wire when it is desired to use several points provisionally as hot junctions. It has however no influence upon the thermo-electric force if the wire be cut in different places and afterward joined in the oxy-hydrogen flame.

Thick lumps along the wires are in the way to be sure, but with a little practice a junction can be made which can hardly be detected.

Thermo-elements as they are now prepared in a diameter of 0.6^{mm} are very homogeneous and over a length of several meters show no differences which could affect the temperature measurement $1°$. Thin wires reduced from the above by drawing are often less uniform. Two wires, for example, which had been drawn down to 0.25^{mm} showed about double this variation, nor could this difference between the thin and thick wires be removed by glowing electrically.

Repeated heating to high temperatures may also under certain circumstances render the wires less homogeneous.

Attention has been called to this before, but now that the accuracy of thermo-electric measurements has been so materially increased it becomes necessary to go into the discussion of this matter in considerable detail.

The changes are such as Holborn and Wien observed in platinum and palladium due to the action of combustion gases or hydrogen in the presence of silicium, and which affect the resistance even more than the electromotive force. Even with electric heating where combustion gases should no longer be feared, the walls of an oven which has not been previously heated to a higher temperature than that to be measured, and kept there until burned through and through, develop gases which pass through platinum.

We have called attention to this in the discussion of a platin-iridium bulb for the gas thermometer in our earlier communication. The thermo-elements are also affected by exposure to these gases but may usually be protected by the use of porcelain, which does not itself develop such gases and is very impervious to them even in its unglazed form. For temperatures below 1100° still further protection is afforded by glazing of course. Tubes of more refractory material intended for temperatures where porcelain can no longer be used and necessarily imperfectly burned we have found to be porous and when first heated to give off gases themselves at high temperatures which in the presence of silicium affect platinum strongly.

If the wires of a thermo element which have been exposed in this way be afterward used for temperature measurement, there is often little or no change in the electromotive force from the normal values *so long as the temperature distribution along the wire remains unchanged*, but considerable variations always appear as soon as it is varied.

We made observations for example with such an element C which had been exposed in an imperfectly burned oven, the hot junction being combined with that of a second element A which had not been so exposed. At 550° the following differences (C−A) in microvolts were observed in an electric oven 40^{cm} long in which the common junction of the two elements could be moved from the center toward either end as described elsewhere :

Hot Junction.		I.	II.
At the center (elements symmetrical)		7	7
6^{cm} to the right { 2 wires heated 14^{cm} {		− 73	+ 3
6^{cm} " left { " " 26^{cm} {		+126	+36

In column I the elements were crossed while in II the two wires of each element passed out of the oven at the same end.

Originally the two elements had agreed for all positions to within 5 microvolts.

Further examples could be cited, but we will confine our-selves to two where efforts were made to restore the damaged elements.

The element C', cut from the same wire as T, was in use for a long time for high temperatures in a long oven and there exposed to these gases, after which the following differences were obtained by comparison with the normal elements G and H under *unchanged* conditions of temperature distribution.

t	H−C'	G−C'
200°	+ 5 MV	+5 MV
300	+ 5	+6
400	+ 4	+5
500	−11	+6
600	−13	+7
700	−14	+7
800	−13	+5
900	−12	+7
1000	−13	+7
1100	−10	+8
1200	− 9	+8

When this same element C' was introduced into the short oven containing melted copper however, where the conditions of temperature distribution were very different, it gave 10,092 MV as the melting temperature while the two normal elements showed 10,212 microvolts.

C' was then taken out, glowed electrically at full white heat for several hours and again introduced into the copper cruci-ble (with proper porcelain protection of course), when it gave 10,170 under the same conditions as before.

The same measurement was then a third time repeated after 5^{cm} had been cut from both wires at the hot junction—the electromotive force now proved to be 10,220 microvolts. In this condition further measurements were made of the melting points of lead, cadmium and zinc with results 2513, 2460 and 3371 microvolts respectively, all in good agreement with the normal values.

As has been said the glowing was done electrically, as much as 17 amperes being sent through the wires. Connected in series both wires glow with equal intensity at high tempera-tures, the smaller resistance of the platinum when cold being offset by its larger temperature coefficient.

An element B of the same group was used several times in an oven tube of a new material which was being tried for the first time, up to a temperature of 1600°.

The platinum wire became very brittle in the vicinity of the junction, broke in several places and was melted together in the oxy-hydrogen flame. Afterward when used for measurement at about 1000° in the long oven (unchanged temperature distribution) its reading was some 90 microvolts too low, but for the melting point of copper (different distribution) it gave only 9840 MV, i. e. some 400 MV too low. Sixteen centimeters were at once cut off at the hot junction, first from the platinum and afterward from the platin-rhodium wire, when the melting temperature of copper was given at 10,290 and 10,150 MV respectively. Further shortening did not serve to change these figures and the pieces cut out were again inserted. In this condition the element was glowed as in the previous case for some hours at white heat, when the melting point of copper was read at 10,216 MV—almost exactly the normal value.

The glowing also restores the bright surface to wires (platinum particularly) which have become dull after exposure to these gases.

It is probable that the standard element which represents the earlier temperature scale of Holborn and Wien and which has been used many times for comparisons up to 1600° has undergone similar changes.

Its indications at 500° are now 10°, and at 1000°, 18° too high, while the new determinations of the melting point of gold differ only 8° from the old. This melting point in particular was well determined at that time, the greatest variations from the mean in twenty-five determinations under varied conditions being ± 4° while the corresponding variations for silver and copper amounted to ± 10°.

At the lower temperatures no fixed points were then determined so that no further comparison is possible.

We draw the conclusion from this that the temperature scale once established can be maintained with certainty only with the help of fixed temperatures such as the melting points given above. It is further advisable to divide a set of thermo-elements into two groups, one to be used only in porcelain tubes up to say 1200° and the other under other conditions or in higher temperatures.

For the latter, up to the present at least, the same accuracy is neither to be obtained nor expected until observations with the gas thermometer have been made and the technical resources much extended.

Especial attention ought again to be called to the fact that in the measurement of the fall in temperature in an extended space such as is necessary for the calculation of its mean temperature, the utmost care should be taken that the elements be in perfect order. The best arrangement is always to measure with two elements having a common hot junction and entering the space in question from opposite sides, as has already been described in our former paper.

Furthermore, inasmuch as thermo-elements have come to be employed very generally for technical purposes during the past few years, we take this opportunity to say that it is impossible to give the limits to which the electromotive force can be affected by the breaking of the protecting tube or other accident through which the wires become exposed to combustion gases in the manner above described—the higher the temperature the greater the danger. In case of such an accident the element should at once be tested and restored as elsewhere indicated.

7. *Conclusion.*

By way of conclusion we will bring together the temperatures which we have obtained for the melting points of the metals based upon observations with the gas thermometer and some recent results of other observers employing other methods.

These include the measurements with platinum resistances as carried out by Callendar and by Heycock and Neville[*] which are also based upon the gas thermometer up to 445°, the boiling point of sulphur, and the higher temperatures exterpolated from a formula of the second degree.

The melting points of gold and silver have also been measured with thermo-elements by Berthelot,[†] using the wire method and determining the temperature optically from the decrease in density of an air column in an electrically heated, open, porcelain tube.

	Gas Thermometer.	Platinum Resistance.	Optical Method.
Cadmium	321·7°	320·7°	
Lead	326·9	327·7	
Zinc	419·0	419·0	
Antimony	630·6	629·5	
Aluminium	657·	654·5	
Silver, in air	955·	955·	962
Silver, pure	961·5	960·7	——
Gold	1064·0	1061·7	1064
Copper, in air	1064·9	——	
Copper, pure	1084·1	1080·5	

Charlottenburg, May, 1900.

[*] Most of the figures are taken from Heycock and Neville's paper—loc cit.; Callendar's values—Phil. Mag. V, xlviii, 519, 1899—differ in several cases 1°.
[†] D. Berthelot, C. R., cxxvi, p. 473, 1898.

ART. XIX.— *Notes on Certain Schists of the Gold and Diamond Regions of Eastern Minas Geraes, Brazil;* by ORVILLE A. DERBY.

IN a recent communication in this Journal (May, 1899) brief reference was made to certain schists of the region that were supposed to be in some way related to the curious schistose partings in the quartz veins that were more particularly discussed. The single type analyzed presented such an abnormal composition that it appeared desirable to make analyses, kindly executed by Dr. Hussak, of such material, unfortunately scanty, as was at hand as seemed suitable for that purpose. This material includes, in addition to the rock already analyzed by Hillebrand, two of the rocks mentioned in the above cited paper from the diamond mine of São João da Chapada. To these has been added a perfectly sound schist found loose in a topaz washing near Ouro Preto and presumed to represent the decomposed material in which the nests of topaz occur. In addition four analyses by Gorceix of schists of the Ouro Preto region are given from the Annaes da Escola de Minas de Ouro Preto (vols. 1 and 2, 1882, 1883).

	I.	II.	III.	IV.	V.	VI.	VII.	VIII.
SiO_2	38·32	37·77	47·83	31·18	54·1	47·4	62·4	36·4
Al_2O_3	28·16	21·37	26·75	20·39	27·3	31·2	22·7	26·1
Fe_2O_3	2·24	28·36	8·51	36·41	7·7	6·7	3·9	23·1
FeO	4·02	....	....	....	....	....	....	....
CaO	0·32	tr	tr.	tr.	....	....	...	1·3
MgO	12·04	1·05	2·43	0·79	0·8	2·0	1·3	2·1
K_2O	1·11	8·96	10·42	7·11	3·0	4·5	4·2	6·4
Na_2O	0·16	....	....	tr.	3·6	2·7	1·5	1·0
Loss on ignition	8·01	3·89	5·33	3·83	3·8	5·6	3·4	3·9
Other elements	5·69	....	....	....	....	....	...	....
Total	100·07	101·40	101·27	99·78	100·3	100·1	99·3	100·3

I.—Greenish schist with large crystals of cyanite; Serra do Gigante, north of Diamantina; loose block; analyst, Hillebrand (this Journal, May, 1889). "Other elements" include TiO_2 4·93, ZrO 0·09, MnO 0·16, (NiCo)O 0·04, P_2O_5 0·47 and traces of SrO, Li_2O, S and F.

II.—Bluish schist found loose in the diamond mine of São João da Chapada but presumed to come from a schistose layer in the lower quartzite (itacolumite); analyst, Hussak. TiO_2 occurs rather abundantly but was not separately determined.

III.—Soft greenish schist found loose in the diamond mine of São João da Chapada but presumed to come from a schistose layer in the upper conglomeritic quartzite; analyst, Hussak. A strong trace of P_2O_5 was noted in the analysis.

IV.—Black schist found loose in a small topaz washing near the Caxambú topaz mine, Ouro Preto, and which possibly represents the original matrix of the topaz occurring in nests in a clay similar to what would be given by the decomposition of this rock; analyst, Hussak.

V.—Unctuous greenish schist from Boa Vista near Ouro Preto; analyst, Gorceix.

VI.—Fibrous schist containing at times crystals of pyrophillite and altered pyrite; Boa Vista near Ouro Preto; analyst, Gorceix.

VII.—Whitish unctuous scaly rock from Boa Vista near Ouro Preto; analyst, Gorceix.

VIII.—Compact violet schist in form of dike in limestone at Ganderela, north of Ouro Preto; analyst, Gorceix.

From macroscopic examination, No. I is evidently a chloritic rock with enormous crystals of cyanite up to 5^{cm} or more in length. The portion insoluble in sulphuric acid (43·41 per cent) calculated, according to Hillebrand's suggestion, as normal cyanite, muscovite, quartz and rutile, gives: cyanite 16·46 per cent, muscovite 8·46 per cent, quartz 13·66 per cent, rutile 4·73 per cent, with a doubtful excess of 0·09 per cent referred to zircon, which could not be found in the heavy residue, while a doubtful trace of zircon oxide was noted in the hydrochloric acid solution indicating that this element probably belongs to a soluble constituent. The soluble portion (56·66 per cent) reduced to 100 agrees very closely with Eaton's analysis of corundophilite from Chester, Mass., the most important differences being an excess of 1·88 per cent SiO_2, 1·13 per cent Al_2O_3, and a deficiency of 4·15 per cent of iron oxides. It is interesting to note that this Chester type of chlorite is also associated with an aluminous mineral, corundum, and that the somewhat similar type with the same association from the Culsagee mine, N. C., carries Mn, Ni and Co like the soluble portion of the Brazilian rock. Another element, monazite, revealed in minute quantities by washing, is probably indicated by the phosphoric acid of the analysis.

In the paper above cited an argument for the eruptive origin of this and other schists of the region was drawn from the occurrence of autigenetic monazite, which was considered as an original element that had passed unchanged through the process of metamorphism. This argument must now be withdrawn since, as will be shown in the following paper, evidence has recently come to hand that monazite may be formed by secondary processes; one of the proofs being furnished by the rock in question, which shows rutile included in monazite in the same manner as in the associated chlorite and cyanite. The probable origin of this rock will be discussed farther on.

No. II, as shown by microscopic examination, is essentially a
icaceous rock heavily charged with a fine dust of hematite,
hich is easily removed by hydrochloric acid, leaving a con-
lerable amount of a poorly characterized titanium mineral,
.tile or anatase, which should give a small percentage of
:anium not determined in the analysis. The alumina is nearly
per cent short of the amount required by the potash for
)rmal muscovite and this deficiency is too great to be supplied
' iron. Calculated on the basis of the Hallgarten sericite,
)wever, the deficiency of alumina is reduced to a reasonable
nit, and there can be no doubt that the rock is composed of
) per cent more or less of sericitic mica with perhaps 2–3 per
nt of chlorite and a small percentage of rutile, or anatase,
id probably also of quartz. Washings reveal the merest
ace of autigenetic monazite, which, in a measure, confirms
e identification of this rock with the decomposition product
iicaceous clay rich in iron oxide and with rutile) of a sheared
ke in the vicinity.

No. III is in appearance a purely micaceous rock with no
'idence, even in the heavy residue, of more than the merest
ace of free quartz and hematite. It is so soft as to go easily
a slime in water and is apparently considerably decomposed,
it this appearance is not confirmed by the high percentage of
)tash, which for normal muscovite requires 7·25 per cent
ore alumina than is given by the analysis. The rock prob-
)ly contains over 80 per cent of an iron-bearing sericite with,
:rhaps, 7 per cent, more or less, of chlorite and a small per-
:ntage of quartz and earthy iron oxide. Washings reveal a
iall amount of microscopic tourmaline, of which some grains
)pear to be secondarily enlarged, and worn zircons of a size
id abundance that seem extraordinary in a rock of such fine
·ain and of so purely argillaceous character. A trace of
iosphoric acid was noted in the analysis and a very decided
action was obtained from the heavy residue, so that it is cer-
in that monazite is also present although in its rolled state it
nnot be distinguished from the zircon.

No. IV is very similar to No. II in appearance and composition
ith a somewhat less deficiency of alumina for normal mus-
·vite and a considerably greater percentage of free iron oxide.
trace of phosphoric acid indicates the presence of a phos-
iate, of which, however, no trace could be detected in the
:avy residue, so that it probably is neither apatite nor monazite.
he heavy residue shows a relatively small amount of auti-
netic rutile and tourmaline and in this respect agrees with
at of the decomposition clays of the immediate vicinity with
:sts of topaz, which, so far as can be made out, must have

resulted from the decay of a rock very similar to the one analyzed.

Nos. V, VI and VII were analyzed by Prof. Gorceix for the express purpose of determining whether the denomination "chloritic" or "talcose," ordinarily applied to the predominant schists of the region, was applicable or not. They may, therefore, be presumed to be the most typical unctuous schists that could be found in a sufficiently sound state to be analyzed. All three show a considerable amount of soda, which in No. V is in excess of the potash, so that the micaceous element cannot be a normal muscovite. Alumina is somewhat in excess of what is required to satisfy the alkalies as normal muscovite and it was noted that macroscopic pyrophillite occurs in the bed from which No. VI was taken, so that it is probable that the excess of alumina may, in all three rocks, represent a small percentage of this mineral, or of cyanite, which is also very frequent throughout the region. Free iron oxide was noted in No. VI in the form of altered pyrite and it is probable that No. V also carries a small amount in the same state, or in that of hematite. All three show a considerable excess of silica, doubtless in the form of quartz, and the small percentage of magnesia probably represents a slight admixture of chlorite. Nothing is known of the heavy residues of these rocks or of the details of their mode of occurrence.

No. VIII has almost exactly the amount of silica and alumina required to satisfy the alkalies and magnesia as normal muscovite and chlorite, leaving an excess of about 8 per cent of silica for quartz. As in Nos. II and IV the iron is probably, for the most part, in the state of hematite.

All of these rocks show a low to extremely low percentage of silica and a high to extremely high one of alumina, while all without exception are poor in lime and, except No. I, in magnesia as well. On the other hand, all but No. 1 show tolerably high percentages of alkalies.

From these analyses no certain conclusion as to the original mineralogical composition and mode of origin of these rocks can be drawn. On the somewhat risky assumption that the present chemical composition of metamorphic rocks is, as regards non-volatile constituents, essentially identical with the original composition, no satisfactory comparison with known types can be made for the majority of these rocks. Nos. V and VII agree fairly well in composition with some of the trachytic and phonolitic tuffs cited in Zirkel's Petrographie, vol. iii, pp. 675 and 680. If, as is unusually done with such rocks, they be considered as metamorphosed arkose, the material must have been derived from a syenitic (or porphyritic) type rich in soda rather than from an ordinary granite or gneiss.

Neither of these hypotheses, however, will apply to Nos. III and VI, which, if original clastics, must have been almost purely micaceous sediments with, in the case of the latter, a slight admixture of kaolin. No. III is certainly clastic as proved by the heavy residue and it is presumed to represent a thin, shaly parting in a conglomeritic quartzite derived from an underying micaceous quartzite and for this such an hypothesis is not improbable. Nos. II, VI and VIII also, if clastic, must likewise have been essentially micaceous but heavily charged with limonite. It is, however, improbable that in five representatives, taken by chance, of original argillaceous sediments, all should prove to be essentially micaceous and none essentially kaolinitic, and the hypothesis may be suggested that in the case of marine clays a gain in alkalies may possibly take place from soluble salts imprisoned from the sea water. No. I, regarded as a clastic, must have been a singular mixture of kaolinitic and magnesian clays.

On the other hand, the hypothesis of eruptive origin for any of these rocks, with the possible exception of Nos. V and VII, involves that of unknown and improbable types, or that of an important loss of lime and magnesia with a consequent concentration of alumina, iron and alkalies. In the paper above cited arguments for the possible eruptive origin of Nos. I and II were deduced from the presence of autigenetic monazite, the lack of recognizable allothigenetic elements and, in the case of No. II, from a presumed connection with a decomposed sheared dike in the immediate vicinity as well as of traces of original structure. The first argument has proved fallacious but the others still hold good and the second one is equally applicable to No. IV, while No. VIII (very similar in composition to Nos. II and IV) was taken to be a dike by Gorceix in his field examination, though it is probable that if he had known its composition a more detailed examination of this point would have been made.

In the same paper reference was made to a peculiar rock from near the fall of the river Dattas which presents many analogies with No. II but with more decided eruptive characteristics. Both of these rocks were supposed to be allied to No. I, and when that on analysis proved to be chloritic rather than sericitic, as was at first supposed, it was assumed that this might also be the case with the two rocks in question. A qualitative* analysis of the Dattas rock, however, shows that

---

* As the rock is heavily charged with tourmaline, which is evidently an introduced element, it was considered that its original composition had been so changed that the work of a quantitative analysis would not be compensated by the results.

like No. II, it is very poor in lime and magnesia and that it is composed essentially of a sericitic mica heavily charged with hematite dust. The two rocks are therefore closely similar, chemically as well as mineralogically, and it can safely be affirmed that the principal differences revealed by a complete analysis would be only such as correspond to the presence of the characteristic elements of tourmaline and to a greater abundance of titanium and probably also of iron oxide, both in the free state. These two rocks also agree in presenting on a polished section transverse to the plane of shearing, appearances of original structure that in the Dattas rock are perfectly well-defined but somewhat indistinct in the other. This appearance is that of the mottled aspect of a rock composed of an intimate mixture of white and colored elements, as, for example, a diabase or a basalt. In the Dattas rock the minute white areas are perfectly defined rectangles like those of the feldspar of the rocks mentioned, while in that from São João da Chapada (No. II) they appear to have lost their original sharpness of outline through shearing. Under the microscope, these rectangular areas of the Dattas rock are seen to be composed of a sericitic aggregate free from the iron and titanium dust with which the rest of the section is thickly sprinkled. On dissolving out the iron, the distinction between the white and colored areas disappears and the whole slide presents a sericitic aggregate sprinkled with a dirty white dust of titanium oxide, rutile or anatase, which before had been concealed by the iron. The appearance is that of a basaltic rock composed of a well crystallized white element (plagioclase or melilite) in a groundmass containing bisilicates (pyroxene?) or in a basic glass rich in iron and titanium oxides, which on alteration has settled in the place of the original minerals, or glass, but without invading the areas of the white element. In addition to this secondary iron (hematite) and titanium dust, the rock also contains these two elements as primary constituents in the form of crystals, often of considerable size, of magnetite and of minute octahedrons of an altered titanium mineral that was most probably perofskite.

The peculiar structure and characteristic primary accessories of this rock are strongly suggestive of an original basaltic character and, in view of the perofskite, most probably a melilite-basalt. On this hypothesis, however, the original rock must have been rich in lime and magnesia and its present chemical and mineralogical composition can best be explained by the hypothesis that before metamorphism it had been decomposed and leached *in situ*, the resulting residual clays and oxides retaining their original positions without mingling. A confirmation of this hypothesis is apparently to be found in the peculiar manner in which the rock has been invaded by tourmaline. This mineral besides lining the shear and fracture planes with beautiful and symmetrical microscopic rosettes of radiating needles, also appears in the body of the rock, where, however, it is confined to the white areas, the needles being sharply cut off by the rectangular limits of the original crystals wholly or in part substituted by tourmaline. This invasion evidently took place after the decomposition and shearing but perhaps during the process of the recrystallization of the rock.

There is nothing intrinsically improbable in the hypothesis that some metamorphic schists may result from the alteration of rocks, both massive and clastic, that had been decomposed and leached *in situ,* and thus many of the anomalies of composition that this group presents may be satisfactorily explained. This hypothesis is involved in that presented by Vogt and Van Hise of the derivation of micaceous iron schists of Norway and of the Lake Superior region from original carbonates, and in another place evidence pointing in the same direction will be presented in relation to the extensive itabirite beds associated with the schists here discussed. If, as seems probable, the itabirites are derived from decomposed and leached carbonates, the process must have taken place on a gigantic scale in the region in question and the associated silicate rocks, whether massive or clastic, must have been affected to a greater or less extent.

The hypothesis of leaching cannot, however, be applied satisfactorily to the rock No. I, which, if an original basic eruptive of any of the ordinary types, must have lost heavily in lime and iron while magnesia was retained. This, however, would be contrary to what is supposed to have taken place in the other rocks here discussed and to what is generally observed in the modern cases of decomposition and leaching. Moreover the presence of monazite in this rock is, so far as present experience goes, incompatible with such types. It seems more probable that in this case the composition has remained comparatively unchanged and that the original rock must have been one of low silica, iron and lime, but of high alumina and magnesia associated with a considerable variety of rare elements. No type of rock with these characteristics has as yet been clearly defined, but that such may exist, either independently or as local phases or segregations in other types, is suggested by the occurrence of corundum and aluminous silicates in association with olivine rocks in North Carolina and Georgia, and that of alumina-magnesia silicates such as cordierite, prismatine and sapphirine as segregation (?) masses in the granulite of Saxony and the mica schist of Fiskernäs, Greenland. If, as may be presumed from the scanty information at hand, these latter minerals, with their associates, form independent rock masses, a possible prototype for such schists as No. I can be imagined. The question still remains of the mode of origin of such a prototype if it exists, but the occurrence of corundum in olivine rocks shows that the association of alumina and magnesia in ultra basic eruptives is not an impossibility.

The probable occurrence in this rock of a soluble compound of zirconia is also suggestive of a possible prototype in the nepheline- or augite-syenite groups, in which such compounds have thus far proved to be most frequent and in which rocks with corundum indicating an excess of alumina have lately been recognized. Thus far, however, no member of this group at all approaching in composition the Serra de Gigante rock appears to have been described. That rocks of this group may have existed in the district and have contributed to the metamorphic schist series, is perhaps indicated by the perofskite- and, possibly, melilite-bearing schists from the neighborhood of Dattas.

As to how far the above analyses are typical of the phyllites of the series in question can only be conjectured. These, which as regards thickness are evidently more important than the quartz, iron-mica and calc-schists with which they are associated, present almost universally an aspect that has led to their being generally denominated as chloritic or talcose, that is to say they are predominantly of micaceous texture and of a character that suggests magnesian minerals. True chloritic and talcose as well as amphibolitic schists undoubtedly occur in the series, but it is almost certain that they are subordinate to sericitic schists, agreeing more or less perfectly in character with Nos. II to VIII of the above analyses. So far as can be judged from a superficial examination of material which for the most part is profoundly decomposed, the more quartzose types represented by Nos. V. to VII are the most abundant and characteristic, and these, from their composition and intimate association with the quartz schists, may be presumed to be, for the most part, of clastic origin. If so, however, the original clays must have been sufficiently rich in alkalies to form a sericitic mica as the predominant element of their metamorphosed state. Judging from the above analyses and from the almost universally micaceous character of the phyllites of the region, the proportion of alkalies must have been very uniformly above 5 per cent.

It is interesting to compare with the proportion of alkalies above deduced for these phyllites that given in the extensive list of analyses of ceramic materials in vols. 16 and 18 of the annual reports of the U. S. Geological Survey. Out of 550 nalyses in which the alkalies were determined, only 67 (12 er cent) have over 5 per cent of potash and soda ; 99 (18 per ent) over 4 per cent, and 177 (30 per cent) over 3 per cent. n the 87 that are given as slate, shale or shale-clay, the pro- ortion having over 5 per cent is the same (11=12 per cent), ut below this figure it rises rapidly, the numbers being 22 (25 er cent) with more than 4 per cent of alkalies and 62 (72 per ent) with more than 8 per cent.

In the low proportions of alkalies the difference above noted nay reasonably be attributed to the greater amount of water n the earthy types, but this explanation will not apply in the omparison of the shales with the Minas schists, which, con- idered as original clastics (as a large, and probably the greater, art undoubtedly were), appear to have been uniformly richer n alkalies than the normal clay deposits of more recent times. f this conclusion is correct, these ancient argillaceous deposits nust either have been composed quite uniformly of compara- ively sound felspathic or micaceous material rather than of aolin, or in some way the original proportion of alkalies must ave been increased. The hypothesis above given that some f these schists are of eruptive origin will in part explain the redominance of micaceous rocks in the region, but, for the resent at least, there are no good reasons for supposing that hese constitute more than a small fraction of the whole. The ase of No. III shows that a high proportion of alkalies is not iecessarily indicative of eruptive origin and a number of other iighly micaceous rocks have been washed with the same result f a residue with well-characterized allothigenetic elements.

An excess of alumina over that required by the alkalies to orm mica, and which goes to form aluminous silicates, seems lso to be a common feature of the schists of the region, as yanite is a very common mineral both in the rock outcrops nd in the loose material of the surface, as well as an element, ften predominant, of the concentrates of the gold and diamond lacers. In general, however, it appears to come from the nore quartzose rather than from the more micaceous members f the series and in these cases it may be presumed to represent aolin in original clastic sediments. Other aluminous silicates hat resist decay so as to appear in the washings, as staurolite, re not at all prominent. From some of the above analyses it nay be suspected that pyrophyllite will prove to be a common nd widespread element, but from its liability to be confounded rith mica it can only be detected in perfectly sound rocks, that re rare in the region. As above indicated, this excess of lumina may be attributed to the normal presence of kaolin in riginal clastics, or to the leaching of original eruptives.

A high proportion of iron oxide, more frequently in the state of hematite than of magnetite, is also a very common feature in the micaceous as well as in the quartzose and calcareous rocks of the region. In the latter case and in that of the actinolitic and a part at least of the quartzose iron schists (itabirites), it may be presumed to come from original carbonates. In other cases it may be attributed to iron sand in clastic deposits, to original elements either oxides or silicates, in eruptives, or to a leaching of limonite into rocks of any character. It is almost invariably accompanied by a certain proportion, often high, of titanium oxide as rutile (more rarely as anatase) which after quartz and the iron oxides is the most abundant and constant mineral of the gold and diamond concentrates. This, whether in clastics or eruptives, is probably indicative of original ilmenite. Chrome mica occasionally appears in the quartzites of the region and in one case (a pebble from the conglomerate of the Cavallo Morto diamond mine near Diamantina) this could be traced very satisfactorily to grains, evidently clastic, of chromic iron contained in the same rock and which probably indicate that a peridotitic rock has contributed to the original sandy deposit.

Art. XX.—*Notes on Monazite;* by Orville A. Derby.

Solubility in acids.—In order to test the relative solubility of the mineral in different acids a sample of monazite sand from Prado, Bahia, was freed from all admixture by careful picking and after grinding was submitted in parcels of about half a gram each to the action of 100 centigrams of the ordinary laboratory acids each diluted with one part of water. After standing for 68 hours without heating, phosphoric acid was determined in each solution as follows: in nitric acid solution, 2·51 per cent; in hydrochloric acid, 1·62 per cent; in sulphuric acid, 1·39 per cent.

Magnetism.—In cleaning up residues with the electro-magnet it has been found that monazite can be quite successfully separated from its usual associate, zircon, and that when monazite and xenotime occur together the latter can, by a proper graduation of the points, be almost entirely drawn away from the former. In one case in which the instrument had been used with an opening between the points of about a centimeter, for the separation of pyroxene of the acmite type which may be presumed to carry about 30 per cent of iron oxide, a mixed sample of monazite and zircon was very neatly and quickly separated without reducing the distance between the points, both minerals being equally free from iron staining or inclusions.

Microchemical reactions—A single granule of the mineral, no matter how minute, can be rapidly and securely identified by moistening it with sulphuric acid on a slip of glass and burning off the acid over a spirit lamp. The characteristic crystallization of cerium in double ball-shaped clusters of radiating needles or minute cucumber seed-shaped isolated crystals, can usually be detected after this operation in the ring of evaporated material about the granule, but better after adding a drop of water and allowing it to evaporate in a desiccator. Another drop of water with a slight admixture of ammonium molybdate solution added to the same preparation gives on evaporation a very satisfactory reaction for phosphoric acid. The reagents are best applied by means of a small loop on the end of a very fine platinum wire and an excess should be avoided, especially with the acid, as too large a drop is liable to run in a very annoying manner in the heating. The same reactions are given by the recently discovered cerium-aluminium phosphate, florencite of Hussak and Prior, but this can be distinguished by its form and cleavage when these are recognizable. The microcrystalline forms of cerium and yttrium sulphates are so similar that a confusion with xenotime is also to be guarded against. It has recently been found also that soluble silicates containing zirconium give very similar forms in the sulphuric acid test, but these are readily distinguishable by the absence of phosphoric acid. Confirmatory tests with oxalic acid have not proved uniformly successful and can usually be dispensed with. The Florence test by crystalliza- tions in a blowpipe bead requires from a half a dozen to a dozen grains of the usual size and is more successful with the salt of phosphorus than with the borax bead in which the pre- sence of phosphoric acid appears to exercise a disturbing influ- ence, although with patience the crystals characteristic of cerium can be obtained.

When, as is usually the case, the grains are transparent, the micro spectroscope will also usually give a very satisfactory test on a single grain of the usual size, by means of the absorb- tion band of didymium. This can also be obtained by either reflected or transmitted light with an ordinary hand, or rain- band, spectroscope, when a number of grains can be brought close together in balsam on a microscopic slide.

Natural etching.—The grains of monazite from decomposed pegmatites or muscovite-granites are frequently etched, though not to the point of completely obliterating their original form and faces, but the phenomenon has rarely been observed in the decomposition-products of other types of rocks. In a miner's residue from the Cavallo Morto (Dead Horse) diamond mine near Diamantina the extraordinarily abundant monazite grains (the mineral is lacking or rare in the residues of most of the mines of the district) are extremely fresh in appearance, pre- senting a strong contrast with the well-worn aspect of the associated zircons, and under the microscope this difference is seen to be due to the profound natural etching of the grains. The deposit is reported by a competent observer to be a decomposed metamorphosed conglomerate like others of the vicinity, and in this case the monazite grains should be equally worn with those of zircon. On the contrary, however, they have been completely rejuvenated in appearance and amongst thousands passed in review under the microscope, none show- ing what could be positively identified as original or worn faces were seen except when, as in the case of the latter, these had been protected by the secondary enlargements described below. In similar deposits in the neighborhood in which the cement of the ancient conglomerate was evidently highly argillaceous, a similar etching of the included quartz grains is almost universal and is often very beautiful.

Secondary enlargement.—In the above-mentioned residue from the Cavello Morto mine, many grains show a darker central portion surrounded by perfectly clear material having the same optical orientation, but, like all the grains of the residue, without distinct crystalline form. The appearance is that of a rounded and worn grain with secondary enlargements principally in the form of clusters of spindle-shaped prolongations of the two poles of an original ovate grain. So far as can be made out from the color and from rather unsatisfactory optical tests, the clear outer shell is of monazite material, and this conclusion is apparently confirmed by microchemical tests, since grains in which the central nucleus appeared to be entirely protected from the action of the acid gave very satisfactorily the cerium and phosphoric acid reactions. As there was a possibility that the enlargement might be the newly discovered cerium-aluminium phosphate, florencite, which, when the form is not distinct, is readily confounded with monazite, a small quantity of the residue was tested by wet analysis for alumina, but with a negative result. It thus seems tolerably certain that in the clastic rock from which the residue comes there had been a new formation of monazite such as has been shown for tourmaline and several other minerals in similar rocks. The main objection to this conclusion is that monazite, as an autigenetic element, is thus far only positively known in rocks of eruptive or presumably eruptive origin, but this objection is weakened by evidence presented below indicative of a possible secondary origin.

Inclusions.—The monazite thus far extracted from granites, gneisses and porphyries, and that found in secondary deposits derived from these rocks, is free from inclusions, and in these cases the mineral, together with the associated accessories (zircon, magnetite or ilmenite, or both, and occasionally the phosphates xenotime and apatite), has evidently been one of the first to crystalize out of an eruptive magma. Quite recently, however, evidence has come to hand, that it may also be formed in other ways and probably by secondary processes. In addition to the secondary enlargements above noticed, which were evidently formed in a metamorphosed clastic rock, the following cases have been noted.

From the residues of diamond washings at São João da Chapada and Sopa near Diamantina, and from a gold washing at Bandeirinha in the same neighborhood, the monazite crystals of the peculiar prismatic habit described in a recent number of this Journal (vol. vii, p. 353) are frequently found heavily charged with scales of hematite and more rarely with minute needles and grains of rutile. This type of monazite has been traced home to some peculiar schists of the region that are presumed (in part on the evidence of the perfectly fresh condition of the monazite crystals), to be of eruptive origin. In the schists from Sópa and São João da Chapada, which are almost purely sericitic, the monazite is free from inclusions, but in the chlorite-cyanite schist from the Serra do Gigante the monazite, like the accompanying chlorite and cyanite, is full of minute grains of rutile and no essential difference in the character and mode of occurrence of the inclusions in the three minerals can be detected. Whatever may have been the original nature of this rock, the silicate elements and the rutile are clearly secondary, and there is no escaping the conclusion that the monazite must be so also.

In a small gold-working called Ogó* a few hundred meters away from the great diamond mine of São João da Chapada, monazite of the ordinary type occurs with rutile in the peculiar partings, composed almost exclusively of bright green muscovite, of a quartz vein in diabase. The crystals are here of the type known as turnerite and both they and the associated muscovite and rutile closely resemble those of the microscopic residue obtained from a specimen of the Binnenthal gneiss faced on one side with the well known macroscopic crystals of that locality. In the case of the Brazilian crystals, however, rutile and muscovite are included in them in such a way as to show contemporaneous crystallization. The rock is a peculiar one and in its present condition may be suspected to be a secondary product, but this cannot be proven. A decomposed massive quartz-muscovite rock from the same mine, which in appearance is of the same nature as the partings except for being finer-grained and more quartzose, did not give monazite. The exposure is unsatisfactory and as in the field examination the greenish rock was taken to be a chloritic alteration phase of the diabase, no special attention was given to it beyond taking a specimen that quite unexpectedly proved of great interest. The quartz vein with its peculiar micaceous accompaniment suggests a comparison with the quartz-kaolin (pegmatitic?) veins of the same vicinity and may be suspected to be a special type of granitic apophyses.

If the possibility of a secondary formation of monazite be admitted, a more plausible explanation than that given in the paper above cited (that of an intimate interlamination of eruptive and clastic material), can be given for the micaceous partings and selvages of the Sopa quartz veins that afford perfectly fresh monazite mixed with well-worn zircons.

* Ogó, which is apparently a term of African origin, is the name given to the fine heavy concentrate of microscopic minerals obtained in panning for gold and diamonds, when the predominant color is yellow. It may consist of monazite, xenotime or rutile, or a mixture of all three, the first generally predominating. In panning the miner, seeing the formation of a heavy yellow concentrate, is often deceived and on cleaning up exclaims in disgust "the gold turned to ogó".

Occurrence in basic rocks.—Up to a very recent period the rule seemed to be very firmly established that it was useless to look for monazite as an antigenetic element in any but highly acid rocks. In the scores of rocks examined, both sound and decomposed, it had been found to be almost universal in the muscovite granites, or their porphyritic and gneissic equivalents; frequent in the biotite granites, but lacking in the amphibole-granites and all other more basic rocks that had been examined. The monazite-bearing schists above mentioned from Sopa, São João da Chapada and Serra do Gigante, however, indicate that this rule is not general, though as the type of mineral that they present is peculiar and perhaps of secondary origin, it may be that it only requires to be somewhat modified rather than set entirely aside. The Sopa schist (that one free from zircon) have not been analyzed, but it is evidently composed almost exclusively of sericite and there is no difficulty in considering it as a modified porphyry. That of São João da Chapada, though basic in the low silica (37·77 per cent) and high iron (28·36 per cent, for the complete analysis see No. II of the preceding paper) contents, may, as I attempted to show in the paper above cited, be plausibly considered as a basic phase of a mixed dike of essentially granitic character. No such hypothesis is, however, possible with the Serra do Gigante schist with its 38·32 per cent of silica, 28·16 per cent of alumina and 12·04 per cent of magnesia. As already stated under the head of inclusions, all the elements of this schist are apparently secondary and it might therefore be considered as a metamorphosed sedimentary clay. The high percentage of magnesia is, however, anomalous for such a clay and in the preceding paper in which this rock is more fully discussed the hypothesis of the decomposition and leaching, prior to metamorphism, of an original eruptive rock is suggested. Even so, however, it seems necessary to imagine a type of eruptives that has not yet been clearly recognized though there are indications of its possible existence.

Art. XXI.—*The Spectra of Hydrogen and the Spectrum of Aqueous Vapor;* by John Trowbridge.

It is customary to consider that there are two spectra characteristic of hydrogen—a four-line spectrum, so-called, and another consisting of many lines widely distributed through the spectrum, known as the white spectrum. The four-line spectrum appears when a condenser discharge is employed with what is called dry hydrogen: it is also readily produced in steam and water vapor. From the fact that a condenser discharge seems necessary to excite it in dry hydrogen, it is supposed to indicate a higher temperature than the white spectrum.

The four-line spectrum is found in the atmosphere of the sun; and is a characteristic spectrum of certain types of stars. There are also other lines attributed to hydrogen in the stars which are supposed to indicate conditions of pressure and temperature which perhaps can be imitated and studied in laboratories.

I propose to show, in this paper, that conclusions in regard to temperature and pressure of hydrogen in celestial bodies, deduced from observations on hydrogen enclosed in glass vessels, are untrustworthy; and that electrical dissociation must be considered as a most important element in determining the characteristics of a gaseous spectrum—more important indeed than the question of pressure and apparent temperature.

I shall give my reasons for believing that the four-line spectrum of hydrogen in the atmosphere of the sun is an evidence of aqueous vapor in that atmosphere and therefore is an evidence of the existence of oxygen in the sun. The conviction is forced upon me that the term dry hydrogen is a misnomer, when the gas is subjected in glass vessels to condenser discharges or to sufficiently powerful steady currents of electricity.

The bibliography of the subject of the spectra of hydrogen is so extensive that I must, with due regard to the limits of this article, refer the reader to the reports of the committee of the British Association, and to the recognized bibliographical authorities on this subject, and I do this because I feel that in my experiments I have exceeded the experimental limits of previous investigators; for I have been enabled to employ more powerful electrical discharges than have been hitherto possible. My work, therefore, does not trench in this respect upon that of previous investigators.

The source of the electrical energy I employed was twenty thousand storage cells of the Planté type. The direct current from these cells through a liquid resistance was used to produce the white spectrum ; and a glass condenser consisting of 300 (three hundred) plates of glass—each plate having a coated surface of 16x20 inches—with a total capacity of about 1·8 microfarads—was charged by the cells to produce the four-line spectrum.

I had great difficulty at first in obtaining tubes which would stand such powerful discharges. I began my work with end-on tubes which were closed by plates of quartz luted on by silicate of soda. The electrodes were hollow cylinders of aluminum connected to thick pieces of platinum wires. These

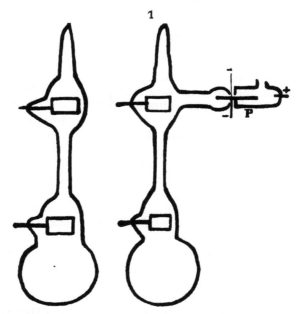

wires passed through the walls of the glass tubes, and were immersed in large vessels of distilled mercury. After considerable experience tubes were constructed which would resist the disruptive and heating effects of the discharges. This form of tube, however, was abandoned for another form, shown in fig. 1. The chief peculiarity of this form of tube is an X-ray bulb in place of the end covered with the quartz plate. The reason of the adoption of this form of tube is this: it was necessary to heat the entire tube to a high temperature for a long period during the process of exhaustion, to drive out the air and aqueous vapor before it was filled with hydrogen, and this heating was impossible with a luted-on end.

Moreover the thin glass of the bulb did not appreciably absorb the ultra-violet rays until the wave length 1800 was reached, and this absorption did not affect my conclusions since my work was confined to the portions of the spectrum studied by astrophysicists. The diameter of the narrowest portion of the tubes was about 1^{cm}. I employed still another form of tube, shown also in fig. 1. At one side of this tube there was an adjunct consisting of a palladium tube P on the outside of the spectrum tube joined hermetically to a platinum wire inserted in the walls of the tube. A glass vessel filled with dilute sulphuric acid slipped over the palladium tube. At one end of this glass vessel was inserted a platinum wire. The palladium wire was made the cathode and the platinum wire the anode of a battery. In this way hydrogen is liberated on the surface of the palladium tube, is then occluded and can be made to pass into the spectrum tube.*

The tubes which were not provided with palladium tubes were filled with hydrogen or nitrogen, through a series of tubes filled with caustic potash, and through a number of tubes filled with phosphoric pentoxide provided with partitions of glass wool. After many attempts I adopted the following arrangement of these tubes, which proved satisfactory. In the first place I found it necessary to discard all gums or other adhesive material and to use ground-glass joints and mercury seals for these joints, having satisfied myself that all other forms of connections consume time and are worthless. The tubes were mounted on a board and after the joints had been adjusted the final adjustment with the gas holder could be readily and safely accomplished by means of a rack and pinion which raised or lowered by slow degrees the gas holder fig. 2. This drying arrangement could be exhausted to a high degree and would hold the rarified gas as long as desired. The method of filling the tubes was as follows: the spectrum tube was exhausted to about ·1^{mm}, having been repeatedly filled with hydrogen. This operation was continued until the hydrogen spectrum appeared. Finally the tube was exhausted to the X-ray stage by long heating and by employing a strong condenser discharge. Then the dried gas was admitted until a pressure of about ·1^{mm} remained in the tube. The vessels with palladium tubes were carefully exhausted to the X-ray stage (and beyond, a spark preferring to jump six inches in air to passing through the tube). When these tubes with the palladium adjunct were connected to the terminals of the battery a slight heating was sufficient to start a discharge and to cause the occluded hydrogen to appear in the spectrum tube.

* A device due to Dr. William Rollins, Boston.

The operation of heating the spectrum tubes is highly important; moreover the discharge from a condenser should be employed in the process of exhaustion and the operation should be continued for several hours. In this subject of the exhaustion of tubes I am much indebted to suggestions of Dr. William Rollins and Mr. Heinze of Boston, who have had long experience in the preparation of X-ray tubes. This experience I believe is indispensable to one who essays to investigate the spectra of gases in glass vessels.

2

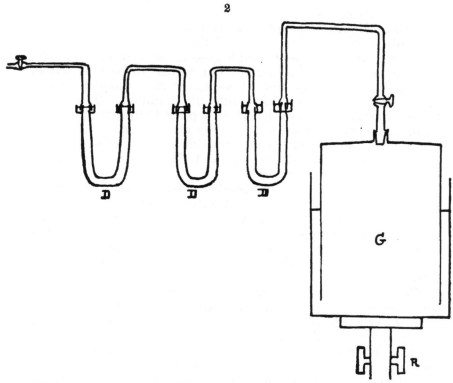

The vacuum pump was a modification of Rapp's pump. It was automatic. There was but one valve and the number of ground-glass joints with mercury seals was reduced to a minimum. The hydrogen was produced by the electrolysis of phosphoric acid and doubly distilled water. The gasholder contained many liters of the gas. The nitrogen was prepared by passing air through ammonia water, then through heated copper gauze, then through tubes of caustic potash; and finally through tubes of phosphoric pentoxide.

The spectroscope consisted of a Rowland grating with a ruled surface five inches by two inches, and a radius of curvature of sixty inches. The dispersion was small on account of the small number of lines to the inch; but it had the advantage of exhibiting the salient features of the spectra on one plate, with a dispersion of the order used by astrophysicists; and thus enabled comparisons to be readily made. The distance between the great H lines in the second spectra was approximately 2^{mm}. The camera was on an arm pivoted at half the distance between the slit and the grating. This arrangement was adopted for compactness. The grating and camera were enclosed in a light-tight box and a lever enabled me to expose the plate when this box was closed.

Various liquid resistances were employed to regulate the strength of the electric current. I finally used distilled water, which was kept running through glass tubes. Since my object was not to measure the current, but rather to keep it within safe limits, I did not find it necessary to employ cadmium terminals in a solution of iodide of cadmium. The direct battery current gave me the white type of hydrogen spectrum, while the second type was given by the powerful condenser discharges. A revolving drum provided with suitable metallic strips enabled me to safely charge and discharge the condensers through the spectrum tubes.

In working with powerful condenser discharges I found that I obtained essentially the same spectrum whether the tube was filled with hydrogen, rarified air or nitrogen, notwithstanding the great care taken in heating the tubes and drying the gases; this characteristic spectrum was that of water vapor—containing lines of hydrogen and of oxygen—with traces of nitrogen lines.

The employment, therefore, of a condenser discharge in glass tubes drives off aqueous vapor from the glass walls, even if extraordinary care is taken to expel such vapor in the process of exhausting such tubes. My experiments entirely confirm Professor Crookes' statement on this point. Glass seems to be of a permeable nature, and in the process of manufacture is doubtless filled with gases which reveal themselves when disruptive electrical discharges are employed. It does not seem possible that dry hydrogen can exist as such in a tube when it is subjected to a condenser discharge; and I therefore believe that the spectrum of hydrogen called the four-line spectrum is an evidence of the presence of aqueous vapor, for it can be produced with the greatest ease when we are certain that this vapor is present. This spectrum, to my mind, is an evidence of the presence of oxygen in the sun's atmosphere.

The true spectrum of hydrogen seems to be the white spectrum. This is developed by a steady current which should

ot exceed a certain strength; for a powerful one drives off
so rarified air from the walls of the tubes and there is a
rocess of occlusion set up if large metallic terminals are
nployed. Such large terminals are necessary if very strong
rrents are used. With aluminum electrodes under the
tion of strong currents, spectra of the combinations of nitro-
n with oxygen and hydrogen are produced in tubes presuma-
y filled with pure dry hydrogen. When the tubes have been
bjected to very strong currents it seems impossible to refill
em with pure hydrogen. In one case I filled a tube with pure
ry nitrogen and passed a current through it sufficiently power-
l to melt down one of the aluminum hollow cylinders which
rmed an electrode. The aluminum surface was filled with
vities as if gases had bubbled from the interior. When this
be was filled with dry hydrogen it gave again spectra of com-
unds of nitrogen and no trace of the hydrogen spectrum.

powerful condenser discharge, however, produced the
queous vapor spectrum. Hydrogen, thus, is extremely elu-
ve when subjected to electrical dissociation in the presence of
her gases in glass vessels with metallic electrodes, and indeed,
steady battery current can be employed at a certain stage of
chaustion to occlude hydrogen to such a degree that the
cuum is apparently raised and no discharge can be forced
rough the tube until it is subjected to external heat. This
clusion or destructive electrical dissociation is very much in
vidence when the terminals of a battery of ten or twenty
ousand storage cells are connected to an X-ray tube through
resistance of several megohms. If this tube is properly
chausted, that is by repeated heating and by the use of electri-
l discharges, no current can be forced through it unless the
be is repeatedly flashed with a powerful Bunsen burner.
uddenly, however, under the action of the heat a blue cloud
ses in the tube. The cathode beam appears, the anticathode
rows intensely hot and the exterior resistance must be quickly
creased to save the tube. The pressure has apparently been
reatly reduced; but the experiment of Dr. William Rollins
ith two connecting X-ray tubes, one of which is heated and
e other not, the heated one carrying a current and the other
sisting an eight-inch spark, shows that this is not the case and
at we have to deal with electrical dissociation and not with
ange of pressure. Let us follow this experiment further.
resently the blue cloud in the tube grows smaller, the cathode
am disappears and the cloud slowly sinks into the positive
ectrode. At that moment the X-rays flash out with great
illiancy; but in a few seconds the light in the tube is totally
ctinguished, and a further heating from an external source
necessary before the phenomena can be repeated. It can be

repeated as often as desired, although the apparent resistance of the tube undoubtedly is raised by a more or less permanent occlusion of what I believe is water vapor. This experiment strongly recalls the familiar one of the glowing of platinum under a stream of non-ignited hydrogen. The electrical dissociation of the aqueous vapor evidently produces an intense heat at the anticathode and the X-rays are emitted during this dissociation.

If we suppose that there are material bodies circulating about the sun and charged negatively—the sun being charged positively—we might conceive of a similar action of a difference of potential on rarified aqueous vapor, which would be competent to produce a corona.

The white spectrum of hydrogen produced by hydrogen coming from palladium is in its main feature similar to that obtained from electrolytic hydrogen which is passed through the drying apparatus I adopted.

There are some bands, however, which need examination to determine whether they arise from impurities conveyed into the tube from the pump. The palladium tube apparently supplies hydrogen more or less continuously to the tube to repair the loss from the process of electrical dissociation, and is therefore more reliable than the form of tube which is filled with hydrogen through drying apparatus which necessarily cannot be subjected to heat, and through which rarified air is conveyed with hydrogen into the spectrum tube.

The study of the effect of powerful electrical discharges on hydrogen led me to endeavor to find the lines in the star ζ Puppis discovered by Professor E. C. Pickering: since their approximate wave lengths satisfy a modification of Balmer's formula, Professor Pickering attributes them to hydrogen. The wave lengths of these new lines are comprised in the spectral region extending from about 4200 to 3700. I have plotted them as short lines contiguous to the normal solar spectrum in fig. 3, No. V. The long lines correspond approximately to the most intense lines or bands in the spectrum of hydrogen produced in the tube provided with the palladium adjunct. The hydrogen spectrum when regarded as a whole on the scale of small dispersion I have employed, seems to be made up of lines spaced according to a certain order, very much as if two sets of lines spaced according to a certain arrangement should be superposed on each other; the fingers of one hand shifted over those of the other. The hydrogen lines are more or less intensified bands or dark accumulation of lines almost obscured by the spectra produced by the compound of nitrogen and aqueous vapor if a very strong battery current is employed in a tube provided with aluminum electrodes. It is possible that certain hydrogen bands may be narrowed and rendered lines in a spectrum of small dispersion such as astrophysicists are compelled to employ, by an electrical dissociation of water vapor in the presence of an excess of oxygen, and that these new lines may thus be evidence of the presence of aqueous vapor in the stars of this new type. To test this theory I filled a tube with

3

oxygen, and submitted it to a powerful condenser discharge. The resulting spectrum was of the general type obtained by these strong discharges in hydrogen, rarified air and nitrogen. It is shown in fig. 3, No. IV. If we compare this spectrum with I, in which the broad bands represent hydrogen lines in hydrogen at atmospheric pressure; with II, the spectrum of rarified air; also with the lower one in III, the spectrum obtained in pure nitrogen, we see that the main hydrogen lines are narrower and certain hydrogen lines are so faint as to be hardly visible even on the negative.

The general character of the spectral lines of hydrogen in respect to breadth and intensity seems not to depend so much upon mere pressure and apparent temperature (temperature deduced from calculations of electrical energy in electrical discharges) as upon electrical dissociation of gases: of aqueous vapor, for instance in the presence of an excess or a small supply of oxygen. I believe, therefore, that any conclusion in regard to temperature and pressure of hydrogen in the stars deduced from laboratory experiments with electrical discharges in glass vessels are misleading, since we have mainly to do with phenomena of electrical dissociation and not of pressure and apparent temperature—that is temperature which can be measured by calorimetric methods or which can be calculated from the electrical dimensions of the circuit which is employed.

It would seem, therefore, that the study of electrical dissociation is necessary for the solution of problems relating to the conditions of gases in the atmosphere of the sun and in that of the stars. This seems to be a logical necessity, since the electromagnetic theory is far reaching and sensible heat is only one of the manifestations of electrical energy.

My conclusions are as follows:—

1. When a condenser discharge is sent through a rarified gas which is confined in a glass vessel, the gas can no longer be considered in the dry state: for aqueous vapor is liberated from the glass. When a sufficiently powerful condenser discharge is employed, dry hydrogen, dry nitrogen and rarified air give substantially the same characteristic spectrum. When a very powerful steady battery current is used to excite the tubes filled with these gases—various compounds of nitrogen and oxygen—nitrogen and hydrogen are formed if aluminum electrodes are employed.

2. The four-line spectrum of hydrogen in the solar atmosphere is an evidence of aqueous vapor, and therefore of oxygen in the sun.

3. Conclusions in regard to the temperature of the stars exhibiting hydrogen spectra are misleading, if based upon conditions of pressure and temperature in glass vessels; for conditions of electrical dissociation of aqueous vapor, for instance, the presence of an excess or lack of supply of oxygen, are the controling ones rather than conditions of the mere pressure of the gases.

4. X-ray phenomena produced by a steady battery current strongly suggest an electrical theory of the origin of the sun's corona.

Jefferson Physical Laboratory, Harvard University.

[XII.—*On a New Effect produced by Stationary Sound-Waves;* by Bergen Davis.

he course of an investigation upon the action of sound-
under certain conditions, I have found that if a small
ʳr, which is closed at one end, is placed in a stationary
ᵥave, it will not only arrange itself perpendicularly to
tion of the wave, but will actually *move* across the wave
ᵢrection perpendicular to the stream-lines. The force
ing this motion is of considerable magnitude and it acts
ly to the closed end of the cylinder, causing it to move
direction of the closed end, i. e., the closed end is driven
the wall of the pipe.

ᵥvestigating this effect I have used an organ pipe for the
tion of the stationary wave, its length being 70^{cm}, and
ᵉadth and thickness 6·85^{cm} by 5^{cm}, respectively. The
ᵥhen giving its first overtone made 358 vibrations per
. From the dimensions above given, it follows that a
ᵥill exist at 22^{cm} from the mouth-piece. and also at the
ᵈ end. One wall of the pipe was made of glass. A
ᵤbber diaphragm was placed across the pipe near the
ext to the mouth-piece, thus converting the remainder
ᵢᵖe into a closed chamber. The air in this chamber
this way kept free from any currents that might arise
lowing the pipe. A diaphragm of this kind at the node,
ot alter the intensity of the sound appreciably. The
ᵥhen blown very hard furnished a very strong overtone.
small cylinders were made from "No. 00 gelatine cap-
such as are used for medical purposes. The hemispher-
ds of these capsules were removed, and flat paper discs
ᵤted for them at one end, the other end being left open.
ᵡf these cylinders were arranged as shown in fig. 1.
ᵥstem of cylinders was supported by a small glass pivot,
was mounted on the point of a fine needle. The needle
tached to a glass rod which was destined to carry the
 into different regions of that part of the pipe located
 the protecting diaphragm.

strength of the air blast was measured by a mercurial
ᵢeter, which was used to obtain the same force of blast
hout a series of experiments. The pipe was placed
lly with the stopped end downward, and the glass rod
ᵧ the system of cylinders was introduced from below
h an airtight opening. It could be moved up and down
pipe, and the system of cylinders placed in any desired
ᵤ in the stationary wave. The plane of the system of

cylinders was perpendicular to the direction of the stream-lines. A scale was placed along the pipe in such a position that the zero was at the stopped end, and the node was at division 48 and the middle of the loop at division 24.

A stroboscopic disc was used to determine the rate of revolution.

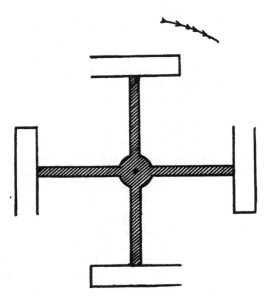

Matters being thus arranged, it was found that the system of cylinders when placed in different parts of the closed portion of the pipe rotated with different velocities. A maximum was obtained at the middle of the loop, viz: at 24 and a minimum at either of the nodes, viz: at 0 and 48. The mean results of a set of experiments are given in the table below.

Position of cylinders in pipe.	Rates of revolution per second.
2	·77
7	3·53
12	6·43
17	7·70
21	8·74
24	9·00
26	8·80
29	8·47
32	8·12
37	7·00
42	4·69
47	·92

The above table is plotted in a curve shown in fig. 2. The abscissæ are distances along the pipe and the ordinates are rates of revolution per second. The full line is the observed curve; the broken line is a sine curve. The agreement of the two curves is sufficiently close to suggest that the rate of revolution at any point in the stationary wave is proportional to the velocity of the vibrating air particles at that point. The

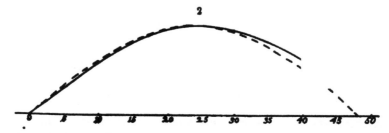

2

curve rises slightly above the sine curve near the dividing membrane, viz: from 30 to 48.

So far as I am aware a rotation of this kind has not before been observed.

The force involved in the above reaction was measured with the aid of a double cylinder supported by a fine wire, which went up through the wall of the organ pipe inside of a glass tube, 20^{cm} long. This double cylinder was carried at the end of a torsion arm and each cylinder was ·757^{cm} in diameter and 3·1^{cm} in length. The area of cross-section of each cylinder was ·45 $^{sq. cm}$. The pipe was placed in a horizontal position during the torsion experiments.

The amount of torsion necessary to bring the cylinders back to their original position was read on a graduated circle at the top of the glass tube.

The mean results of a number of experiments are given below.

Position of cylinders in pipe.	Torsion.
2	12°
7	72°
12	190°
17	355°
21	480°
23	555°
27	490°
32	315°
37	165°
41	72°
45	14°.

The force exerted upon the cylinders increases very rapidly from the nodes to the middle of the loop. The table is plotted in a curve shown in fig. 3. The ordinates are the *square roots* of the plate readings. The curve corresponds not badly to a sine curve. Experiments with cylinders of the same length but of smaller diameters gave almost an identical curve.

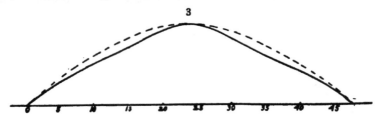

Experiments made in various gases.—The body of the pipe between the diaphragm and the stopped end forms a closed chamber and into this chamber various gases were introduced. The pipe was still set into action by an ordinary *air* blast. The sound wave within the pipe was transmitted through two media, air and the gas, the diaphragm forming the division between them. The force of the air blast was kept constant throughout the experiments. The torsional measurements were made only at the middle of the loop. Although the number of vibrations executed by this composite mass of gas changed very much, the position of the node at the diaphragm was but slightly altered.

Several readings were taken in air, before and after each experiment with the gases. A fresh supply of gas was introduced into the pipe before each reading. The mean results of these experiments in different gases are given in the annexed table. The readings in air, as well as its density are taken as unity.

	Torsion deflections.	Density of gas.
Air	1·	1·
CO_2	1·47	1·52
Illuminating gas	·77	·75
Hydrogen	·064	·069

As will be seen, the force exerted is proportional to the density of the gas.

The rates of revolution of the system of cylinders mentioned above were also taken in the different gases and the mean result of several experiments are given below, the rate of revolution in air being taken as unity.

Air.	Illum. gas.	CO_2.	Hydrogen.
1·	1·	1·04	·59

The rate may be considered as approximately a constant for all the gases, although the force producing this rotation varies with the density of the gas. This can be explained by the circumstance that the resistance to motion is also proportional to the density. The deviation from this result in the case of hydrogen is explained by the fact that the friction on the supporting pivot was not an inconsiderable quantity compared to the total resistance experienced by the cylinders, and this factor is proportionately larger in the case of hydrogen.

Prof. Hallock first suggested to me the explanation of the effect described in this paper.* He pointed out that Daniel Bernoulli has shown that a gas in motion is virtually less dense than the same gas at rest. The air in the capsule cylinder is at rest, while that on the outside of it is in motion. The energy exerted is due to the difference in density on the two sides of the closed end of the cylinder.

An inspection of the curve in fig. 3 shows that the force is approximately proportional to the square of the velocity of the vibrating air.

I have applied this effect to the problem of determining the amplitude of the vibrating air particles in the middle of the loop of the stationary wave. It is evident from the formula given below, that if the force per unit of area on the closed end of the cylinder is obtained experimentally, the change of density may be found, and the amplitude of vibration calculated from it.

Prof. R. S. Woodward has kindly assisted me in applying the proper principles of hydrodynamics to the problem. The relation between the velocity, pressure and density of a moving fluid may be expressed by Bernoulli's equation, since, for this case the average state of the fluid only is under consideration.

$$\int \frac{dp}{\rho} = R - \tfrac{1}{2} u^2$$

R is the potential due to external forces, and is negligible in this case ; hence

$$\int \frac{dp}{\rho} = - \tfrac{1}{2} u^2$$

The adiabatic gas relation gives, $p = c\rho^r$, where c is the gas constant.

---

* This effect is somewhat similar in nature to the acoustic attractions and repulsions of Guthrie and the rotating mill of Dvořak that is, they all depend on the change in density arising from motions of a fluid. The effect herein described is of a different order of magnitude than that of the rotating resonators B D.

By substitution and integration I obtain,

$$\frac{1}{n}\frac{p_2}{\rho_2}\left[1-\left(1-\frac{p_2-p_1}{p_2}\right)^n\right]=\tfrac{1}{2}(u_1{}^2-u_2{}^2),$$

where $n=\dfrac{\gamma-1}{\gamma}$.

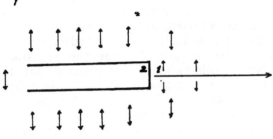

The cylinder is in the moving air as shown in fig. 4; u_1 is the velocity of the vibrating air outside of the cylinder, u_2 is the vilocity inside the cylinder near the closed end. P_2 and p_1 are the pressures on the two sides of the closed end of the cylinder. If the air in the cylinder next to the closed end is at rest, then u_2 becomes zero.

By expansion and reduction, the above equation becomes;

$$u_1{}^2=2\frac{p_2-p_1}{\rho^2}$$

p_2-p_1 is the quantity measured on the torsion balance. The absolute value of the pressure as determined from the experiments gave a force of 21 dynes on an area of one sq.^{cm}. This gives a linear velocity to the vibrating air of 187^{cm} per second. The amplitude so obtained is 2·61^{mm}. This value for the amplitude is somewhat less than a corresponding value obtained by means of the sound-wave anemometer recently devised by myself, an account of which will be shortly published.

Physical Laboratory, Columbia University,
June 1st, 1900.

ART. XXIII.—*On some Interesting Developments of Calcite Crystals;* by S. L. PENFIELD and W. E. FORD.

1. CALCITE FROM UNION SPRINGS, CAYUGA COUNTY, N. Y.

THE crystals under consideration were found during the summer of 1899 by Mr. J. M. Clarke, Director of the Geological Survey of the State of New York, and were sent to New Haven for examination. Mr. Clarke had observed that the crystals presented certain features of unusual interest, and it was his wish that they should be described and that the specimens should be deposited in the Yale Collection. The crystals occur in the Onondaga limestone, in a region where slight tectonic disturbances have taken place, giving rise to fissures in which calcite has deposited as vein material. The most interesting feature presented by the crystals is their diversity of habit, shown often on a single hand specimen, and due to different methods of twinning together with peculiarities in the development of certain crystal faces.

Most of the crystals were not well adapted for measurement with the reflecting goniometer, but, using one of the smaller ones, about 5^{mm} in length and 2^{mm} in diameter, it was possible to identify the prominent forms by means of their angles. The small crystals are quite highly modified and their development is represented by fig. 1. The terminal faces are the brightest and best developed, and are those of the common scalenohedron v, $21\bar{3}1$. There was meas-

ured for the identification of this form r (cleavage) $\wedge v$, $10\bar{1}1 \wedge 21\bar{3}1 = 28° 56'$, calculated $29° 1' 30''$. In the zone r, v, and making a very small angle with v is the scalenohedron v_1, $7·4·\bar{11}·3$, which is especially prominent on the crystals from this locality. This form was identified by von Bournon on crystals from Derbyshire and the Dauphiné Alps, and appears as form No. 37, plate 31, of his *Traité de Minéralogie*, published in 1808. The form was identified by its position in the zone r, v and the measurement $v \wedge v_1 = 3° 23'$, calculated $3° 55'$. On the crystals under consideration the faces of the scalenohedron v_1 have a vicinal development, and thus the contrast between them and the better developed faces of the scalenohedron v is generally quite marked. A negative rhombohedron, f_1, truncates the edges of v_1 and appears always as a narrow face with vicinal development from which no reflection could be obtained. A rhombohedron in this position would have the symbol 0·12·$\overline{12}$·5, and is a little steeper than the common rhombohedron f, 02$\overline{2}$1, which truncates the pole edges of the scalenohedron v. The pyramid of the second order γ, 8·8·$\overline{16}$·3, was identified by the measurement 8·8·$\overline{16}$·3 $\wedge$. 8·8·$\overline{16}$·3 = 25° 40′, calculated 24° 46′, and, further, by its being truncated by the positive rhombohedron M, 40$\overline{4}$1. This rare pyramid was first identified by vom Rath[*] on crystals from Andreasberg in the Harz, and, as pointed out by the authors,[†] this same pyramid is the prevailing form of the siliceous calcites from the Bad Lands of South Dakota. On crystals from Union Springs there is a tendency for the upper and lower faces of the pyramid γ to round into one another, owing to vicinal development, and because of this rounding it was impossible to obtain an accurate measurement between the upper and lower γ faces.

On the majority of the specimens the crystals are not so highly modified as the one just described, but as already stated the variation in habit due to twinning and the unequal development of certain faces, gives to the specimens a peculiar interest. All the types to be described occur on a single specimen having a surface about half the size of one's hand covered with crystals. The crystals on this specimen were not suitable for measurement and therefore no angles will be given, but the forms were evidently like those identified on the small crystal previously described.

Scalenohedral type.—The scalenohedron v_{I}, 7·4·$\overline{11}$·3, fig. 2, is apparently very common at the locality. It should be stated

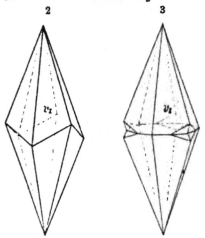

that this form has the same middle edges as the rhombohedron r, 10$\overline{1}$1, and the common scalenohedron v, 21$\overline{3}$1, but is somewhat steeper than the latter form. A twinning about the basal plane, fig. 3, is quite common.

[*] Pogg. Annalen, cxxxii, p. 521, 1867.   [†] This Journal, ix, p. 352, 1900.

Twins, with the rhombohedron e, $01\bar{1}2$, *as twinning plane.*—
The habit resulting from this kind of twinning is like that of
the well known Guanajuato calcites, fig. 4, described by Pirsson,[*] and it should also be stated that as early as 1837 Lévy[†]
also described and figured calcite twins of this same type from
Streifenberg, Nertschinsk, Siberia. Fig. 4 is analogous to the
figures of Pirsson and Lévy, though drawn in a different position, and represents the common scalenohedron v, $21\bar{3}1$, drawn
with the twinning plane vertical and having a position like
that of the side face of a cube, or the pinacoid 010 of any of
the three axial systems. This position has been adopted for
representing the twin crystals as it gives the best idea of their
peculiar development. Fig. 5 represents the scalenohedron v_I,

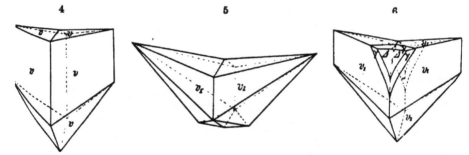

$7\cdot4\cdot\bar{1}1\cdot3$ twinned without distortion, a type which has not been
observed on any of the specimens, but the figure is introduced
in order to show how, by the extension of the two lettered
faces in front and the parallel faces behind, together with the
suppression of the four small faces below, the Guanajuato
type, fig. 4, results. Most of the Union Springs crystals of the
Guanajuato type show in addition to the scalenohedron certain
modifications at the reëntrant angle, fig. 6. The faces forming
the reëntrant angle are the pyramid of the second order γ,
$8\cdot8\cdot\bar{16}\cdot3$, and a rhombohedron designated as Δ, apparently in
the zone with v_I and γ, which would cause it to have the symbol $80\bar{8}3$. The surfaces forming the gash or reëntrant angle,
however, are curved to such an extent that exact symbols cannot be assigned to portions of them.

Twins with the rhombohedron f, $02\bar{2}1$, *as twinning plane.*—
The rhombohedron f is one of the rare twinning planes of calcite,
and the habit presented by the crystals from Union Springs is
very striking. The scalenohedron v_I, $7\cdot4\cdot\bar{1}1\cdot3$, twinned about
f, and drawn with the twinning plane vertical, as previously

* This Journal III, xli, p 61, 1891.
† Description d'une Collection de Minéreaux formée par H. Heuland, vol. i, p.
10, fig. 5, Plate 1.

described, is represented by fig. 7. In the Union Springs crystals representing this twinning law the reëntrant angle at the top wholly fails, and a peculiar, pointed, spear-head devel-

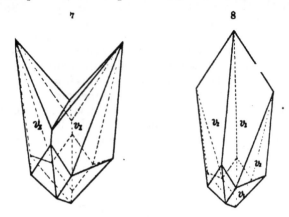

opment, fig. 8, results from the extension of the two front lettered faces of fig. 7 and the corresponding faces at the back. The crystals observed have always been attached at the lower end. Several crystals of this peculiar type were observed on the specimens sent by Mr. Clarke, and they are said to be quite common at the locality. On a crystal with a broken point the reëntrant angle measured from the rhombohedral cleavages was found to be 35° 38′, calculated 35° 27′.

Lévy, in fig. 6 of the atlas to his work already cited, gives a representation of a crystal from Kongsberg in Norway, of identically the same habit as fig. 8 of this article; however, the habit is apparently a very unusual one, and it is interesting to record it at a new locality.

On a single specimen or even at one locality, as a rule, all crystals of a certain mineral have the same or nearly the same habit, resulting undoubtedly from crystallization under uniform conditions, and therefore it seems a matter of more than usual interest to note on a single hand specimen from the Union Springs locality, the occurrence of simple scalenohedrons, fig. 2, and of three distinct types of twinning, figs. 3, 6 and 8. The calcite crystals seem to be all of one generation. Associated with them are a few crystals of dolomite, apparently of later growth.

2. Butterfly Twins from Egremont, Cumberland, England.

The so-called butterfly twins from Egremont are well-known and are figured in many mineralogies. Lévy in his work, already cited, gives three figures of them, No. 17, 68, and 69 of his atlas. A few words concerning them and new figures are introduced in the present article for the sake of comparison with the two types of rhombohedral twinning previously described. The twinning plane in these crystals is the rhombohedron r, 10Ī1, and the common scalenohedron v, 21Ī1, thus twinned and drawn, as in previous cases, with the twinning plane vertical, is represented by fig. 9. Fig. 10 represents a

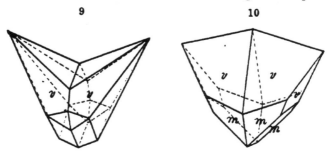

crystal of the butterfly twin type in the Brush collection, and, by comparison with fig. 9, it may be seen that the upper faces of the crystal result from the development of the two front, lettered faces of fig. 9 and corresponding faces behind, to the complete obliteration of the reëntrant angle. The faces at the lower extremity of fig. 10 are those of the prism m, 10Ī0.

It is a matter of interest to observe how the scalenohedron when twinned as described according to the three rhombohedral laws gives apparently simpler shapes by distortion, or unequal development of some of its faces, than if the distortion had not taken place.

3. Crystals from Pallaflat, Cumberland, England.

A feature of the crystals from this locality, as represented by specimens in the Brush collection, is the prominent development of the negative scalenohedron x, 13Ī1. This form, as shown by fig. 11, has its shorter pole edges bevelled by the common scalenohedron v, 21Ī1, and has the same middle edges as the negative rhombohedron, f, 02Ī1. Fig. 11 was drawn by Mr. W. Valentine of the Sheffield Laboratory. It presents nothing new, and is practically identical with fig. 674 of von Bournon's *Traité de Minéralogie* published in 1808. The figure is introduced in the present article, because by understanding its simple zonal relations, the same forms can be easily identified as they occur on a twin crystal to be described. Fig. 12 represents the development of two beautiful twin crystals in the Brush collection, both occurring on the same hand specimen. The twinning plane is the unit rhombohedron, and the development is analogous to that of the butterfly twins from Egremont, fig. 10. A prominent feature of the twins is the vertical zone r, f and x of the individual to the

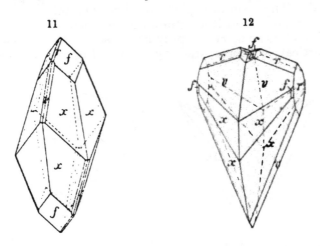

11   12

right, extending over the twinning plane to x, f, and r of the left hand individual, and so on around the crystal. Thus with this method of twinning four x faces, two in front and two behind, form as it were a vertical prism, analogous to the prism formed by four faces of the scalenohedron v, fig. 4, when the flat rhombohedron e, $01\bar{1}2$, is the twinning plane.

In figs. 4 and 12 the rhombohedral symmetry is not apparent, and the habit is like that of twin crystals of the monoclinic system, having the vertical faces v and x, respectively, as prisms and a pinacoid as twinning plane. The twin crystals represented by fig. 12 are so attached that only a portion of the lower x and v faces are visible.

4. Crystals from the Stank Mine, Lancashire, England.

This locality is represented in the Brush collection by a suite of eleven groups of crystals. A form which appears on all of the crystals, and which seems to be especially characteristic for the locality, is the negative scalenohedron B, $2\cdot8\cdot\bar{10}\cdot3$, having the same *middle edges* as the rhombohedron f, $02\bar{2}1$. On all of the specimens the faces of the scalenohedron have a decided vicinal character. Some of the specimens are simply groups of scalenohedral points without modifications. Generally, however, as shown by fig. 13, the flat negative rhombohedron e, $01\bar{1}2$, is slightly developed, and long narrow faces of the positive scalenohedron v, $21\bar{3}1$, modify the pole edges, while some portions of the negative rhombohedron s, $05\bar{5}1$, may be seen near the attachment of the crystals. Another prominent type which seems to be common is represented by fig. 14. In this type of crystal the rhombohedron *s* is prominent and is bevelled by the scalenohedron *v*; the pole edges of *v* are truncated by the rhombohedron *f*, and *B* and *e* appear as slight modifications only. The type of crystal, however, which it is desired especially to call attention to is represented by fig. 15, there being several almost ideally developed crystals of this type in the collection. The peculiar feature presented by the crystals consists in the termination of the prism *m*, 10$\overline{1}$0, by twelve faces of almost equal size and shape, thus giving the appearance of a prism terminated by a dihexagonal pyramid. Six of these terminal faces belong to the positive scalenohedron *v*, and the remaining six to the negative scalenohedron *B*, while the negative rhombohedron *e* appears at the end of the crystal. The distinction between the *v* and *B* faces is at

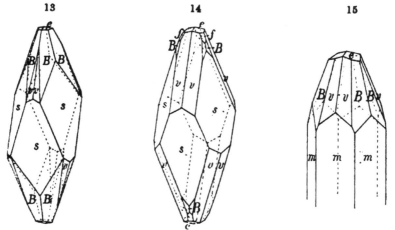

13   14   15

first glance not at all apparent, but the forms may be distinguished by their relative positions with reference to the rhombohedron *e* and by slight differences in their surface markings, the *v* faces being finely striated parallel to their intersection with the unit rhombohedron, while slightly undulating, smooth, vicinal surfaces characterize the form *B*.

5. CRYSTALS FROM ELÆOLITE-SYENITE, MONTREAL, CANADA.

The crystals under consideration are small tabular individuals, fig. 16, measuring about 8^{mm} in diameter, and were found by Mr. W. F. Ferrier of Rossland, B. C., at the Corporation Quarry, Outremeut near Montreal. Only a few crystals were found, and one of these was presented to the Brush collection by Mr. Ferrier. A tabular habit is not at all uncommon for calcite, but the crystals deserve brief men-

16

tion because of the peculiar combination of forms.   Besides the basal planes only two forms are present, the negative rhombohedron f, 02$\bar{2}$1, a very common form, and the positive scalenohedron F, 42$\bar{6}$1, a rare form for the species.   The scalenohedron F has the same parameter relation on the horizontal axis as the common scalenohedron v, 21$\bar{3}$1, but is twice as steep, F intersecting the vertical axis at 6 and v at 3.   Fig. 16 was drawn by Dr. H. E. Gregory while a student in the laboratory.

Sheffield Laboratory of Mineralogy and Petrography,
    Yale University, New Haven, May, 1900.

ART. XXIV.—*A Method of Measuring Surface Tension;* by
JAMES S. STEVENS.

In the methods of measuring surface tension which have
formerly been employed there seems to be two difficulties :
the weights applied to break the tension cannot be added in
continuous increments, and they afford a certain accelerating
force beside their own mass when they are dropped on to a
scale pan ; and secondly, it is difficult to apply the pull
exactly at the center of inertia of the body used to break
through the surface skin. The second difficulty was overcome
in a method described by T. P. Hall, where filaments of glass
were detached from a liquid surface by weights applied to a
scale pan which formed the opposite end of a balance.

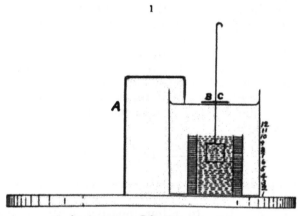

1

A, hook gauge; BC, disk; D, sinker.

The method described here is free from difficulties of the
above nature. An iron wire is bent into the form of a T (fig.
1) and the cross allowed to float on the surface of water. The
body of the T projects a little distance into the center of a
coil of copper wire through which a current may pass. The
part of the T forming the plunger is adjusted so that its weight
is nearly sufficient to break through the liquid surface. A
slight pull towards the center of the coil due to the passage of
a current will then accomplish this. The system is raised and
lowered by a rack and pinion so that it may be placed on the
liquid surface as gently as possible.

By means of a continuous resistance the current may be
applied in small increments, and the amount required to break
through read from an ammeter. When the current is applied
the plunger immediately moves to the center of the coil and

the pull is then steadily applied to the center of inertia of the system.

The magnetic force may be calibrated at leisure by use of a balance or it may be calculated from the constants of the coil. This in proper units must be added to the weight of the system.

In some experiments already made by this method a circular disk of thin ferrotype plate was used and a rather heavy sinker of soft iron fastened to its center. The variation of surface tension with temperature in case of distilled water, the effect of adding a drop of oil, the surface tension of a soap solution and the results with disks of different areas have been tested.

In order to test the law of inverse squares a hook gauge was employed so that the height of the water surface, and hence the floating disk, could be adjusted very closely. By means of a scale fastened to the side of the vessel containing the liquid, the weight of the surface could be changed by successive centimeters. If we assume that the force required to break the tension is constant, and if we call the distance between the center of attraction of the coil and the magnetic center of the system attracted, when they are nearest together, x; then $x+1$, $x+2$, etc., will represent other distances used. Let c', c'', etc., stand for currents corresponding to these distances and we have

$$\frac{c'}{x} = \frac{c''}{x+1} = \frac{c'''}{x+2} \text{ , etc.}$$

Substituting values and solving we have for a given experiment $x = 6^{cm}$.

Applying the values to other measurements we have 1·88, 1·81, and 1·95 as ratio numbers expressing the value of the same force, namely that required to break through the surface tension.

Very little work has been done with this apparatus and this paper is designed to describe the method rather than announce results.

Orono, Maine.

SCIENTIFIC INTELLIGENCE.

I. Geology and Mineralogy.

1. *The Glacial Gravels of Maine and their associated deposits;* by George H. Stone. U. S. Geol. Survey, Monograph XXXIV, pp. 1–499.—Ever since the early writings of Agassiz on the glacial geology of New England, the "horsebacks" and terraces of Maine have been generally known, but the great extent and remarkable perfection of the glacial deposits is a recent discovery. Prof. Stone's work, extending over a period of thirteen years, has resulted in a series of maps showing the distribution of the water-laid sediments of glacial time over the whole state. The deposits from each glacial stream have been mapped and grouped into some sixty gravel "systems." Each "system" is composed of trunk stream (usually an esker) with tributaries and distributaries (delta branches). Under separate heads a detailed description is given of composition, shape, etc., of each group. The gravel "systems" are usually 10–25 miles in length ; the Houlton "system" is 115 miles long and the Katahdin stretches out for 125 miles. The longer "systems" extend for many miles like an artificial roadway carefully constructed, and with the exception of Sweden constitute the finest examples of gravel ridges in the world.

This reconstruction of the river systems of glacial time has been done with great labor, for there are no adequate maps of Maine and much of the region is a succession of swamps and forests. The fifteen maps accompanying the monogram show at once the difficult nature of the work and the remarkable development of the forms of drift in the state. The chapters on Classification and Genesis are valuable not so much because of the theoretical conclusions drawn as for the great mass of detailed description and discussion of the abundant material at hand. Much has been done in glacial geology since the manuscript for this monograph was completed (1889), but we know of no more important contribution to the general discussion of eskers. Besides the treatment of eskers space is given to descriptive and theoretical matter in regard to kames, coast gravels, valley drift and tills. A study of the till and the marine clays was made to determine whether there was an interglacial period in Maine as in the Mississippi Valley, but the evidence was found to point to a single glacial period. H. E. G.

2. *The Illinois Glacial Lobe;* by Frank Leverett. U. S. Geol. Survey, Monograph XXXVIII, pp. 1–797, with 18 maps and 15 illustrations.—This is the most elaborate volume on Glacial Geology yet published and renders accessible the mass of facts gathered during many years of research by a force of geologists under the direction of Prof. Chamberlain. The problem before Mr. Leverett was the investigation of the complicated phenomena

arising from the movements and deposits of the successive advances of the ice front in the Lake Michigan region. Not only detailed study but more refined methods of investigation were demanded, and the development of these methods will be useful in all future glacial work. The various drift sheets are described in detail under the following heads : 1. The Illinoian Drift Sheet—the most extensive in the region, which was preceded by an interval of deglaciation (Yarmouth) and followed by the Sangamon interval of deglaciation. 2. The Iowan Drift Sheet, in the discussion of which the whole loess problem is gone over and conflicting views compared, without however arriving at a definite general conclusion as to the origin of the loess. The Iowan deposits are overlaid in places by the "Peorian soil" of another interval of deglaciation. 3. Two Wisconsin Drift Sheets complete the series.

Perhaps the most generally interesting part of the monograph are chapters xi and xii, dealing with beaches of Lake Chicago and the Chicago outlet and with the many cases of reconstructed drainage throughout the drift-covered area. Each stream is treated separately and the past and present conditions of its drainage basin is discussed. In Chapter xiv all the reliable well records of Illinois are given and conclusions drawn as to character of soil, relation of drift to water supply, etc. From the well data and from comparison of the drift areas to adjoining driftless areas the average thickness of the Illinois drift sheet is given as 100 to 130 feet. H. E. G.

3. *The Geography of the Region about Devils Lake and the Dalles of the Wisconsin with some notes on its Surface Geology;* by ROLLIN D. SALISBURY and WALLACE W. ATTWOOD. Wisconsin Geol. and Nat. Hist. Survey, Bulletin No. 5, pp. 1–146. 85 plates and illustrations.—This bulletin of the Wisconsin Survey begins an educational series designed primarily for use in the schools. It marks the beginning of a plan to make the work of the geological surveys directly useful to all intelligent people of a state instead of to the few specialists. The volume was evidently prepared by skillful teachers, for the region chosen for description is clearly marked and typical and interesting and the plan of presentation is that of the most successful instructors. Part I is a description of the topography of the area—the quartzite ridges and the sandstone plains. Part II deals with the history of the topography. The origin of the rocks which show themselves at the surface is explained and their scenic features which result from erosion are described. The glacial period and its effect on topographic form is treated of easily. In connection with each division of the subject there is given just enough theoretical matter to put the abundance of local facts in proper perspective. The volume contains no geological facts regarding this interesting region which have not been published before, but it is timely and valuable as a plan for presenting a geographical subject to schools and as a suggestion to state surveys. H. E. G.

4. *The Moraines of Southeastern South Dakota and their Attendant Deposits;* by James E. Todd. U. S. Geol. Survey, Bulletin 158, pp. 1–168.—This report is a description of the Pleistocene deposits of the Dakota - Nebraska region. The deposits are described as a first or outer moraine which extends down to the Missouri River, a second moraine bordering a region farther north, and various forms of drift outside the moraines. The drift region beyond the outer moraine is largely loess and Prof. Todd describes at some length the distribution character of these interesting deposits. The Missouri River is believed by the author to have been largely excavated since the ice age and, furthermore, that the river is still deepening its trough.

H. E. G.

5. *Preliminary Report on the Copper-bearing Rocks of Douglas County, Wis.;* by U. S. Grant. Wisconsin Geol. and Nat. Hist. Surv. Econ. Series, No. 3, Bulletin VI. Pp. 55, maps and plates. Madison, 1900.—This work has for its purpose a study of the copper-bearing rocks of the Keweenawan series in the region southeast of Duluth and Superior, the so-called Douglas Copper Range, to ascertain the character and value of the copper deposits. The rocks and the occurrence of the ores are quite similar to what is found in the well-known region of Keweenaw Point. The promise of valuable deposits is not, however, very great, the copper being scattered so much as to make in general very lean ores. While the value of the work is chiefly local, it will also prove of interest to all who are interested in the origin and occurrence of ore deposits.

L. V. P.

6. *Bulletins of the Geological Survey of Finland, No. 9. Esquisse hypsometrique de la Finlande;* by J. J. Sederholm. Pp. 17, 8vo. Helsingfors, 1899.—This is a short description of a relief map of Finland on a scale of 1 : 2,000,000, in which the main features of the topographic relief are brought out and described.

No. 10. *Les depots Quaternaires en Finlande,* by J. J. Sederholm (pp. 28, 1899). This little work contains a brief but very good description of the glacial deposits of Finland and is accompanied by an excellent map on the same scale as the preceding one. The differences of relative level of land and sea at different times, the movement of the ice and the remarkable and celebrated osar, colossal glacial embankments crossing the country in long lines and well shown on the map, are described and discussed.

No. 11. *Neue Mitteilungen über das Ijolitmassiv in Kuusamo,* by V. Hackmann (pp. 45, 8vo, 1900). Ijolite is a name given by Ramsay and Berghell to the rock forming a mountain mass composed of pyroxene and nephelite and of granular structure. Recently Hackman has studied this occurrence in the field and gives a full description, showing it to be a complex of alkaline rocks ranging from ijolite through essexite into pyroxene-syenite. The ijolite is the main type. Several new analyses are given and the whole is an interesting contribution to the petrography of the alkaline rocks.

L. V. P.

7. *Analyses of Rocks, Laboratory of the U. S. Geological Survey*, 1880 *to* 1899; F. W. CLARKE. Bull. 168, U. S. Geol. Surv., 1900, pp. 308. This is a new edition of Bulletin 148 of the Survey publications, containing a large number of new analyses, chiefly executed by W. F. HILLEBRAND and H. N. STOKES. All told, some 1,400 analyses of rocks, clay, soils, etc., are given. The order of the analyses has been changed, the more important constituents being given first, as is usual with many petrographers. This makes the analyses much more convenient for use. As before, the arrangement is geographical. The former chapter on rock analysis by Hillebrand is omitted and it is stated that this, in an enlarged form, will soon appear as a separate bulletin. One cannot but be impressed by the enormous amount of painstaking, careful and skillful work represented by these analyses, so complete and so detailed. From no other source could such a mass of similar analyses be gathered, and they doubtless represent the most important work done in American chemical geology.

L. V. P.

8. *The Bjurböle, Finland, Meteorite.*—Portions of this meteorite are now on exhibition in the Finland pavilion at the Paris Exposition. The fall took place March 12, 1899. The mass, which had a total weight of about 340 kilos, fell into the Gulf of Finland, making a hole about nine yards in diameter in the ice (which had a thickness of about 28 inches) and penetrating the submarine clay to a depth of six yards. The removal of the mass has been a matter of considerable difficulty and was accomplished only in sections. The portions on exhibition at Paris are the largest secured and weigh 83 and 22 kilos respectively. The interior of the stone is gray with yellowish spots. The exterior is covered with a black crust. Prof. Stanislas Meunier gives in a recent issue of "Le Naturaliste" the results of a detailed study of the meteorite. He describes it as friable in texture and in structure of the variety to which he applies the term öolitic. The mineralogical composition is largely chrysolite and pyroxene, with which are associated grains of nickel-iron, iron sulphide and chromite. Sp. gr. $= 3\cdot54$. Chondrules varying in diameter from $\cdot5$ to $2\cdot5^{mm}$ are easily separated. They are of several types, which are fully figured. The most remarkable are those of a fibrous, probably feldspathic, mineral often enclosed in a metallic shell, and those containing porphyritic crystals of augite. The meteorite is classed by Meunier as belonging to his Type 38, Montrejite. The stones of Hessle, Aussun, Prairie Dog Creek and Searsmont will be remembered to be other examples of this type.

O. C. F.

OBITUARY.

Professor G. H. F. ULRICH, F.G.S., Director of the School of Mines of Otago University, Dunedin, New Zealand, died on May 26, at the age of seventy years. By his active investigations, first in Victoria and later in New Zealand, he contributed much to the knowledge of the Mineralogy and Geology of those countries.

ART. XXV.—*Notes on the Colorado Canyon District;* by
W. M. DAVIS.

A VISIT to the district of the Colorado Canyon leaves the
traveler with a deepened impression of his indebtedness to the
three explorers from whom an understanding of that marvelous
region has come. Newberry, crossing the plateaus south of
the canyon in 1858, recognized the deep erosion of the fundamental crystallines before the deposition of the Paleozoic
series, and the extensive denudation of the plateau uplands
where the retreating escarpments succeed each other in order
of age; he made explicit statement that not only the canyon
but the escarpments of the plateaus " belong to a vast system
of erosion, and are wholly due to the action of water."[*]
Powell's adventurous expedition down the river in 1869, justly
classed with the most daring explorations of the continent,
added an account of the division of the plateaus into huge
blocks by faults and monoclines trending about north and
south, in contrast to the cliffs of erosion which in a general
way trend east and west; he clearly stated that " the cliffs of
erosion are very irregular in direction, but somewhat constant
in vertical outline; and the cliffs of displacement are somewhat regular in direction, but very inconstant in vertical outline."[†] Dutton's more elaborate surveys of the canyon and of
the plateaus to the north of it in 1879 and 1880, aided by
Holmes' wonderful drawings, have made the canyon district
classic ground, the type of all that is gigantic in displacements
and denudation, the region to be cited, at home and abroad, as

[*] Ives Expedition: Report upon the Colorado River of the West, Washington,
1861. Part III, Geological Report, 45.
[†] Exploration of the Colorado River of the West, Washington, 1875, 191.

the standard example, best studied, best described, and best illustrated, of river work on a grand scale.* It should be added that the topographical maps made by Bodfish and Renshawe in 1879, although of small scale and large contour interval, are of great value in giving the location and altitude of the more important features.

The wish that I have had for many years to visit the canyon was fortunately brought to fulfilment in June. Our party consisted of Prof. R. E. Dodge of Columbia University, Dr. H. E. Gregory of Yale University, Mr. R. L. Barrett of Chicago, Mr. Richard Wetherill of Pueblo Bonito, New Mex., Dr. Tempest Anderson of York, England, and the writer. Great assistance in plans and outfit was given by Mr. F. E. Hyde, Jr., of New York, who was unhappily unable to go with us. We left the Sante Fe Pacific Railroad at Flagstaff, Arizona, on June 4, and followed a very irregular route northward, partly with wagons, partly with horses and pack train, and on June 26 reached Milford, Utah, whence the Oregon Short Line carried us to Salt Lake City. Postponing a fuller account of our observations to a later date, I desire to make brief statement here of certain points on which other conclusions than those announced in the Survey reports seem to be admissible.

The unconformities in the Canyon.—The Kaibab section of the canyon discloses the nearly even floor on which the horizontal Paleozoic strata rest. The floor is of complex structure. The fundamental schists with granitic dikes, which are exposed in the western Kaibab section, are overlaid in the eastern section by the heavy Unkar and Chuar series, dipping eastward and measuring according to Walcott about two miles in thickness.† The wedge in which the tilted formations terminate westward is a most remarkable geological structure, alike for its distinctness and for its significance. It is easily recognized near the bottom of the northern cliffs of the canyon when viewed from the southern rim at Hance's or at Cameron and Berry's; these primitive stopping places being at the ends of the road from Flagstaff, which forks a few miles before reaching the rim. The wedge is still better seen by descending the Grand View trail from Cameron and Berry's to a promontory of the Red Wall limestone north of the "copper mine." The crystalline floor of the Unkar series is of extraordinary evenness; no inequalities in it were detected in the view from the Red Wall promontory, except a small pre-Paleozoic fault (see

* Tertiary History of the Grand Canyon District, with Atlas, U. S. Geol. Survey, Monogr. II, Washington, 1882.
† 14th Ann Rept. U. S. G. S., 1894, 508–512.

diagram by Gilbert, fig. 52, in Walcott's article, above referred to). Close to the base of the Unkar is a sheet of basalt, described as a contemporaneous flow by Walcott (l. c., 516–518), but to our view seeming rather to be an intrusive sill, inasmuch as it broke from one layer to another, instead of lying conformably between the under and overlying beds.

The floor of the Paleozoic series is somewhat less regular than that of the Unkar. A few miles down the canyon there is a mound of schists that rises rather higher than the basal Tonto sandstone, and the harder layers of the lower Unkar also interrupt the Tonto for a mile or more up stream from the apex of the wedge on the northern side of the canyon. It is notable that the deformed schists beneath the Unkar series descend by a continuous and very steep slope to the river; hence their whole mass must be about equally resistant; but west of the apex of the wedge and for a number of miles down

FIG. 1. Diagram to illustrate the section and profile of the canyon in the Kaibab

stream, the schists show a distinct bench just beneath the Tonto before steepening to their precipitous lower slope; hence here the upper part of the schists must be weaker than the lower. It occurred to me that this weakness may be ascribed to a pre-Tonto subærial weathering of the schists. If so, the gently undulating sub-Tonto floor may be, with renewed confidence, regarded as an ancient lowland of subærial denudation, as described by Powell, only slightly modified by marine erosion previous to Tonto deposition; while the sub-Unkar floor may have been planed down by the sea to smooth and firm rock from some unknown but probably moderate measure of relief, before the Unkar series was laid down on it.

The erosion of the plateaus.—The retreating cliffs of Permian, Triassic and other formations, on the north and east of the Grand Canyon district, have been taken to prove extensive erosion over the plateaus in which the canyon has been cut. Dutton concludes that this erosion was accomplished during a lower stand of the region ; in other words, that the plateaus are parts of a vast peneplain produced in a cycle of erosion anterior to that in which the canyon has been eroded, the two cycles being separated by an uplift of several thousand feet. It has been pointed out by other observers (although I believe that no statement of their opinion has been published) that the supposed peneplain surface coincides roughly with the resistant Upper Carboniferous strata over large areas, and hence that the plateaus might be explained as a surface of halting or at least of hesitating erosion in a single cycle, after the stripping of the weaker overlying formations. It is not easy to make definite choice between these two explanations, but I believe Dutton's to be the more probable of the two, for such reasons as the following. There is a mesa in the valley of the Little Colorado, southeast of the road from Flagstaff to Tuba, capped by a horizontal sheet of lava which rests unconformably on a bevelled surface of tilted Permian shales ; but the shales are elsewhere denuded to a lower level. Again, there are numerous and extensive landslides at the base of the Echo and Vermilion (Triassic) cliffs south and west of Lee's Ferry; the slides seem to indicate a revival of sapping after a long pause. Further, the considerable length of many streams that flow against the dip of the strata suggests the need of the most favorable conditions for their development, such as would be associated with two cycles of erosion rather than with only one. Finally, the strong contrast between the gigantic young ravines of the steep canyon walls and the shallow mature valleys of the plateaus seems beyond explanation as dependent on the resistance of the Upper Carboniferous strata during a single cycle of erosion. But much more work should be done before this question can be definitely settled.

The Esplanade.—The canyon, where traversing the Uinkaret plateau and the greater part of the adjacent Kanab, consists of a broad and flat-floored upper valley, beneath which a deep and narrow chasm has been cut. Dutton terms these two parts the upper (or outer) and the inner canyons, and gives the name, esplanade, to the well developed floor of the upper canyon. He explains the esplanade as the result of valley-broadening during a pause in the general uplift in consequence of which the plateaus have been dissected : in other words, after a first cycle in which the plateaus were broadly denuded, a second cycle was introduced by general uplift of the region ; but this uplift was interrupted by a pause long enough for the erosion of the upper canyon (l. c., 121). Other observers have suggested that the esplanade is a structural bench determined by the resistance of the Red Wall limestone, from which the overlying weak lower Aubrey sandstones and shales have been stripped. Certainly the surface of the esplanade, as seen in the magnificent prospect from Vulcan's Throne, where the canyon cuts the Toroweap valley, coincides closely with the top of the Red Wall series. Further up the canyon, the topographical maps show the esplanade in Kanab canyon on the north and in Cataract canyon on the south, but its distinctness lessens eastward and it disappears before entering the Kaibab. Where we saw the canyon in the Kaibab, its cross section was altogether different from that where the esplanade is well developed in the western Kanab. In the Kaibab there are two structural benches, as shown in fig. 1: one is determined by the Red Wall limestone, the other by the Tonto sandstone ; and of the two the latter is rather more pronounced than the former. The change in form from the Uinkaret to the Kaibab seems to result from a weakening of the strata that are included between the firm limestones of the Red Wall and the firm sandstones of the Tonto. This change in structure is fully recognized by Dutton, and he explains the difference of form between the Uinkaret and the Kaibab sections by the greater altitude of the Kaibab during the erosion of the canyon (l. c., 257, 258). He does not, however, explain the absence in the Kaibab section of all signs of the pause during the general uplift by which he explained the esplanade further west. To my view, the esplanade, like the Kaibab benches, seemed to be entirely of structural origin ; it does not seem to be necessary to postulate a pause during uplift in order to account for it.

The Toroweap is a broad valley that has been eroded along the fault that divides the Kanab and the Uinkaret plateaus, north of the canyon. A similar valley, which I have designated in my notes as the South Toroweap, is seen on the southern side of the canyon. Both of these valleys are flat-floored and open on the canyon close to the level of the esplanade. Dutton concluded that the Toroweap is the broadened valley of an ancient river which disappeared by reason of change from humid to arid climate before the inner canyon was cut. He recognizes that the valley has been somewhat shallowed by lava flows from the volcanoes of the Uinkaret on the west, but does not attach much importance to this fact (l. c., 92, 99). It does not seem necessary to adopt Dutton's conclusion as to climatic change, for the lavas in the floor of the Toroweap appeared to be in sufficient volume to account for its shallowness in contrast to the depth of Kanab and Cataract canyons a little further east.

Date of faulting.—The displacements by faults and monoclines, resulting in the division of the plateau region into a number of huge blocks, Shiwits, Uinkaret, Kanab, Kaibab, etc., have been variously dated by different observers. Powell states that when the great denudation began, there were no faults and no benches; the first uplift was broad and even, without differential displacements; the displacements began later, and so slowly as not to affect the course of the original streams (l. c., 200, 201). Dutton dates the Hurricane fault as later than the earlier lava flows and earlier than the later ones (hence later than the great denudation and before the later canyon cutting); the Toroweap fault was not begun until the upper canyon had attained very nearly its present form, so that the floor of the esplanade as a topographic feature was dislocated by the fault; the Kaibab began to have a distinct existence only after the great Miocene denudation, when the Colorado was beginning the erosion of the upper canyon; the Echo cliffs monocline is rated as very nearly coeval with the East Kaibab monocline, and the latter began with the Pliocene (l. c., 117, 94, 192, 205). Walcott places the uplift of the Kaibab before the completion of the great (Miocene) denudation,[*] while there were still some heavy overlying strata upon it, because the layers of the monoclines are flexed without manifest fractures. The dating of the displacements therefore still offers an interesting subject for discussion.

While subordinate movements on the fault lines and monoclines may have occurred during the erosion of the inner canyon, the chief movements seem to be much older. This conclusion is enforced by several observations. First, there is the arrangement of the Permian and Triassic escarpments with respect to the northern extension of the fault lines: had the faults been of recent date, later than the great denudation by which the escarpments were pushed back for scores of miles from the original extension of their strata, the borders of the several formations as shown on a map ought not to be significantly out of line on the two sides of a displacement; for in the post-displacement period there would not have been time for much retreat of the cliffs on one side of a fault in excess of the retreat on the other side. But as a matter of fact, the Permian and Triassic cliff-makers have been greatly denuded with respect to and subsequent to the displacement of the blocked plateaus. The Permian cliffs retreat northward as they approach the Hurricane fault south of Toquerville; the Permian and the Triassic cliffs are out of line by several miles on the two sides of the fault that passes near Pipe Spring, west of Kanab. This style of arrangement was recognized by Powell, who stated that the higher blocks have been more eroded than the lower ones, and hence "the cliffs of the higher blocks stand further back from the axis of upheaval than those of the

* Bull. Geol. Soc. Amer., i, 1890, 60, 64.

lower blocks " (l. c., 191). Secondly, the Vermilion cliffs of the Trias make a great detour around the northern end of the Kaibab arch, such as could not be accounted for if the uplift of the Kaibab were as recent as the erosion of the canyon. Thirdly, the Trias margin along the Echo cliffs is in close accordance with the monocline of the same name; this proves that time enough must have elapsed since the bending of the monocline for the Triassic sandstones to retreat from whatever irregular front they may have had while they still lay horizontal, to their present well-defined structural alignment. The time that sufficed for the erosion of the canyon by the active Colorado does not seem long enough for the recession of the Triassic cliffs from an irregular to a regular front under the attack of dry-climate weather alone. Fourthly, the relatively small retreat of the Aubrey cliffs from the fault lines, as along the Hurricane and the Toroweap faults, may be explained by the relatively recent uncovering of these structures, long ago faulted.

The origin of the drainage system.—Both Powell and Dutton attribute an antecedent origin to the larger members of the drainage system in the plateau region. It is possible to conceive of the conditions necessary for the maintenance of such a drainage system through all the movements of upheaval and displacement that the plateau region has suffered; but it seems to me doubtful if these necessary conditions have been actually provided; and it is very difficult to find decisive proofs of an antecedent origin. During the great denudation by which thousands of feet of Mesozoic strata were stripped from the plateaus, there must have been opportunity for many spontaneous rearrangements of river courses. This opportunity would have been increased if there had been several successive cycles or partial cycles of denudation, separated by movements of elevation, as is entirely possible: the opportunity would have been still further increased if, as is eminently probable, displacements with slight tilting occurred during the denudation, for, as Hayes has shown, the greater rearrangements of drainage are dependent rather on gentle tilting movements of elevation than on inequality of rock resistance. The denudation of the region is usually dated as beginning in early Eocene time, because it is postulated that late Eocene strata once stretched all over the canyon district; but if a true-scale section be drawn, it does not appear improbable that the cañon district may have been slightly uplifted so as to undergo some denudation while the Eocene deposits were accumulating in basins on the north and east. The whole stretch of Tertiary time, with repeated uplifts and displacements, would certainly offer abundant opportunity for the development of new stream courses, even if the drainage of the region had been chiefly antecedent at the beginning. Moreover, some of the valleys actually follow the strike of weak monoclinal structures, and are therefore to be explained as of subsequent rather than as of antecedent or consequent origin; other valleys slope against the dip of the strata, and may come to be classed as obsequent instead of antecedent. Altogether, this problem is so complicated that its solution seems out of reach at present. The main trunk of the Colorado may be antecedent, but such an origin seems improbable for the side streams.

Climatic changes.—The change from a moist Miocene climate to an arid Pliocene climate, already referred to in connection with the Toroweap valley, does not seem to be essential to the development of existing forms. House Rock valley, for example, is explained by Dutton as having been eroded by a member of the original antecedent drainage system, whose waters dried up at the time of the change from the moist Miocene to the dry Pliocene period; but may this valley not be equally well explained by headward erosion under a persistently dry climate along the weak Permian monoclinal strata that lie between the resistant Carboniferous strata of the Kaibab arch on the west and the heavy Triassic sandstones of the Paria plateau on the east? The valley seems to belong to the class of subsequent valleys above referred to; the depth of its floor having been determined by the sill of resistant Carboniferous strata in the platform of the Marble canyon on the southeast. The valley is not yet significantly deepened in response to the uplift which has permitted the erosion of the main canyon by the full-bodied Colorado.

The ravines which dissect the flanks of the Kaibab and the uplands of the Paria plateaus are ascribed by Dutton to the increased rainfall of the Glacial period, by which the aridity of Pliocene time was interrupted (l. c., 196, 202, 228); but in view of the active although intermittent erosion and transportation now going on in at least some of the flanking ravines of the Kaibab, it seems inadvisable to limit the origin of ravines of this class to any special period of time. Effective work is done on them to-day when thunderstorm torrents flush their channels, and similar work has probably been done ever since the uplands were stripped of the weak overlying Permian strata and exposed to erosion.

It may be added that the longitudinal summit valley of the Kaibab, classed by Dutton with House Rock and Toroweap valleys as the work of Miocene streams that became extinct with the coming of a dry Pliocene climate (l. c., 193–195), seems explicable as a result of underground drainage. Sink holes of various sizes are found on the Kaibab uplands, and Dutton reports that large springs issue from the walls of the Kaibab canyon (l. c., 138, 196).

It was impressed upon me throughout our excursion over the plateaus that a multitude of problems connected with the history of the canyon district await solution. The earlier explorers gathered a great harvest there, and presented their results so vividly that newcomers may enter the region as if already acquainted with it. But the gleanings that the earlier explorers left are abundant; local structural studies as well as such considerations as are involved in dates of faulting, origin of stream courses, cycles of erosion and past climatic conditions, offer plentiful material to investigators for years to come.

Harvard University, July, 1900.

ART XXVI.—*On the Determination of Minerals in thin Rock-sections by their maximum Birefringence ;* by L. V. PIRSSON and H. H. ROBINSON.

IN the determination of minerals in thin sections of rocks, the common species possess certain diagnostic characters by which they are as a rule readily recognized. In doubly refracting minerals the interference colors or the birefringence between crossed nicols is probably the most generally useful character for diagnosis and this combined with the refractive index is sufficient in many cases to determine the mineral at a glance, when taken in connection with its color, cleavage, etc.

In the case of the less common or rare minerals, the approximate determination of the maximum birefringence becomes in many cases of the highest importance and in combination with other characters may be sufficient to definitely determine the species.

It is thus a matter of the greatest value to the beginner in microscopical petrography that he should learn the values of the maximum birefringence exhibited by the common minerals and associate these in his mind with the colors exhibited between crossed nicols. To do this he should have a simple and relatively accurate method. The more experienced petrographer has of course learned these relations for the common minerals by practice, though generally without any definite method.

In the case of birefringent minerals the strength of the interference colors shown by sections between crossed nicols depends on three conditions : (1) on the orientation of the section with regard to the ellipsoid of elasticity, (2) the thickness of the plate and (3) the difference in values between the axes of greatest and of least elasticity.

If the axial ellipsoid be cut perpendicular to an optic axis, and this in uniaxial crystals is of course a basal section, the section of the ellipsoid has the form of a circle. Hence for parallel light the elasticities are alike in all directions, the plate is isotropic and shows no interference-colors between crossed nicols. As soon, however, as the ellipsoidal cross-section is inclined to this position it takes an elliptical form, the light elasticities are not alike in all directions and hence interference colors are shown between crossed nicols. As the cross-section is inclined more and more the eccentricity of its elliptical form continually increases and with it the strength of the interference colors until the maximum of both is reached ; in uniaxial crystals this is in the direction of the prismatic zone parallel to the vertical axis and in biaxial crystals in a direction parallel to the plane of the axes of greatest and least elasticities, that is to say the plane of the optic axes. In these cases the difference between the indices of the axes of greatest and of least elasticity is the maximum birefringence, since it indicates the maximum eccentricity that can be obtained in any elliptical section of the ellipsoid of elasticity ; for this, therefore, the strongest interference colors are shown. We have prepared a table of the values of maximum birefringence which is given on page 264. These values, which have been carefully collated from the best and most recent determinations, are given in descending order and the list embraces the great majority of rock-forming minerals.

The recognition of sections of a mineral in a rock slide which exhibit the maximum birefringence is somewhat inadequately treated in most handbooks, it being generally assumed that the student will form his own method from the optical data given. It involves the determination of the "order" of the interference color displayed. It rarely happens that the bounding walls of a mineral section in the slide are exactly perpendicular to the plane of the object and cover glasses ; on the contrary they are apt to be inclined or in many cases have the form of a thin wedge. From the thin edge of such a wedge, especially by the aid of a high power and where the colors are not interfered with by the overlapping of some other mineral, as at the edge of the section, or near a crack or hole, or against some isotropic mineral or one cut nearly normal to an optic axis, it is often possible to count the successive colors in ascending order, finally arriving at the top and noting the color given by the full thickness of the section. This is essentially the same process as moving up the thickening quartz wedge and noting the colors it yields. If the ascending slope to the top of the section is extremely steep the stronger colors of the first order may be so crowded as to appear almost like a dark line. If by any chance this process cannot be used, then the section must be brought to extinction by the quartz wedge, the section then removed and the color in the wedge noted and its order ascertained by slowly withdrawing the wedge and noting the successive orders of colors down to zero.

The colors also vary with the thickness of the section. Good sections as made to-day should not vary sensibly in thickness in different portions, that is the two surfaces should be parallel. They should also average about 0.03^{mm} or 0.001 inch in thickness, or the maximum color exhibited by quartz should not be above clear white or at most pale straw-yellow.

With this thickness for the slide the majority of rock-making minerals show maximum colors in the first and second orders, colors easily remembered and determined.

In a rock slide containing a number of sections of the same mineral, the ellipsoid of elasticity will be cut in various directions and some of them must approach the maximum of birefringence, the probability increasing with the number of the sections. While it will rarely happen that one is thus oriented exactly parallel to the axial plane or optic axis, sensible divergences may exist without perceptibly altering the birefringence or vitiating the method to be presently described. In a number of haphazard sections of a mineral then, the one showing the maximum of birefringence may be assumed to be nearly parallel to the optic plane or axis thus giving approximately the highest birefringence of the mineral. This determination may be confirmed by the fact that such sections should not yield any loci of optic axes or axial bars or bisectrices in convergent light. The color in plain light, the pleochroism with one nicol, the cleavage and crystal form often help, in addition, to confirm the orientation.

If now we consider only those sections of minerals which give the maximum of birefringence, we may eliminate the factor of the orientation of the section in considering the strength of the interference colors and this will then depend on the other two factors, the thickness of the plate and the value of birefringence. There is thus a constant relation between them and if we know the thickness of the plate and the value of the maximum birefringence, we may readily calculate the interference color; or, given the thickness and the color, the maximum birefringence may be found; or, finally, with this and the color, the thickness of the plate.

These relations have been put into graphic form by Michel-Lévy in the beautiful colored plate in "Les Minéraux des roches" by Michel-Lévy and Lacroix. We have not found them expressed in graphic form in any other handbook with which we are acquainted.[*] The advantage of having these relations shown in a diagram where they can be readily seen, followed and used is very great. It is probable that the uncertainty, difficulty and cost of the colored plate may also have deterred some authors from using it. It is, however, by no means necessary that the plate should be colored, the diagram is the essential part and this we have thought would be of service to petrographers and students in this country, many of whom are unacquainted with the colored plate of Michel-Lévy, or are deterred from using it by its cost and inconvenient form. We have put it into compact shape and have simplified the color-names, by indicating the place of the distinct and well-

---

[*] Such as Rosenbusch, Rosenbusch-Iddings, Zirkel, Luquer or Harker. The last named, however, gives some very useful practical hints in this direction.

known colors rather than the transition tones. If anything is sacrificed to accuracy by this method, it is more than compensated for by simplicity and convenience in use.

In using this diagram (p. 265) for the determination of an unknown mineral, the thickness of the rock-section is first determined. This is done by taking some well-known mineral, such as quartz, for instance, which furnishes a number of sections in the slide and observing the highest color given by any of them. The maximum birefringence for quartz is 0·009 as seen in the table and by following this line down toward the left-hand lower corner until it intersects the vertical line of the highest interference-color observed, and by then following out to the edge on a horizontal line the thickness of the section can be told. The numbers represent hundredths of millimeters.

For the determination of an unknown mineral the highest color given in numerous sections is observed and the thickness having been determined as above, by means of the diagonals the numerical value is noted which corresponds to the given color in a section of the determined thickness. The maximum birefringence of the unknown mineral having been determined the table of birefringences is referred to and the mineral usually found to be one of a group of several having approximately the same values. Which particular one it may be, is in the vast majority of cases readily told by its comparative refractive index in plain light, cleavage, color and other optical properties.

Thus, for example, in a slide having numerous quartz sections the highest color shown is a pale yellow. The birefringence of quartz is 0·009. The diagonal 0·009 crosses the line of pale yellow at about 0·035 which gives the thickness of the section. In the slide are numerous grains of an unknown mineral, colorless, of strong refraction and whose highest color, given by several of them, is an orange-red of the second order. But there passes through the intersection of the thickness 0·035 and the line of orange-red of the second order the diagonal 0·029 which is the approximate value of the birefringence of the mineral. Referring to the table this is found to approximate to that of allanite, chondrodite, diopside cancrinite and tremolite. The refraction, color, cleavage, etc., show at once that of these, diopside is the only one which corresponds and thus the mineral is determined.

Sheffield Petrographical Laboratory,
  Yale University, New Haven, Conn., April, 1900.

BIREFRINGENCES (MAXIMUM).

0·287	Rutile.	0·028	Tremolite.	0·011	Clinochlore.
0·179	Dolomite.	0·025	Actinolite.	0·010	Dumortierite.
0·172	Calcite.	0·024	Anthrophyllite.	0.010	Staurolite.
0·126	Aragonite.	0·024	Diallage.	0·010	Topaz.
0·121	Titanite.	0·024	Hornblende,	0·009	Bronzite.
0·107	Cassiterite.		common.	0·009	Labradorite.
0·072	Hornblende,	0·023	Gibbsite.	0·009	Quartz.
	basaltic.	0·023	Tourmaline.	0·008	Albite.
0 062	Zircon.	0·022	Glaucophane.	0·008	Andesine.
0·061	Anatase.	0·021	Augite.	0·008	Cordierite.
0·055 ?	Biotite, black.	0·021	Barkevikite.	0·008	Corundum.
0·055	Epidote, dark.	0·021	Brucite.	0 008	Kaolin.
0·055	Astrophyllite.	0·021	Sillimanite.	0·008	Oligoclase.
0·050	Talc.	0·019	Pargasite.	0·007	Soda-orthoclase.
0·049	Aegirite.	0·018 ?	Tridymite.	0·007	Soda-microcline.
0·045	Monazite.	0·016	Cyanite.	0·006	Orthoclase.
0·043	Anhydrite.	0·015	Dipyre.	0·006	Microcline.
0·038	Muscovite.	0·015	Ottrelite.	0·005	Eudialyte.
0·037	Epidote, light.	0·014	Wollastonite.	0·005	Melilite.
0·036	Meionite.	0·013	Hypersthene.	0·005	Nephelite.
0·036	Olivine.	0·013	Serpentine.	0·005	Zoisite.
0·035	Biotite, light green.	0·012	Anorthite.	0·004	Apatite.
0·032	Allanite.	0·012	Natrolite.	0·002	Leucite.
0·032	Chondrodite.	0·012	Seybertite.	0·002	Vesuvianite.
0·029	Diopside.	0·011	Andalusite.	0·001	Chlorite.
0·028	Cancrinite.	0·011	Antigorite.		

INDICES OF REFRACTION (MEAN).

2·711	Rutile.	1·669	Bronzite.	1·566	Brucite.
2·534	Anatase.	1·667	Sillimanite.	1·56	Biotite,
2·53	Brookite.	1·654	Seybertite.		light green.
2·38	Perofskite.	1·65	Dumortierite.	1·559	Labradorite.
2·029	Cassiterite.	1·645	Apatite.	1·553	Andesine.
1·952	Zircon.	1·644	Glaucophane.	1·551	Quartz.
1·930	Titanite.	1·644	Anthophyllite.	1·551	Talc.
1·812	Pyrope.	1·641	Hornblende,	1·55	Dipyre.
1·811	Monazite.		common.	1·545	Nephelite.
1·791	Aegirite.	1·638	Andalusite.	1·542	Oligoclase
1·78 +	Allanite.	1·635	Tourmaline.	1·542	Gibbsite.
1·770	Almandine.	1·633	Aragonite.	1·541	Cordierite.
1·765	Corundum.	1·632	Melilite.	1·54	Serpentine.
1·751	Epidote.	1·627	Actinolite.	1·54	Kaolin.
1·747	Grossular.	1·626	Wollastonite.	1·54 +	Canada balsam.
1·741	Staurolite.	1·624	Pargasite.	1·535	Albite.
1.721	Vesuvianite.	1·622	Tremolite.	1·526	Microcline.
1·721	Augite.	1·622	Dolomite.	1·523	Orthoclase.
1·720	Cyanite.	1·622	Chondrodite.	1·507	Cancrinite.
1·719	Hornblende,	1·62	Topaz.	1·507	Leucite.
	basaltic.	1·609	Eudialyte.	1·496	Hauyne.
1·718	Ottrelite.	1·601	Calcite.	1·487	Analcite.
1·715	Spinel.	1·591	Biotite, black.	1·486	Natrolite.
1·705	Astrophyllite.	1·59	Clinochlore.	1·483	Sodalite.
1·701	Barkevikite.	1·587	Anhydrite.	1·476	Tridymite.
1·70	Zoisite.	1·585	Muscovite.	1·46	Nosean.
1·699	Hypersthene.	1·583	Meionite.	1·44	Opal.
1·688	Diallage.	1·580	Anorthite.	1·434	Fluorite.
1·683	Diopside.	1·577	Chlorite.		
1·679	Olivine.	1·567	Antigorite.		

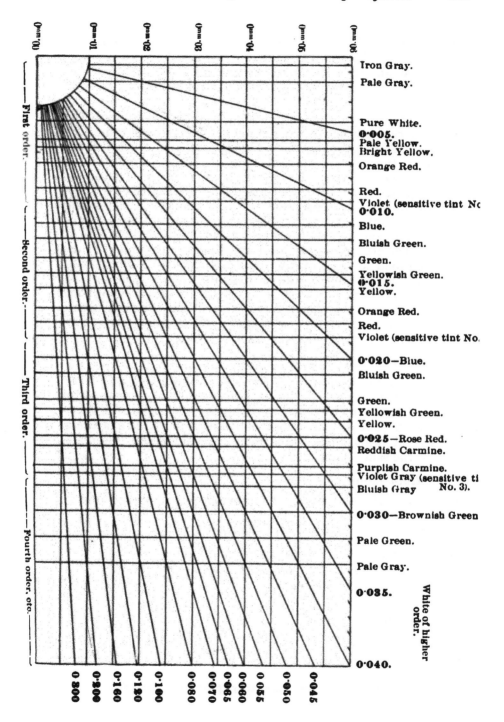

. ART. XXVII.—*Studies in the Cyperaceæ;* by THEO. HOLM.
XIV. On a collection of *Carices* from Alaska with remarks
upon the affinities of *Carex circinata* C. A. Mey. and *C.
lejocarpa* C. A. Mey. (With figures in the text.)

AMONG the very extensive botanical collections, which Dr.
Walter II. Evans brought from Alaska during the summers of
1897 and 1898, were several *Carices*, which were kindly
donated to the writer, and upon which we have prepared the
following paper. The Alaskan coast with adjacent islands has
for many years been known to possess an unusually rich vegetation of *Cyperaceæ*, especially *Carices*, and a considerable
number of highly interesting species are recorded in the works
of Boott, Hooker, Ledebour, Macoun and Meyer.* Several of
these species are absolutely local and their geographical range
is thus confined to some of the small islands or to a few localities on the main land; some others are found, also, on the
Asiatic side of Bering Sea; a few are distributed throughout
the northern part of Asia, Europe and this country, while
some of these species are even arctic and circumpolar. The
following species were collected by Dr. Evans:

Carex macrocephala Willd.

Many gigantic specimens with spikes 7^{cm} in length and over
4^{cm} in thickness; the species is considered as diœcious by
Franchet, who examined a large number of specimens from
eastern Asia, but among those collected by Dr. Evans there
were several individuals with purely staminate and pistillate
spikes on the same plant.

var. *bracteata* Holm nov. var.

The bracts of the inflorescence developed into green leaves,
about 10^{cm} in length.
Common in sand along shore, Kussiloff.

C. Macloviana d'Urv. var. *Haydeniana* (Olney).

On flat at Feeny's ranch, five miles from Kadiak.

C. canescens L.

Common about Sitka.

var. *polystachya* Bail.

In low places, Unalaska.

* For references consult the bibliography appended to this article.

C. cryptantha Holm nov. sp. (fig. E).

Roots very slender; rhizome short, cespitose; culms numerous until 45ᶜᵐ in length, triangular, scabrous above; leaves shorter than the culm, about 25ᶜᵐ in length, narrow, flat, scabrous along the margins and on the lower surface, their sheaths short, pale brown, thin in texture, not fibrillose; inflorescence very short, from 1½ to 2ᶜᵐ in length, consisting of 2 or 3 contiguous, sessile, reddish-brown spikes, mostly androgynous with a few female flowers at base; bracts scale-like, obovate, brown, membranaceous, the midrib sometimes excurrent as a bristle-like awn; scales of staminate and pistillate flowers (fig. F) broadly elliptic, acute, shining reddish brown, pointed, the margins hyaline; utriculus (fig. G) rather broadly elliptic, attenuated at both ends, pale brown, glabrous, coriaceous, with several, distinct nerves, the beak short, entire; caryopsis dull and brown, roundish in outline, almost flat; stigmas 2.

In dense mats along small streams between Kussiloff and Kenai. Its nearest ally is *C. heleonastes* Ehrh., from which it differs in its narrow inflorescence and shining reddish brown scales, which are larger and much broader, almost entirely concealing the utricle.

C. læviculmis Meinsh.

This species has been known for many years, but has not been well understood; it was referred to *C. elongata* by Bongard; to *C. remota* by Treviranus, and as "potius *C. glareosa*" by Ruprecht, while L. H. Bailey described it as a variety *sparsiflora* of *C. Deweyana* Schw. It was not until 1893 that Meinshausen segregated it as a species, and his diagnosis reads as follows: "Radix cæspitosa, fibrosa; culmis tenuissimis acute triquetris, 1–1½ ped., flaccidis lævissimis inferne foliatis; foliis linearibus planis, infimis brevissimis, summis dimidio culmi subæquantibus; spica elongata, spiculis 4–8 subglobosis remotius interruptis, infima sæpe valde remota, ferrugineo—viridibus, basi masculis; bracteis squamæformibus, ima setacea longe aristata rarius foliacea spiculæ 2–3—plo. excedente; perigyniis ovatis rostro acuminato, ore minimo integro, plano convexis marginibus acutis scabris, utrinque rubro nervatis patulis, apice paulo incurvatis demum ferrugineis, squama ovata acuta v. acuminata ferruginea acute carinata paulo breviora."

The specimens examined by Meinshausen were from Sitka and Kamtschatka (legit Mertens). Dr. Evans collected it on edges of bogs on cleared lands, Wrangell; along the beach above high tide, Howkan; in woods, common, Kussiloff.

The species has, furthermore, been collected by Thos. Howell in marshes by Yes Bay, Alaska, and by John Macoun in the vicinity of Comox, Vancouver Island.   Rev. G. Küken-

Carex lugens (A–D) and *C. cryptantha* (E–G) natural size.

thal informs us (in litteris) that he has, also, seen specimens from Idaho, Washington and Oregon, which he considers identical with this species.

C. polytrichoides Muhl.

Edge of marsh, Kussiloff.

C. tenella Schk.

In marsh, not abundant, Kussiloff.

The specimens are somewhat shorter and more robust than the typical plant from Scandinavia and Rocky Mountains, but the structure of the spikes and utriculus is the same.

C. circinata C. A. Mey.

On rocky hilltops, 900 to 1200 feet alt., Kadiak.

This species is one of the very rarest, and has only been found a few times before, perhaps not very far from where Dr. Evans collected it. Meyer, who is the author of the species, gives the locality only as "in rupibus Unalaschkæ," while Prescott, who described the Cyperaceæ for Bongard, states that it was found on "the island of Sitcha"; according to Ledebour, Redowsky is credited with having collected the plant somewhere in Kamtschatka, though with a query. In later years Professor John Macoun has reported the species as having been found by Barclay near Sitka, but these specimens, which we have seen, do not belong to *C. circinata*, but to *C. pyrenaica* Wahlbg.; the species occurs, however, on the Shunagin Islands, where it was collected by Mr. Harrington in 1871–72, and of which a few specimens are preserved in the National Herbarium. But we find no record of it in Trautvetter's works, in Kjellman's or Franchet's, nor is it included in James Macoun's list of plants from the Pribilof Islands. The species must be very local, and seems to be confined to the Alaskan coast with adjacent islands.

C. Gmelini Hook.

Among rocks on beach, Sitka; on flats, Kussiloff; abundant at Homer, Cook inlet.

C. nigella Boott.

A very few specimens were collected on Kadiak, where it grew together with *C. circinata*.

C. vulgaris Fr.

Growing in small tufts in wet places, Wrangell.

C. interrupta Bœckl.

Abundant at Wrangell, grows in dense tufts.

C. lugens Holm, nov. sp. (fig. A).

Roots thick, very hairy; rhizome ascending, stoloniferous; culm from 30 to 45^{cm} in height, slender, but stiff, sharply triangular, scabrous; leaves shorter than the culm, narrow and flat, scabrous along the margins and the lower surface; spikes 3 to 4, contiguous: one terminal staminate, 2 or 3 lateral pistillate; staminate spike peduncled, about 2^{cm} in length, the scales (fig. B) obovate-oblong, obtuse, dark brown with hyaline margins, the midrib pale, not excurrent; pistillate spikes erect, very dark, short, only 1 to 2^{cm} in length, densely flowered, the upper sessile or nearly so, the others peduncled; bracts not sheathing, very short and filiform; scale of pistillate spike (fig. C) obovate and acutish, black with very narrow hyaline margins, midrib not excurrent, barely visible; utriculus (fig. D) glabrous, a little longer than the scale, thin in texture, round obovoid with a minute entire beak, purplish above, pale green below, nerves barely visible; caryopsis obovate, flat; stigmas, 2.

Common in drier places on bay, Kussiloff.

Related to *C. cæspitosa* Good., but is at once distinguished from this by its stoloniferous habit and darker colored spikes. *C. cæspitosa* grows always in dense tufts and the culms are at the base surrounded by purplish, bladeless sheaths, besides that the utricle is pale green.

C. cryptocarpa C. A. Mey.

Abundant near villages, Sitka; on sands with *Elymus*, Kussiloff; in wet places near beach, Wrangell and Howkan.

C. macrochæta C. A. Mey.

In bogs at Sitka, 1000 feet above sea-level; on mountain top about 1200 feet, Kadiak.

C. salina Wahlbg. var. *cuspidata* Wahlbg.

On flat in rich soil, Kussiloff. The specimens were low, from only 10 to 17^{cm} in height, but agree very well with the Norwegian plant; it is very interesting to notice that the species exhibits exactly the same variation in Alaska as in Norway, thus the var. *reducta* (Drej.) has been collected by Mr. James M. Macoun on St. Paul Island, while the var. *hæmatolepis* (Drej.) was collected on the same island by Mr. Thos. H. Kearney Jr.

C. stygia Fries.

Common on flats near edge of bog, Kussiloff; in bogs and wood meadows, Howkan, Sitka.

C. livida Willd.

In rather dry places, Sitka. Specimens very tall, reaching until 30^{cm} in height.

C. pauciflora Lightf.

Abundant in bogs, Sitka.

C. utriculata Boott.

On bog, Sitka.

C. physocarpa Presl.

Common at Kadiak on rather dry mountain side at an eleva·tion of 500 feet. The specimens seem better referable to this species than to *C. compacta* R. Br., but neither this or *C. physocarpa* is well understood, thus the identification is very uncertain.

In considering the geographical distribution of some of these species, the following are circumpolar: *C. canescens, C. vulgaris, C. salina* and partly, also, *C. livida; C. cryptocarpa* is very common on the western and eastern coast of Bering Sea, and extends as far south as Japan and China, but it is otherwise known only in Greenland, Iceland, Farœ Islands and a few places on the Norwegian coast; *C. stygia,* which seems to be very abundant in several parts of Alaska has not, so far, been recorded with certainty from other countries than Finmark in Norway. *C. tenella,* which has a very wide distribution in North America, from the Atlantic to the Pacific Ocean, does not occur in Greenland, is rare in Scandinavia and Arctic Russia, has not yet been reported from Arctic Siberia, and is first met with again in Kamtschatka. *C. macrocephala* and *C. Gmelini* are only known from the eastern and western coast of Bering Sea but both extend as far south as Japan, the former even to China; *C. macrochœta* has been collected in many places on the Alaskan coast, but not, so far, on the Asiatic side, unless Kjellman's specimens from Konyam Bay may be referred to this rather than to *C. podocarpa* R. Br. *C. Macloviana* shows an exceedingly wide distribution ranging from Alaska southwards to Magellan, besides that it occurs, also, on Sandwich Islands, in Greenland and Arctic Europe. *C. nigella* is only known from Alaska and the Rocky Mountains in middle Colorado; *C. interrupta* is only known on this side of Bering Sea: Alaska and Oregon.

The very considerable material of the rare *C. circinata,* which was placed at our disposal, induced us to study its affinities, inasmuch as we received, also, a number of excellent specimens of its nearest ally, the equally rare *C. lejocarpa* C. A. Mey., for which the writer is indebted to Mr. James M. Macoun, who, some years ago, detected this species on the Pribilof Islands in Bering Sea.

These two species resemble each other very much, but are, nevertheless, readily distinguished by the following characters:

The rhizome is horizontal and creeping in *C. lejocarpa*, but cespitose in the other; the leaves are flat and straight in the former, but almost filiform and "circinate" in the other; the utricle has a hyaline bidentate beak in *C. circinata*, but a merely truncate in the other. As regards the distribution of the sexes, the spike was found to be androgynous in all the specimens, several hundreds, which we have examined of *C. circinata;* in *C. lejocarpa*, on the other hand, we have found specimens that were truly *diœcious*, and others in which the spike was androgynous, the latter being quite scarce. The number of stigmas varies in both, and we observed two and three in the same spike of *C. circinata*, three being, however, the commonest; in *C. lejocarpa* we noticed two stigmas in only two specimens.

In making our first disposition of these species we did not hesitate to place them among the "*Carices genuinæ*," not only on account of the number of stigmas being mostly three, but especially because the external structure and coloration of the spike suggested affinities to certain species of *Carex* proper rather than of *Vignea*. However it is often very difficult to decide where to place some of these little, monostachyous species, and one is often most inclined to consider the number of stigmas as much more important than it really is. The distigmatic monostachyæ are thus generally placed among the "*Vigneæ*" as representing small sections of their own, as, for instance, the *Capituligeræ* of Rev. G. Kükenthal, to which are referred *C. nardina*, *C. capitata*, *C. pulicaris*, and others, yet the *C. nardina* of the Rocky Mountains in Colorado has most often three instead of two stigmas, and as we have stated above a similar variation exists also in both *C. circinata* and *C. lejocarpa*. We should prefer to compare the structure of utriculus besides the shape and relative size of the scales and bear in mind that the formerly so-called "*Psyllophoræ*" in most cases may be referable to the higher developed "greges" not only among "Carices genuinæ," but also among "Vigneæ," as illustrated by Drejer and Boott in various instances. And in regard to our two Alaskan species, no Caricologist can possibly deny that their spikes, considered by themselves, show a striking resemblance to those of *C. sempervirens* Vill. and its allies. We have, also, but so far in vain, tried to establish some connection between them and *C. nigricans*, *C. pyrenaica*, *C. macrostyla* and *C. pulicaris;* the affinities of these species with their deciduous squamæ and reflexed utricles may, perhaps, be sought among some of the higher forms with squarrose spikes.

In our further disposition of *C. circinata* and *C. lejocarpa* we have considered them as "*formæ hebetatæ*" of that section

o which *C. sempervirens* is supposed to belong, the affinities
of which may naturally be sought among such boreal or alpine
forms as *C. misandra* R. Br., *C. frigida* All., *C. ferruginea*
Scop., etc. The systematic position of these species is not
defined in Drejer's work, but he makes some allusions to a possibly existing affinity between *C. circinata, C. Davalliana* on
he one side and *C. sempervirens, C. misandra* and *C. frigida*
on the other, while he enumerates these as "*formæ desciscen-
es*" under "*Melananthæ.*" However, on his table of affini-
ies III Drejer places these between "*Sphæridiophoræ* and
Lamprochlænæ," and he evidently considered them as represent-
ng a section or grex of their own; the work, however, which
his author began in such an admirable manner was, unfortunately, never finished. In Tuckerman's "Enumeratio methodica" we find some of these species classified as "*Fuliginosæ*"
C. fuliginosa and *C. frigida*) and as "*Ferrugineæ*" (*C. ferruginea, C. sempervirens, C. brachystachys* and *C. firma*), while
he does not seem to have known *C. lejocarpa* or *C. circinata.*
A similar disposition is, furthermore, to be noticed in Dr.
Christ's Catalogue, where these same species are arranged in
one section "*Frigidæ* of Fries," together with *C. mucronata*
All. and *C. brachystachys* Schrank. The *Frigidæ* are taken
up again by Rev. G. Kükenthal, placed between *Sphæridiophoræ* and *Hymenochlænæ.* Still another, but most unsuccessful classification is proposed by Professor Bailey, who includes
all the *Frigidæ* in *Microrhynchæ*, which he, furthermore, combines with *Æorastachyæ,** *Melananthæ* and *Trachychlænæ* (*C.
glauca* Scop.) of Drejer, while *C. circinata* and *C. lejocarpa*
are placed under a new section *Leptocephalæ* Bailey including
C. polytrichoides! It does not seem necessary to discuss further such combination, which is in no respect in accord with
the system proposed by Drejer, to whom Professor Bailey
refers as author of these various sections.

It appears as if the *Frigidæ* may really constitute a special
section, as proposed by Dr. Christ and Rev. G. Kükenthal, but
we do not know how far Mr. Kükenthal extends this section,
since this author has only treated the South American *Carices,*
of which only one, *C. Vallis pulchræ,* is enumerated. However, by examining the European and boreal species of this
section, we have noticed that certain and, indeed, very striking
analogies exist between them; moreover, the utriculus may be
well referable to one single type in all of them. This organ is
in these species more or less trigonous in transverse section;
it is membranaceous, usually glabrous or, sometimes a little

---

* The sectional name *Æorastachyæ* (αἰώρα) as proposed by Drejer is by Professor Bailey constantly cited as "*Ærostachyæ.*"

scabrous above; it is generally attenuated at both ends, and the orifice of the beak is mostly bidentate with erect teeth, or entire, oblique. The nerves are seldom distinct, and the coloration may vary from green to deep purple or brown; thus the spikes in some species attain the same dark aspect as is characteristic of the *Melananthœ.* Another character common to these European and boreal species is the position of the pistillate spikes, being more or less remote and nearly always borne on very conspicuous peduncles, in some species of quite considerable length and nearly filiform. The sheaths of the bracts are well developed, and in regard to the general habit, these species exhibit but a few divergences, which, however, may be considered as analogous deviations, such as we know exist in some of the other sections. The terminal spike is sometimes gynæcandrous or androgynous in *C. brachystachys* and, as it seems, constantly gynæcandrous in *C. misandra ;* in all the other pliostachyous, which we have examined, the terminal spike is purely staminate. Such varied distribution of the sexes is also met with among the *Melananthœ,* the *Hymenochlœnœ* and the *Spirostachyœ.* And if we, moreover, consider *C. lejocarpa* and *C. circinata* as members of the *Frigidœ,* we then perceive a still more gradual variation among the species themselves: from the diœcious *C. lejocarpa* to *C. circinata* with androgynous spike, and to *C. sempervirens* and its very nearest allies with unisexual spikes, borne on the same culm, to *C. brachystachys* with occasionally androgynous or even gynæcandrous terminal spike and *C. misandra* with its invariably *gynæcandrous.* Abnormally decompound pistillate spikes, as these are frequently met with in several of the other sections, have also been noticed in some of the *Frigidœ ;* Boott records such cases from *C. brachystachys,* and we have, furthermore, found these in *C. ablata* Bailey and in one single inflorescence of *C. circinata.*

In looking over the various sections, as these are described by Drejer, we have not been able to find any under which the *Frigidœ* might be properly arranged. They seem to possess some characters common to the *Hymenochlœnœ,* yet lacking the essential character of these: " perigynia plerisque nervata, rostro apice hyalino bilobo—bifidove, vel distincte bifido," besides " squamæ membranaceæ albidæ" and " color plerumque læte viridis." And in comparing them with the *Spirostachyœ* certain analogies seem to exist; for instance, the darker color of the spikes, the mostly erect and remote, pistillate spikes, while " perigyniis distincte bifidis " and " spicis densifloris quasi squarrosis " are not possessed by the *Frigidœ.* We have not, on the other hand, been able to detect any intergradating forms between these three sections, by which the *Frigidœ* might be connected as "lesser" or "higher" developed types; on the contrary, it seems as if we in the *Frigidæ* themselves might trace some of their earlier types represented by, for instance, *C. lejocarpa*, that we might count *C. sempervirens* and *C. frigida* among their central forms, while *C. misandra* may be looked upon as a "*forma desciscens*," yet possessing the most essential characters of the section.

While thus fully recognizing the maintenance of these species as representing a section of their own, we should prefer not to adopt "*Frigidæ*" as the sectional name but one in Greek which, moreover, would be in stricter conformity with those suggested by Drejer for the other sections. And when we suggest "*Stenocarpæ*" as a more proper name, this refers not only to the characteristic shape of their utriculus, but also to one of the central forms, "*Carex stenocarpa*" of Turczaninow, the Asiatic homologue of *C. sempervirens*. The principal characteristics of the section, drawn from the central forms, are as follows:

Carices stenocarpæ.

Terminal spike staminate; lateral ones pistillate, distant, peduncled, erect or finally drooping. Bracts distinctly sheathing, mostly filiform, short. Utricle membranaceous, glabrous or a little scabrous above, narrowly attenuated at both ends; orifice of beak bidentate with erect teeth or obliquely cut, often hyaline; stigmas 3, achenium trigonous. Boreal or alpine species with mostly dark-colored spikes. The lesser developed, monostachyous types (*C. lejocarpa* and *C. circinata*) remind of certain *Vigneæ*, the central of *Spirostachyæ*; at the limits of the section are types (*C. misandra, C. brachystachys*) with androgynous or gynæcandrous terminal spikes, which remind of certain *Melananthæ* (*C. atrata, C. Mertensii*).

Among the species which represent this section may be enumerated:

Hebetatæ {	*C. lejocarpa* C. A. Mey.
	C. circinata C. A. Mey.
	C. sempervirens Vill.
	C. frigida All.
Centrales {	*C. ferruginea* Scop.
	C. ablata Bail.
	C. firma Host.
	C. hispidula Gaud.
Desciscentes {	*C. brachystachys* Schrank.
	C. misandra R. Br.

To these may possibly be added some others, of which we have seen no specimens, but judging from the diagnoses and illustrations, we suppose that they may be regarded as members of this section: the monostachyous *C. hakkodensis* Franch., *C. rhizopoda* Maxim., *C. heteroclita* Franch. and *C. grallatoria* Maxim. all from Japan; *C. Vallis pulchræ* Phil. from South America, and *C. acicularis* Boott from New Zealand; finally the pliostachyous *C. hæmatostoma* Nees, *C. psychrophila* Nees and *C. macrogyna* (Turcz.) from northern India. Boott himself considered *C. macrogyna* as being an ally of *C. sempervirens*, but states that in the former the uppermost spikes are sometimes androgynous. In regard to *C. hirtella* Drej., *C. hæmatostoma* and *C. psychrophila* Boott compares, also, these with the *Ferrugineæ* Tuckm., as representing on the mountains of northern India those analogous species found on the Alps of Europe. *C. grallatoria* is, according to Maximowicz, a close ally of *C. lejocarpa*.

The *Stenocarpæ* constitute a small, very limited section, and the types are either alpine, arctic or confined to the coasts of the Pacific from Bering Sea to Japan. *C. misandra* exhibits the widest range in geographical distribution; besides being circumpolar, this species occurs on the mountains of both northern and middle Europe and on the Rocky Mountains as far south as middle Colorado. The central forms, excepting *C. ablata* from Vancouver Island, are confined to the European Alps or, if we include those cited by Boott, to the mountains of India. Finally, as mentioned above, there is a species in South America and one in New Zealand, both with a single, androgynous spike, which Mr. Kükenthal considers as belonging to this same section, with him the *Frigidæ* of Fries.

There is, thus, considered from a geographical viewpoint, a wide gap between the "*hebetatæ*" and "*centrales;*" yet, as we have stated above, there is one of the latter which inhabits Vancouver Island; moreover by including the Japanese representatives of "*hebetatæ*" these are not so very remote from their higher developed allies on the mountains of northern India. The Pyrenees and middle European Alps possess, on the other hand, types which are very isolated and are only connected with their northwest American allies through *C. misandra*. The most gradual development of the section may, thus, be traced in Asia and in the northwestern corner of America, where all the types occur: *hebetatæ, centrales* and *desciscentes*. But we have no species of this section in Europe, which as *C. lejocarpa* may point towards some fundamental types of the recent time, and the *formæ centrales* of the European mountains, *C. sempervirens*, etc., may be considered as representing the remnants of an old center of which the earlier types have disappeared long ago, but leaving a group of species, the highest developed in the section.

While thus the *Stenocarpæ* represent a very natural little group of *Carices* in respect to their external, morphological, haracters, we have, furthermore, examined the anatomy of ome of these species, which seems to be very uniform. Vhether uniformity in anatomy, however, may be considered s being an absolute necessity for the establishment of sections f *Carices* to which morphological peculiarities are in common, is by no means certain. We will, no doubt, meet with 1any exceptions by extending our studies to several of the ther and larger sections, as for instance the *Melananthæ* and *Microrhynchæ*, the members of which exhibit so many diverse ypes from the extreme north and south, and living under very ifferent conditions as to climate and soil. It is, also, very ossible that several of the smaller sections may be found to ossess a like structure in anatomical respects, yet being apparntly distinct when considered from a morphological view· oint. This is readily noticed from the literature on this subect, the works of Mazel and Lemcke. The former of these uthors has examined the structure of the root, the stem and he leaf of 43 species of *Carex*, nearly all from Europe, and is final conclusion is thus expressed : " On ne peut en aucun as se baser sur des caractères anatomiques pour grouper sysématiquement les espèces dans le genre *Carex*" ; yet it does ot seem as if this author has made any attempt to deduct classification from his anatomical results. The other author, Alfred Lemcke, who has examined the rhizome and the aboveround stem of about 160 species of *Carex* from Europe and his country, believes, on the contrary, in the possibility of classiying *Carices* from the structure of their rhizomes, and considers Dr. Christ's system as the most practical for this purpose. The *Frigidæ* Fr. including *C. mucronata* All. are, thus, conidered as being closely related to each other, while we find *C. circinata* enumerated under *Psyllophoræ* among such pecies as: *C. gynocrates*, *C. capitata*, *C. polytrichoides* and ven *C. Fraseri*. However we doubt very much whether ny conclusion may be drawn from the structure of the hizome alone or from the stem so as to establish sections or maller groups of *Carices*. Nevertheless the works of these wo authors constitute a valuable contribution to our knowldge of the anatomy of *Carex*, even if their results do not ear directly upon the classification.

In regard to our *Stenocarpæ* we are, thus, well aware of the ossibility that the anatomical characterization which we resent as supplemental to the morphological, may also conain several points that are common to some of the other secions. These anatomical notes may, nevertheless, become seful to further studies of the genus, at least as a contribution to similar investigations. As regards the material which we have examined, the specimens were collected in the following localities:

C. circinata C. A. Mey., rocky hilltops, 360–375ᵐ alt., Kadiak; *C. lejocarpa* C. A. Mey., grassy banks, St. George Island, Bering Sea, and marshes at Yes Bay, Alaska; *C. sempervirens* Vill., Alps of Switzerland, Tyrol; *C. ferruginea* Scop., along streams, 2000ᵐ alt., *C. frigida* All., 2300ᵐ alt., *C. firma* Host., 2000ᵐ alt., *C. hispidula* Gaud. 2100ᵐ alt. and *C. brachystachys* Schrank, 1260ᵐ alt., all from the Alps of Switzerland; *C. ablata* Bailey, 600ᵐ alt., Vancouver Island; *C. misandra* R.Br., St. Mathew Island, Bering Sea; along mountain streams on Gray's Peak, 4000ᵐ alt., Colorado; on dry, grassy mountain slopes or swamps, West Greenland; Rendalen, Spitzbergen; in swamps on rocks, Nova Zembla; Alps of Tyrol at 3000ᵐ alt.

The root.

In the species examined we found the roots very strongly built with a more or less thick-walled hypoderm inside the epidermis. The outermost strata of the cortex show a similar thickening and constitute thus a firm sheath around the inner bark, which is thin-walled and usually collapsed throughout to the innermost stratum that rests on endodermis. In *C. sempervirens* the outer six strata of the cortex were especially heavily thickened, also in *C. hispidula*, in which even the innermost four or five strata exhibited a like thickening, while in *C. firma* the entire cortical parenchyma was developed as a solid mass of stereïds. Endodermis is thick-walled in all the species but not to the same extent; thus an U-endodermis may be found in *C. lejocarpa*, *C. circinata*, *C. misandra*, *C. hispidula* and *C. sempervirens*, while it appears as a V-endodermis in the remaining species; the thickening is especially heavy in *C. firma* and *C. sempervirens*. The pericambium is most often thin-walled, or in some roots of *C. firma*, *C. frigida* and *C. sempervirens* the cell-walls may be observed as being somewhat thickened. It consists of only one stratum, which forms a closed ring in *C. sempervirens* and sometimes also in *C. firma*, but is, in the other species, interrupted by the proto-hadrome vessels in a more or less regular manner. We have noticed the following cases: In *C. circinata*, *C. lejocarpa*, *C. misandra*, *C. frigida* and *C. ablata* the pericambium was constantly interrupted by all the proto-hadrome vessels, there being mostly four or five pericambium cells between each two of these vessels. In *C. hispidula* we found in some roots the pericambium interrupted by all the proto-hadrome vessels or by only five out of twenty-five, and these roots were from the

ıe individual; the number of pericambium-cells between
h two vessels varied in this species from four to seven, but
ɪ and five was the commonest. This peculiar case was,
reover, noticed in other specimens from a still higher eleva-
ı (2800ᵐ Mt. Ryssel), in which sometimes most of the
ɪto-hadrome vessels were situated inside the pericambium,
l only a very few bordered on endodermis. In *C. firma* a
ıilar and very irregular position was noticed, since the peri-
ɪbium was either not interrupted at all or it was broken by
ɪ majority of the vessels, by 14 out of 15. In *C. brachy-*
chys we found no case where the pericambium was not
errupted, but the position of the proto-hadrome was very
ɪgular even in the same root, cut at different places. The
ɪnber of interruptions varied thus between 8 and 13, while
ɪ distance between the places where the sections were made
s about 5ᶜᵐ; in most of the roots of this species, however,
observed that the majority of the proto-hadrome vessels
re located inside the pericambium, and only in one root did
notice that fourteen out of nineteen bordered on endoder-
ɪ. In *C. ferruginea* Scop. we observed that all the proto-
ɪrome vessels bordered on endodermis in roots of very dif-
ent thickness, but in a single root the following position
s noticed: 25 out of 26 vessels had broken through the
ricambium, but at the other end of the same root there were
ly 18 of these 25 vessels that bordered on endodermis.—
hispidula, C. ferruginea and *C. brachystachys* illustrate
ıs the singular fact that the pericambium is very irregularly
errupted by the proto-hadrome, but, on the other hand,
found no cases in these species where it was not inter-
ɔted at all. *C. firma*, on the contrary, possesses roots in
ıich the pericambium may sometimes form a closed ring or
ɪay be broken by nearly all the proto-hadrome vessels. This
gular instance is, moreover, to be observed in roots of *C.*
ɔina Wahlbg. in some specimens which we collected in West
eenland; the species belongs, however, to another section,
ɪdently to the *Lamprochlænæ* Drej., which we intend to
cuss in a subsequent paper.—But we have not so far noticed
ɣ such variation in any of the other *Cyperaceæ* which we
ve heretofore examined. In regard to the other vessels in
ɪ roots of these species we have not found any occupying
ɪ center of the central-cylinder, with the exception of small,
eral roots, which usually possessed only one central vessel.
e conjunctive tissue which thus occupies the innermost part
the central-cylinder is distinctly thick-walled in all the
cies, with the exception of *C. ablata*, where it is thin-walled
ɪn in such roots of which the endodermis shows the cell-
lls to be very considerably thickened. The leptome, includ-

ing the proto-leptome, was very well developed in all the species and was located between each two proto-hadrome vessels.

In examining the above-ground stem we find that it does not exhibit such striking divergences as was noticed in the roots. Sections taken from about the middle part of the culm showed the outline to vary from cylindrical to sharply triangular as follows: it was found to be terete and nearly glabrous in *C. circinata, C. sempervirens* and *C. brachystachys;* pentagonal and almost glabrous in *C. ferruginea;* obtusely triangular and glabrous in *C. ablata;* triangular and nearly glabrous in *C. lejocarpa, C. misandra* and *C. frigida,* while triangular and very scabrous in *C. firma* and *C. hispidula.* The cuticle is quite thick and perfectly smooth, not wrinkled in these species. The epidermis shows generally a distinct thickening of the outer wall, especially in *C. sempervirens* and *C. ablata;* the radial cell-walls are a little thickened in *C. ferruginea, C. frigida* and *C. ablata.* Cone-cells were found outside the hypodermal stereome in all species; stomata occurred frequently in that part of epidermis which covered the bark and showed the same structure as in the leaves, which will be described later. A somewhat peculiar epidermal structure is represented by *C. hispidula,* where the outer wall in nearly all the cells was noticed to be extended into roundish papillæ, especially near the stomata, although not covering these. The cortical parenchyma is mostly developed as short palisades, radiating towards the center of the stem, and contains lacunes, sometimes of quite considerable width in *C. circinata, C. lejocarpa* and *C. hispidula.* The stereome occurs as hypodermal groups, covering the larger mestome-bundles, but is also to be found on the hadrome side of these, bordering on the pith. The smaller mestome-bundles have usually only a minute support of stereome, this being merely developed as a few cells on either face of these. This tissue, the stereome, is very thick-walled, at least on the leptome-side, in *C. circinata, C. sempervirens, C. hispidula, C. firma, C. frigida* and *C. brachystachys,* much less so in the remaining species, and it is generally rather open on the hadrome-side of all the mestome-bundles.

The mestome-bundles are developed as larger, in transverse sections oval, or as smaller, nearly orbicular, and are arranged in almost regular alternation with each other. They constitute in most of the species only one peripheral band, but in *C. hispidula, C. ablata* and *C. brachystachys* we noticed an inner band of a few, 3 to 5, large bundles, almost entirely imbedded in the pith. The parenchyma-sheath is invariably thin-walled and seems often to contain chlorophyll; a mestome-sheath with the inner walls distinctly thickened is also noticeable in all the species; but we found no trace of the inner green sheath, which we remember is known to occur in many genera of the *Cyperaceæ*, but not so far known from any species of *Carex*.

There is a central, thin-walled pith in all the species, and this seems to break down, leaving a wide central cavity in all the species examined, with the exception of *C. circinata*.

The leaf.

The leaf exhibits a somewhat greater variation than is noticeable in the stem. The narrowest leaf-blade is possessed by *C. circinata* and *C. brachystachys*, but if we consider the sections taken from the middle of the blade, this is in none of these so narrow that it might be described as semi-cylindrical. The surface of the leaf is very smooth in most of the species, but in *C. hispidula* the blade is somewhat constricted between the mestome-bundles, besides that the epidermis in this species is very scabrous on either face. In *C. brachystachys* the lateral parts of the blade are much thicker than the mediane, and very large prickle-like projections are developed on the upper surface on each side of the bulliform-cells. The leaf-surface is thus mostly smooth with no deep furrows, but it is scabrous from prickle-like projections in *C. sempervirens*, *C. ferruginea* and *C. brachystachys*; in *C. hispidula* both surfaces are quite scabrous from the numerous papillæ, which also characterized the stem of this species. Bulliform-cells occur on the whole upper face of *C. circinata* and *C. lejocarpa*, but these cells are in the other species restricted to a single group just above the midrib; the outer cellwall of epidermis is very thick in some species, for instance, *C. hispidula*, *C. frigida* and *C. firma*, but not so in *C. ablata*. Some certain variation exists also in the relative length and breadth of the epidermis-cells outside the stereome, and this is especially distinct when we examine the leaf in superficial sections; we notice, for instance, in *C. ferruginea*, *C. brachystachys* and *C. misandra* that these cells are narrower, but not shorter, than the surrounding epidermis, while in *C. sempervirens*, *C. frigida* and *C. hispidula* they appeared to be much shorter but not narrower. In *C. firma* and *C. ablata* these same cells, covering the stereome, were distinctly both narrower and shorter than the stomatiferous strata.

In regard to the stomata, these are in the *Stenocarpæ* confined to the lower surface of the leaf, and show a very uniform structure; they are free in all the species, slightly projecting in *C. misandra*, *C. firma* and *C. ablata*, but level with epidermis in the others; in *C. hispidula*, as mentioned under the stem, the stomata are surrounded by mostly four papillæ, which, however, are quite short and not bent over the stoma, thus this is almost free as in the other species. This structure of the subsidiary cells with their papillæ in *C. fimbriata* has already been described by Schwendener in his paper on the stomata of the *Gramineæ* and *Cyperaceæ.**

The mesophyll is either represented as a homogeneous tissue throughout the leaf, as in *C. circinata* and *C. lejocarpa*, or it is differentiated into a more or less distinct palisade-tissue on the upper face and a more open, pneumatic on the lower, as in *C. ferruginea;* in the other species the palisades are short and not so regular, but appear nevertheless to be placed vertically on the blade, with the exception of *C. ablata*, in which the palisade-cells show a tendency to radiate towards the center of the mestome-bundles. Lacunes were observed in all the species, one between each two mestome-bundles, and they were especially wide in *C. circinata*, *C. lejocarpa*, *C. misandra* and *C. hispidula*. The stereome occurs as in the stem on both faces of the mestome-bundles and as hypodermal in the larger of these; it is, moreover, to be observed as a small, isolated group in each of the two leaf-margins. It is very thick-walled in *C. sempervirens* and *C. ferruginea*, moderately so in the other species; it is rather weakly developed in *C. ablata*.

The mestome-bundles lie in one plane, and the midrib is generally somewhat larger than the others and more promi-nent. We find in the leaf as in the stem two forms of these bundles, larger and smaller, which show the same alternation in regard to their arrangement and the same structure as we have described under the stem.

* The position of the stomata in *Carex*, whether these are free and exposed or sunk below the surrounding epidermis and sometimes partly covered by papillæ, does not seem to be of any importance in regard to the classification of these species. On the contrary, Mazel has shown that species from the various sec-tions of *Carex* may exhibit the same structure and position of stomata, and there does not appear to be any relation between stomata, being free or not, and the local environment, climate and soil. This author has shown, for instance, that *C. Davalliana*, *C. Pseudocyperus* and *C. hirta*, all from swamps, possess super-ficial stomata, while these are sunk in *C. paniculata*, *C. vulgaris*, *C. paludosa*, etc., from similar wet localities. In species which inhabit drier stations, woods and hills, a similar variation exists; thus the stomata are superficial and perfectly free in, for instance, *C. divulsa*, *C. brizoides*, *C. præcox*, *C. sylvatica*, etc., but not so in *C. maxima*, *C. glauca*, etc. A still more peculiar case may be illustrated by *C. misandra*, which we have examined from very different and remote localities: arctic and alpine regions of America and Europe, still maintaining the same structure and free position of the stomata, while in *C. capitata*, which we found growing almost side by side with this species, the stomata are protected by long and almost ramified papillæ.

It does seem as if the position of the stomata is not always an indication of the conditions of the surroundings under which the plants live, and it is, no doubt, one of the many inherited characters which are explainable only in a few species to which the natural surroundings are unchanged.

Utriculus.

The anatomical structure of this little organ appears to be very uniform in these species of the *Stenocarpæ;* it is very thin on account of the very little mesophyll, which forms here only one stratum between the two mestome-bundles, while it may be represented as two or three in the immediate vicinity of these. The outer wall of epidermis, on the dorsal face, is, however, thick in these species, while that on the ventral face is invariably very thin, like the radial walls. Stereome is also developed in the utricle of these species, but it is not very thick-walled and does not occur in groups of any considerable size either; it accompanies the two mestome-bundles, and it occurs besides as isolated groups between these, but not in all the species. In *C. misandra* and *C. frigida* the stereome was restricted to the mestome-bundles, but in the other species we found from 5 to 14 isolated groups between these; the utricle of *C. brachystachys* and *C. ferruginea* appear to possess the strongest mechanical support of these species.

In comparing these anatomical details of the *Stenocarpæ*, the species examined seem to possess a very uniform structure in spite of the fact that they were collected from very remote stations, Alaska, Colorado, Middle-Europe and the arctic region, and in the case of *C. misandra*, the specimens from Alaska exhibited the same anatomical characteristics as those from the alpine slopes of Rocky Mountains (Colorado) and Switzerland besides from the arctic region. The internal structure considered by itself is very much in conformity with that of a number of arctic plants of various orders which we have had opportunity to examine, and which seems to indicate that these species lived under conditions that were influenced by a moist rather than by a dry atmosphere. The development of the *Stenocarpæ* seems thus to have taken place, at least, in recent time, in alpine or boreal regions, where they are yet in existence, and if we combine the geographical distribution, including the nature of the surroundings, with the external and internal peculiarities of these *Carices*, we see no objection to consider *C. circinata* and *C. lejocarpa* as "*formæ hebetatæ*" of this section.

Brookland, D. C., April, 1900.

Bibliography.

Bongard: Observations sur la végétation de l'ile de Sitcha. (Mém. de l'Acad. Imp. d. sc. de St. Pétersbourg, Ser. 6, vol. ii, 1831, p. 168.)

Clarke, C. B.: *Cyperaceæ* in Flora of British India by J. D. Hooker, vol. vi, 1894, p. 699.

Franchet, A.: Les Carex de l'Asie orientale. (Nouv. archives du Muséum d'hist. nat., Ser. 3, vols. viii–x, Paris, 1896–98.)

Lemcke, Alfred: Beiträge zur Kenntniss der Gattung *Carex* Mich. Inaug. diss. Königsberg in Pr. 1892.

Macoun, John: Catalogue of Canadian plants, vol. iv, Montreal, 1888, p. 110.

Macoun, James M.: A list of the plants of the Pribilof islands. (The Fur Seals and Fur-seal islands of the North Pacific ocean. Washington, 1899, part iii, p. 559.)

Mazel, Antoine: Etudes d'anatomie comparée sur les organes de végétation dans le genre *Carex*. Thesis. Genève, 1891.

Maximovicz, C. I.: Diagnoses plantarum Asiaticarum VI. (Mélanges biologiques d. Bull. de l'Acad. Imp. d. sc. d. St. Pétersbourg, vol. xii, 1886, p. 560.)

Meyer, C. A.: Cyperaceæ novæ descriptionibus et iconibus illustratæ (Mém. de l'Acad. Imp. d. sc. de St. Pétersbourg, vol. i, 1831, p. 195. "Des savants étrang.")

Meinshausen, K.: Ueber einige kritische und neue Carex-Arten der Flora Russlands (Bot. Centralbl., vol. lv, No. 7, 1893).

Schwendener, S.: Die Spaltöffnungen der Gramineen und Cyperaceen. (Sitzungsber. d. K. Preuss. Akad. d. Wiss. Berlin, 1889, p. 65.)

ART. XXVIII.—*Experiments on High Electrical Resistance:* Part I; by OGDEN N. ROOD, Professor of Physics in Columbia University.

IN making electrical measurements of high resistance, if the " direct deflection method " is used, under ordinary circumstances a resistance of 5000 megohms can be measured with some degree of certainty, and if the conditions are very favorable, 50,000 megohms may be reached. The "loss of charge method " is limited by the insulation of the condenser employed, this insulation itself being equal only to about 10,000 megohms. In the method described in this paper 5000 megohms is small resistance, and it is not till 15,000 or 20,000 megohms has been reached that the manipulation becomes particularly simple. After that, the only limit is found when the resistance is so great that the electrometer contrived by me refuses to give any indications, even with electricity of very high potential. Just what this limiting resistance is, I cannot at the present moment state, being cut off by the dampness of the air from such determinations; meanwhile, it is certain that the figure is quite high.

In the older methods of measuring resistance, a flowing current of electricity is made to act on a system of astatic needles, but in the method here proposed the electricity is allowed to slowly accumulate, till it reaches a certain potential: in the older methods the electricity is treated as though its velocity were infinite, but the present plan is based on the fact that when the resistance is very great the electricity may require ten minutes or even an hour before raising the potential of the electrometer to the required degree. Thus, while in the older methods of work the electricity in its flow resembles a quick-moving incompressible fluid, in the one here described it may be compared to that of a thick pitchy fluid, moving slowly in a long open tank, and attaining its level at the farther end of the tank after the lapse of a longer or shorter time. After its level has been attained, if a leak is made anywhere in the floor of the tank, over this leak there will be a depression of the fluid; if the hole is made in the middle of the tank, some of the fluid drawn off and the hole then stopped, the pitchy fluid will flow to the depression from both sides, but it may take minutes before the old level is once more established. Again, if we imagine the walls of the tank to be cold and the pitchy fluid to be at a higher temperature, some of it will harden on the walls, and to a certain extent obstruct the flow of the fluid by contracting their cross-section. This illustration corresponds to a polarizing action which often sets in, and gradually diminishes the flow of the electricity, through or over a substance of very high resistance, and the greater the resistance the larger will be this counter action. What has been said so far applies to the cases of glass, mica, paraffin paper, silk and most artificial insulating substances.

If we now imagine our tank to be quite narrow, offering thus great resistance to the passage of the pitchy fluid, then at the end where the fluid is introduced, the height or potential will equal that of the supplying source, but as the distance from this end increases, the potential falls and finally becomes zero. This case corresponds to the surface conduction of electricity by black sealing wax, jade and gutta percha. They all furnish to the electrometer only small quantities of electricity, which can only be drawn from the neighborhood of the reservoir, and here the polarization is very marked.

Finally, imagine the walls of the tank to be still farther contracted and that the pitch-like fluid refuses to enter them in perceptible quantities, and we have the cases of ebonite, ordinary rosin and exfoliated mica (in very dry weather).

Evidently, if two tanks of the same length and cross-section deliver equal amounts of the pitchy fluid in equal times, then their conducting powers should be set as equal, and in general, if there be no loss in transit, the amounts delivered per minute, by tanks differing from each other, will be a measure of their relative conducting powers. Finally, if the tanks almost refuse to deliver up any of the pitchy fluid, still, if their potentials or the heights of the fluid at various points can be measured, in a manner analogous to that employed in the study of the conduction of heat by solid bodies that have reached the stationary condition, then it becomes possible to determine their relative conducting powers, and by comparison with somewhat better conductors, an estimate of their resistance in standard units can be formed.

Apparatus.—I modified a single leaf electroscope, so that it became a serviceable electrometer for this kind of work. Its plan may be seen in the diagram. A brass rod AR was supported on a bar of ebonite, which was placed on two rods of the same substance fastened on a base of ebonite. The rods and the bar were painted with melted rosin. The rod AR carried an aluminum leaf, bluntly pointed below, and attached above to the point R by a gold leaf hinge. Aluminum leaf was used, it having been found that under the action of the minute but very numerous electrical discharges, the gold leaf slowly wasted away at its corners, which it always presented to

plate D with an undesirable twist. D is a plate of sheet
ss faced with platinum foil, and receives the charge from
aluminum leaf when its potential has been sufficiently
sed. The distance of this plate from the leaf can be varied
a micrometer screw, M, and thus the sensitiveness of the
paratus regulated. When a Leyden jar was used as the
rce of the electricity, the plate D was connected with the
th ; when the street current
s employed, D was connected
h one of the electric wires,
sistance of two or three thou-
d ohms being interposed, and
only thousands of megohms
e dealt with, this amounted
nothing. An arrangement by
ich a slight vibration could be
imunicated to D was added,
l by a single motion D was
rated and the electrometer
charged by contact with A.
en the electrometer gives a
ke the leaf usually remains
iched to the plate D, which is
reason of the above men-
ied contrivance. At A, vari-
adjuncts can be placed, bind-
screws, etc. The whole ar-
gement amounts to a kind of

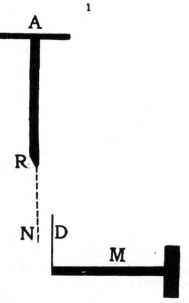

t jar which discharges itself when charged to a certain
ential. Over it was placed a glass case to avoid air currents,
the case was in contact only with the base of the apparatus.
o such electrometers were made by me, one rather large for
rden jar experiments; the other was smaller, especially in
acity, and was used only with currents of 110 to 112 volts.
ne small electroscopes on the above mentioned plan were
made; they could be set to indicate certain potentials in
lies either in contact with them or at fixed distances from
m. Their capacities were quite small. I also made a
menberger electroscope with its dry pile, the delicacy of
ich could be varied at will by moving the brass plates con-
ted with the terminals of the pile. It was used only in
study of good insulators. The source of electricity was
letimes a Leyden jar charged to a potential of 12,000 volts,
in the measurements given in megohms a current of only
i to 112 volts was used. This same current was employed
he preparation of standard resistances. It is to be under-
od that the experiments were performed in dry winter

weather, the hygrometer having a range from 12 to 30 per cent of moisture. At the end of this paper certain determinations that were made in damp weather are added.

Preliminary experiments.—The case of a sealed glass tube 1^m 30^c in length, will illustrate partially what has been said about the behavior of electricity moving over a bad conductor. When the tube was made to establish connection between a Leyden jar charged to a potential of 12,000 volts and the electrometer, the electricity advanced slowly over its entire length, the electrometer not being at all affected for some time. In about five minutes it gave regular strokes; the tube had assumed the stationary condition and had nearly the same potential over its whole length, except near the electrometer, where for 3 or 4^{cm} it was lower. For half an hour it was allowed to deliver charges to the electrometer with much regu-larity, the rate being observed with a stop-watch. The elec-trometer was then connected with the middle of the tube; at first the rate of discharge was greatly increased, as the poten-tial was higher here than at the drained end, and the electricity was flowing into the electrometer from both sides, but in about five minutes the flow had fallen nearly to the old rate and become quite regular. An air space of a centimeter was now made between the tube and the electrometer; the latter continued to give discharges, but more slowly; the potential of the middle part that had been somewhat drained was rising and acted inductively on the electrometer, and this process continued for some minutes, till the original level had been established. The electrometer was then again connected with the farther end of the tube and matters allowed to get into a stationary condition. The strokes having become regular, the middle of the tube was connected with the ground by wet paper and wire. The electrometer strokes however continued for a considerable time, but with diminishing frequency, till the supply of electricity at the farther end of the tube had exhausted itself. In this and in all similar experiments it is to be understood that the two ends of the conducting body were provided with metallic armatures, tinfoil fastened with thread. A great many experiments were made on conductors of this class without any armatures at all; the tinfoil however facili-tates matters, and makes the flow more regular.

All this illustrates the case of a glass tube with rather good insulating properties, and I pass on now to more ordinary samples. A sealed tube of hard Bohemian glass, 1^m 30^c in length, was washed in water, dried in sunshine and connected with jar and electrometer. In a second or two the electro-meter began its discharges, the rate being about three seconds per stroke. The entire tube was then heated quite hot with a

Bunsen burner; for two minutes the flow of the electricity was stopped, but it then began again, the intervals between the strokes being 8, 7, 4, 3, 3, 3 seconds. Two days afterwards the experiment was repeated with a result of 10 strokes in 32 seconds. Ten days later, its rate was 10 strokes in 42 seconds. A sealed tube of soft German glass of the same length that had been carefully cleaned and heated gave 10 strokes in 20 seconds; heated again for three minutes and cooled, its rate was 10 strokes in 36 seconds and on the following day this had fallen to 10 in 15 seconds. Ten days afterwards it gave 10 strokes in 20 seconds. These slight irregularities are however thrown into the shade by an observation made on a tube of the same kind of glass, where the resistance of one-half of it was twelve times as great as that of the other, and it remained in this condition for two weeks, or until it was cleaned; then the whole of the tube had the lower resistance. A tube made from the glass that is now used in Germany for Leyden jars, was found to have a high resistance, so that the electricity did not seem to reach its farther end at all. After being in contact for three-quarters of an hour with a Leyden jar having a potential of 12,000 volts, at a distance of 16^{cm} from the jar, the potential of the tube was only 600 volts, and an additional exposure for the same length of time extended this distance only to 22^{cm}. At the time these experiments were made I had not constructed a set of units of high resistance, or it would have been easy to have measured the resistance of these tubes in megohms when traversed by a current of 12,000 volts. Later on I measured the resistance of a tube of soft German glass, 27^{cm} in length, which had been for some days in a box dried with sulphuric acid. This at first proved to be 250,000 megohms, then under the influence of the current it increased to 380,000 megohms. On this day the hygrometer indicated 43 per cent of moisture. Some time afterwards, the hygrometer indicating 53 per cent, the resistance of the tube had fallen to 180,000 megohms. The current here employed, as in all cases where actual figures are given, had a potential of only 110–112 volts.

Silk.—A double thickness of rich brown silk, 45^{cm} long, 2^{cm} broad, conducted electricity at a potential of 12,000 volts very well; the electrometer strokes began immediately, and continued hour after hour. A band of white silk was then taken, a meter in length and 3^{cc} in breadth; the discharges were at the rate of about one per second. The silk was then boiled twice in water, and dried, without affecting its rate. It was then boiled in alcohol and dried; the discharges were a little slower at first, one in four seconds, but the rate soon increased to one in two and a half seconds. Two threads of white sewing silk failed to affect the electrometer, but with the electroscope it was found that the charge had extended over 15^{cm}. When twenty-four threads of this silk, half a meter in length, were used, the electrometer gave slow strokes needing three minutes each. A single thread of this silk was heated in a bath of one part gutta perch, three parts rosin ; a large quantity of vapor bubbles escaped, and it became a good non-conductor. When it was connected with the Leyden jar, charged *positively* to 12,000 volts, not only did it not conduct, but became *negatively* electrified, with a potential of about 200 volts. Negative electrification under such circumstances will be referred to later on.

Paraffin paper.—Ordinary commercial paraffin paper, viz : writing paper which had been passed through melted paraffin wax, conducted quite well. A strip a meter in length, in fifteen minutes had acquired over its whole length a potential of about 12,000 volts. Some measurements were made of short strips of this paper provided with metallic armatures, a potential of only 110 volts being employed.

Length.	Breadth.	Resistance in megohms.
25^{mm}	18^{mm}	80,000
63^{mm}	"	125,000
132^{mm}	"	300,000

It is evident that the resistance of the connecting armatures has influenced these results.

Mica.—Using a potential of 12,000 volts, a strip of mica (muscovite) 30^{cm} long and 2^{cm} broad, was compared with a strip of paraffin paper of the same dimensions ; the resistance of the mica was sixty times that of the paraffin paper. The hygrometer at the time indicated 30 per cent of moisture. Two strips of mica from the same specimens were prepared ; the length of each was 27^{cm} the breadth 2^{cm}. Both were heated till their edges had become somewhat white, and on the following day their relative resistances were determined ; these turned out to be nearly as 1 to 2. Knowing that mica when heated gives off traces of water and fluorhydric acid, a piece of this mica was heated in a Bunsen flame till its structure was destroyed, and on the following day, having been provided with a tin-foil armature, it remained for twenty minutes in contact with a Leyden jar charged to a potential of 12,000 volts. It was then removed and presented to an electroscope capable of indicating 100 volts, but failed to show any charge.

In the case of all the substances that thus far have been mentioned except two, it was found that when they were connected with the Leyden jar, that within at least fifteen minutes the part farthest from the jar had assumed its potential, provided that no electricity was conducted off from it.   This is not true of the substances that follow ; their case is similar to that of rods conducting heat and slowly attaining the stationary condition.

Jade, gutta percha, black sealing wax.—There was placed at my disposal a fine bar of jade, 30^{cm} long with a square cross section of 12^{mm}.   This was simply washed in water and sun-dried.   Afterwards it was exposed to 12,000 volts for fifteen minutes.   The electricity had traversed its whole length with a diminishing potential, that of the farther end being only 600 volts.   This would have been much more than enough to have furnished electrometer strokes, but the rate of travel and the rate of delivery was so slow, that the electrometer was not affected.   The following experiment was then made with the jade, and also later, with black sealing-wax and gutta percha. The end of the jade in contact with the Leyden jar had of course a tin-foil armature to facilitate conduction from the jar to the surface of the jade ; at a distance of two centimeters from this armature the jade was wrapt with a strip of tin-foil, and this was brought into metallic connection with the electrometer placed at some distance.   This arrangement assumes that the electricity will leave the Leyden jar, traverse two centimeters of naked jade, and then reach the electro-meter by a connecting wire, and that no electricity will pass through the air by convection, and reach the movable conduct-ing strip of foil and hence the electrometer.   This supposition appeared to be practically justified by repeated experiments, where the knob of the Leyden jar retained its short distance from the movable armature, but was disconnected from the fixed terminal armature of the jade.   I repeatedly watched for ten minutes at a time without finding that the disconnected jar produced any effect at all on the electrometer.   This fact makes it possible to study cases which otherwise would offer difficulty.   Using two centimeters of naked jade in this way, the electrometer furnished strokes needing ten seconds. Using four centimeters, the time of the first stroke was not 20 seconds but 47 seconds, the next following stroke needing 54 seconds, indicating the polarization usually observed in cases of surface conduction in connection with high resistance. Gutta percha and black sealing wax were now treated in exactly the same way ; the movable armature being at 4^{cm} ; they both refused to affect the electrometer, but with a distance of 2^{cm} the times needed for electrometer strokes were as fol-lows ·

	Gutta percha.	Sealing wax.
1st stroke	1' 20"	2' 2"
2d "	2' 9"	5'
3d "	2' 25"	no stroke.

On this day there was 30 per cent of moisture in the air.

In all the cases thus far examined the substances in question conducted sufficiently well to affect the electrometer arranged with a medium sensitiveness, but there were other insulators having still higher resistances, at least in dry weather.

Ebonite.—One of the most interesting of these is ebonite or hard rubber. It being very doubtful according to my previous experiments whether ebonite conducted electricity at all in dry weather, a strip of this material, 40^{cm} long and 1·5^{cm} broad, was provided with tin-foil armatures and then tested with the Bohnenberger electroscope for accidental charges, and afterwards connected for half an hour with a jar charged positively to 12,000 volts. It was then removed from the jar, the tin-foil discharged, and passed over the Bohnenberger electroscope. It was found to be *negatively* charged for about 17^{cm} from the jar end, the potential being a little under 110 volts, and gradually rising near the jar end, probably not much above 200 volts. This negative charge, I take it, is probably due to a separation of the ions, owing to the prolonged inductive action of the strong field. The final result is, that if the ebonite conducted at all in the regular way, such conduction was masked by the inductive effects.

Gutta percha.—Exactly the same experiment was made with a similar strip of gutta percha: it showed a charge of positive electricity that extended 16^{cm} from the jar-end of the strip; beyond this the electrification was negative, with a potential of from 100 to 200 volts. There seemed to be a neutral zone dividing the two opposite electrifications.

Amber.—The amber was 3^{cm} in length; one-half of it was wrapt in tin-foil fastened down by thread. Being supported by *insulating* silk thread, it was left for half an hour in contact with the knob of a Leyden jar charged positively to 12,000 volts. On removal and discharge of the foil armature, it was found charged positively, with a potential of 300 or 400 volts.

Rosin.—A piece of freshly cut rosin of the same dimensions and treated in the same way, became negatively electrified. Other experiments given below tend to show, that taking dry and damp weather together, plain rosin is the best insulator that was examined.

Insulating silk thread.—Ordinary thick white sewing silk was passed through a bath of one part of gutta percha melted with three of rosin. Four centimeters of this was treated as above and became feebly *negatively* electrified, with a potential of about 200 volts. The temperature of this bath is rather high, and one composed of wax and rosin probably would answer all practical purposes.

Is the conduction of electricity by glass entirely on the surface, or does the electricity penetrate a little below the surface?—One experiment bearing on this question is here given : a sealed tube of soft German glass conducted quite well ; a portion of it was heated till it melted black sealing wax, and in this way a few centimeters of it were coated with the wax, but when it was cold it conducted about the same as before. An application of boiled linseed oil produced no effect on its conducting power; finally, a few centimeters of the surface were ground away on a stone, but after washing, drying and the application of heat, it conducted just as before.

Experiments in damp weather.—The results of a few experiments on the effects produced by dampness of the air on the resisting power of certain insulators may be of interest, and some are here given.

Ebonite.—Hygrometer at 78° ; a piece of the same sample previously mentioned was used ; the distance from armature to armature was 17^{mm}; the breadth 13^{mm}, thickness 2^{mm}; it had been exposed to the air for several days. Employing 110 volts at first, it conducted quite well, its resistance being about 5000 megohms. It was then warmed, and while cooling showed a resistance as high as 1,200,000 megohms. In three minutes this fell to 15,000 megohms and then remained constant. Two hours later, the same figure was obtained.

Gutta percha of the same dimensions with the ebonite, without being warmed or dried in any way, had a resistance of 12,000 megohms.

Rosin, with armatures of tin-foil, the distance between them being only 9^{mm}, had a resistance of 1,500,000 megohms. It had been exposed to the damp air for days, and was not warmed or dried. These experiments were made on the same day, but on the following day when the air was somewhat drier, the hygrometer standing at 65°, the resistance of the ebonite was 340,000 and that of the gutta percha 440,000 megohms. The resistance of the rosin was too great to be measured without altering the distance of the plate P, which at the moment was undesirable. Freshly cut natural beeswax also appears to be a good insulator in damp weather. A few days later, the hygrometer standing at 60°, the resistance of the ebonite was again measured and turned out to be 1,100,000 megohms. It had in the meanwhile been exposed to the air which had been damp. Wishing to improve the resisting power of ebonite against moisture, I painted a similar piece with boiled linseed oil, and allowed it to dry for two weeks. It was not improved as shown by my electrometer, and to avoid all uncertainty, I requested Mr. H. C. Parker to measure its resistance by the "direct deflection method" with a galvanometer; it turned out to be only 5300 megohms. It was then gently heated by a spirit lamp and while warm measured; the resistance had been raised to 90,000 megohms; it fell at first quite rapidly, and afterwards slowly, finally reaching, in four minutes, 6000 megohms. It was then heated to a temperature of 100° C. for four hours, and after free exposure to the air for three days, it was found that its resistance had been increased fifteen fold, the hygrometer standing at 54° when it was measured.

One more experiment may be given; on a somewhat damp day the hygrometer stood at 56°, the thermometer at 80° F.; pieces of mica and thin ground glass were prepared with metallic armatures; the pieces were 13^{mm} broad and the distance from armature to armature was 55^{mm}. The resistance of the glass was found to be nearly one thousand megohms, that of the mica 50,000 megohms.

In the preliminary measurements that have been detailed in this article, it has been assumed that during the transit of the electricity there is practically no loss by convection nor by leakage in the electrometer, so that the time consumed by a given charge in traversing a conductor and charging the electrometer to a given potential, would be directly proportional to the resistance of the conductor, and the few experiments that I have been able to make seem to justify this conclusion, but my set of high resistances was not completed in time to give it a rigorous examination. In dry weather the Leyden jar lost only two-thirds of its charge in twelve hours, although its necessary construction was far less favorable for insulation than was the case with all of the other arrangements.

The measurements given in this article were made with the help of certain standard units that I have recently constructed on a new plan. A considerable number were made, the resistance in each case not exceeding 5000 megohms. These were measured separately by Mr. H. C. Parker with a galvanometer and the "direct deflection method." Afterwards, they were joined together, and furnished suitable standards

It may finally be remarked that almost all of the experiments described in this article are quite easy of execution, and many of them are well adapted for class illustrations.

June 18th, 1900.

RT. **XXIX.** — *On two new Occurrences of Corundum in North Carolina;* by JOSEPH HYDE PRATT.

WHERE formerly corundum was supposed to be rare in its :currence and to be found in quantity only in the basic mag-
:sian rocks, it is now known to occur in various types of
·cks and in quantity, in syenites, gneisses and schists. Many
·w occurrences have been discovered during the past few
:ars, some of which give indications of being of considerable
:onomic importance, while others are only of scientific value.
i the present paper two new occurrences are to be described,
iat have been observed in North Carolina, one in an amphi-
)le-schist and the other in a quartz-schist.

Corundum in Amphibole-Schist.

At the Sheffield mine in Cowee township, Macon County,
·orth Carolina, corundum has been mined in a saprolitic rock
: various times for a number of years. While sinking a shaft
ght feet square to penetrate the depth of the corundum-bear-
ig saprolite, the solid unaltered rock was encountered. The
iaft was 87 feet deep and showed the following sequence
)wnward. The first 12 feet was through the saprolitic rock,
i which there were seams, a few inches wide, of kaolin; the
2xt two feet were corundum-bearing; from 14 to 28 feet the
.me saprolite was encountered and then another two feet that
as corundum bearing, followed by another ten feet of the
.prolite and two more feet of the corundum-bearing rock;
·om 42 to 65 feet the rock began to be less decomposed and
·om 63 to 66 feet another seam that was corundum-bearing
as encountered. From this point the rock became more and
iore solid, until at 77 feet the fresh rock was encountered.
hese various seams in the rock are very pronounced and are
ipping 30° toward the west near the top, but become nearly
orizontal nearer the bottom of the shaft. The seams of
ecomposed feldspar observed near the top of the shaft become
·ss and less kaolinized downward, until in the solid rock the
:ams are of a pure plagioclase feldspar. In the hard rock
kposed there are two seams of corundum similar to those
bove, although in the fresh rock the corundum seams are not
; pronounced as in the saprolitic rock. There is often consid-
rable of the feldspar bordering the seams of corundum. The
eneral trend of the rock is about N. 5–10° E.

From what could be seen of the solid and the saprolitic rocks
ie corundum occurs in seams a few feet in width at intervals
i the rock, and while the corundum may be ten or more per
ent in these veins, its percentage in the rock, that it would be
ecessary to mine, would not probably be over three or four.
'he actual width of the dike is not known, but the saprolitic

rock has been cut across for nearly 100 feet in a direction about at right angles to the strike.

The fresh rock at the bottom of the shaft is somewhat varied in appearance and while it all does not show any definite gneissoid structure, the more finely divided portions are distinctly so. There are streaks, a few inches thick in the rock, that are composed almost wholly of a plagioclase feldspar. Some portions of the rock are decidedly porphoritic and contain phenocrysts of a light gray amphibole, a centimeter in diameter, in a groundmass of feldspar. A large part of the rock is made up, however, of small roughly-outlined prismatic crystals of an amphibole, probably hornblende, and irregular fragments of plagioclase feldspar. The hornblende is almost black in color but in thin splinters it has a bronze luster and a deep resinous color. Biotite of a deep brown color occurs sparingly, and a pink garnet is rather abundant. It is this part of the rock that is of a gneissoid structure and in which the corundum occurs. The corundum is of a light to a purplish pink color and in nodules up to two or three centimeters in diameter. There are some streaks in the rock that are very highly garnetiferous, composed essentially of the garnet and plagioclase feldspar, or of the garnet and biotite. Chalcopyrite occurs very sparingly in these portions of the rock. Small particles of graphite have been observed in the coarsely crystallized portions.

Prof. L. V. Pirsson has kindly made a microscopical examination of this rock, the results of which are embodied in the following paragraphs.

In thin section the microscope disclosed the minerals, hornblende, plagioclase feldspar, garnet, biotite, muscovite, staurolite and rutile. Hornblende is the most common, forming about two-fifths of the section, while of the remainder, plagioclase and garnet occur in about equal quantities and the others in comparatively insignificant amount.

"The hornblende is formless but tends to irregular columns almost invariably extended in the plane of schistosity; it has very rarely a somewhat stringy tendency in its cleavage but is usually homogeneous in broad plates. Its color is a clear olive-brown and it is somewhat pleochroic but not strongly so. It is everywhere dotted by the small grains of garnet, which rarely show good crystal form. The garnet occurs associated also with the plagioclase."

"The plagioclase occurs twinned according to the albite law only. In sections perpendicular to 010, the lamellæ show extinctions as great as 30° and the plagioclase is therefore rich in lime and as basic as labradorite, which it probably is. It shows strong evidence of shearing movement in the rock; it is often broken, exhibits rolling extinctions and the albite lamellæ are curved and bent. It runs along the planes of schistosity between the feldspars and forms a mosaic of angular broken grains."

"Staurolite was found in rather broad irregular grains, and the rutile in small irregular grains and well crystallized prisms."

Prof. Pirsson has indicated that the character and structure of this rock, composed chiefly of amphibole, labradorite and garnet, suggests most strongly that it is a metamorphosed igneous rock of the gabbroid family. During metamorphism, the augite of the gabbro would be converted into the brown hornblende and any iron ore that was present would be taken up by the hornblende and garnet. The rutile would have resulted from the titanic acid that is a regular component of the iron ores in these gabbro or diabase rocks. Staurolite is rather naturally expected, as it is usually a mineral of metamorphism, and its natural home is in the schistose rocks. The feldspar has suffered the least (except the corundum) chemically and shows only the shearing of dynamic processes.

The corundum does not occur in crystals but in small fragments and in elongated nodules, which are cracked and seamed, and appear to have been drawn out by the shearing processes. The general character and shape of the corundums would indicate that they were original constituents of the igneous rock and were not formed during its metamorphism.

The exact classification of this rock is not easy, but it will probably be nearer correct to bring it under the head of an amphibole-schist.

Corundum in Quartz-Schist.

In the crystalline rocks of the southwestern part of North Carolina and the northeastern part of Georgia an interesting occurrence has recently been observed, namely that portions or bands of these are corundum-bearing. These corundum-bearing bands are first encountered on the head waters of Tallulah river, in the northern part of Rabun County, Georgia, and can be followed in a northeasterly direction to the Yellow Mountain in Clay County, North Carolina. They are near the top of the Blue Ridge, at an elevation of from 3000 to 4000 feet.

The composition of these rocks vary from those that are a normal gneiss to those that contain no feldspar and can best be described as quartz-schist, composed of biotite mica and quartz. Some portions of the rock are rich in garnet, while others are almost entirely free from this mineral, and occasionally there are small bands of white quartz. They are distinctly laminated and are frequently intersected by granitic dikes, some of which are coarsely crystallized and of a pegmatitic character, that are often parallel with the beddings of the schists, although many of them are cutting irregularly through them. Where these dikes are parallel to the bedding of the schists, the laminated structure of the latter is more apparent. The general strike of these crystalline rocks is N.E.-S.W. and with a dip of about 30° to the N.W.

Portions, or bands, of these schists are corundum-bearing, but they are irregularly defined and gradually merge into the normal rock. They have a similar relation to the normal schists that the garnet-bearing bands of a gneiss have to the normal gneiss in which they occur. They are not veins in any sense of the word, but are simply portions of the same mass of crystalline rocks in which corundum occurs as a constituent of the rock. These bands vary in width from a foot or two to 12 or 15 feet, but in these wider ones the corundum-bearing portion is not continuous but is intercepted by streaks of barren rock and granitic dikes.

These bands can be traced for a distance of five or six miles in a N.E.-S.W. direction, sometimes outcropping continuously for nearly a mile. There are at least two of these corundum-bearing bands which are parallel to each other and about two miles apart. The only variation that has been observed in them is the percentage of corundum and garnet, otherwise they are identical. The percentage of corundum is never high, and from determinations made on samples from various parts of the deposits, it varies from two to five per cent.

The corundum occurs for the most part in small particles and fragments that have no definite shape and are of a gray, white, and bluish-white color to almost colorless. It is also in crystals from minute ones to some that were observed two and a half inches long and about one-half an inch in diameter, which are usually fairly well developed in the prismatic zone.

It is probable that these schists are the result of the metamorphism of sandstones and shales formed from alluvial deposits of many thousand feet in thickness, that were formerly the bed of the ocean. By lateral compression these have been folded and raised into the mountain ranges of this section. That these were much higher than at the present time is very evident from the granitic dikes that are of deep-seated origin. By decomposition and erosion the mountains have been worn down to their present condition, thus exposing the schists in contact with granitic dikes which have aided in their thorough metamorphism. The shales were rich in alumina which not improbably was in the form of bauxite, and during their metamorphism the excess of the alumina crystallized out as corundum. This mineral has crystallized out along the planes of lamination so that during the subsequent weathering of the rock the corundum has been left in knotty nodules, studding the surface of the rock, giving it the appearance of *containing* a high percentage.

ART. XXX.—*On the Products of the Explosion of Acetylene, and of Mixtures of Acetylene and Nitrogen;* by W. G. MIXTER. Second Paper.

[Contributions from the Sheffield Laboratory of Yale University.]

THE bomb described in the first paper* was again used in the work unless otherwise stated. The figure of it is here reproduced (fig. 1) for convenience. The acetylene gas for the experiments was made by thrusting lumps of calcium carbide through the tubulure of a glass gas holder previously filled with water. The gas was dried as before by passing it through

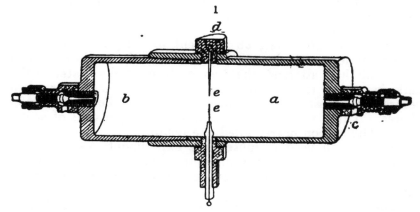

a cylinder containing a kilo of small sticks of caustic potash. The determinations of acetylene were made as follows: The gas to be tested was passed into a eudiometer through a stop cock at the top of it. The acetylene was absorbed by an ammoniacal solution of cuprous chloride and the ammonia was finally washed out with water. The temperature and pressure of the gas remaining after the absorption of acetylene was made the same as before. This method is applicable only in absence of oxygen and carbonic oxide. The latter will be present when the acetylene exploded was mixed with water vapor or oxygen. Repeated tests for carbonic oxide failed to show the presence of that gas in the products of the acetylene exploded.

Experiment 36.—Gas, 99 per cent of acetylene; pressure, 2 atmospheres. The explosion was prompt. It extended through a brass tube $1\frac{1}{2}^{mm}$ in diameter and 1 meter in length to the manometer. The determinations of acetylene in the residual gas gave 4, 4·3 and 3·7 per cent.

* This Journal, vol. ix, 5.

Experiment 37.—Gas, 98 per cent of acetylene ; pressure, 1289^{mm}; temperature, 12° ; electrodes, 2^{mm} apart. The gas was sparked repeatedly for several seconds. Carbon separated about the electrodes, but there was no explosion and the gas after the sparking was found to contain 98 per cent of acetylene.

Experiment 38.—Gas, 98 per cent of acetylene; pressure, 1312^{mm}; temperature, 15°. The electrodes were 3^{mm} apart. There was no explosion and no change in pressure during the sparking.

Experiment 39.—Gas, 96 per cent of acetylene; pressure, 1472^{mm}; temperature, 14°. The pressure after the explosion at 14° was 1425^{mm}, showing a condensation of 3·2 per cent. Two determinations of acetylene in the residual gas gave 3·4 and 3·3 per cent.

Experiment 40.—Gas, 98 per cent of acetylene; pressure, 5·1 atmospheres, which was the same after the explosion when the bomb had attained the temperature of the water in the tank in which it was immersed. The gas tested was taken before the pressure in the bomb had fallen below 4 atmospheres. Three tests yielded 2·3, 2·4 and 2·4 per cent of acetylene. The carbon nearly filled the bomb.

The nitrogen used in subsequent experiments was prepared from sodium nitrite and ammonium chloride. The mixtures of acetylene and nitrogen were dried as described by caustic potash.

Experiment 41.—Gas, 89·3 per cent of acetylene and 10·7 per cent of nitrogen; pressure 3·2 atmospheres before and after the explosion. Two estimations of acetylene in the gas after the explosion gave 1·4 and 1·4 per cent. In order to determine the amount of hydrocyanic acid, the gases (933^{cc} reduced to 0° and 760^{mm}) from the bomb were passed slowly through a dilute solution of pure potassium hydroxide. This solution was then titrated with a decinormal solution of silver nitrate, requiring 5·2^{cc} to produce a permanent precipitate of silver cyanide. This corresponds to 0·028 gram of hydrocyanic acid and 23^{cc} of the compound in the gaseous state. The bomb had a capacity of 1$\frac{1}{10}$ liters, but for convenience the result may be given as the amount of hydrocyanic acid in a volume of 1 liter in the bomb. It was 0·083 gram and 63·3^{cc}. The potash solution after the titration was acidified with hydrochloric acid, boiled, cooled, then an excess of potash was added and the solution was again boiled. The escaping vapor was found to be free from ammonia. Nor was any found in another test similar in all respects except that the titration with silver nitrate was omitted.

Experiment 42.—Gas, 85 per cent of acetylene and 15 per cent of nitrogen; pressure, 3061^{mm}; temperature, $17·8°$. The pressure after the explosion when the gas had cooled to $17°·8$ was the same as before the explosion. Each of two tests for acetylene in the residual gas gave 0·25 per cent. The residual gas was passed slowly through an absorbing tube containing a dilute solution of potassium hydroxide until the pressure in the bomb had fallen to that of the atmosphere; 20^{cc} of a deci-normal solution of silver nitrate were used in titrating the solution of potassium hydroxide. Next 10^{cc} of a concentrated solution of potassium hydroxide were poured into the bomb to absorb the hydrocyanic acid remaining in it. The mixture of carbon and potassium hydroxide was thoroughly shaken and then washed on a filter. The filtrate required $14·4^{cc}$ of a decinormal solution of silver nitrate, making the total amount of the solution of silver used $34·4^{cc}$, corresponding to 0·1858 gram of hydrocyanic acid, or 0·1688 gram and $139·6^{cc}$ (at $0°$ and 760^{mm}) per liter in the bomb.

Experiment 43.—Gas, 76·6 per cent of acetylene and 23·4 per cent of nitrogen; pressure 3818^{mm} at $14·8°$. 1·7 per cent of acetylene was found in the first portions of gas drawn from the bomb and 2·5 per cent when the pressure was little more than that of the atmosphere. The attempt to estimate the hydrocyanic acid remaining in the bomb was unsuccessful. An alcoholic solution of potash was used, as this wets the finely divided carbon better than water. The filtered solution became turbid on adding silver nitrate and the titration was so unsatisfactory that the result was rejected. The solution of potassium hydroxide used to absorb the gas drawn from the bomb required in titrating $32·3^{cc}$ of a decinormal solution of silver nitrate corresponding to 0·1747 gram of hydrocyanic acid. As $\frac{4}{5}$ths of the gas in the bomb were taken for this determination we may assume that $0·1747 \times \frac{5}{4} = 0·2187$ gram was the total amount of hydrocyanic acid in the gas in the bomb and not condensed on the carbon. This result calculated for a volume of 1 liter gives 0·200 gram and 165^{cc} (at $0°$ and 760^{mm}) of hydrocyanic acid. The gas after the explosion was alkaline to litmus and the solution of potassium hydroxide gave off ammonia on warming.

Experiment 44.—Gas, 87 per cent of acetylene and 13 per cent of nitrogen. The nitrogen was collected over an alka-line solution of pyrogallic acid to absorb traces of oxygen, and then was mixed with the acetylene over a solution of ferrous sulphate in order to free the gases from possible traces of ammonia. As the bomb leaked after the explosion, it was only possible to determine approximately the ratio between the quantities of hydrocyanic acid and ammonia formed. The potassium hydroxide solution through which the gaseous products of the explosion were passed required 5^{cc} of a deci-normal solution of silver nitrate to form a permanent precipitate of silver cyanide, 5^{cc} more of the silver solution were added in order to remove the cyanides from the solution, and the precipitated silver cyanide was filtered off. The filtrate was then cautiously distilled into hydrochloric acid. The acid solution was finally evaporated to dryness. The residue was 0·0035 gram of ammonium chloride, equivalent to 0·001 gram of ammonia. As 5^{cc} of the silver solution correspond to 0·027 gram of hydrocyanic acid, the ratio between the ammonia and hydrocyanic acid formed is apparent.

In order to determine whether or not the gas after the explo-sion contained cyanogen, the alkaline solution from which ammonia was distilled was acidified with hydrochloric acid, boiled, then made alkaline with potassium hydroxide. This last solution gave when distilled no ammonia. If cyanogen had been present in the gas, there would have been formed in the solution of potassium hydroxide through which the gaseous products of the explosion were passed, cyanide and cyanate. The latter salt, as is well known, yields an ammonium salt when treated with an acid.

Experiment 45.—The mixture of acetylene and nitrogen containing 28 per cent of the latter gas was allowed to stand 24 hours over a solution of ferrous sulphate mixed with ferrous hydroxide in order to free the gas from oxygen and possible traces of ammonia. The pressure of the gas before the explosion was 3781^{mm}, a little less than 5 atmospheres, at 6·4°. The gas drawn from the bomb after the explosion when the pressure was nearly 5 atmospheres contained 3·2 per cent of acetylene, and that portion drawn at 1·3 atmospheres contained 5·6. The hydrocyanic acid estimated as in previous experiments was found to be ·170 gram and 140^{cc} per liter in the bomb. The ammonia found in the same volume of gaseous products was ·0047 gram. Here we have a ratio of 1 of ammonia to 36 of hydrocyanic acid in the products of the explosion.

Experiment 46.—The mixture of acetylene and nitrogen containing 10 per cent of the latter was purified as in the previous experiment. It was sparked when at a pressure of 3856^{mm}. The first portion of the gas drawn from the bomb was found to contain 1·7 per cent of acetylene and that taken at a pressure of 2 atmospheres 2·2 per cent. The hydrocyanic acid found corresponded to 0·145 gram and 120^{cc} per liter in the bomb. Likewise the ammonia was 0·0046 gram. Here the ratio of the ammonia to the hydrocyanic acid is 1 to 30.

Experiment 47.—Two liters of a dried mixture of acetylene and nitrogen containing 10 per cent of the latter gas were passed through a combustion tube which was heated to a temperature sufficient to decompose the acetylene. The gaseous products were passed through a solution of potassium hydroxide to absorb hydrocyanic acid. But not a trace was found by the prussian blue reaction. The same result was obtained a second time.

Experiment 48.—A mixture of acetylene and air was exploded in an open liter jar. Much soot separated and the flame passed slowly into the jar. No hydrocyanic acid was detected in the products, and none in several repetitions of the experiment.

Experiment 49.—The sides of a jar were moistened with strong ammonia water and the air in the jar was mostly displaced by acetylene. On applying a flame to the mouth of the jar there was a slight puff and an abundant separation of soot in the jar. Hydrocyanic acid was found in the products of four experiments, and it was also formed rather abundantly when illuminating gas was substituted for acetylene.

Experiment 50.—Acetylene was passed through concentrated ammonia water. then through a tube heated to dull redness by one burner. Hydrocyanic acid was formed.

The experiments 19 to 28 described in the first paper (loc. cit.) were made in U-tubes having a diameter of 17 to 20^{mm}. The residual gas contained much acetylene and the amount of condensation products was considerable. In order to learn if the character of the products is due to the expansion of the gas when exploded, the next two tests were made in an iron U-tube $7 \cdot 5^{cm}$ in diameter. To the ends of this tube were attached the parts a and b of the iron bomb (fig. 1, p. 299). The apparatus was designed for observing the effect of explosions under the same conditions as in the bomb except that the gas was not subjected to constant volume. About 50 kilos of mercury were required in the apparatus and 1800^{cc} of dry acetylene measured at atmospheric pressure were used. The U-tube when charged for an explosion was hung in a tank of water in order to cool it quickly after an explosion. When acetylene is fired in such a U-tube the products cool more rapidly than in the bomb.

Experiment 51.—Gas, 99 per cent of acetylene; pressure, 2060^{mm} at 18°. The electrodes were 2^{mm} apart. The explosion was prompt and the pressure after it was 2097^{mm} at 18°. Two determinations of acetylene in the gas after explosion gave 3 per cent.

Experiment 52.—Gas, 99 per cent of acetylene ; pressure, 2300^{mm}. The electrodes were 4^{mm} apart. Two estimations of acetylene in the gas taken at 3 atmospheres gave each 1·5 per cent, and when the pressure in the apparatus had fallen nearly to that of the atmosphere the portions tested were found to contain 1·9 and 1·8 per cent of acetylene.

The next two experiments were made in the iron U-tube in order to find the effect that the expansion accompanying the explosion has on the formation of hydrocyanic acid. The mixtures of acetylene and nitrogen were kept for a day over a solution of ferrous sulphate in which was suspended ferrous hydroxide.

Experiment 53.—About 1·5 liters, measured at atmospheric pressure, of gas containing 15 per cent of nitrogen were used. The mixture was sparked when under a pressure of 2·5 atmospheres. The residual gas contained 1·25 per cent of acetylene. 800^{cc} of it were passed through a solution of potassium hydroxide. This was then titrated with a decinormal solution of silver nitrate, requiring 0·4^{cc} to produce a permanent precipitate, a result not indicating satisfactorily the presence of cyanides.

Experiment 54.—The mixture contained 13·5 per cent of nitrogen. Two liters of gas were taken and subjected to a pressure of 2·5 atmospheres. The explosion was prompt. The estimation of acetylene in the residual gas was not satisfactory but it showed there was less than 3 per cent present. 1000^{cc} of the gas left by the explosion were passed slowly through a solution of potassium hydroxide. One drop of a decinormal solution of silver nitrate gave to the solution a permanent brown precipitate. Next, 200^{cc} of the gas were passed through a solution of potassium hydroxide, made from the metal. Finally a few drops of a solution of ferrous ferric sulphate were added and the mixture was warmed, cooled, and made acid with hydrochloric acid. No prussian blue was formed, a proof that no hydrocyanic acid had been formed.

These results in the iron U-tube prove conclusively that the large amounts of acetylene and condensation products in the earlier experiments in glass tubes are not attributable to expansion at the time of the explosion.

In order to learn whether the peculiar explosive **wave** of experiments 27 and 28 of the first paper can be transmitted to gas in a larger vessel and there produce the same effect, the next two experiments were made in the apparatus shown in fig. 2. The tube *a* is glass and has a diameter of 18^{mm}.

riment 55.—Gas, 98 per cent of acetylene ; pressure,
ospheres. The electrodes were 3^{mm} apart and less than
m the glass. The sparking at first produced a filament
n between the electrodes. A much stronger secondary
was then applied and after some seconds there was a
glow and carbon was deposited
tly in a length of 10^{cm} of the
The explosion did not extend to
in the bomb. The residual gas
d 97·5 per cent of acetylene.
riment 56.—Gas, 98 per cent of
e ; pressure, 2316^{mm} (3·3 atmos-
at 15°. The electrodes were 4^{mm}
d 3^{mm} from the glass. A moment-
ing of the primary circuit pro-
n explosion. The pressure when
had cooled to 15° was the same as
he explosion. There were in the
be brilliant black rings of carbon
d by spaces less densely coated.
timations of acetylene in the re-
as taken at 2·7 to 3·3 atmospheres
ch 1·3 per cent, and two portions
hen the gas in the bomb was at 1·5
eres and less gave 1·7 and 1·6 per

e peculiar explosive wave started
tube extended to the gas in bomb
d almost complete decomposition
cetylene.
he acetylene used thus far in the
as not free from traces of water, the following experi-
vere carried out in U-tubes such as shown in fig. 3 filled
s containing 98·8 per cent of acetylene. Some phos-
pentoxide was placed in the tube and also a clean piece
phorus ; the latter was to remove traces of oxygen.
riment 57.—The volume of the gas diminished slowly
ding indicating an absorption, but there was no appear-
a tarry product. Five days after the tube was filled,
measuring 60^{cc}, was condensed to 19^{cc} and then sparked.
plosion was prompt and violent. The condensation was
nt and the acetylene in the residual gas was 4 per cent
gas taken.
tube was filled again with acetylene and the phos-
pentoxide was made to coat the upper part of it. A
f phosphorus was placed in the gas. The volume

diminished slowly and a reddish liquid appeared. A little water was introduced in order to find if phosphoric acid will condense acetylene. The change if any was slight in two days. Owing to an accident the experiment was not completed, but the absorption observed suggests doubt as to the value of phosphorus pentoxide as a drying agent for acetylene. In the remaining experiments no such absorption was observed.

Experiment 58.—After 8 days the phosphorus was removed and the pressure increased to 3·3 atmospheres and the gas was then sparked. The condensation was 8 per cent and the acetylene remaining was 8 per cent.

Experiment 59.—A smaller quantity of phosphorus pentoxide was used than before. However, some remained unaltered and after 8 days the pressure was made 3 atmospheres and the gas was sparked. The explosion was prompt and apparently as violent as any in the glass U-tubes. 6 per cent of the gas had formed condensation products and the acetylene in the residual gas was 24 per cent of that taken.

Experiment 60.—The acetylene stood over a solution of calcium hydroxide 3 days before it was put into the U-tube. The phosphorus pentoxide used filled a tube 2^{cm} long and 1^{cm} wide used to hold it. The phosphorus was also placed in a short tube. When the apparatus was filled the volume of the gas was noted and it did not change appreciably in the two weeks the gas was left to dry and to give up traces of oxygen. The gas, 57^{cc}, was then condensed to 19^{cc} and sparked by closing the primary for an instant. The explosion was prompt and violent. The condensation was 5 per cent and the acetylene in the residual gas was 21 per cent of the gas taken for the test.

The results obtained with dry acetylene indicate that water is not essential to the explosion of the gas, and that the wide variations noted in the first paper in the products of explosions in glass tubes are not to be ascribed to the presence of water.

The following table gives the number of cubic centimeters of acetylene and hydrocyanic acid per liter of gas in the bomb after explosion.

No. of Exp.	Pressure in atmospheres.				Per cent of C_2H_2 in gas after explosion.	C. C. at 0° and 760mm of C_2H_2 per l. in bomb.	C. C. at 0° and 760mm of H C N per l. in bomb.
29	3				4	120	
30	3				4·1	123	
36	2				4·	80	
39	1·95				3·3	66	
40	5·1				2·4	122	
41	3·2	N	10·7	%	1·4	45	63
42	4·	N	15	"	0·25	10	140
43	5·	N	23	"	2·1	105	165
45	5·	N	28	"	4·4	220	140
46	5·		10		2·	100	120
51 in U-tube	2·7				3·	81	
52 "	3·				1·7	51	
53 "	2·5	N	15	"	1·3	32	0
54 "	2·5		13	"	Not estimated.		0

The above are all of the results obtained with the bomb, fig. 1, and with the iron U-tube having the same diameter as the bomb, excepting experiments 31 to 35 described in the first paper. The amounts of acetylene and hydrocyanic acid found in the residual gas drawn from the bomb do not include quite all of these compounds, as there was some condensation on the carbon. This is evident from the fact that gas taken at 4 or 5 atmospheres contained less acetylene than that taken at lower pressures. The above data derived from the experiments are not sufficient to yield the relation between the pressure and the quantity of acetylene in the residual products. It is obvious, however, that at the lowest pressure at which explosion will occur the total quantity of acetylene left in the bomb is much less than when the initial pressure is 3 or more atmospheres. The small amounts found in experiments 51, 52 and 53 show that a more rapid cooling in the large iron U-tube than occurs in the bomb does not increase the acetylene in the residual gas. This fact supports the view expressed in the first paper that the acetylene in the gas after explosion is not an undecomposed portion of the original gas.

There is, however, no question about the synthesis of hydrocyanic acid in the explosions of mixtures of acetylene and nitrogen in the bomb. But in the U-tube the nitrogen did not combine. In the former instances the molecules of nitrogen either acquired energy adequate for combination or were dissociated, while in the latter they remained in the inert condition characteristic of the element. The velocity of explosion of acetylene is 1000 meters and higher per second,[*] and in the iron U-tube the expanding gas must move a column of mercury about 80^{cm} in length, that is, the pressure on $1^{sq\ cm}$ must move one kilo of mercury. It is highly improbable that this mass is displaced materially before the explosion is completed. If this assumption be correct, the temperature at the instant of explosion is the same in the U-tube as in the bomb. In the former the expanding gas cools more rapidly than in the latter. In the bomb there is a sufficient number of collisions to impart to the molecules of nitrogen the energy adequate for combination or to dissociate them into atoms. In the U-tube the volume of the gas increases after the explosion and the collisions are fewer than in the bomb and the nitrogen does not combine. Thus we may conclude that, even at the temperature of exploding acetylene, a sufficient frequency of collisions is requisite to cause nitrogen to combine.

The amount of hydrocyanic acid remaining in the bomb does not appear to depend upon the proportion of nitrogen in the mixture exploded, for the result was the same with 15 per cent of nitrogen in the mixed gases at a pressure of 4 atmospheres as with 28 per cent at 5 atmospheres. It may be that a fixed quantity of hydrocyanic acid per unit of space remains when the products of the explosion cool and assume a state of equilibrium.

Acetylene and ammonia, as shown by experiments 49 and 50, yield hydrocyanic acid at a much lower temperature than is required to cause nitrogen to combine. It may be that ammonia is the first compound of nitrogen formed in the bomb, but the fact that a little ammonia is found in the products of an explosion is not conclusive, as this may have resulted from the decomposition of hydrocyanic acid. Dewar[†] passed a mixture of hydrogen and nitrogen through a carbon tube heated externally by an electric arc and obtained hydrocyanic acid. Here the conditions were similar to those in bomb, namely, carbon, hydrogen and nitrogen at a very high temperature. Dewar made no statement regarding the presence or absence of ammonia in the products of his experiment, and Berthellot[‡] considered that ammonia played no part in the synthesis of hydrocyanic acid effected by sparking a mixture of acetylene and nitrogen. The results in the bomb indicate that less acetylene remains when nitrogen is present than when pure acetylene is exploded, but they are not conclusive. Berthellot (loc. cit.)

[*] Berthellot and Chatelier, Comptes Rendus, cxxix, 427 ; Ann. Chim. et Phys., xx. 15.
[†] Proc. Roy Soc , xxix, 188, and xxx, 85.
[‡] Comptes Rendus, lxvii, 1141.

observed that a mixture of acetylene and nitrogen diluted with hydrogen did not deposit carbon when sparked, and that all of the acetylene or nitrogen, an excess of one or the other being present, was removed by prolonged sparking of the mixed gases over a solution of caustic potash to remove the hydrocyanic acid. He considered that nitrogen united directly with acetylene and represented the reaction by the equation

$$CH_2 + N_2 = 2HCN.$$

As acetylene was not decomposed by the sparks, it must have combined directly with nitrogen. The reaction is of great interest, as it shows that the nitrogen molecules are rendered chemically active by conditions which do not dissociate endothermic acetylene.

ART. XXXI.—*Scapolite Rocks from Alaska;** by J. E. SPURR.

THE following rocks containing scapolite as an essential constituent were collected by the writer in the summer of 1898,
while making a geological reconnoissance in southwestern
Alaska.

Andesine-oligoclase-scapolite-biotite rock.

This rock was found on the Yentna River (a branch of the
Sushitna) some 12 or 15 miles above the junction with the
Sushitna, where it forms large masses of uniform appearance,
which are probably great dikes cutting an older rock. Specimens of this older rock, taken in different places, proved to be
hornblende syenite and hornblende diorite. The scapolite
rock is coarse-grained and granitoid in appearance, the bluish-
gray of the scapolite suggesting quartz at first glance. Under
the microscope the structure is hypidiomorphic granular, grad-
ing to panidiomorphic. The essential minerals are feldspar,
scapolite, and biotite, abundant in the order of naming. The
feldspars are mostly twinned according to the albite law, and
one crystal examined by the Fouqué method showed on a section perpendicular to the negative bisectrix an angle of 70°
between the plane of the optic axes and the albite twinning,
proving it to be an andesine-oligoclase. The scapolite is both
idiomorphic and hypidiomorphic, like the feldspar. Some sections are always dark, and give the dark cross of uniaxial
minerals in convergent light. The cleavages, which are often
broken, intersect at right angles and belong to the faces 100
and 010; another less perfect cleavage evidently belongs to
the prism 110. The relief of the mineral is small, like quartz,
while the double refraction is higher, giving a light-yellow
color; it is optically negative. The biotite is fresh, and like
the other minerals is idiomorphic or hypidiomorphic in habit;
it contains grains of apatite. Small crystals of zircon, showing
the prism and pyramid faces, are occasionally found.

Analysis of andesine-oligoclase-scapolite-biotite rock.
[Analyst, Dr. H. N. Stokes.]

	Per cent.		Per cent.
SiO_2	62·78	BaO	·04
TiO_2	·56	SrO	trace
CO_2	——	CaO	4·84
P_2O_5	·15	MgO	2·32
S	·02	K_2O	2·15
Cl	trace	Na_2O	4·11
Al_2O_3	17·16	Li_2O	trace
Fe_2O_3	1·96	H_2O at 100°	·24
FeO	2·31	H_2O at 110°	·88
MnO	·06		
		Total	99·58

* Published by permission of the Director of the U. S. Geological Survey.

The above analysis bears out the results of the optical investigations and throws some additional light on the probable variety of scapolite. The high percentage of silica in the rock and the low percentage of lime indicate that the scapolite is probably of the variety dipyre. The rock therefore presents some analogy to the so-called dipyre-diorite of the Norwegian geologists.

Microcline-scapolite rock.

This is from the same locality as the rock just described and is doubtless a variation produced by differentiation. In the outcrop it has the appearance of a vein, forming a distinct tongue of light-colored material between walls of mottled rock, which appear to be essentially the same as the scapolite rock previously described, but of finer grain. This finer-grained biotite-scapolite rock grades off distinctly into the coarser-grained lighter-colored vein, which is about two inches thick. Under the microscope the structure is very coarse and the constituent grains are intergrown. Among the minerals microcline prevails, with considerable scapolite, which is largely altered to calcite. The scapolite gives in the basal section uniform darkness and the interference cross; it has abundant irregular cleavages and cracking, and shows yellow of the first order as an interference color.

Quartz-scapolite porphyry.

A rock called quartz-scapolite porphyry, for want of a better name, comes from a point far distant from the rocks just described, being found on the Kuskokwim River at the mouth of the Holiknuk, where it occurs among the light-colored dikes which cut the Cretaceous shales and shaly limestones. Under the microscope the rock shows phenocrysts of quartz, which are of small size and are bounded chiefly by the dihexahedral planes, but sometimes have the prism sparingly developed and sometimes show the faces of pyramids of two orders; the edges are often frayed by magmatic resorption. There are frequent large idiomorphic phenocrysts of scapolite, more or less broken down by decomposition. These phenocrysts give a uniaxial figure in convergent light. The outlines of the crystals show that they are of the tetragonal system, and they are characterized by low single and double refraction; they are optically negative. There is good cleavage parallel to the two lateral pinacoids, while that parallel to the prism is present but not so well developed. Irregular fractures are common. The mineral has decomposed along the margins and cleavages, partly to calcite, but mainly to an opaque kaolinic substance which is colored brown with iron oxide. There are abundant microscopic inclusions arranged in zones within the crystal. Judging from the paucity of calcite among the decomposition products and also from the apparent small single and double refraction of these scapolites, the mineral is probably a soda scapolite (marialite end of the series). By exception, some almost entirely altered phenocrysts now consist chiefly of calcite, but what remains seems to be scapolite, and the outlines of the phenocrysts corroborate this inference. These decomposed scapolites are probably lime scapolites, or meionites. A few small phenocrysts, mostly altered to calcite, muscovite, etc., seem to have been originally soda-lime feldspar. The groundmass is a fine-grained aggregate, consisting chiefly of quartz, orthoclase, and muscovite. In the fresher scapolite phenocrysts the cross sections are sometimes rectangles, but the favorite occurrence is in the form of a penetration twin of two such crystals along the composition plane 110; and the corners of the pinacoids whose faces constitute the rectangles are truncated by the prism. There are no true longitudinal sections in the slide, all being of small double refraction and having approximately the same orientation.

Occurrence of Scapolite Rocks elsewhere.

The scapolite rock from Norway in the vicinity of Oedegården has been described by Brögger and Reusch. This rock is essentially a mixture of scapolite and amphibole with accessory titanite, and was regarded by the describers as a facies of the gabbro (hyperite) in the neighborhood of veins of apatite which traverse the gabbro. The observers concluded, moreover, that the scapolite rock was formed by the alteration of the normal hyperite or gabbro, since they found remnants of diallage in the hornblende, as if the former were the original mineral, and of plagioclase in the aggregates of scapolite, as if the scapolite had formed from the plagioclase. The alteration, according to Brögger's view, was due to gaseous or pneumatolitic action.[*]

In Canada, also, granular rocks containing scapolite have been described from a number of localities.[†] In one of these rocks the principal constituents are pyroxene, hornblende, and scapolite, with accessory epidote, enstatite, pyrrhotite, and rutile, and the rock is classified as a scapolite diorite, it having a granular structure. Another rock is made up of an aggre-

[*] See Rosenbusch's Mikroscopische Physiographie der Massigen Gesteine, vol. i, p. 331.

[†] On some Canadian rocks containing scapolite, etc., by Frank D. Adams and Andrew C. Lawson: Canadian Rec. Sci.

gate of plagioclase, scapolite, and green hornblende, with accessory pyroxene, quartz, epidote, and pyrite, and this rock is termed a plagioclase-scapolite diorite. It is stated in the pamphlet referred to that "although the derivation of at least a part of the hornblende of these rocks from pyroxene is well-nigh certain, the derivation of the scapolite from plagioclase, which, as already stated, has been pretty clearly proved in the case of the Norwegian rocks, is not so evident in these similar rocks from Canada." Also, Dr. A. P. Coleman reports from Ontario a scapolite rock consisting of hornblende, plagioclase, and scapolite, in which the scapolite was evidently primary.*

Finally, scapolite rock occurs in the white limestone of New Jersey at many localities and in large amounts, the rock consisting in one locality of scapolite, hornblende, pyroxene, and sphene, and in another locality of the same minerals with the addition of a little plagioclase. These dikes are evidently igneous, and fumarolic or gaseous action has been suggested as taking part in their formation.†

In addition to these granular scapolite rocks, scapolite gneisses and amphibolites have been described from various places.‡

Origin and Classification of the Alaskan Scapolite Rocks.

In the locality on the Yentna River which has been described the main intrusive rock of the country was noted in the field to be a biotite granite, but of this, unfortunately, no specimen was retained. Dikes of quartz-feldspar rock or alaskite,§ which were found cutting the granite, have been carefully examined microscopically, and connected with this alaskite were rocks transitional into the granite, and quartz veins sometimes containing tourmaline. The specimen of scapolite rock was from a large mass forming bluffs 40 feet high, and seemed typical; it is certain that this rock occurs in notable quantity, but in just what proportion to the other dikes is not certain. All these dikes cut an ancient igneous rock, which was shown in one case to be a hornblende syenite and again a hornblende diorite.

In the second case which has been described, namely, the quartz-scapolite rock from the Kuskokwim River, the scapolite dikes were associated with a number of other siliceous dikes. Those which were examined microscopically, besides the scapolite porphyry, are an alaskite porphyry and a granite porphyry.

---

* See Zirkel's Lehrbuch der Petrographie, vol ii, p. 783.

† F. L. Nason, Annual Report of the Geological Survey of New Jersey, 1890, p. 33.

‡ Zirkel, op cit., vol. iii, p. 339.

§ See author's paper on Classification of Igneous Rocks according to Composition; Am. Geol., xxv, April, 1900, p. 210.

In both these cases the scapolite is regarded as original. There is no evidence of its derivation from feldspar, but rather in every case of contemporaneous formation. Certainly the mineral is not the product of weathering, for it is nearly always in a process of decomposition, showing itself as unstable under atmospheric conditions as are the feldspars. It is exceedingly probable, however, that in the formation of this, as in other occurrences of scapolite rock, gases have played an important part. The scapolites contain chlorine, otherwise they have essentially the composition of the soda-lime feldspars. The marialite scapolite corresponds to oligoclase, and the meionite scapolite to anorthite, with a series between like that of the feldspars, so that when scapolite is found instead of feldspar we may suppose chlorine gas to have been present at the time of the formation of the rock. This is, however, no good reason for considering the rock as of secondary origin, for in most igneous rocks the gases play some part in the formation and in many a very weighty part, the rocks nevertheless belonging to the class of original igneous rocks. The writer is inclined, therefore, to give the scapolite rocks a place among the primary igneous rocks wherever the scapolite shows evidence of having formed contemporaneously with the other rock minerals. In cases where it evidently is an alteration product from some other rock, the classification will of course be different.

In case classification is attempted, the scapolites might be considered, for the purposes of classification, as equivalent to the feldspars, when they occur in rocks where feldspar is present in an equal or greater amount. Thus the biotite-scapolite rock from the Yentna River might be called a biotite scapolite-belugite,* since the feldspar belongs to the andesine-oligoclase series, and the equally important scapolite seems to be of a variety containing much lime and corresponding to anorthite in the feldspar series. The scapolite-belugites might then be considered a group of the belugite family.

The porphyritic scapolite rock from the mouth of the Holiknuk on the Kuskokwim might be called a scapolite-adamellite porphyry, adamellite being the name adopted by Brögger for the quartz-bearing monzonites. In this case the scapolite appears to belong mainly to the soda end of the soda-lime series, and thus is the equivalent of oligoclase-andesine. The groundmass appearing to be mainly orthoclase, the feldspathic constituents (considering scapolite as such) would be on the whole intermediate between the granite and diorite families, or in the monzonite family. In this case the rock might be con-

* See writer in Am. Geol., xxv, April, 1900, p. 233.

sidered to belong to what might be called the scapolite-ada-
mellites, which would be a group under the monzonite family
where the scapolite partly takes the place of the feldspar.

It would probably be best, however, to separate the scapolite-feldspar rocks into a distinct class, and to give the groups
characteristic names, the analogy with the feldspar rocks being
expressed by writing in the tabulation each group laterally
opposite the corresponding group of feldspar rocks. Thus the
rock first described may be called yentnite, from the Yentna
River, instead of scapolite-belugite, but it may be written
opposite the belugite group; the type would then be a biotite
yentnite. Similarly the quartz-scapolite porphyry from the
Kuskokwim* River might be called kuskite, instead of scapolite-adamellite; the kuskite group of the scapolite-feldspar
class of rocks could be written opposite the adamellite (or
quartz-monzonite) group of feldspar rocks, and the type might
be called a kuskite porphyry.

* Eskimo *Kuska*, derivation uncertain, *kwik*, genitive *kwim*, river.

ART. XXXII.—*On the Qualitative Separation of Nickel from Cobalt by the action of Ammonium Hydroxide on the Ferricyanides;* by PHILIP E. BROWNING and JOHN B. HARTWELL.

[Contributions from the Kent Chemical Laboratory of Yale University—XCVII.]

SOME years ago F. W. Clarke* suggested a method for the separation of nickel from cobalt depending upon the solvent action of ammonium hydroxide upon the precipitated ferricyanides. The method may best be described by quoting from the original article : " To the slightly acid solution containing the two metals, I first add an excess of ammonium chloride. This causes the cobalt precipitate, which otherwise would run through the filter, to fall in a denser state, and also of a much darker color, often nearly black. I then add the potassium ferricyanide until the precipitation is complete, and afterwards agitate strongly with a considerable excess of ammonia. Upon filtering, all the cobalt remains upon the filter, being recognized by the characteristic color of the precipitate, and the nickel is readily detected in the filtrate, by means of ammonium sulphide. If, upon filtering, the portion at first running through is turbid, it may be disregarded, or returned to the filter, that which filters through subsequently being almost invariably clear."

In making a study of this method we found two serious objections; first, the practical impossibility of obtaining a good filtration from the cobalt ferricyanide, even in the presence of the ammonium chloride, and, second, the large amount of sulphur thrown down when ammonium sulphide was added to the filtrate containing the nickel with the excess of ferricyanide.

Our first attempt was to secure, if possible, a complete separation of the precipitated cobalt ferricyanide and the dissolved nickel by filtration. This we were able to accomplish by the addition of a small amount of a solution of an aluminum salt to the original solution which held back the cobalt, and, as experiment showed, allowed the complete solvent action of the ammonium hydroxide upon the nickel salt. Amounts of nickel as small as 0·0001 grm. were detected, when mixed with the aluminum salt, by precipitating as ferricyanide, extracting with ammonium hydroxide, and testing in the manner to be described.

On turning our attention to a possible improvement in the method for the detection of the nickel, it was found that when the ammoniacal solution of the nickel ferricyanide was treated

* This Journal, xlviii, 67.

with strong sodium or potassinm hydroxide solution, in the presence of an excess of potassium ferricyanide, a black flocky precipitate formed which gave no test for ferro or ferricyanide, and gave every indication of being nickelic hydroxide. This reaction we found to afford us a most delicate test for nickel.

The method as modified by us may be described as follows: Dissolve not more than 0·1 gm. of the salts of the two elements in about 5$^{cm^3}$ of water, add a few drops of a saturated solution of alum, destroy any free mineral acid by neutralizing with ammonium hydroxide, and make faintly acid with acetic acid. To this solution add about 0·5 grm. of potassium ferricyanide and agitate to effect the solution of the ferricyanide and the complete precipitation of the nickel and cobalt salts. Then add about 5$^{cm^3}$ of strong ammonium hydroxide and filter. To the filtrate, which should have no reddish color, add a piece of sodium or potassium hydroxide about the size of a pea and boil. The appearance of a black precipitate, in case of very small amounts of nickel showing first as a dark coloration, indicates nickel.

The tables following give a record of the experimental results. With the precautions indicated, this method may be applied very satisfactorily.

TABLE I.

	CoSO$_4$7H$_2$O grm.	NiSO$_4$. 7H$_2$O grm.	KAl(SO$_4$)$_2$ saturated sol. cm^3	K$_3$FeC$_6$N$_6$ grm.	NH$_4$OH (conc.) cm^3	NaOH, solid, about size of a pea.	Result.
1.	----	0·0100	2	0·5	5·	"	Heavy ppt.
2.	----	0·0050	"	"	"	"	"
3.	----	0·0010	"	"	"		"
4.	----	0·0003				"	Distinct
5.	----	0·0001	··	··			Plain

TABLE II.

1.	0·10	----	"	"			None
2.	0·10	0·0100		"		"	Heavy
3.	0·10	0·0050				"	Distinct
4.	0·10	0·0030				"	Very faint
5.	0·05	----					None
6.	0·05	0·0100				"	Heavy
7.	0·05	0·0050	··	··		"	Distinct
8.	0·05	0·0030	··	··			Plain
9.	0·05	0·0010*	"	"	··		Faint

* Equivalent to 0·0002 of the metal.

SCIENTIFIC INTELLIGENCE. .

I. Physics.

1. *Radio-activity of Uranium.*—Sir William Crookes has recently made an investigation in regard to Becquerel's rays, the most important result of which is the fact that he has succeeded in purifying uranium salts to such an extent that they are photographically inactive. There can be no doubt, therefore, that the radio-active rays are not a specific property of pure uranium, as has been previously supposed.

In the course of his experiments Crookes has tested a large number of specimens of barite and witherite from various localities without obtaining the slightest photographic action from them. He then tested every mineral in his extensive collection with the result that the following were found to be active: Pitchblende, uranite, autunite, orangite, thorite, euxenite, samarskite, alvite, bröggerite, monazite, xenotime, arrhenite, sipylite, fergusonite, chalcolite, hielmite. These minerals, which are arranged in the order of their intensity of action, all contain either uranium or thorium. Pitchblende, the most active mineral showed great variations in the activity of specimens from different localities, but roughly speaking the variation was proportional to the percentage of uranium present. A layer of powdered pitchblende a quarter of an inch in thickness gave as strong an action as one of the same material two inches thick, showing that the action does not pass through much thickness of the active material. The action was shown to be proportional to the time of exposure. Several salts and oxides of uranium were tested for their comparative activity upon the photographic plate with the result that they did not show much difference. A sample of metallic uranium from M. Moissan, however, showed a slighter action. With the intention of using it as a standard for comparison the author prepared a very pure specimen of uranium nitrate by treatment with ether and subsequent repeated crystallization. To his surprise this gave no action upon the sensitive plate. It was then shown by a series of experiments that no modification of physical or chemical conditions materially affects the radio-activity of a uranium compound when, to begin with, the salt experimented on possesses it; other similar experiments showed that, starting with an inactive uranium salt, nothing that can be done to it will cause it to acquire activity. It was found that when ordinary crystallized uranium nitrate is dissolved in ether a heavy aqueous liquid separates containing uranium nitrate that is much more active than that which remains dissolved in the ether. Systematic recrystallization from water showed a concentration of active material at the soluble end. Fractionation carried out by heating uranium nitrate until a certain amount of basic salt was produced, proved that the active body gradually umulated towards the basic end. By treating a solution of y active uranium nitrate with ammonium carbonate in excess, mall flocculent precipitate was obtained which showed great io-activity ; it was found, however, that the active body was wholly insoluble in ammonium carbonate. The precipitate t mentioned gave in five minutes as strong an action upon the sitive plate as is given in 24 hours by ordinary uranium rate. The author calls the active substance provisionally rX." Thus far he has not observed any special spectrum for substance. He decides that UrX is certainly distinct from olonium," because the rays emanating from it penetrate glass ile polonium rays do not. It is not so easy to settle whether X is distinct from radium, although many arguments point to not being radium.

Experiments have been commenced by the author, which tend show that thorium may be separated into an active and inactive dy.

It is the author's opinion that in the present state of our owledge of radio-active bodies it is safest to retain an open, or n a slightly sceptical mind. He calls attention to the fact t we recognize them mainly by photographic and electrical ts—reactions which give strong results even when the active dy is present in too small a quantity to be detected by the ctrum—one of the most delicate of tests. He remarks that radiographic test is cumulative. If no action is apparent at end of an hour, one may be shown after twenty-four hours. a day's exposure will show nothing, try a week's. The article ncludes with the following sentence : " Considering my most ive UrX does not contain sufficient of the real material to w in the spectrograph, yet is powerful enough to give a good pression on a photographic plate in five minutes, what must be dilution in compounds which require an hour, a day, or a week give an action?"— *Chem. News*, lxxxi, 253, 265.   H. L. W.

2. *Electrical conductivity of Gases traversed by Cathode rys.*—Starting from the results reached by J. J. Thomson and itherford in regard to the conductivity of gases traversed by ntgen or uranium rays, owing to the production of positive d negative ions within them, J. C. McLENNAN shows that thode rays impress a similar condition upon a gas. The con- ctivity produced by the cathode rays is explained as due to a motion of the ions, positive and negative, produced in the s by the radiation. The ionization by cathode rays was found be about 300 times that due to intense Röntgen radiation. n experimenting with different gases, the result was established at to determine the relative ionization produced in two gases cathode rays of the same intensity it is sufficient to deter- ne the absorbing power of the two gases for the same rays. is also probably true that the ionizations produced by rays of nstant intensity traversing different gases at the same pressure proportional to the densities of the gases. The calculated

values expressing the relation between the ionization by cathode rays in different gases (oxygen, nitrogen, etc.) compared with that of air ($= 1$) was found to be approximately the same as those observed for Röntgen rays, except in the case of hydrogen. —*Proc. Roy. Soc.*, No. 431, p. 375.

3. *Kleiner Leitfaden der Praktischen Physik* von FRIEDRICH KOHLRAUSCH. Pp. xix, 260; 8vo. Leipzig, 1900 (B. G. Teubner). —The many excellent features of the work on Practical Physics by Dr. Kohlrausch, now before the public for some twenty-five years and in its eighth edition, are well known. The same author has now prepared an elementary volume on the same lines, which is designed especially for beginners in Physics and which has a unity character impossible in the larger work. His long experience has enabled him to write a book eminently fitted to fill the place for which it is designed; it is indeed a model both in arrangement and in clearness of statement and conciseness of presentation. It should have a wide sphere of usefulness not only at home, but wherever laboratory work in physics is carried on.

4. *Photometrical Measurements and Manual for the General Practice of Photometry with special reference to the Photometry of Arc and Incandescent Lamps;* by WILBUR M. STINE, Swarthmore College. Pp. vii, 270. New York, 1900 (The Macmillan Company).—The introduction of new forms of illumination in recent years has much increased the importance of photometrical measurements and extended the range of their application. Hence the value of the present volume, which gives in compact form the physical basis of photometry, the various forms of photometers, the light standards in use and finally the particular application of photometrical measurements to electric lamps of the two types. The author has not only brought together a remarkable amount of material within small compass, but in the frequent and well selected references has practically placed before the reader the important part of the original literature of the subject.

II. GEOLOGY AND MINERALOGY.

1. *Status of the Mesozoic Floras of the United States.* First Paper: *The Older Mesozoic;* by LESTER F. WARD, with the collaboration of Wm. M. Fontaine, Atreus Wanner, and F. H. Knowlton. From the Twentieth Annual Report of the U. S. Geological Survey, 1898–99, Part II, General Geology and Palaeontology, pp. 211–748, with 158 plates. Washington, 1900. —This extended and timely collaboration brings into convenient form for reference the principal known data concerning the Triassic and Jurassic floras of the United States, including many new descriptions. The salient facts dealt with are here enumerated.

The structural determination of several new genera and species

conifers from Triassic and Jurassic horizons is by Dr. Knowl-
. The new genus *Pinoxylon* may be specially mentioned as a
rassic pine from South Dakota.

The Triassic flora of North Carolina, as based upon the orig-
l collection of Emmons and most of his types, is the work of
fessor Fontaine. The Emmons collection of fossil plants
de some fifty years since, and long supposed to have been lost,
s recently found in the geological collection of Williams Col-
e by Professor T. Nelson Dale. Their redescription accom-
ied by new figures, and made in the light of a knowledge of the
eral other Triassic plant-bearing horizons since discovered in the
ited States, is especially satisfactory. The Triassic horizons of
rth Carolina being among the earliest investigated in this
ntry, represent classic ground. Moreover, from the wooded
racter of the plant-bearing region, and from the fact that there
few natural products to induce excavation, specimens of this
a are now seldom to be obtained, save from a few localities prin-
ally in Chatham County, where, as at Egypt and Deep River,
ermittent coal mining is still being carried on in the same horizons
ich yielded the important vertebrate fossils *Dromatherium*, and
croconodon, and also *Rhytidodon rostratus* Marsh. The Deep
ver locality was visited by the reviewer a few years since, but
ortunately the mines and culm heaps were on fire and could
be examined at that time, though being again put in order
further work.

A rich new Triassic flora from York County, Pennsylvania is
much interest. This was discovered by Mr. A. Wanner, and
n large part figured and described by him, Professor Fontaine
o having determined the several forms. While some new
nts occur, the essential agreement of this flora with that of
rth Carolina gives to its description a fundamental value.

A brief notice of the Triassic flora of Virginia, made by Fon-
ne the subject of Monograph VI of the U. S. Geological Sur-
y publications, is given by Professor Ward, who notes the
eement between this and the Triassic flora of Lunz in Austria
er studied by Stur and referred by him to the Keuper.

A characteristic Jurassic flora from Oroville, California, is
cribed and illustrated by Professor Fontaine.

The Jurassic cycadean trunks from Carbon County, Wyoming,
ming the new genus *Cycadella* with twenty species, are
cribed by Professor Ward. A review of his preliminary
cription of these cycads was given by the present reviewer
this Journal for April, 1900, pp. 383–387, and need not be
eated here. The 104 plates now devoted to the illustration of
se fine fossils forms the principal feature of the present
ume, as well as a most interesting addition to the series of
tes showing various forms of isolated cycadaceous leaves so
minent in the several Triassic and Jurassic floras mentioned
ve.

n addition to these Wyoming cycads Professor Ward also

describes as *Cycadeoidea nigra* a remarkably well preserved and handsome new species of cycadean trunk of supposed Jurassic age from the vicinity of Boulder, Colorado. This trunk is of more particular interest as extending the geographical range of the genus *Cycadeoidea* one step further west. While the distinctness of the present species is quite certain, it might be noted that externally it bears a strong resemblance to *Cycadeoidea Uhleri* from the Potomac formation of Maryland, the latter being in all probability a close specific, or even varietal relative. There is likewise a close resemblance to *Cycadeoidea* (Raumeria) *Masseiana* Capellini, from the scaly clays of Italy.

If in the present invaluable contribution there is anything open to criticism, it is perhaps the absence, in a work otherwise so profusely illustrated, of maps and sections of the more important localities and areas dealt with, such for instance as appear in the present author's contributions on the lower Cretaceous of the Black Hills in the XIXth Annual Report of the U. S. Geological Survey.   G. R. W.

2. *La Flore Wealdienne de Bernissart;* par A. C. SEWARD. Memoires du Musée Royal d'Histoire Naturelle de Belgique, Vol. i, 1900.—The plants described in this memoir are the unfortunately somewhat fragmentary remains of the Flora occurring in the Wealden beds of Bernissart near the French frontier between Mons and Tournay, famous for the discovery in 1877 of numerous remarkably preserved Iguanodon skeletons.

The writer's conclusions are translated in part as follows: The plants of Bernissart are sufficient to demonstrate a closer resemblance to the Wealden facies than to any other Mesozoic type of vegetation.

I have elsewhere remarked that from the botanical point of view there exists an intimate resemblance between the Wealden type of Flora wherever well developed, as in England and Northern Germany, and the Flora of the inferior Oölite. There is not in fact a well marked break in the palaeobotanical continuity between the Jurassic and Wealden Floras of itself justifying the term Wealden as applied to plant beds such as those of Hastings on the coast of Sussex; comprised in the lower Wealden by the English geologists.

The composition of the flora of Bernissart is interesting for the marked preponderance of ferns, the apparently total absence of cycads, and the rarity of conifers. In other regions, for example in England, Portugal, and Germany the cycads play a preëminent role in Wealden vegetation. A collection containing many conifers and some ferns has recently been made in the Wealden strata at Bracquegnies, 30 kilomètres east of Bernissart, whilst specimens from Baume, 8 kilomètres further east, consist principally of conifers.

It is also remarkable that the specimens of *Weichselia Mantelli* and *Laccopteris Dunkeri* are more numerous in the Bernissart sediments than all others. Perhaps the deduction may be made that the land drained by the rivers which laid down the plant debris in the Bernissart sediments was to a great extent covered by the two species first named and that the remaining vegetation also consisted principally of ferns. The aspect of the specimens and this abundance of ferns likewise suggests the idea that these plant-bearing sediments were derived from a low region chiefly occupied by ferns to the relative exclusion of larger plants or trees. These probably grew in the higher background or in the localities beyond the reach of the water which deposited the Bernissart strata. It is interesting, again, to note that the plant debris contained in these strata furnishes no facts favoring the presence of Angiospermous species. As we know from richer floras of the same age, it appears most probable that the higher classes of Phanerogams were not represented, or at least occupied a very secondary position in the vegetation of the Jurassic and Wealden periods. G. R. W.

3. *Notes on Some Jurassic Plants in the Manchester Museum ;* by A. C. SEWARD, M.A., F.R.S., from Volume xliv, Part III of "Memoirs and Proceedings of the Manchester Literary and Philosophical Society, Session 1899–1900.—This communication deals with Inferior Oölite species from the Gristhorpe plant beds, in part figured by Lindley and Hutton. It adds certain important data to our knowledge of Jurassic vegetation. G. R. W.

4. *" The Maidenhair Tree "* (*Gingko biloba* L.); by A. C. SEWARD, F.R.S., and Miss J. Gowan, Newnham College. Cambridge, Annals of Botany, Vol. xiv, No. LIII, March 1900 (pp. 100–124, with plates viii–x.—This beautiful tree is here described with a completeness and precision which will be most pleasing and timely to every student of Botany or Palaeobotany.

Engler's subdivision *Gingkoaceæ* is adopted for the monotypic genus *Gingko,*—distinguished by the possession of motile male cells as well as by other characters of more or less importance. Following an historical sketch and diagnosis, the vegetative structures and fructification are described, reference always being had to the literature on the several subjects. A sketch of fossil *Gingkoaceæ* is appended, which adds greatly to the value of the contribution. G. R. W.

5. *New minerals from Greenland.*—A recently published account[*] of minerals collected in 1897 by G. FLINK in the Julianehaab district of southern Greenland, contains descriptions of a considerable number of new species and also of other rare known species associated with them. This important paper consists of two parts. Part I (pp. 7–180), by G. Flink, is on the minerals of Narsarsuk on the Tunugdliarfik fiord in southern Greenland. These are found only in a very limited area of the prevailing syenite, where the rock has in spots a pegmatitic character. The prominent minerals are microcline and ægirite, which often occur as large crystal individuals; in drusy cavities between these minerals the rarer species occur in successive generations. Of

[*] Meddelelser om Grönland, xxiv, 1899.

the more interesting known species of which full descriptions are given, may be mentioned the following : Parisite, eudidymite, epididymite, ægirite, arfvedsonite, catapleiite, neptunite, elpidite.

The new species described are nine in number. The descriptions of these leave nothing to be desired in the way of fullness and accuracy, but they can only be briefly characterized here.

CORDYLITE is a *barium-parisite.* It occurs in minute hexagonal crystals, club-like in form; they are related in angle to parisite; hardness 4·5; fracture conchoidal; sp. gravity 4·31; color wax-yellow. The composition is expressed by the formula $Ce_2F_2BaC_3O_9$.

ANCYLITE occurs in minute orthorhombic crystals with strongly curved faces. Color light yellow to orange or resin-brown; hardness 4·5; no cleavage; sp. gravity 3·95. In composition it is a hydrated carbonate of cerium and strontium for which the formula deduced is : $4Ce(OH)CO_3 + 3SrCO_3 + 3H_2O$.

SPODIOPHYLLITE resembles a chlorite; it occurs in crystals belonging to the rhombohedral system, combinations of the hexagonal prism and base; cleavage basal, micaceous; hardness about 3 ; sp. gravity 2·633; color ash- to pearl-gray. In composition it is a metasilicate related to ægirite; formula $(Fe,Al)_2 (Mg,Fe,Mn)_3 (Na_2,K_2)_2 (SiO_3)_6$.

TAINIOLITE ('Tæniolite) is a kind of mica occurring in elongated colorless crystals with sp. gravity = 2·86. An analysis by Mauzelius on 0·1 gram gave: SiO_2 52·2, Al_2O_3 2·7, FeO 0·6, MgO 19·1, K_2O 11·5, Na_2O 1·8, Li_2O 3·8, loss 8·7 = 100. The loss is referred to water (hydroxyl) and fluorine.

LORENZENITE occurs in needle-like orthorhombic crystals, colorless to violet or brown; luster adamantine; hardness 6; sp. gravity 3·42. An analysis (by Mauzelius) gave SiO_2 34·26, TiO_2 35·15, ZrO_2 11·92, Na_2O 17·12, K_2O 0·37, H_2O 0·77 = 99·59. The formula deduced is $Na_2 (Ti,Zr)_2 Si_2O_9$.

LEUCOSPHENITE occurs in wedge-shaped monoclinic crystals; color white inclining to grayish blue; luster vitreous, on some faces pearly; hardness 6·5; sp. gravity 3·05. An analysis (Mauzelius) gave : SiO_2 56·94, TiO_2 13·20, ZrO_2 3·50, BaO 13·75, Na_2O 11·14, K_2O 0·56, H_2O 0·31 = 99·40. The formula deduced is $Na_4Ba (TiO)_2 (Si_3O_8)_3$. The mineral seems to be related to eudidymite.

NARSARSUKITE occurs in tabular tetragonal crystals ; color honey-yellow to brownish gray; hardness a little above 7 ; sp. gravity 2·751. An analysis (Christensen) gave : SiO_2 61·63, TiO_2 14·00, Fe_2O_3 6·30, Al_2O_3 0·28, MnO 0·47, MgO 0·24, Na_2O 16·12, F 0·71, H_2O 0·29 = 100·04·

CHALCOLAMPRITE occurs in small regular octahedrons, resembling pyrochlore; color dark grayish brown inclining to red; hardness 5·5; sp. gravity 3·77. An analysis (Mauzelius) gave : Nb_2O_5 59·65, SiO_2 10·86, TiO_2 0·52, ZrO_2 5·71, Ce_2O_3 (etc.) 3·41, Fe_2O_3 1·87, MnO 0·44, CaO 9·08, K_2O 0·38, Na_2O 3·99, H_2O 1·79, F 5·06 = 102·76 (deduct O) = 100·63. It is regarded as consisting of equal amounts of $RNb_2O_6F_2$ and $RSiO_3$.

ENDEIOLITE is related to the preceding species and like it occurs in regular octahedrons, at first taken for pyrochlore. Color dark chocolate-brown; hardness 5, sp. gravity 3·44. An analysis (Mauzelius) gave (assuming the loss to be SiO_2): Nb_2O_5 59·93, SiO_2 [11·48], TiO_2 0·76, ZrO_2 3·78, Ce_2O_3 4·43, Fe_2O_3 2·81, MnO 0·37, CaO 7·89, K_2O 0·43, Na_2O 3·58, H_2O 4·14, F 0·69 = 100·29 (deduct O) = 100. The formula deduced is $RNb_2O_6(HO)_2$ + $RSiO_3$.

Part II (pp. 181–213) contains descriptions of minerals from the nephelite-syenite of Julianehaab, by O. B. Boeggild and Chr. Winther. Three new species are included here and also an account of steenstrupite.

EPISTOLITE is a silver-white mineral described by Boeggild. It resembles brucite in its tabular crystals and pearly luster on the basal cleavage; it is, however, monoclinic in crystallization. Color white; hardness 1 to 1·5; sp. gravity 2·885. An analysis (Christensen) gave: SiO_2 27·59, Nb_2O_5 33·56, TiO_2 7·22, FeO 0·20, MnO 0·30, CaO 0·77, MgO 0·13, Na_2O 17·59, H_2O 11·01, F 1·98 = 100·35 (deduct O) = 99·52. No definite formula can be obtained, as the material used was somewhat altered.

BRITHOLITE, described by Winther, occurs in pseudo-hexagonal crystals, related in composite form to aragonite. Color brown, opaque; luster greasy to vitreous; hardness 5·5; sp. gravity 4·446. An analysis (Christensen) gave: SiO_2 16·77, P_2O_5 6·48 $(Ce,La,Di)_2O_3$ 60·54, Fe_2O_3 0·43, CaO 11·28, MgO 0·13, Na_2O 1·85, H_2O 1·27, F 1·33 = 100·08.

SCHIZOLITE, also described by Winther, is characterized as a manganese pectolite. It occurs in columnar masses and prismatic crystals, elongated parallel the *ab*-axis; cleavage parallel to two faces in this zone. Color pink to brown; hardness 5 to 5·5; sp. gravity 3·089. An analysis (Christensen) gave: SiO_2 51·06, TiO_2 0·68, Ce_2O_3 1·47, FeO 2·79, MnO 12·90, CaO 19·48, Na_2O 10·71, H_2O 1·36 = 100·45. The formula deduced is $4R_2O.10RO.15SiO_2$, which does not agree very closely with that of pectolite.

OBITUARY.

JAMES EDWARD KEELER. The Director of the Lick Observatory, Professor James E. Keeler, died at San Francisco after a brief illness on August 12th. Few of his friends had a suspicion that his health was otherwise than in the excellent condition which was believed characteristic, until the widely published announcement of his death met their eyes.

James Edward Keeler, born at La Salle, Illinois, on September 8th, 1857, where he also received his early schooling, doubtless owed much of his intellectual alertness to his father whose career had been singularly varied and interesting. A paymaster in the navy during the civil war, his intimate connection with the monitor class of ironclads, including a participation in the famous fight between the Monitor and the Merrimac and subsequent experience in the East, strengthened in him the innate taste for mechanical arts which he shared and enjoyed with his distinguished son. The removal of the family to Mayport, Florida, while the future astronomer was still quite young, profoundly modified his course of development in replacing an ordinary career of schooling by an employment of all his energies in adapting himself to a new life and a new environment. But even in this active life his natural inclination towards astronomy asserted itself. By the exercise of much self-denial he was able to purchase an achromatic telescope objective of about two and a half inches diameter; this he mounted and supplied with a thoroughly serviceable stand, and with the product of his ingenuity he commenced the acquiring of his remarkable store of knowledge of physical astronomy. This was followed by a singular consequence. Mr. Charles H. Rockwell, who has ever remained a kind and sympathetic friend, learned by accident of the young astronomer in Florida and, eager to advance scientific learning in every way, secured for him an opportunity to gain the advantages afforded by one of our greater universities. As Mr. Keeler's schooling was not of a character which fitted him immediately for entrance to university standing, he entered as a special student in the Johns Hopkins University, then in the initial stages of its evolution. Here it was that the writer became intimate with him and learned his extraordinary capacity for fruitful work. Shortly after securing his bachelor's degree here in 1881, he became an assistant to Professor Langley, aiding him most efficiently in his delicate researches with the bolometer, at the same time establishing a reputation for scientific ability which practically secured his future. With this experience, broadened by two years of study at Berlin and Heidelberg and further increased by four or five years passed at Mt. Hamilton as an astronomer of the Lick Observatory, his subsequent career is wholly natural. In 1891 he accepted the position as Director of the Allegheny Observatory, a position exceptionally congenial to him because of the many friends gained during the time in which he acted as assistant to his predecessor Professor Langley. The opportunity for more important work, however, which came to him in 1898, with an invitation to become Director of the Lick Observatory, obliged him to sacrifice this agreeable environment and to enter upon a life of great fruitfulness and activity which has been, so unfortunately for science, interrupted by his death.

This brief note does not offer the place to review the scientific achievements of Mr. Keeler; but it is gratifying to find that not only those engaged in allied branches of science recognized his merits, but more popular appreciation was evinced by his election as an Associate of the Royal Astronomical Society in 1898, and as a member of the National Academy at its last meeting. Quite fitting is it, however, that the writer should take this occasion to record the grief of a great number of his personal friends through whom his life was enriched in a degree never attained by men less selfish and less loyal than he.   C. S. H.

AMERICAN JOURNAL OF SCIENCE

[FOURTH SERIES.]

ART. XXXIII.—*Elaboration of the Fossil Cycads in the Yale Museum;* by LESTER F. WARD. (With Plates II–IV.)

THE visitor to the Peabody Museum of Yale University, if he penetrates to the basement of that building, finds himself literally "in the woods,"—a petrified forest of Mesozoic cycads. A few of the finest specimens may be seen in the exhibition halls on the second floor, and many more will be ultimately placed there, but at present the bulk of the collection is undergoing elaboration below, in close association with the gigantic dinosaur bones that have made the names of Marsh and of Yale so justly celebrated in the history of science. In a very literal sense, cycads are to the vegetable kingdom what dinosaurs are to the animal, each representing the culmination in Mesozoic time of the ruling dynasties in the life of that age. Professor Marsh saw this, and, as the last act of his life, had the sagacity to make the Yale Museum for all time the Mecca for all who shall wish to gain a realizing sense of the fauna and flora of America in a period now forever closed.

It has chanced to be my fortune or misfortune to be situated, as it were, in the storm track of cycadean investigation on this continent, and upon me has devolved the duty of roughly blocking out the general line of study of that wonderful extinct vegetation that has been coming to light in this country in such rich profusion. Science demands a terminology and a nomenclature, and whatever may be said of the superficial character of all systematic work, it is now true and always has been true, nay, it must always be a need, that the systematist precede the structuralist and provide him with a language and a framework for his finer researches. This is all I profess to have attempted, and not only am I keenly aware of the superficiality and defectiveness of my work, but I have urged on all occasions the importance of the exhaustive study of the wonderful structure of these cycadean trunks that reveals itself to the lens and even to the naked eye. I presented this aspect of the case in the strongest form I could command to Professor Marsh, and pointed out to him the "unlimited possibilities" of such a study of the great Yale collection. I am happy to record his warm appreciation of the fact, which led him actually to inaugurate it some months before his death, by inducing Mr. George R. Wieland to undertake it and by placing at his disposal every possible facility for the prosecution of this work. Professor Marsh's successor, Dr. C. E. Beecher, with the approval of the Trustees of Yale University Museum, has allowed no interruption in these valuable researches, so creditable to the institution, and has placed them on a permanent basis.

I am only to speak here of my own work, much of which I have already recorded and need only allude to,[*] confining myself to the additional results that have been reached by recent study. As stated in the Nineteenth Annual Report of the U. S. Geological Survey, Part II, pp. 546 and 547, I worked up all the cycad material from the Black Hills in the Yale Museum, in the months of March and June, 1898, and the 22 species described in that report were based on the material in the U. S. National Museum, and on 126 specimens in the Yale Museum, contained in two invoices, the first of 87 specimens and the second of 39. These were illustrated in 97 plates, and that paper constitutes the basis for subsequent and future investigations. It was known at that time that other

---

[*] For the benefit of any who may be specially interested, I herewith refer to the following papers of mine relating in whole or in part to fossil cycadean trunks: Fossil Cycadean Trunks of North America, with a Revision of the Genus Cycadeoidea Buckland Proc Biol. Soc. Washington, vol ix, Washington, April 9, 1894, pp. 75–88; The Cretaceous Rim of the Black Hills, Journal of Geology vol. ii. No. 3, Chicago, April–May, 1894, pp. 250–266; Recent Discoveries of Cycadean Trunks in the Potomac Formation of Maryland. Bull. Torrey Bot Club, vol xxi, No 7, July 20, 1894, pp 291–299; Some Analogies in the Lower Cretaceous of Europe and America, Sixteenth Annual Report, U. S. Geological Survey, 1894–95, Pt I. Washington, 1896, pp. 463–542, pls. xcvii–cvii; Descriptions of the Species of Cycadeoidea, or Fossil Cycadean Trunks, thus far discovered in the Iron Ore Belt, Potomac Formation, of Maryland. Proc. Biol Soc Washington, vol. ix, March 13, 1897, pp 1–17; Descriptions of the Species of Cycadeoidea, or Fossil Cycadean Trunks, thus far determined from the Lower Cretaceous Rim of the Black Hills, Proc U. S. Nat. Mus., vol. xxi (No. 1141), Washington, 1898, pp. 195–229; The Cretaceous Formation of the Black Hills as indicated by the Fossil Plants, Nineteenth Annual Report, U. S. Geological Survey, 1897–98, Pt. II, Washington, 1899, pp 521–946, pls lvii–clxxii; Description of a New Genus and Twenty New Species of Fossil Cycadean Trunks from the Jurassic of Wyoming Proc Washington Acad Sci, vol i. pp. 253–300, pls xiv–xxi; Status of the Mesozoic Floras of the United States, First Paper The Older Mesozoic, Twentieth Annual Report, U. S. Geological Survey, 1898–99, Pt. II, pp. 211–748, pls. xxi–clxxix.

invoices were on their way, but it was impossible to wait for them to arrive and be described before going to press with the Nineteenth Annual Report. They came, however, during the summer, and added 44 specimens to the Yale collection, which I studied in November of that same year, but the results were not then published. In fact, the investigation was purposely left incomplete, because in the meantime, viz., in October of that year, I had been over the entire cycad-bearing area of the Black Hills in company with Mr. H. F. Wells, who had collected all the cycads for Professor Marsh, and had seen on the ground such an immense number of trunks and fragments that I had determined to make every effort to have these added to the collections already made. I had understood from Mr. Wells that Professor Marsh had declined to purchase any more, and I made a strong effort to induce the authorities of the U. S. National Museum to secure them on the extremely reasonable terms for which Mr. Wells offered to do the work. Failing in this, I appealed to Professor Marsh to secure them for Yale, which he did promptly, and before spring they had all arrived.

To give an idea of the extent and wealth of these latest accessions, it is only necessary to say that while, even after the arrival of the third invoice in the summer of 1898, the entire collection numbered only 170 specimens, it now numbers 731 specimens! It is true that many of these are fragments broken from larger trunks, but this was precisely what was needed to complete and perfect the collection. From a scientific point of view, fragments are often more valuable than perfect trunks, since they reveal the internal structure and throw light on the entire nature of the plants. Then again many of these fragments and *disjecta membra* are found to belong to specimens previously received, and add directly to their value. For example, the largest specimen in the collection, which, when the parts were gotten together and weighed, proved to be the largest trunk known in the world, came in different invoices and in four pieces. One branch, No. 145, was in the third invoice, and I had left it unassigned, well knowing that it was incomplete, and hoping that the remainder might ultimately be found. It did in fact arrive, and the parts have been brought into position and mounted in the exhibition hall on the second floor, where it may be seen of all men. Happily it proves to belong to the great branching species which I dedicated to Professor Marsh, and will stand forever as a tribute to his labors in this field—the *Cycadeoidea Marshiana.*

As was naturally to be supposed, and as I fully expected, the greater part of these numerous accessions have been found to belong with more or less certainty to one or other of the 22 species already described, but not only do they complete and supplement the previous collections in the manner above pointed out, but they greatly enrich it by adding many and often much finer specimens than any that existed before. First of all, they have added two species to the Yale collection which were formerly only represented in that of the U. S. National Museum. These are *Cycadeoidea excelsa* and *C. occidentalis.* In the second place, the beautiful *C. pulcherrima,* of which the type is at the National Museum, is now represented at Yale by at least one almost equally fine specimen, while the great *C. Jenneyana,* of which there were only a few broken pieces, now numbers its representatives by scores, some of them very fine trunks. The species which the new accessions have perhaps most richly endowed is the rare *C. Wellsii,* to which three or four huge and superb trunks are now added. Every species known from the Black Hills is now represented in the Yale collection, which was not previously the case, and it is safe to say that that collection constitutes the largest and finest assemblage of fossil cycadean trunks in the world.

Besides the specimens from the Black Hills, the Yale Museum possesses two specimens from the Jurassic cycad bed of the Freezeout Hills of Wyoming, treated by me in the paper above cited in the Proceedings of the Washington Academy of Sciences, and still further illustrated in the Twentieth Annual Report of the U. S. Geological Survey. These are Nos. 127 (*Cycadella Reedii*) and 128 (*C. Beecheriana*), for the completion of which Prof. Wilbur C. Knight of the University of Wyoming, at my suggestion, has given to Yale the complementary fragment, No. 500.54, of his much larger collection from the same bed. There have also been discovered among the forgotten collections in the Yale Museum two fragments of cycads from the Iron Ore Clays of Maryland. When these trunks were refound, Professor Marsh at once remembered that he had secured them in 1867, from Mr. Philip Tyson, the original discoverer of cycads in Maryland. I have examined these specimens, and find them both to belong to the predominant species of that region, *Cycadeoidea marylandica.* They bear the numbers 729 and 730.

There have also now been found in the Yale Museum collections, 7 specimens of cycadean trunks from the original Purbeck forest beds of the Isle of Portland, that overlie the Portland stone. These are in two lots. The first consists of a single specimen, the gift of Dr. Gideon Mantell to Prof. Benjamin Silliman shortly after the attention of paleobotanists had been called to these objects, so long known to quarrymen as "crow's nests." It bears Mantell's label, *Mantellia nidi-*

formis Brongniart, with a reference to the figure in his Wonders of Geology, 1839, p. 365, fig. 70 (or is it to his Medals of Creation, in the later editions of which, but not in the earlier, the same figure occurs?). As the *Mantellianidi-formis* of Brongniart is the *Cycadeoidea megalophylla* of Buckland, it is evident that Dr. Mantell supposed the specimen to belong to this species, but it does not agree with his own figure, and seems rather to represent the small-scarred species, *Cycadeoidea microphylla* Buckland. This specimen has been numbered 732 of the Yale collection.

The other lot consists of 6 specimens which were received on Sept. 24, 1898, from Mr. A. N. Leeds, the collector and donor. They have now been given the numbers 733–738. Three of them (Nos. 733–735) belong without doubt to *Cycadeoidea microphylla* Buckland. No. 736 probably belongs to the same species. Nos. 737 and 738 have the large scars of *C. megalophylla*, and undoubtedly represent that species.

I made my fourth visit to the Yale Museum in May, 1900, and commenced work on the new material on the third of that month. My first work was to identify as many of the specimens as possible with previously described species. In many cases this was easy of accomplishment, but in dealing with the immense number of fragments and the imperfect, immature, dwarf, and depauperate specimens that necessarily accompany a complete collection, and which, in spite of the trouble they give, ought always to be collected, doubts constantly arise and cannot be removed. In the list which follows I have expressed such doubts by interrogation marks, and they are taken to indicate the tentative and incomplete character of the work. The future will doubtless greatly change matters in these respects, and the reader is asked to make due allowance for the unsettled state of our knowledge of these objects. A mere list of species with numbers attached is admitted to be unattractive reading, but it nevertheless affords an idea of the relative abundance of the species, and will have value as a guide or catalogue of the collection for any who may wish to study or examine it. An alphabetical arrangement of the species will probably prove more convenient than any attempt at systematic classification. The following are the 29 species thus far known, with the specimens in the Yale collection that have been assigned to each:

Cycadeoidea aspera, No. 104.
Cycadeoidea cicatricula, No. 118.
Cycadeoidea Colei, Nos. 12, 20, 25 ?, 28 ?, 48, 52 ?, 57, 68, 80 ?, 224 ?, 240 ?, 246 ?, 291, 321, 433, 444, 476 ?, 539.

Cycadeoidea colossalis, Nos. 2, 7, 10, 13, 17, 37, 40, 55, 133, 238, 333 ?, 351, 354 ?, 438 ? ?

Cycadeoidea dacotensis, Nos. 1, 3, 5, 6, 30, 39 ?, 43, 54, 61 ?, 62, 63, 95 ?, 106 ?, 134, 212, 213, 214, 215, 232, 307 ?, 324, 328, 353 ?, 355, 356, 358, 369, 372, 373 ?, 384 ?, 398, 425 ?, 435, 503 ?, 504 ?, 505 ?, 508 ?, 509, 510, 511 ?, 512, 513, 516 ?, 518, 523 ? ?, 525, 526, 528, 530, 538, 542, 543, 546, 547, 548, 549, 550, 716 ?, 719 ?, 720 ?

Cycadeoidea excelsa, Nos. 236, 239, 248, 370 ?, 453, 455, 461 ?, 479, 481, 498.

Cycadeoidea formosa, Nos. 89, 611, 615 ?, 639 ?, 641 ?

Cycadeoidea furcata, Nos. 18, 60, 718.

Cycadeoidea heliochorea, Nos. 722, 723, 724, 725, 726.

Cycadeoidea ingens, Nos. 92, 94, 99, 100, 103, 110, 117, 122, 123, 208, 554 ?, 557 ?, 560 ?, 561, 562, 563, 565, 566, 568, 572, 577, 578, 579, 580, 581, 582 ?, 584 ?, 592, 607, 609, 610, 612, 614 ?, 624, 625, 626, 628 ?, 633, 642 ?, 644, 646, 650, 653 ?, 656, 657, 661, 670, 671, 675, 676 ?, 678, 684, 690, 697, 705, 706 ?, 710.

Cycadeoidea insolita, Nos. 50, 64 ?

Cycadeoidea Jenneyana, Nos. 81, 87, 88, 90, 91 ?, 93, 96, 97, 98, 101, 102, 108, 109, 111, 112, 113 ?, 114, 115, 116, 120 ?, 121 ?, 124 ?, 125, 126, 171 ?, 173 ?, 174 ?, 176 ?, 192 ?, 193, 194, 195, 196, 197, 198, 199, 200, 201 ?, 202, 203, 204, 205, 206, 207, 551, 553, 555 ?, 556 ?, 558, 559, 564 ?, 567 ?, 569 ?, 570 ?, 571 ?, 573, 574 ?, 575, 576, 583, 585, 586, 587, 588, 589, 590, 591, 593, 594 ?, 595 ?, 596 ?, 597, 598, 599 ?, 600, 601 ?, 602 ?, 603 ?, 605 ?, 606, 608 ?, 613 ?, 616, 617, 618, 619, 621, 622 ?, 632 ?, 634 ?, 636 ?, 638 ?, 643, 645, 647, 648 ?, 649, 652, 654 ?, 655 ?, 658, 659, 660 ?, 663 ?, 665, 666 ?, 667 ?, 668 ?, 669, 672, 673, 674, 677, 680 ?, 681, 682 ? ?, 683 ?, 685 ?, 686, 688, 689 ?, 691 ?, 692 ?, 693, 694, 695, 696, 698, 699, 700, 701, 702, 703 ?, 708, 710, 721.

Cycadeoidea McBridei, Nos. 8, 19, 23, 26 ?, 27, 29 ? ?, 38, 42 ?, 46, 73, 76, 179, 180, 189 ?, 225 ?, 227, 228 ? ?, 231, 233 ?, 235, 245, 250, 252, 257 ?, 258 ? ?, 262, 268 ?, 271 ?, 288, 289, 298, 299, 301 ?, 305, 314 ?, 320 ?, 325, 326 ?, 329 ?, 338 ?, 339 ?, 340, 343 ?, 346 ?, 361 ?, 375, 379, 380, 381, 383 ?, 387 ?, 388, 407, 415, 422, 429, 431, 436 ?, 437, 439 ?, 449, 456, 462 ?, 467, 469 ?, 471, 472, 489 ?, 490, 492, 494, 495 ?, 496, 497 ?, 522 ?, 544 ?

Cycadeoidea Marshiana, Nos. 4, 9, 11, 33 ?, 44, 79 ?, 129, 143, 145, 161, 164, 167, 169 ?, 216, 219 ?, 221 ?, 256, 267, 276, 281, 283 ?, 285, 300, 392, 405, 410, 411, 418 ?, 430, 502, 541, 711.

Cycadeoidea minima, Nos. 53, 149, 150, 152, 153, 154, 155, 156, 157, 168, 426, 468, 474, 478, 714.

Cycadeoidea minnekahtensis, Nos. 14, 22, 24, 34 ?, 41, 47, 70, 71, 72, 83, 86, 140 ?, 142, 144, 148, 158, 160, 162 ?, 181, 182 ?, 184, 186 ?, 210, 211, 217, 220, 229 ?, 230, 242, 244 ?, 249 ?, 259 ?, 264, 265, 266 ?, 273 ? ?, 277, 279 ?, 286 ?, 290 ?, 292, 309, 310, 318 ?, 327 ?, 331, 332, 344 ?, 348 ?, 349, 350, 357, 362 ?, 365 ?, 366, 367 ?, 396 ?, 397, 401, 403 ?, 404, 412, 420 ? ?, 432 ?, 440, 445, 473 ?, 480, 491, 506 ?, 507, 520, 532, 533, 534, 535, 536 ?, 540, 728.

Cycadeoidea nana, No. 84.

Cycadeoidea occidentalis, No. 234.

Cycadeoidea Paynei, Nos. 52 ?, 58 ?, 69 ?. 132, 135, 163 ?, 165, 188 ?, 247 ?, 263 ?, 272, 280 ?, 284 ?, 293 ??, 316, 319 ??, 334, 336 ?, 337 ?, 364, 376, 386, 395 ?, 399, 413 ?, 423 ?, 428 ?, 434 ?, 447, 448, 451 ?, 452, 493 ?, 712, 713.

Cycadeoidea protea, Nos. 32, 185, 187, 241, 253, 296, 297, 303, 315, 359, 382, 414, 457, 458, 463, 466, 487, 499, 521, 529.

Cycadeoidea pulcherrima, Nos. 78, 159, 545.

Cycadeoidea reticulata, Nos. 254, 282, 287, 335, 342, 377, 378.

Cycadeoidea rhombica, Nos. 620, 623, 627, 629, 630, 631, 640.

Cycadeoidea Stillwelli, Nos. 16, 36, 56 ?, 105, 107 ?, 119, 175 ?, 226 ?, 515 ??, 517 ?, 552 ?, 651 ??, 662 ??

Cycadeoidea superba, Nos. 137, 146, 147, 218, 717.

Cycadeoidea turrita, Nos. 15, 35 ?, 45, 49, 51, 65, 66, 67, 74, 75, 82, 85, 139, 141, 151 ?, 166, 183, 190 ?, 191 ?, 223 ?, 255 ??, 261 ?, 269, 270 ?, 278, 295 ??, 304 ?, 313, 323, 330, 352 ?, 360, 368, 374, 394, 402, 406, 408 ?, 409, 446 ?, 450 ?, 482, 483, 484, 486, 519 ??, 731.

Cycadeoidea utopiensis, No. 727.

Cycadeoidea Wellsii, Nos. 21, 59, 130, 136, 138, 222, 243, 322, 391, 400, 500, 501, 537.

Cycadeoidea Wielandi, Nos. 77, 131, 393, 424.

Indeterminable, Nos. 31, 170, 172, 177, 178, 209, 237, 251, 260, 274, 275, 294, 302, 306, 308, 311, 312, 317, 341, 345, 347, 363, 371, 385, 389, 390, 416, 417, 419, 421, 427, 441, 442, 443, 454, 459, 460, 464, 465, 470, 475, 477, 485, 488, 514, 524, 527, 531, 604, 635, 637, 664, 679, 687, 704, 707, 709, 715.

The specimens marked as "indeterminable" are largely fragments, i. e., small pieces broken away from the trunks to which they belonged, and it is to be hoped that an exhaustive comparison with all the specimens in the collection may result in the restoration of a considerable number to the trunks now in the collection. As an aid in this work I have indicated in many cases the species to which they are most likely to belong, and given other hints as to where to look with the greatest prospect of success.

Considerable search has already been made to bring such fragments together, and it has been successful in about twenty cases. As a natural result this has changed a few of the assignments previously made and published; thus, No. 9, doubtfully referred to *C. colossalis*, proves to be a part of the trunk, No. 33, of *C. Marshiana;* No. 13, referred to *C. dacotensis*, belongs to Nos. 17 and 40, constituting a fine trunk of *C. colossalis;* No. 47, thought perhaps to be a branch of *C. Marshiana*, fits No. 142, and both together make up a good specimen of *C. minnekahtensis;* and No. 70, which resembles

C. turrita and was so referred, when joined to No. 480, to which it belongs, proves to represent *C. minnekahtensis.*

Other alterations made are as follows:

In one case, viz., that of No. 110, I have changed the assign-ment previously made and published in the Nineteenth Annual Report of the U. S. Geological Survey (pp. 614, 615, pls. xcix and c), referring it to *C. ingens* instead of *C. McBridei.* This change is important because this specimen was the only one from the Blackhawk region that I had referred to the latter species. After an examination of the much larger collection now in hand from that region, I am satisfied that this specimen belongs to *C. ingens*, and that *C. McBridei* is not represented in that section of the Black Hills.

The small specimen, No. 53, which I regarded (loc. cit., p. 615) as perhaps a "miniature" or "undeveloped" form of *C. McBridei*, I now make the type of a new species, *Cycadeoidea minima*, represented by 17 specimens (see infra, p. 341), and I do the same with another small specimen, No. 32, which I doubtfully referred (loc. cit., p. 608) to *C. minnekahtensis* as a dwarf representative similar to the specimen No. 19 of the U. S. National Museum. This new species is now represented by 20 specimens, all but one of which are in the Yale Museum, and may bear the name *Cycadeoidea protea* (see infra, p. 343).

After all possible assignments had been made and the inde-terminable material separated out, there remained 60 specimens which, while exhibiting good specific characters, were not refer-able to any of the species described. These 60 specimens, however, proved to belong to very different types, and a care-ful classification of them shows that they constitute 7 distinct specific groups, or, in other words, 7 new species. The sys-tematic description of these new species will therefore be our next and final task.

Cycadeoidea superba n. sp.

Trunks large (30–40^{cm} high, 30–50^{cm} in diameter, with a girth of over one meter), short-conical or somewhat globular, little compressed, unbranched; rock soft but not fragile, red-dish brown, of low specific gravity; organs of the armor horizontal except near the summit where they are ascending and pass into a large terminal bud, which, however, is some-times wanting, and a cavity, or crow's nest, occupies the sum-mit; scars definitely arranged in spiral rows around the trunk, the angle made by the rows with the vertical axis diminishing above, those from left to right making an angle of 20–35°, those from right to left from 60–75°; leaf scars subrhombic, high in proportion to their width, the vertical angles usually

ınded, 30–40^{mm} wide, 20–25^{mm} high; leaf bases present,
ʋally 1–3^{cm} below the surface, sometimes filling the scars or
ɛn projecting, soft, porous or spongy in structure; vascular
ndles occasionally visible in a row some distance from the
ırgin, appearing either as depressions or elevations; walls
8^{mm} thick, very distinct, lighter colored than the leaf bases,
ooth but more or less grooved on the surface, sometimes
ʋided by a commissure in two plates; reproductive organs
ınerous and well developed, sometimes tending to arrange
ɛmselves in vertical rows, raised above the surface, usually
ɡe, 4×7^{cm}, or even 6×9^{cm} in diameter, surrounded by numer-
ʋ subtriangular bract scars covering much of the surface,
ɛ central portion relatively small, usually solid and amor-
ous, of a spongy consistency; armor 7–8^{cm} thick, obscurely
ached to the axis; woody zone 5–8^{cm} thick, showing 2 or 3
ıgs; cortical parenchyma 3^{cm} thick; fibrous zone 3–5^{cm} thick,
one specimen consisting of two rings, the outer 12^{mm} thick,
ɔwing both longitudinal and radiate structure, the inner 8^{cm}
ick, apparently subdivided into three subordinate rings;
ɛdulla 10^{cm} or more in diameter, of a homogenous structure,
 nally decayed at the base.

This fine species is represented in the Yale collection by
specimens, the numbers of which, with their weights and
ıte of preservation, are as follows:

> No. 137, 48·35 kilograms, nearly complete.
> No. 146, 45·36   "   "   "
> No. 147, 39·90   "   a hemisphere.
> No. 218, 56·12   "   nearly complete.
> No. 717, 9 75   "   incomplete.

Nos. 137 and 146 must have grown close together, as each
ʊ a flattened area near the base, and these two surfaces
actly fit together. They are, however, covered by the
ʋpressed leaf scars and had no organic attachment. More-
ɛr, each has its own independent base and axis, and there is
ʋ proof that they were connected otherwise than mechani-
lly. No. 147 also has some similar flattened areas, and it
ɛms to have been the habit of this species to grow in clumps
clusters.

Its nearest affinities are with *C. McBridei*, but it differs in
ʊ globular, symmetrical form, in the more open scars, and in
eir smooth well-defined walls, also in the finer structure.
ʊ next nearest relationship is with *C. Wellsii*, but it lacks
e thick walls and most of the other essential features of that
ecies.

All the specimens are from the Minnekahta region.

On Plate III may be seen side views of Nos. 137 and 146,
owing also the terminal bud.

Cycadeoidea rhombica n. sp.

Trunks of medium size (40–50^{cm} high, 20–30^{cm} in diameter), subcylindrical, tapering upward, more or less laterally compressed, unbranched ; rock soft on the immediate surface, hard and flinty within, reddish brown on the weathered surfaces, whitish and somewhat chalcedonized in the interior, with rather high specific gravity; organs of the armor horizontal or somewhat descending; phyllotaxy clear and well marked, the spiral rows from left to right forming an angle with the axis of 25°, those from right to left of 60°; left scars strictly rhombic with parallel sides and all angles, vertical as well as lateral, sharp, 15^{mm} wide, 9^{mm} high, and very uniform ; leaf bases present usually filling the scars, square across the top but lying at different levels, rough-granular and somewhat porous, without apparent bundle scars ; walls very thin, rarely 1^{mm}, consisting of a white flinty substance, the surface smooth and longitudinally grooved or pitted, often not rising to the summit of the leaf bases, thus forming cracks between them, often split into two, more or less equal, thin plates; reproductive organs present but not specially numerous or conspicuous, fairly well developed, usually projecting or forming elevations, but occasionally decayed, leaving cavities, small, 15×25^{mm} in diameter or smaller, the involucral bract scars inconspicuous, the central portion solid and amorphous externally; armor very thin, 15–20^{mm}, joined to the axis by a more or less irregular line sometimes appearing definite; woody zone about 4^{cm} thick, usually showing three distinct layers ; cortical parenchyma 12–15^{mm} thick, often conspicuously marked by the thick vascular bundles passing across it and curving upward and outward to the leaf bases, its outer wall marked by shallow, longitudinal grooves, 6–10^{cm} long, pointed at their extremities, lying side by side but overlapping one another; fibrous zone in two distinct rings, the outer 10–15^{mm} thick, the inner 12^{mm} thick, separated from the outer by a definite line appearing on the fractures as a seam or crack, its inner wall also definite, both rings appearing longitudinally striate on radial fractures, but both, and especially the outer, showing on transverse fractures a radiate structure with woody wedges; medulla 6–8^{cm} thick, porous in its outer, and cherty or flinty in its inner, portion.

This is one of the best-defined species in the collection, although none of the specimens are complete. It consists of Nos. 620, 623, 627, 629, 630, 631, and 640, all from the Blackhawk region. These specimens all came in the same box with a number of others of different species (*C. Jenneyana* and *C. ingens*), and it seems probable that they were found close together. Still a comparison of them shows that they ɔbably represent at least three different trunks. Nos. 620 d 623 are the bases of two different trunks. Nos. 627 and 0 fit together, and Nos. 629 and 631 probably belong to that ɪnk. No. 640 is a small fragment clearly showing the same aracters, but not known to belong to any of the other speci-ᴇns.

The only affinities shown by this species with any other are th *C. Stillwelli*, and here the resemblance is confined chiefly one specimen, No. 105, yet the scars of that specimen are t only larger but have more or less curving sides, and the .f bases are not so squarely truncated as in these forms. It also more strictly cylindrical. Further material may tend assimilate these specimens; if so the effect will be to remove ɪt specimen from *C. Stillwelli* and enlarge the scope of ɪs species.

The specific name relates to the exactly rhombic scars. On ate II a view of the fragment No. 629 is represented, which ɔws the rhombic scars to good advantage.

Cycadeoidea heliochorea n. sp.

Trunks very large, the largest probably 50^{cm}, ellipsoidal or arly globular, flattened at the summit where a small terminal d is set in the center of a broad surface occupied by the ɪall, spirally arranged scars of the upper leaves, laterally com-ᴇssed to an unknown extent, all the specimens probably rep-ᴇnting the broader sides, unbranched; rock soft, reddish own, rather fragile, of low specific gravity; organs of the ːnor ascending in all the specimens, but these all belong to ᴇ upper part; phyllotaxy not traceable; leaf scars obscurely own, subrhombic or rhombic, the upper vertical angle as ɑrp as the lower, 25^{mm} wide and 15^{mm} high in the larger ɑmples, diminishing to the size of bract scars; leaf bases ɑgh and structureless, often sunk 2–3^{cm} below the surface, owing in a few cases faint traces of bundles in the form of ɔjections from the bottom of the scars or ribs in the sides ove the bottom; walls very thick and irregular, of the same ɤture as the leaf bases, presenting a rough, jagged, and .even surface with scarcely any longitudinal arrangement or bdivision into plates; reproductive organs very numerous d prominent, nearly covering the surface to the exclusion of ᴇ leaves, usually projecting 3 or 4^{cm} as decorticated cones, ɱetimes absent, having fallen out, leaving deep, bowl-ɑped cavities, usually large, attaining a diameter of 4×7^{cm}, t smaller ones are also common, surrounded by numerous d conspicuous, concentrically arranged, involucral bract scars, ᴇ inner bracts themselves still present forming involucres to the projecting spadices which they invest, the central portions consisting of obovate, longitudinally striate fruits having definite and peculiar pits and markings; armor 8^{cm} thick, the leaf bases appearing to blend with the large curving vascular strands that are seen passing through the cortical parenchyma; fibrous zone and medulla not represented.

This well-marked species consists, so far as known, of the 5 fragments Nos. 722–726, collected by Mr. Wells ten miles west of Sundance in Wyoming, 90 miles northwest of the Minnekahta localities. These specimens therefore possess an especial interest from the point of view of distribution, the locality lying between the old one and the Hay Creek region from which Professor Jenney made his fine collection of fossil plants. Nos. 722–724 consist of broad portions of the armor at the upper part of three different trunks, the first two including the terminal bud and broad, flat surrounding area. No. 722 is the largest specimen, measuring 28^{cm} the longest way. They all show a curvature on both sides, which if carried all the way round would make an immense trunk. It is altogether probable that they may represent the flattened sides of much compressed trunks. It is greatly to be hoped that more material may be obtained from this locality.

The specific name, from the Greek words for sun and dance, alludes to Sundance, the only name with which the locality has been associated.

Plate IV shows the outer surface of No. 722, with the small scars spirally arranged around the small terminal bud.

Cycadeoidea utopiensis　n. sp.

Trunk small (21^{cm} high, 17 × 22^{cm} in diameter, with a girth of 63^{cm}), irregularly short-conical, elliptical in cross section but apparently not forcibly compressed, showing one small branch and a cavity from which a second has been removed, the surface well preserved, but having an area near the summit covered by what appears to be an outer coating of ramentum as in the genus Cycadella, more or less obscuring the organs and definitely broken away on three sides including the apex; terminal bud well developed, 5^{cm} high, 7 × 9^{cm} in diameter, rounded towards the summit where there is a depressed area 3^{cm} in diameter occupied by small rhombic scars surrounding the somewhat projecting axis 12^{mm} in diameter; base preserved in great part, somewhat hollowed out; rock soft except the central portion which is hard and fine grained, light drab color on the surface with a somewhat calcareous appearance, but containing no lime, jet black at the center, of rather low specific gravity, weighing 8·17 kilograms; organs of the armor nearly

rizontal in the middle part of the trunk, declined near the
ᵉ and progressively ascending above to the erect terminal
ves; phyllotaxy faintly traceable over some small areas near
ᵇ base, indicating that the spiral rows from left to right
ke an angle of about 50°, and those from right to left one
about 60°, with the axis; leaf scars subrhombic, the vertical
gles rounded, the upper often reduced to a curve, only those
ar the base normal or distinct, the rest much reduced and
torted, crowded and massed together so as to present a
ᶜuliar wrinkled and gnarly appearance, small and very nar-
ᵛ in proportion to their width, normal ones 2ᶜᵐ wide, 1ᶜᵐ
ᵍh; leaf bases always present, of an open structure, with
ᵉp, variously-shaped pores separated by thin partitions, some
ᵖjecting, others depressed, often showing the leaf bundles,
ich are arranged in a row all the way round some distance
ᵐ the margin, thus appearing near the center always as
ge distinct pits; walls 2–5ᵐᵐ thick, white, flinty and fine
ined, much grooved and divided longitudinally, those of
ᵇ central and upper parts of the trunk consisting of two
ooth plates separated by a deep furrow, rising above and
rounding the small leaf bases so as to enclose them and form
merous shallow cups or short tubes, the whole giving to the
face a peculiar wrinkled appearance; reproductive organs
ᵐmon but some distance apart and not the cause of the gen-
ᵃl distortion, normal and fairly well developed, rising a little
ᵒve the surface, variable in size, some having a diameter of
×40ᵐᵐ, others only of 15×25ᵐᵐ, surrounded by very numer-
ᵇ subrhombic involucral bract scars resembling and gradually
ᵇsing into the leaf scars, the central portion showing mark-
ᵍs that represent the essential organs; armor about 25ᵐᵐ
ᵗck, somewhat definitely joined to the axis; woody zone
4ᶜᵐ thick, the parts incapable of being distinguished or
scribed, but consisting in large part of loose open structure;
ᵉdulla 6×10ᶜᵐ in diameter, black within, hard and fine
ined.

This species is represented by the single specimen, No. 727,
rchased by Mr. Wells from a dealer in Hot Springs, who
ted that it was obtained from some unknown person who
imed to have found it "50 miles west of Hot Springs in
yoming." If the direction were due west, this point would
l in longitude 104° 30′ west from Greenwich and 12 miles
st of the line of the Cretaceous border on the maps of the
ack Hills. As the foot hills end about 20 miles west of Hot
rings and are succeeded here as in all parts of the Hills by
ᵉ Upper Cretaceous, and these again by higher deposits, one
ᵘld naturally suppose that this locality would fall far out on
ᵉ Tertiary terrane. It therefore seems more probable that

the direction was northwest from Hot Springs, and this might locate it in the Lakota formation some distance north of Cambria and in the general region of the Newcastle coal field. This, could it be proved, would be interesting as supplying another link in the cycad chain which is fast encircling the Hills, and would partially close the wide gap between the Minnekahta and the Sundance regions. But, as it now stands, everything is in doubt, and it is greatly to be hoped not only that we shall ultimately learn the true locality, but that other specimens may be found. The patch of ramentum, if such it be, near the summit of the specimen raises the suspicion that it may belong to the genus Cycadella, and as all the specimens of that genus thus far known have come from the Jurassic, it is possible that the horizon of the bed holding this specimen may be lower than that of the other Black Hills cycads.

The specific name alludes to the alleged locality, which, as we have seen, would be a sort of geological nowhere.

Plate III shows a view of the best-preserved side of the specimen.

Cycadeoidea reticulata n. sp.

Trunks small (9–15^{cm} high, 10–20^{cm} in diameter, with a girth of 40–50^{cm}), globular or oblate-spheroidal, vertically compressed and also elliptical in cross section, unbranched, usually hollowed at the summit, but in one specimen showing the worn bases of the terminal leaves concentrically arranged, depressed at the base with a concave center surrounded by a groove between the armor and medulla; rock generally soft, but sometimes hard in the interior, reddish brown or lighter on the weathered surfaces, drab or dark within, of rather low specific gravity, the specimens all weighing less than 4 kilograms; organs of the armor normally horizontal; phyllotaxy more or less distinct, the rows of scars from left to right making an angle of 45°, those from right to left of 60°; leaf scars subrhombic, 15–20^{mm} wide, 6–9^{mm} high; leaf bases soft, porous or spongy, sunk 1–5^{mm} below the surface, occasionally with indistinct bundle scars near the center; walls very thin, often less than 1^{mm}, rarely exceeding 2^{mm}, hard and fine grained, presenting a smooth white surface contrasting strongly with the leaf bases, rising above them in plates or dikes or separating them with a delicate network of fine lines, thus giving the whole surface a reticulate appearance, sometimes divided by a commissure into two plates, or by two such into three; reproductive organs not abundant or conspicuous, rarely wanting, their places occupied by cavities, usually small, but variable in size, vaguely defined, averaging 15x20^{mm} in diameter, having scarcely any visible involucral bracts, the central portions porous or open, the hollow interior in one instance containing a collection of small round bodies of doubtful nature; armor 2–3^{cm} thick, somewhat definitely joined to the axis; woody zone 2–4^{cm} thick; cortical parenchyma 1–2^{cm} thick, generally of a loose porous structure; fibrous zone 1–2^{cm} thick in two rings, the outer thicker, dark, hard, and fine grained, the inner thin (5–7^{mm}), partitioned off by medullary septa and bounded on both sides by scalloped lines; medulla 5x8^{cm} in diameter, of a coarse and soft structure.

This species embraces 7 specimens, the numbers of which, with their weights and state of preservation, are as follows:

No. 254, 1·02 kilograms, incomplete.
No. 282, 3·42   "   nearly complete.
No. 287, 3·86   "   "
No. 335, 3·18   "   "
No. 342, 1·25   "   incomplete.
No. 377, 3·42   "   "
No. 378, 2·61   "   "

These all came from the original Minnekahta locality. Their nearest affinities are with *C. turrita*, and they resemble certain of the branches or turrets of that species, but are all entire trunks and show no signs of branching.

The specific name refers to the reticulate appearance produced by thin white walls that form a network over the surface.

A side and top view of No. 342 are given on Plate IV. This specimen shows the leading characters as clearly as any example.

Cycadeoidea minima n. sp.

Trunks diminutive, the smallest known (6–12^{cm} high, 8–14^{cm} in diameter, with a girth of 20–39^{cm}), ovoid or obovoid, eccentric or oblique, sometimes vertically compressed or flattened at the top, simple, with a terminal bud or corresponding depression, the base usually hollowed out, but sometimes downwardly projecting; rock generally soft and porous, reddish brown or drab colored, of low specific gravity; organs of the armor horizontal except near the summit; phyllotaxy traceable in several specimens, the rows of scars from left to right making an angle with the axis that varies from 30–50°, those from right to left making an angle in all cases of 75°; leaf scars subrhombic to nearly rhombic, 15–20^{mm} wide, 7–9^{mm} high; leaf bases nearly filling the scars, sometimes rising above the walls, presenting a rough, spongy appearance, often having a pair of large pits near the center, and occasionally showing indistinct marginal rows of bundle scars; walls thin, often less than 1^{mm}, hard and flinty, light colored or white, wrinkled, striate, or grooved on the surface, sometimes with a visible commissure; reproductive organs few and obscure, small and inconspicuous, never exceeding 3^{cm} in diameter, the bract scars very obscure and the central portion mostly solid; armor poorly exposed, $2-4^{cm}$ thick, separated from the axis by a somewhat definite line; woody zone $2-3^{cm}$ thick; cortical parenchyma 1^{cm} thick, traversed by the distinctly visible strands; fibrous zone $1-2^{cm}$ thick, not differentiated, somewhat definitely bounded on both walls, the outer often a scalloped surface, the inner marked by longitudinal ridges forming narrowly rhombic meshes; medulla $2-5^{cm}$ in diameter, porous or spongy, homogeneous and structureless.

This species embraces 15 specimens of the Yale collection and one from that of the U. S. National Museum, which are the smallest complete cycadean trunks known. The numbers, weights, and state of preservation, are as follows;

No. 53,	1·57	kilograms, complete.	
No. 149,	1·22	"	"
No. 150,	1·39	"	"
No. 152,	0·79	"	
No. 153,	0·45	"	
No. 154,	0 45	"	
No. 155,	0·34	"	
No. 156,	0·30	"	"
No. 157,	0·23	"	nearly complete.
No. 168,	2·15	"	complete.
No. 426,	0·30	"	fragment.
No. 468,	0·45	"	nearly complete.
No. 474,	0·45	"	incomplete.
No. 478,	0·11	"	fragment.
No. 714,	1·36	"	complete.

U. S. Nat. Mus. specimen, 0·62 kilograms, nearly complete.

When I made my first study of the 87 specimens then in the Yale collection, I found No. 53 of this group, and rather than create a new species for it I doubtfully referred it to *C. McBridei*, as previously stated. Mr. Wells's next shipment, which I studied in November, 1898, contained 9 specimens so much like it that I was then certain that they constituted a specific group. The later accessions have added 5 others that belong to the same group.

In October of that year, when in the field with Mr. Wells, I picked up a small perfect specimen in the region southeast of Minnekahta station, and on account of its diminutive size I brought it back with me to Washington in my valise. It also belongs in this group. It now bears the number 2248 of the locality catalogue of fossil plants of the U. S. Geological Sur-

7, and will form a part of the cycad collection of the U. S.
.tional Museum. Its specific gravity is considerably higher
ın that of the other specimens, but beyond this there are no
ential differences.

On Plate II is represented the original specimen, No. 53,
ich is typical of the species, but was not figured in the
neteenth Annual Report of the U. S. Geological Survey.

Cycadeoidea protea n. sp.

Trunks small (8–12^{cm} high, 10–20^{cm} in diameter), low, sprang-
g, and of all shapes, often contracted at the base, the dis-
tion probably only slightly due to compression, nsually
ıch branched, or consisting entirely of several somewhat
nal systems or branches, often with a well-preserved termi-
l bud at the summit of each branch, usually nearly complete
d little broken or worn; rock hard, firm, and fine grained,
ldish brown on the surface with lighter stripes, darker
thin, of high specific gravity; organs of the armor wholly
der the influence of the subordinate systems or branches,
ially erect with reference to the terminal buds; phyllotaxy
o relating wholly to the branches, around which and the
en flattened summits the leaves are spirally arranged, some-
ıes with great regularity; leaf scars, where visible, of nearly
rmal subrhombic shape, but very variable in this respect due
the exuberant branching, often reduced to mere slits; very
all, 6–13^{mm} wide, averaging 9^{mm}, 1–6^{mm} high, averaging 4^{mm};
f bases porous or showing a columnar structure but of a
rd substance, usually sunk some distance below the surface,
asionally exhibiting a few large pits; walls very thin, often
s than 1^{mm}, hard and flinty, of a lighter color than the leaf
ses and divided into several plates the edges of which present
striate or grooved appearance, the middle plate sometimes
rided by a commissure; reproductive organs small, few, and
scure, but certainly present and apparently functional, some-
ıat raised, 12×15^{mm} in diameter, surrounded by scars that
rcely differ from the leaf scars, the central portion hetero-
neous, apparently showing the ends of the essential organs;
nor where exposed about 2^{cm} thick, but certain leaf bases
ve a length of 4–6^{cm}, joined to the axis by a somewhat defi-
e line; wood rarely exposed, appearing about 3^{cm} thick and
rided into several rings, the outermost ring (cortical paren-
yma) thin, the others showing no structure, the innermost
ll (exposed in one specimen) marked with broad shallow
oves terminating in pits or scars and alternating with one
other; medulla of a coarse sandy consistency, contrasting

with the surrounding hard tissues, often projecting downward below the armor, varying in size with the specimen.

I have already mentioned the specimen No. 32, and the similar one No. 19 of the U. S. National Museum Collection, both of which I took for immature forms of *C. minnekahtensis.* The latest invoices contained 19 additional specimens so like these as to render that theory no longer tenable, and I feel entire confidence in erecting this large and very distinctive group into a new species. In fact, notwithstanding the anomalous character of these forms and the consequent difficulty in finding terms by which to describe them, there is scarcely a species known to me that has less confusing relationships with other species than has this one. Though truly protean, as I have attempted to imply by the specific name chosen, none of its many forms is at all the same as those of any other species.

It now consists, as we have seen, of 21 specimens, 20 of which are in the Yale collection. The numbers, with their respective weights and state of preservation, are as follows:

No. 32,	2·61	kilograms,	complete.
No. 185,	0·94	"	nearly complete.
No. 187,	0·14	"	"
No. 241,	1·59	"	"
No. 253,	1·02	"	fragment.
No. 296,	1·25	"	nearly complete.
No. 297,	0·34	"	fragment.
No. 303,	1·59	"	nearly complete.
No. 315,	0·91	"	"
No. 359,	1·81	"	"
No. 382,	1·25	"	
No. 414,	3·52	"	"
No. 457,	1·47	"	complete.
No. 458,	1·47	"	incomplete.
No. 463,	0·56	"	fragment.
No. 466,	0 68	"	incomplete.
No. 487,	3·18	"	nearly complete.
No. 499,	0·68	"	" "
No. 521,	0·79	"	" "
No. 529,	0·22	"	" "

No. 19, U. S. Nat. Mus., 1.81 kilograms, nearly complete.

A careful study of all these specimens, and a reëxamination of No. 19 of the U. S. National Museum, have tended to strengthen the impression which the last-named specimen, considered alone, made upon my mind, and which I noted at the bottom of page 607 of my memoir on the Cretaceous Formation of the Black Hills, viz., that these small, gnarled, and branching forms have a decidedly rootlike appearance, and suggest that the trunks, like those of the "coontie," may have developed entirely below the surface.

I have selected for illustration of this species the somewhat exceptionally regular, but still fairly characteristic specimen, No. 457, represented on Plate IV. This is a chiefly top view, showing the three nearly equal systems or branches, each with its apical scars and central bud perfectly preserved.

We now have, therefore, 29 species of fossil cycadean trunks from the Black Hills, all of which are represented in the Yale collection. To any one interested in the extinct floras of America this collection can scarcely fail to appeal in an especial manner. But we are probably only at the threshold of the subject. Not only will the raw material continue to accumulate, but as we penetrate deeper into the inner structure it is safe to predict that the results will be such as even the botanists proper cannot afford to ignore.

EXPLANATION OF PLATES.

PLATE II.

No. 629.—*Cycadeoidea rhombica* Ward; type; side view. From near Blackhawk, South Dakota.

No. 53.—*Cycadeoidea minima* Ward; type; side view. Minnekahta, South Dakota.

(The numbers refer to the catalogue of the Yale Collection.)

PLATE III.

No. 727.—*Cycadeoidea utopiensis* Ward; type; side view. From west of the Black Hills, Wyoming.

Nos. 137, 146.—*Cycadeoidea superba* Ward; type; side view. Minnekahta, South Dakota.

PLATE IV.

No. 722.—*Cycadeoidea heliochorea* Ward; type; side view. From ten miles west of Sundance, Wyoming.

No. 342.—*Cycadeoidea reticulata* Ward; type; top view. Minnekahta, South Dakota.

No. 457.—*Cycadeoidea protea* Ward; type; top view. Minnekahta, South Dakota.

Art. XXXIV.—*On the Chemical Composition of Turquois;* by S. L. Penfield.

Through the kindness of Mr. Ernest Schernikow of New York City, the writer has recently received a suite of turquois specimens from deposits in Los Cerillos Mts., New Mexico, and the Crescent Mining District, Lincoln Co., Nevada, and one fragment of exceptionally fine quality from the last named locality was presented with the special request that it should be used for chemical analysis. The material was very fine-grained, of a beautiful robin's-egg blue color, and broke with a smooth fracture. A thin section of the material appeared translucent and almost colorless, and when examined under the microscope, the turquois seemed to be perfectly uniform, showing no evidence of being made up of two substances, such, for example, as an aluminium phosphate, mixed with a copper salt as coloring material. The material was so fine-grained that no clue as to its crystallization could be made out, other than that it acted somewhat on polarized light. The specific gravity, taken by suspension in the heavy solution, was found to be 2·791.

In considering the chemical composition of turquois, it should be borne in mind that analyses have been made of only massive, cryptocrystalline fragments, and although they may be selected ever so carefully, no such guarantee of the purity of the material can be given as when, for example, a well crystallized mineral is analyzed. In order to show, however, that turquois is a material of nearly uniform composition, the new analysis is given below in connection with analyses made by other investigators. Analyses have not been included which show a large proportion of foreign constituents other than silica. The analyses are as follows:

	I. Lincoln Co., Nevada. Penfield.	II. Nichabour, Persia. Church *	III. Karkaralinsk, Russia. Nicolajew.†	IV. Fresno Co., California. Moore ‡	V.	VI.	VII.
					\multicolumn{3}{l}{Los Cerillos, New Mexico. Three analyses by Clarke.§}		
P_2O_5	34·18	32·86	34·42	33·21	31·96	32·86	28·63
Al_2O_3	35·03	40·19	[35·79]	35·98	39·53[b]	36·88	37·88
Fe_2O_3	1·44	2·45[a]	3 52	2·99	....	2·40	4·07
CuO	8·57	5·27	7·67	7·80	6·30	7·51	6·56
H_2O	19·38	19·34	18·60	19 98	19·80	19·60	18·49
Insol.	0·93	---	----	----	1·15	0·16	4·20
X	....	MnO ·36	----	----	CaO ·13	CaO ·38	...
	99·53	100·47	100·00	99·96	98 87	99·79	99·83
Sp. gr.	2·79	2·75	2·89	2·86		2·80	

* Given as 2·21 per cent FeO.   b Includes some Fe_2O_3.

* Chemical News, x, p. 290, 1864.   † Kokscharow's Min. Russland, ix, p. 86, 1884.
‡ Zeitschr. Kryst., x, p. 247, 1884.   § This Journal, III, xxxii, p. 212, 1886.

In the new analysis the iron was found to exist wholly in the ferric condition, and therefore the iron in Church's analysis, given as FeO in the original article, has been calculated to Fe_2O_3 to agree with the observations of the author and other investigators.

It is evident from an examination of the foregoing analyses that turquois is a material which is quite uniform in its chemical composition, so uniform in fact that it does not seem reasonable to consider it as an accidental mixture of an aluminium phosphate and a copper phosphate. The presence of the bivalent element copper, however, in somewhat variable amounts, is not so easily accounted for if we are to consider a copper phosphate as isomorphous with an aluminium phosphate. The small amount of iron is probably isomorphous with the aluminium, and it is to be expected that the iron phosphate would have little effect upon the color of the stone, for the hydrated ferric-phosphate, strengite, and the hydrated ferric-arsenate, scorodite, are both light-colored minerals. The idea that the iron is present as the hydrated oxide, limonite, can scarcely be entertained.

Clarke,[*] in discussing the composition of turquois, states that if the alumina is combined with the phosphoric acid and water to form a molecule $2Al_2O_3 \cdot P_2O_5 \cdot 5H_2O$, there remains an excess of phosphoric acid and water which forms with the copper a salt of the composition $2CuO \cdot P_2O_5 \cdot 4H_2O$. Turquois is considered therefore by him as consisting of variable mixtures of the foregoing salts. He regards normal turquois as the aluminium salt, $2Al_2O_3 \cdot P_2O_5 \cdot 5H_2O$, which he also expresses as $Al_2HPO_4(OH)_4$, and "the copper salt, to which the mineral owes its color, is to be considered merely as an impurity." By means of ratios it is quite easy to apply Clarke's theory to the analyses as tabulated on page 346. Taking Al and Fe as a basis, and establishing a ratio of $P : Al+Fe : H = 1 : 2 : 5$, as demanded by Clarke's formula for normal turquois, $Al_2HPO_4(OH)_4$, the ratio of *the excess of Phosphorus : Cu : the excess of Hydrogen* can then be found. The results of the calculation are as follows:

		I.	II.	III.	IV.	V.	VI.	VII.
Normal turquois, Clarke.	P	·352	·408	·373	·371	·387	·376	·396
	Al+Fe	·704	·816	·746	·742	·774	·752	·792
	H	1·760	2·040	1·865	1·855	1·935	1·880	1·980
Residues.	P	·130	·054	·111	·097	·063	·088	·008
	Cu	·108	·066	·096	·098	·080	·094	·083
	H	·394	·108	·201	·365	·365	·300	·074

* Loc. cit.

Taking copper as a basis, the ratios of $P : Cu : H$ in the residues are as follows :

		I.	II.	III.	IV.	V.	VL	VII.
Residues,	P	1·20	0·82	1·15	0·99	0·78	0·93	0·09
copper	Cu	1·00	1·00	1·00	1·00	1·00	1·00	1·00
as unity.	H	3·65	1·63	2·09	3·72	4·56	3·19	0·90

It is to be taken into consideration that it is a very severe test of a formula to throw all of the errors resulting from possible impurities in the materials and inaccuracies of the analyses upon a single constituent, in the case in hand on the supposed copper salt; but still the ratios of $P : Cu : H$ are so variable that it cannot be considered that turquois is a mixture of an aluminium salt, $Al_2HPO_4(OH)_4$, and a hydrated copper phosphate having the definite composition $2CuO. P_2O_5. 4H_2O$ as suggested by Clarke. A compound having the composition $2CuO. P_2O_5. 4H_2O$ demands a ratio of $P : Cu : H = 1:1:4$.

An important factor to be taken into consideration is that the hydrogen in turquois is to be regarded as representing hydroxyl and not water of crystallization, for water is not expelled from the mineral at a low temperature; hence hydroxyl radicals may be considered as playing a part in the chemical composition of the mineral. Considering copper as an essential constituent of turquois and not as an impurity, two theories naturally suggest themselves : one, that the bivalent copper is isomorphous with, and replaces the bivalent aluminium-hydroxide radical $[Al(OH)]''$; the other, that the univalent copper-hydroxide radical $[Cu(OH)]'$ is isomorphous with the univalent aluminium-hydroxide radical $[Al(OH)_2]'$. The first of these ideas has led to no satisfactory solution of the problem; the second, however, reveals a constancy in the chemical relations of the mineral which can scarcely be regarded as due to accident. The relations in question are shown by combining aluminium and iron with two hydroxyls to form the groups $[Al(OH)_2]$ and $[Fe(OH)_2]$, respectively, and copper with one hydroxyl to form the group $[Cu(OH)]$, and then finding the ratio between the phosphorus and $[Al(OH)_2]' + [Fe(OH)_2]' + [Cu(OH)]' +$ Excess of hydrogen. The relations are shown by the ratios derived from the several analyses tabulated on page 346, as follows :

		I.			II.			III.	
P		·482			·462			·484	
Al(OH)₂	·686			·788			·702		
Fe(OH)₂	·018		1·450	·028		1·332	·044		1·320
Cu(OH)	·108			·066			·096		
H	·638			·450			·478		

	IV.	V.	VI.	VII.
P	·468	·450	·464	·404
Al(OH)$_2$	·706 ⎫	·774 ⎫	·722 ⎫	·742 ⎫
Fe(OH)$_2$	·036 ⎬ 1·478	· · · ⎬ 1·426	·030 ⎬ 1·428	·050 ⎬ 1·262
Cu(OH)	·098 ⎪	·080 ⎪	·094 ⎪	·083 ⎪
H	·638 ⎭	·572 ⎭	·582 ⎭	·387 ⎭

Considering $[Al(OH)_2]' + [Fe(OH)_2]' + [Cu(OH)]' + H'$ as playing the role of a univalent radical R', the ratios of P : R in the several analyses are as follows:

I, P : R = ·482 : 1·450 = 1 : 3·01
II, " " = ·462 : 1·382 = 1 : 2·88
III, " " = ·484 : 1·320 = : 73
IV, " " = ·468 : 1·478 = : ₁·6
V, " " = ·450 : 1·426 = : 17
VI, " " = ·464 : 1·428 = : ·08
VII, " " = ·404 : 1·262 = : 12   Average = 1 : 3·02

The author can vouch for the purity of the material analyzed by him, as far as it is possible to do so in the case of a cryptocrystalline mineral, and can also testify as to the accuracy of the analysis, and the very close approximation to the exact ratio 1 : 3, between the phosphorus and the sum of the univalent radicals plus the hydrogen, is very suggestive. The ratios in the other analyses approximate as closely to 1 : 3 as might be expected when the character of the material is taken into consideration, and the average of all the ratios is almost exactly 1 : 3. The ratio 1 : 3 is that of phosphorus to hydrogen in ortho-phosphoric acid, H_3PO_4. Turquois may therefore be regarded as a derivative of ortho-phosphoric acid in which the hydrogen atoms are to a large extent replaced by the univalent radical $[Al(OH)_2]$, $[Fe(OH)_2]$ and $[Cu(OH)]$. There seems to be no fixed ratio between the radicals $[Al(OH)_2]$, $[Fe(OH)_2]$ and $[Cu(OH)]$, nor between the sum of the hydroxyl radicals and the hydrogen. In some cases, however, there is an approximation to the ratio 2 : 1 between the sum of the hydroxyl radicals and the hydrogen, as follows:

	$[Al(OH)_2] + [Fe(OH)_2] + [Cu(OH)]$:	H		
II,	·882	:	·450 = 2 : 1·02
III,	·844	:	·478 = 2 : 1·13
VII,	·875	:	·387 = 2 : 0·89

In cases like the foregoing, the composition of turquois might be considered as a mixture of an aluminium salt, $H[Al(OH)_2]_2PO_4$, with the isomorphous molecules $H[Fe(OH)_2]_2$PO$_4$ and $H[Cu(OH)]_2PO_4$. The molecule $H[Al(OH)_2]_2PO_4$ is equivalent to Clarke's formula for "normal turquois," $2Al_2O_3.P_2O_5.5H_2O$, which he also writes $Al_2HPO_4(OH)_4$. Adopting

Clarke's suggestion that turquois contains very finely†
admixtures of iron and copper phosphates as impurities, †
his formula for the pure mineral (*normal turquois* of
Groth* expresses the composition as $PO_4Al_2(OH)_3.H$
suggests, however, that the formula is perhaps $PO_4H[Al$

In conclusion it may be stated that it is the author
that copper and the small amounts of iron are to be r
as constituents of turquois, rather than as impurities.
port of this idea the constant occurrence of copper, as
by all the published analyses, may be cited.   Furthe
finely pulverized turquois is only partially dissolved by
in a test tube with hydrochloric acid; hence, if the n
contained copper phosphate as an impurity, it would ve
expected that the copper phosphate would dissolve readily,
leaving the basic aluminium phosphate as a pure white residue,
while in tests which have been made the insoluble residues
have remained blue, from beginning to end of the experiments.
Considering the existence in turquois of the univalent radicals
$[Al(OH)_2]$, $[Fe(OH)_2]$ and $[Cu(OH)]$, the composition of the
mineral, as shown by the published analyses, may be expressed
as a derivative of ortho-phosphoric acid, as follows:

$$[Al(OH)_2,Fe(OH)_2,Cu(OH),H]_3PO_4.$$

The $[Al(OH)_2]$ radical always predominates, but is not
present in fixed proportion. Some analyses (II, III, and VII)
conform closely. to the formula $[Al(OH)_2,Fe(OH)_2,Cu(OH)]$,
HPO_4.

Disregarding the iron, the calculated composition of turquois for two special cases of isomorphous replacements are
given below:

	$[Al(OH)_2, Cu(OH),H]_3PO_4.$ $Al(OH)_2 : Cu(OH) : H = 7:1:6.$	Analysis I, page 346.	$[Al(OH)_2, Cu(OH)]_2HPO_4,$ $Al(OH)_2 : Cu(OH) = 12 : 1.$	Analysis II, page 346.
P_2O_5	34·64	34·18	32·13	32·86
Al_2O_3	37·32	36·47*	42·61	42·64*
CuO	8·28	8·57	5·52	5·27
H_2O	19·76	19·38	19·74	19·34
	....	Insol. 0·93	....	MnO 0·36
	100·00	99·53	100·00	100·47

* Include the Fe_2O_3.

Considering that turquois is not a crystallized mineral, the
agreement between theory and the analyses is certainly as
close as could be expected.

* Tabellarische Uebersicht der Mineralien, 1898, p. 97.

Sheffield Laboratory of Mineralogy and Petrography,
Yale University, New Haven, June, 1900.

ART. XXXV. — *Quartz-muscovite rock from Belmont, Nevada; the Equivalent of the Russian Beresite;** by J. E. SPURR.

Occurrence.

THE rock which will be described in this paper occurs in a large dike just east of Belmont, the seat of Nye County, Nevada. It affords an interesting study in view of its peculiar mineralogic composition; its relation to other siliceous rocks, into which it grades in the same dike; its identity with the Russian rock, beresite; and its connection with ore deposits. The dike in which the quartz-muscovite rock is found cuts black, limy slates and gray, fine-grained, saccharoidal, crystalline limestones, which stand vertically. The formation is preminently a slaty one and often becomes schistose, this condition being due to the metamorphism occasioned by the dike under consideration and by other connected dikes. The main dike is half a mile wide and runs in a north and south direction; on one side (the east) it is definitely bounded by the slates, but on the west these slates form only a narrow band, a few hundred yards in width, separating the dike from a large mass of coarser intrusive siliceous granite which lies west and south of Belmont. While the body of rock under consideration is technically a dike, therefore, yet it is perhaps also to be regarded as the marginal facies of the main intrusion.

Near the contact of the dike, the shaly limestones become transformed into jasperoid, as microscopic examination shows, and the development of mica has occurred along certain planes, so that the rock passes into a micaceous schist. The jasperoid itself is often schistose and contains small bunches of yellow and red metallic oxides which give it the aspect of a knotted schist (Knotenschiefer). In the unaltered slates near here Mr. Hilbert[†] found graptolites which mark the rocks as Silurian. According to Mr. Walcott,[‡] the horizon probably corresponds with part of the Upper Pogonip formation at Eureka. In this limestone are found quartz veins which carry rich antimonial silver ores.

Composition of the Dike.

The composition of this dike is not at all uniform, although as a whole the rocks are moderately fine-grained and siliceous. The essential constituents are quartz, feldspar and white mica, but the proportions of these vary in the different localities by

---

* Published with the permission of the Director of the U. S. Geological Survey.
† U. S. Geographical Surveys, vol. iii, Geology, p. 180.
‡ Monograph VIII, U. S. G. S, p. 2.

the increase of one mineral to the exclusion of the others, the changes being gradual and irregular; thus in some places the rock becomes mainly quartz, in others mainly feldspar. Quartz veins are abundant, but are of irregular form and are evidently segregational, being a part of the results of crystallization contemporaneous in a general way with the crystallization of the rest of the rock; these veins often contain considerable muscovite. Biotite is sparingly present in many of the rock types observed. Only two miles to the south this great, relatively fine-grained dike runs into a mass of coarse, siliceous, biotite granite with which it is apparently continuous, although it may be that the dike is a slightly later intrusion connected with the granite. The chemical and mineralogic composition of this coarse biotite granite is much the same as those of the finer-grained dike rocks above referred to, but the texture is very different. Further south again this coarse granite is overlain by flows of massive biotite rhyolite, which has nearly the same chemical and mineralogic composition as the granite, so far as can be discerned.

It is probable that the coarse-grained granite and the rocks of the finer-grained, siliceous, variable dike above described, together with the quartz veins of the vicinity, are all nearly contemporaneous, or closely consecutive allied phenomena, representing a single period of intrusion and a single magma, the variations in composition and period of injection being due to differentiation. It is probable, also, that the similarly constituted biotite rhyolite is connected with the intrusive rocks, which may possibly represent the roots or feeders of the volcanic out-pourings.

Microscopic examination.

Biotite quartz-monzonite.—This specimen is of a type which is very abundant in the dike, particularly near its central portion. It is fine-grained and holocrystalline in the hand-specimen and contains occasional small phenocrysts of feldspar, quartz and muscovite in a granular, saccharoidal groundmass. There is a slight gneissic structure apparent in the hand specimen and the fracture follows this, but it does not appear in the thin section; it is evidently not a cataclastic phenomenon but probably a flow structure. Under the microscope the groundmass is seen to be relatively fine-grained, granular and allotromorphic. It consists of a mosaic of quartz and feldspar, with some biotite, the proportion of the different minerals being, in the order named, about $15:10:1$. The feldspar is both striated and unstriated. The unstriated feldspar was determined once by the Fouqué method as labradorite-bytownite. The striated feldspars were determined in one case to be andesine;

another, oligoclase-albite. There is probably also consider-
ole orthoclase. The rock, therefore, appears to belong to the
ionzonitic family and to the quartz-monzonite group.

Siliceous muscovite-biotite granite.—This rock has in general
tine grain, like that of the quartz-monzonite just described.
1 general, however, it is much lighter colored than this rock,
nce the mica which it contains is sporadic, leaving the rest of
ie rock composed of quartz and feldspar. Much of the rock
bserved in the field contains practically no mica, so that it
asses into alaskite. The specimen studied is remarkable as
ontaining frequent bunches of pure granular quartz, about a
uarter of an inch in diameter, and resembling exactly vein-
uartz. In thin section the rock is seen to be in general of
ledium grain, while it varies locally from very coarse to very
ne. The habit of the minerals is allotriomorphic granular.
ust as the texture varies so enormously, so the arrangement
f the constituent minerals is irregular, some areas being
ntirely of quartz in irregular interlocking grains, while others
onsist of quartz and feldspar. The feldspar is mostly
ntwinned and was determined as orthoclase, while some
winned feldspar was determined as albite. There are in the
ction occasional tiny accessory grains of muscovite and bio-
te, and a large broken crystal of zircon. The feldspar in
ne section encloses primary muscovite, while in another case .
ie muscovite encloses quartz. This would seem to make the
rder of crystallization quartz, muscovite, and feldspar, but the
regular intergrowth of the whole section shows that the
rystallization of all these minerals was essentially contempo-
aneous. The quartz and feldspar are often intergrown in
iicrographic fashion. The feldspar is slightly kaolinized and
1 places there are very small flakes of secondary muscovite,
hich must be held separate from the rest of the muscovite in
ie rock, which is plainly primary.

Quartz-muscovite rock.—This rock is the one of chief
iterest in the dike, and occurs in large masses. It changes
radually and irregularly into the alaskite or muscovite biotite
ranite above described, and from its field relations is evi-
ently a variation of this rock. The specimen examined,
hich is typical, has the exact appearance of a white or light-
ray, medium-grained, micaceous quartzite. It is found espe-
ially near the margins of the dike and cuts into the sedimentary
ates, which are noticeably metamorphosed into jasperoid and
iica-schist near the contact. Under the microscope the rock
seen to be medium-grained, with a fairly even, allotrio-
iorphic, granular structure. The chief minerals are quartz
id muscovite, the former predominant. The rock is evi-

dently what it seems to be from field relations—a variation of the alaskite or muscovite-biotite granite, the variation consisting in the substitution of muscovite for feldspar. The proportion of muscovite to quartz in this rock is such that the rock is evidently nearly a chemical equivalent of the muscovite granite. Tiny grains of feldspar are found, generally striated, and seem to be albite.

Both in the hand-specimen and in the section this rock appears perfectly fresh and all the constituents primary. The freshness of the albite particles seems to remove all question of decomposition, for there is hardly any trace of kaolinization. A more thorough investigation, however, shows that muscovite occurs in two distinct habits. In the first habit it forms large irregular grains intergrown with the quartz, and sometimes enclosing smaller grains of quartz—poikilitic structure. In the second habit it occurs either in medium-sized blades or in small sheaf-like flakes, often spherulitic in arrangement, which are intergrown with quartz grains.

Concerning the muscovite found with the first described habit there is no question as to its primary nature. The last described habit, however, suggests a secondary origin, in spite of the freshness of all the minerals of the rock. Finally, a thorough investigation has shown that these aggregates of muscovite and quartz are plainly derived from orthoclase, some fragments of which may yet be found in the midst, with blades of muscovite and quartz grains penetrating them. It is plain that the alteration of the feldspars has been confined to the orthoclase and has not attacked the albite, which remains fresh.

The amount of primary orthoclase which the aggregates of secondary muscovite and quartz represent is so considerable as to put the primary rock among the fine-grained muscovite granites and to make it almost exactly similar to the muscovite granite just described, with which it is so closely connected in the field.

Biotite granite.—The biotite granite, into which this great fine-grained dike appears to merge south of Belmont, may be briefly described. The rock is porphyritic, carrying numerous crystals of orthoclase up to two inches in length. Under the microscope these orthoclase crystals are found to be intergrown with shreds of orthoclase differently oriented from the main crystal, but uniformly oriented among themselves. The crystals also contain shreds of muscovite. The groundmass is very coarse, and is remarkable for containing blotches of pure quartz in granular aggregates similar to that described above for the fine-grained muscovite-biotite granite of the great dike. In this case the blotches of quartz are from one-third of an inch to one-half an inch in diameter, and each sists of a mosaic of intergrown grains. Besides this quartz, groundmass consists of an allotriomorphic, granular inter-wth of quartz and feldspar, with subordinate biotite. The lspar is mostly orthoclase. A faintly striated feldspar was ermined to be anorthoclase, or microcline-anorthoclase. The lspars sometimes show zonal structure.

Differentiation of the Great Dike.

'his has not been carefully studied, but in general the ker colored rock, which has been shown in one specimen to ı quartz-monzonite, occupies the central portion of the dike, le the rock near the margin consists of the siliceous mus-ite granite, or quartz-muscovite rock. This seems to indi-:, at this locality at least, that the borders of the dike are re siliceous than the center.

Origin of Quartz-muscovite rock and connection with ores.

Although the quartz-muscovite rock is evidently a product alteration, yet it is not due to surface weathering, since the dition of the rock is quite fresh and hard, and the albite ws no trace of decomposition. The process must be arded as one of endomorphism and as connected and prob-y contemporaneous with the exomorphism indicated by the :ration of the siliceous limestone of the wall-rocks to jas-oid and mica schist. In both the intrusive and the intruded k the result of the metamorphism has been the same, pro-:ing quartz and muscovite at the expense of the orthoclase the one hand, and of the calcite and subordinate minerals the other. In the case of the wall-rock the metamorphism, ng apparently from its distribution dependent upon the ~usion, evidently took place after this intrusion and was ught about by the solutions which accompanied the igneous k, or were residual from its solidification. Within the dike similar alteration was probably contemporaneous with that :he country rock.

n the immediate vicinity of this intrusive mass are ore iosits, which in the time of Nevada's prosperity made this ion one of considerable wealth, although at present the ing industry is perfectly dormant. The writer had not iortunity to study these, but according to Mr. Emmons[*] ores generally occur in white quartz veins which are often eral feet in width. These quartz veins are probably con-iporaneous with those already described as occurring in ʒgular form within the dike rock itself, and as evidently representing the final product of the residual solution of the general magma. In these quartz veins the metallic minerals are scattered in bunches or disseminated particles, rarely in banded form. These metallic minerals consist principally of stetefeldtite, which is an argentiferous ore of antimony, containing besides silver, lead, copper and iron. Dana* notes that this mineral in Peru has been regarded as probably arising from the decomposition of chalcostibite, a sulphide of antimony and copper. Chalcostibite occurs at Wolfsberg in the Hartz in nests embedded in quartz.

The metallic minerals being, from their habit, plainly contemporaneous with the quartz veins which enclose them, it is evident that the deposition of these minerals, the formation of the quartz veins, the metamorphism of the country-rock to jasperoid and muscovite schist, and the endomorphism of the muscovite granite to quartz muscovite rock were contemporaneous occurrences, all brought about by the same agencies, which were the solutions representing the end product of the differentiation of the granitic intrusive rock.

The identity of the quartz-muscovite rock with the beresite of Russia.

The rock called beresite occurs in the vicinity of Beresovsk in the Urals, where it is intimately connected with veins of auriferous quartz. The beresite itself forms distinct dikes, varying from two to twenty meters, and reaching forty meters in width. These veins have a general north and south direction, but vary locally and interlace. They were first described by G. Rose† in 1837. He describes the constituents of the rock as orthoclase, plagioclase, quartz, and muscovite in exceedingly varying proportions. The feldspar diminishes and often completely disappears, leaving the rock composed of quartz and muscovite. Some varieties are like sericite schists; others like micaceous sandstones. Rutile is occasionally found as accessory. Rose was inclined to regard the Beresovsk veins as apophyses from the neighboring granite of Schartassh,‡ which is a fine-grained rock of semiporphyritic nature consisting chiefly of quartz and feldspar (the latter in part albite), with small flakes of biotite. The exact relation of the beresite to the neighboring granite, however, has not been ascertained.§ The beresite is intrusive into vertical or highly dipping schistose strata, either chloritic (listvenites), or talcose and argillaceous.

* System of Mineralogy, sixth edition, 1896, p. 204.
† Reise nach dem. Ural, vol. i, p. 186; vol. ii, p. 557.
‡ Op. cit., vol. i. p. 189.
§ Op. cit., vol. ii, p. 559.

se schists strike north and south parallel to the beresite
s; they often contain serpentine. The beresite had been
named by the miners of the district previous to Rose's
stigation, and had been hunted for as the surest index to
l. The dikes are cut obliquely by highly inclined or ver-
l auriferous quartz veins. Ordinarily these veins do not
nd beyond the beresite, but sometimes they enter the
ntry-rock, and even extend to the next dike. The beresite
lf, where it consists chiefly of quartz, muscovite and pyrite,
ibly near the river Tchéremchanka, contains, according to
researches of A. Sokolow, 50 drachms of gold to the ton.
ely, native gold has been found in it.* A peculiarity of
beresite, which makes it difficult of investigation, is its
found decomposition, fresh portions being very rarely met
h. In the upper horizons it is often altered to a light col-
l clayey mass, while the neighboring schists of the country-
k are altered to a red clay.

After the thorough description of beresite by Rose no other
lies were made until Karpinsky published his results in
6.† Karpinsky investigated what appeared to him to be
h pieces of the beresite, and these showed themselves to be
from feldspar and to be composed of muscovite and quartz,
h a little iron-pyrite. He came to the conclusion that the
esite is a feldspar-free rock, and, contrary to Rose's opinion,
to be connected with granite. In his second paper, Kar-
sky admitted that orthoclase is present in beresite from
ther locality, and separated the rock into a feldspar-free
a feldspar-bearing variety, which are connected with one
ther by transitions. Besides the beresites at Beresovsk, a
nber of other Russian localities have been described by
e,‡ while Karpinsky and Arzruni§ have added still others.

What appears to be the most thorough and clear-sighted
ly of the beresite was made in 1885 by Arzruni ‖ He
ribes the beresite as a fine-grained dike-rock of coarser or
r texture and often semi-porphyritic structure, the varieties
this rock being so varied in their peculiarities that it seems
per to describe the different occurrences separately. By the
inishing of one or the other of the mineral constituents,
n as the feldspar or the mica, special types are presented.

A. Karpinsky, Guide des Excursions du VII Congrés Geologique Interna-
l, No. 5, p. 41.
Protocol of the Geologic-Mineralogic section of the Natural Science Society at
etersburg meeting on the 3d of May, 1875, and on the 9th of September,
German summary of the Russian text given by Arzruni, Zeitschr. d.
sch. Geolog. Gesell., vol. xxxvii, 1885, p. 867.
teise, etc., vol. i, pp. 294, 302, 321, 436; vol. ii, pp. 34, 36, and 557.
Zeitschr. d. Deutsch. Geolog. Gesell., vol. xxxvii, 1885, p. 870.
p. cit., pp. 865–896.

Arzruni found, however, that those beresites which consist entirely of quartz and muscovite, and which had been regarded by Karpinsky as fresh feldspar-free rocks, actually contained feldspar in their original condition, although the feldspar has undergone a complete transition to mica and quartz. It appears, also, that the alteration has affected only the ortho-clase, while the plagioclase has remained unaltered. Since in the different parts of the rock the relative proportion of ortho-clase and plagioclase is very variable, it results that the varieties rich in plagioclase contain feldspar, even when the orthoclase is altered to quartz and muscovite, while the varieties poor or lacking in plagioclase can reach a transitional stage where they can be regarded as feldspar-free beresites. As the result of the study Arzruni concludes that beresite is a genuine musco-vite granite with a tolerably constant mineralogic constitution.

The rocks described from Belmont are identical in nearly every particular with those of Beresovsk, except that in Bel-mont the decomposition has not affected the rocks to any extent, permitting a much more accurate study.

Analysis of Belmont rock.

A sample of the quartz-muscovite rock from Belmont proved to have the following composition :

Analyst, Dr. H. N. STOKES.

SiO_2	84·15
Al_2O_3	9·67
Fe_2O_3	·51
FeO	·07
MgO	·04
CaO	·53
Na_2O	2·65
K_2O	1·57
H_2O-	·21
H_2O+	·74
TiO_2	trace
F	·02
MnO	trace
SrO	trace

100·14

A rough computation of this analysis shows the proportion by weight of albite to muscovite to be about 5 : 6. The speci-men then is one which contains much more unaltered feldspar than some of the other varieties. Those varieties which origi-nally contained orthoclase, but no albite, would now be rocks free from feldspar.

ART. XXXVI.—*The Volumetric Estimation of Copper as the Oxalate, with Separation from Cadmium, Arsenic, Tin, and Zinc ;* by CHARLES A. PETERS.

[Contributions from the Kent Chemical Laboratory of Yale University—XCVIII.]

IT is a well known fact that copper oxalate is insoluble in water and scarcely attacked by moderate amounts of dilute nitric acid.* Upon this fact Bournemann† has recently based a method for the separation of copper from cadmium by precipitating copper as the oxalate in the presence of nitric acid, filtering hot, and estimating the copper after ignition, by any of the well known gravimetric methods. Six to ten grams of copper, as the oxide, were used for a single determination, and the errors were large. Bournemann does not recommend this process as an accurate analytical method. Classen‡ describes a method for the separation of metals as oxalates by adding to the solution of the salt of the metals a dilute solution of the potassium oxalate (1 : 6) and concentrated acetic acid to 80 per cent of the total volume. Regarding copper salts in particular, Classen states that precipitation takes place only in dilute solution and then not completely.

It has been the experience of the writer, that the precipitation of copper oxalate from solutions containing at least 0·0128 gm. of the oxide and saturated with the oxalic acid is practically complete. The filtrate in such cases gives no blue color with ammonia, looking down on a column of liquid in a test tube, and only a faint brown color is developed when the filtrate is neutralized, made acid with acetic acid, and tested with potassium ferrocyanide. It is the object of this paper to show that moderate amounts of copper may be determined quantitatively as the oxalate by precipitation with oxalic acid and titration of the precipitate by potassium permanganate, and also to show that moderate amounts of copper may be separated from other metals in the presence of nitric acid, by the addition of considerable amounts of oxalic acid.

Before attempting the quantitative separation of copper from solution by the addition of oxalic acid a few qualitative experiments upon the precipitation of varying amounts of copper sulphate by varying amounts of oxalic acid were tried at different dilutions. In all the experiments the mixtures stood 16–20 hours, and were filtered from 2 to 4 times through four filters folded together, and the filtrates were tested

---

* Storer, Dictionary of Chemical Solubilities, p. 463.
† Chem. Ztg., xxiii, 565.
‡ Ber., x, b, 1316.

AM. JOUR. SCI.—FOURTH SERIES, VOL. X, No. 59.—NOVEMBER, 1900.
24

both with ammonia and with potassium ferrocyanide. In cases where the filtrate gave no blue color with ammonia and only a slight precipitate with ferrocyanide the precipitation was considered practically complete and the conditions were regarded suitable for the trial of the method quantitatively. In the following table is recorded the work upon the precipitation of copper sulphate by 0·5 gm., 1·0 gm., and 2·0 gms. of oxalic acid in 50$^{cm^3}$ of solution.

TABLE I.

Dilution 50$^{cm^3}$.

	CuO taken as CuSO₄ gms.	Oxalic acid added in solution.		Oxalic acid added in crystalline form.	
		Filtrate treated with NH₄OH	Filtrate treated with K₄FeC₆N₆	Filtrated with NH₄OH	Filtrate treated with K₄FeC₆N₆
2·0 gms. oxalic acid present.	0·018	blue color	abundant ppt.	blue color	abundant
	0·031	trace "	" "	trace "	evident
	0·051	....	evident "	....	"
	0·064	....	trace "	....	trace
1·0 gm. oxalic acid present.	0·018	blue color	abundant ppt.	blue color	abundant
	0·031	"	" "	trace "	evident
	0 051	trace "	" "	....	"
	0·064	....	evident "	.....	"
	0·094	....	trace "	....	trace
0·5 gm. oxalic acid. present.	0·018	blue color	abundant ppt.	blue color	abundant
	0·031	"	" "	trace "	"
	0·051	trace "	" "	....	evident
	0·064	....	evident "	....	trace
	0·094	....	trace "	....	"

It will be seen readily by comparison of the right and left hand sides of the table above that somewhat smaller amounts of copper may be precipitated completely by the addition of crystallized oxalic acid than by the same amount of oxalic acid already in solution. Thus, when dissolved oxalic acid is added to the solution of 50$^{cm^3}$ amounts of copper sulphate less than 0·040–0·050 gm. are not precipitated completely, while under conditions otherwise the same excepting that the oxalic acid is added in crystalline form, the precipitation of amounts as small as 0 030 gm. is practically complete. The amount of oxalic acid in solution necessary for the complete precipitation (after 16 to 20 hours) of this minimum amount of copper, 0·031 gm. of copper oxide taken as the sulphate, appears, as shown in Table II, A, which follows, to be about 3·5 gms. in 50·0$^{cm^3}$. If the amount of oxalic acid is increased to 5 gms., making the solution saturated for that substance, using the same volume of liquid, the minimum amount completely precipitable is reduced to 0·0128 gms. but not to one-half that amount.

TABLE II.

CuO taken as CuSO₄ gms.	Oxalic acid added in solution. gms.	Volume at precipitation. cm³.	A Filtrate treated with NH₄OH.	Filtrate treated with K₄FeC₆N₆.
0·031	0·5	50	blue color	abundant ppt.
"	1·0	"	"	" "
"	2·0	"	trace "	" "
	3·0	" slight "	"	"
"	3·5	"	--------	evident "
0·0128	5·0		no blue	trace "
0·0064	5·0		blue color	abundant "
		B		
0·0064	0·5	20	faint blue	abundant ppt.
"	"	15	"	" "
"	"	10	"	faint "
"	"	5	--------	trace "
0·0003	0·5	5*	--------	--------
"	0·1	1†	--------	--------

It appears from the experiments of Table II, B, that the volume of liquid in which precipitation takes place influences the complete precipitation of the copper oxalate. Thus the precipitation of 0·0064 gm. of copper oxide taken as the sulphate by 0·5 gm. of oxalic acid is complete in 5^cm³ of liquid. The precipitate which falls from 0·0003 gm. of the oxide taken as the sulphate dissolves in 5^cm³ of liquid, but remains visible in 1^cm³.

As a result of the preliminary experiments, it may be said that the presence of a certain minimum amount of copper, varying with the conditions, is essential to complete precipitation. Thus, at a dilution of 50^cm³ a saturated solution of oxalic acid will precipitate with practical completeness copper taken as the sulphate in amounts exceeding the equivalent of 0·0128 gm. of copper oxide; that 2·0 gms. of oxalic acid will precipitate almost completely for the same volume of solution the equivalent of 0·03 gm. of copper oxide; and that 1·0 gm. or 0·5 gm. of oxalic acid will precipitate the equivalent of 0·064 gm. of the oxide.

In the quantitative separation of copper as the oxalate the method of treatment was in general as follows. Copper sul-

---

* Precipitate redissolved.
† Precipitate remained.

phate in 50^{cm^3} of water was thrown down by the addition of dry oxalic acid to the hot solution, and, after standing over night, the precipitate was filtered on asbestos, washed two or three times with small amounts of cold water. The precipitate, still in the crucible, was returned to the beaker in which precipitation took place, 5 or 10^{cm^3} of dilute sulphuric acid (1 : 1) were then added together with a convenient amount of water, and, after heating the liquid to boiling, the oxalic acid was titrated with permanganate, the oxalate of copper dissolving readily as fast as the excess of oxalic acid is removed by the permanganate. The precipitate may also be dissolved in 10^{cm^3} of strong hydrochloric acid,* 0·5 gm. manganous chloride added and titrated at 30°–50°. Experiments 4 and 5 were conducted after this manner. In Table III, A, which follows, are recorded results of the quantitative tests of the method.

TABLE III.

	CuO taken as CuSO₄. gms.	Oxalic acid. gms.	Volume at precipitation. cm³.	CuO found. gms.	Error. gms.
			A		
1	0·0372	0·15	100	0·0286	—0·0086
2	0·1860	0·50	125	0·1831	—0·0029
3	0·0398	"	50	0·0376	—0·0022
4	0·1860	1·0	150	0·1834	—0·0026
5	"	0·5	50	0·1864	+0·0004
6	"	"	"	0·1866	+0·0006
7		"	"	0·1866	+0·0006
8	"	1·0		0·1866	+0·0006
9	0·0398	"	··	0·0391	—0·0007

In experiments 1–4, deficiencies are found in the amounts of oxalate precipitated at different degrees of dilution and by different amounts of the precipitant which are in agreement with the results obtained in the preliminary work; the results of experiments 5–9, in which 0·5 grm. and 1·0 grm. of oxalic acid act in a total volume of 50^{cm^3}, show the precipitation to be essentially complete under these conditions.

To study the insolubility of the copper oxalate in nitric acid the experiments in Section B of the table were made.

In experiments 10–13 amounts of oxalic acid varying from 0·5 gm. to 3·0 gms. appear to precipitate the copper completely in the presence of 5^{cm^3} of strong nitric acid. In experiment 14 the amount of oxalic acid used was not sufficient to throw down all the copper in the presence of 10^{cm^3} of nitric acid, but

* Gooch and Peters, this Journal, vii, 461, 1899.

TABLE III (continued).

	CuO taken as CuSO₄. gms.	Oxalic acid. gms.	HNO₃ sp. gr. 1·40. cm³.	Volume at precipitation. cm³.	CuO found. gms.	Error. gms.
			B			
10	0·1860	0·5	5·0	55	0·1859	—0·0001
11	"	"	"	"	0·1860	±0·0000
12	0·1990	2·0	"	"	0·1989	—0·0001
13	"	3·0	"	"	0·1990	±0·0000
14	"	2·0	10·0	60	0·1971	—0·0019
15		3·0	"	"	0·1987	—0·0003
16		"	"	"	0·1985	—0·0005
17		5·0	12·0	130	0·1977	—0·0013
18		"	"	"	0·1975	—0·0015
19		"	25·0	"	0·1837	—0·0153
20			"	"	0·1831	—0·0159
21			5·0	"	0·1983	—0·0007
22			"	"	0·1988	—0·0002
			C			
23		2·5	5·0*	65	0·1971	—0·0029
24		2·0	"	"	0·1981	—0·0019

the copper does come down completely in the presence of the large amount of the nitric acid upon the addition of more oxalic acid, as seen in experiments 15 and 16. In experiments 17 and 18 with a larger volume of water and a larger absolute amount, though approximately the same percentage, of nitric acid present as in experiments 10–13, there is a slight loss of copper; but in experiments 21 and 22 when the amount of nitric acid is reduced to 5^{cm³} in the larger total volume the results are normal. Experiments 19 and 20 show the increased loss when still larger amounts of nitric are present. These facts would make it seem best to limit the absolute amount of nitric in solution to about 5^{cm³}.

One observation may well be noted here; namely, that while one-half gram oxalic acid is all that is needed for the complete precipitation of the copper in the presence of 5^{cm³} strong nitric acid, still the oxalic acid may be added up to the point of saturation of the solution. More than this causes difficulty owing to the fact that a large amount of water is necessary to wash the precipitated oxalate. About 2·0 gms. of oxalic acid to 50^{cm³} of water is a convenient proportion.

In experiments 23 and 24, 5^{cm³} of nitric acid were neutralized with ammonium hydroxide before adding the 5^{cm³} strong

* About 9 gms. of ammonium nitrate present in addition to the 5^{cm³} of nitric acid.

nitric acid in excess. The results show the solubility of copper oxalate in ammonium nitrate and exclude the possibility of such a procedure in this work.

Some experiments were made to show the time necessary for the complete precipitation, both in the presence and absence of nitric acid. Following is the record of such work.

TABLE III (continued).

	CuO taken as CuSO$_4$. gms.	Oxalic acid. gms.	HNO$_3$ sp. gr. 1·40. cm³.	Volume at precipitation. cm³.	CuO found. gms.	Error. gms.	Details of filtration.
				D			
3	0·1990	2·0	....	50	0·1984	—0·0006	Filtered ho immediat
4	0·2030	"		"	0·2025	—0·0005	Filtered ho immediat
5	0·1990	1·0	....	"	0·1990	±0·0000	Filtered al cooling; stood 15 minutes
6				"	0·1987	—0·0003	Filtered af cooling; st 15 minute
				E			
7	"	2·0	5·0	55	0·1943	—0·0047	Filtered a cooling; stood 15 minutes
8				"	0·1969	—0·0021	Stood 2½ ho
9				"	0·1973	—0·0017	Stood 6 ho
0	..	..	..	"	0·1989	—0·0001	Stood 16 ho

The results in section D would seem to show that a solution containing copper may be precipitated hot as the oxalate and filtered either hot or after cooling with a very slight loss. Tests of the filtrates made with potassium ferro-cyanide confirmed these results. When nitric acid is present, however, the mixture must stand after the addition of the precipitant. In section E the gradual decrease of the minus error is noticed, as the time of standing is extended, the precipitation being practically complete upon standing over night.

Separation from Cadmium.

Bournemann* has used nitric acid for a rough separation of copper from cadmium. This method was tried for a quantitative separation in the presence of 6–10 per cent strong nitric acid. The results are found in section F of the table to follow.

* Loc. cit.

Experiments 33–35 stood six hours before filtering. Experiments 36 and 37 stood over night. Copper is separated from more than twice its weight of cadmium, and the results are accurate.

Separation from Arsenic, in Both Conditions of Oxidation.

For the separation from arsenic, arsenious oxide dissolved in sodium carbonate, and di-hydrogen sodium arseniate were the forms of arsenic used. The results are accurate and are given in sections G and H of the table. In experiments 38–40 and 44 and 45 no nitric acid was added. While the presence of the nitric acid is not necessary for the separation of the copper from the arsenic; still the filtration in the absence of the nitric acid is so slow as to be objectionable. The presence of the nitric acid causes the precipitate to come down in a coarser condition, and in such condition it filters easily and is capable of being washed quickly.

Separation from Tin, in Both Conditions of Oxidation.

For the separation of copper from tin a preparation of stannous chloride (20$^{cm^s}$ giving 0·3746 gm. metallic tin by the battery) containing sufficient hydrochloric acid to prevent deposition of oxy-salts was used. The solution of stannic chloride contained 1·0 gm. metallic tin to every 10$^{cm^s}$, and was used without hydrochloric acid. The results of the work are found in sections I and K of the table. The experiments go to show that while copper may be separated from small amounts of tin as stannous chloride yet there is a limit to the amount of tin which may be present. One-tenth of a gram of metallic tin is the largest amount that can be present, with 0·15 gm. copper oxide taken as the sulphate, without significant error. Practically the same statement can be made of the separation of copper from tin taken as stannic chloride. Experiment 57 shows a greater loss of copper when the nitric acid is omitted.

Separation of Copper from Iron.

A solution of ferric nitrate was used for the work on the separation of copper from iron. Low results were obtained when a solution of ferrous or ferric sulphate was used as the source of iron. The results of the experiments are recorded in section L of the table, and show that 0·20 gm. copper oxide as the sulphate may be separated from 0·2–0·3 gm. iron oxide taken as the nitrate. In experiment 64 a good result was obtained when no nitric acid was present, save that added in combination with the iron. A comparison of experiments 63 and 65 shows that it is best to avoid the use of large amounts

TABLE III (continued).

	CuO taken as CuSO₄. gms.	Element from which copper was separated. gms.	Oxalic acid. gms.	HNO₃ sp. gr. 1·40. cm³.	Volume at precipitation. cm³.	CuO found. gms.	Error. gms.
		CdO taken as CdSO₄.		F			
33	0·1990	0·10	2·0	5·0	60	0·1983	−0·0007
34	"	0·20	"	"	65	0·1987	−0·0003
35	"	0-30	"	"	70	0·1987	−0·0003
36		0·40	"	"	75	0·1994	+0·0004
37		0·50	"	"	80	0·1996	+0·0006
		As₂O₃ taken as Na₂AsO₃.		G			
38		0·10	"	----	55	0·1991	+0·0001
39		0·20	"	----	60	0·1987	−0·0003
40		0·50	"	----	75	0·1986	−0·0004
41		0·10	"	5·0	60	0·1994	+0·0004
42		0·20	"	"	75	0·1992	+0·0002
43		0·60	"	"	85	0·1995	+0·0005
		As₂O₃ taken as H₂KAsO₄.		H			
44		0·10	"	----	60	0·1985	−0·0005
45		0·20	"	----	70	0·1990	±0·0000
46		0·10	"	5·0	65	0·1990	±0·0000
47		0·20	"	"	75	0·1992	+0·0002
48	"	0·30	"	"	85	0·1985	−0·0005
49	0·2030	0·30	3·0	"	85	0·2026	−0·0004
	Cu taken as CuSO₄.	Sn taken as SnCl₂ + HCl.		I		Cu found.	
50	0·1590	0·0468	2·0	5·0	55	0·1581	−0·0009
51	"	0·0936	"	"	60	0·1603	+0·0013
51a	"	"	"	"	"	0·1591	+0·0001
52		"	"	"	"	0·1594	+0·0004
53		0·1873	"	"	65	0·1603	+0·0013
54		0 2809	"	"	70	0·1914	+0·0324
55		"	3·0	"	75	0·1988	+0·0398
		Sn taken as SnCl₄.		K			
56		0·10	2·0	"	55	0·1581	−0·0009
57		0·10	"	----	"	0·1565	−0·0025
58		0·20	"	5·0	"	0·1577	−0·0013
59	"	0·50	"	"	60	0·1562	−0·0028
	CuO taken as CuSO₄.	Fe₂O₃ taken as Fe(NO₃)₃.		L		CuO found.	
60	0·1990	0·136	2·0	5·0	60	0·1987	−0·0003
61	"	0·272	"	"	"	0·1983	−0·0007
62	"	0·364	"	"	"	0·1988	−0·0002
63		0·544	"	"	65	0·1971	−0·0019
64		0·272	"	----	60	0·1995	+0·0005
65		0·544	"	2·0	"	0·1998	+0·0008
66	"	0·218	"	"	65	0·1999	+0·0009
		ZnO taken as ZnSO₄.		M			
67	"	0·028	"	5·0	60	0·2007	+0·0017
68		0·057	"	"	65	0·2008	+0·0018
69		"	"	"	"	0·2008	+0·0018
70		0·085	"	"	70	0·2035	0·0045

nitric acid when the larger amounts of ferric nitrate are
sent.

'or a practical application of the above separation of copper
n iron a convenient amount of finely ground chalcopyrite
gm.) was roasted 2–3 hours in a porcelain crucible until
sulphur was driven off, washed into a beaker, strong nitric
l about 5$^{cm^3}$ was added and, with the beaker covered,
wed to evaporate slowly on a hot plate, nearly to dryness.
little dilute nitric acid was added, the solution was filtered,
residue was washed with water containing dilute nitric
l, the filtrate, about 50$^{cm^3}$ in volume, was precipitated with
gms. oxalic acid, and the precipitate was estimated after
iding 12–16 hours, as previously described. The washing
h water acidified with nitric acid is important because the
ly ground ferric oxide remaining undissolved passes through
filter when washed with water alone, but gives no trouble
he water be acidic. The results of two estimations are here
en.

alcopyrite. gms.	Copper found by battery.	Copper found by oxalate method.	Difference.
0·5000	31·00%	30·92%	−0·08%
1·0000	"	31·25	+0·25

Separation of Copper from Zinc.

'he separation of copper from zinc was not altogether suc-
ful owing to the tendency of the zinc oxalate to come
rn with the copper oxalate. Some experiments are given
ection M of the table.

'he separations of copper from bismuth and antimony were
nccessful.

'he work may be briefly summarized as follows: Copper
eeding in amount the equivalent of 0·0128 gm. of the
le to 50$^{cm^3}$ of solution as the sulphate may be separated
ipletely, even in the presence of a moderate amount of
ng nitric, by the addition of sufficient amount of oxalic
l.

'opper may be separated from cadmium, arsenic, iron, and
ll amounts of tin, when precipitated by oxalic acid in a
ime of 50·0$^{cm^3}$ containing 5$^{cm^3}$ strong nitric acid. Inasmuch
he completeness of precipitation of the copper depends
n the presence of a certain minimum amount of the cop-
salt this method is not applicable when the amount of
per falls below 0·0128 gm. of the oxide to 50$^{cm^3}$ of solution.
'he author wishes to thank Professor F. A. Gooch for much
d help given in the preparation of this paper.

ART. XXXVII.—*Synopsis of the Collections of Invertebrate fossils made by the Princeton Expedition to Southern Patagonia;* by Dr. A. E. ORTMANN.

IN the February number (1900) of this Journal, Mr. J. B. Hatcher has given a general account of the sedimentary rocks of Southern Patagonia, frequently (pp. 98, 101–104, 108) referring to the present writer's studies on the fossils of this region. Since these studies have now come to an end, it seems well to give a preliminary report on the chief results obtained, in order to give an idea of the richness of the material at hand, and to enable the scientific public to get a more correct understanding of the beds in question than has hitherto been possible. A final report on the Tertiary Paleontology of Southern Patagonia will be given by the writer in a volume of the "Princeton Expedition to Patagonia," which is in course of preparation, and will be freely illustrated by figures of all species represented, drawn by Mr. F. van Iterson. It is hoped that this volume will be ready by the end of this year or the beginning of next.

The present article is intended to treat only of the so-called "Patagonian formation." As Mr. Hatcher has already pointed out (l. c., p. 101), we have sufficient reason to believe, that the different marine horizons distinguished by F. Ameghino and accepted—at least in part—by H. von Ihering (the Patagonian formation, divided into a "Piso Juliense" and "Piso Leonense," and Suprapatagonian beds or lower part of the Santa-Cruz-formation) are identical, i. e., belong in one and the same continuous series of marine deposits underlying the Mammiferous Santa-Cruz-formation, and we retain the old term *Patagonian formation* for this series, which is certainly a paleontological unit, and belongs, as the writer is now fully satisfied, in the *Lower Miocene.* The fauna of this series is represented in our collection by over 140 species, many of which are new to science.

I shall give here first a list of the *new species,* with short diagnoses and other remarks that are necessary to recognize them, and then I shall add a list of species known from other localities, but *not found previously in Patagonia,* and shall conclude this article by some remarks on the *synonymy,* etc., of known Patagonian forms.

Diagnoses of new species from the Patagonian formation.

ECHINODERMATA.

1. *Cidaris antarctica* sp. nov. Plates with a moderately large, perforated central tubercle, the neck of which is slightly crenulated. Scrobicule large, surrounded by a circle of small tubercles, between which there are still smaller ones. Spines subcylindrical, often slightly compressed, neck somewhat constricted. Articular surface finely striated, with a deep articular groove. Surface of spines closely covered with fine, rounded granules, forming irregular longitudinal rows.

Only isolated spines and plates, San Julian, Santa Cruz, Upper Rio Chalia, Lake Pueyrredon.

2. *Toxopneustes præcursor* sp. nov. Test suborbicular. Ambulacral and interambulacral spaces with 4–8 vertical rows of tubercles of subequal size, those of the ambulacral spaces being somewhat smaller. Poriferous zone moderately broad. Pores in three pairs, the two outer vertical rows separated from the inner row by a small tubercle. All the primary tubercles surrounded by small secondaries and miliaries. Actinostome sunken, and lower surface concave, the actinal cuts comparatively slight.

This species differs from the known recent species of *Toxopneustes* chiefly in the more crowded tubercles. The most closely allied form seems to be: *T. pileolus* (Lmck.).

San Julian; Shell Gap (Upper Rio Chico).

3. *Cyrtoma posthumum* sp. nov. Test subcircular-elliptic, depressed. Apex central, upper side covered with very fine tubercles. Ambulacra petaloid, open, lanceolate, subequal, extending about two-thirds from the apex toward the periphery, the posterior ones closer together than the others. Anus situated on the upper surface, in a deep depression, of a pyriform shape, narrow above and suddenly widening toward the periphery. Lower surface of test concave, covered with larger, more widely separated tubercles. Mouth subcentral, surrounded by a floscelle. Diameter ca. 110^{mm}, height ca. 28^{mm}.

The peculiar shape of the anal depression brings this species into the genus *Cyrtoma* of McClelland (Calcutta Journ. Nat. Hist., 1840), a synonym of which is *Stigmatopygus* of d'Orbigny. It is the first Tertiary representative of this so far exclusively Cretaceous genus.

Lake Pueyrredon.

VERMES.

4. *Serpula patagonica* sp. nov. Tubes solid, calcareous, cylindrical, irregularly contorted and vermiculate, growing

upon shells, stones, etc.   Outer surface transversely rugose. Diameter, 3^{mm}.

San Julian.

5. *Terebella magna* sp. nov.   Large cylindrical tubes, isolated or growing in groups of two or three, straight or slightly curved.   Walls composed of large and irregular fragments of shells.   Inner surface of tubes smooth, outer surface very rough.   Diameter of inner tube, $12–15^{mm}$.

Systematic position of these tubes, apparently built by a worm, doubtful.

San Julian.

BRYOZOA.

6. *Melicerita triforis* sp. nov.   Zoarium foliaceous, lobate. Zooecia hexagonal, disposed quincuncially on both surfaces of the zoarium.   Orifice crescentic, large, about in the middle of each cell.   Besides there is an ovarian opening on the summit of the cell, and two (? avicularian) openings on the side of the mouth.

The three openings in the upper part of the cell distinguish this species from all the rest.

Upper Rio Chalia.

7. *Reticulipora patagonica* sp. nov.   Closely resembling *R. transennata* Waters (Quart. Journ. Geol. Soc., vol. xl, 1884, p. 689), and differing only, if at all, in the branches of the zoarium being a little stronger, and the zooecial openings being more crowded.

Santa Cruz.

8. *Tennysonia subcylindrica* sp. nov.   Closely resembling the only known species of the genus, *T. stellata* Busk (Cat. Mar. Pol. Brit. Mus., 3, 1875, p. 34), but differing by the more slender branches of the Zoarium, which are subcylindrical, and the slightly prominent orifices of the cells.

Santa Cruz.

PELECYPODA.

9. *Modiola andina* sp. nov.   Shell small, elongated.   Apex near anterior end.   Both valves convex, with a blunt ridge running down from the apex to the posterior and inferior end. This ridge is slightly curved, concave toward the lower margin. Upper margin almost straight in its anterior part, forming a blunt angle with the straight posterior part, which passes in a regular curve into the posterior margin.   Ventral margin concave.   Surface of shell in the upper half (above the oblique ridge) finely radially striated.   Lower part of surface smooth, only near the anterior end, below the apex, with a few fine striae.   Length of shell, 24^{mm}; height, 9^{mm}.

Lake Pueyrredon.

10. *Nucula reticularis* sp. nov. Shell small, moderately convex and moderately thick, subovate, oblique. Posterior and anterior dorsal margin slightly convex, ventral margin strongly arcuate. Surface with very fine concentric ribs, which are irregular and often bifurcate. These ribs are crossed by still finer radial striæ, which give a beautifully reticulated appearance to the shell. Ventral margin finely crenulated on inner side. Hinge teeth fine, both parts of the series forming an obtuse angle, anterior part with ca. 9, posterior with ca. 18 teeth. Length, 7·5^{mm}; height, 6^{mm}.

The sculpture of this species is of the type of that of the Oligocene *N. chasteli* Nyst.

Santa Cruz and Mt. of Observation.

11. *Crassatella quarta* sp. nov. Shell elongated-ovate, comparatively thin, not very convex. Apex only slightly prominent. Anterior end rounded, posterior hardly angulated and hardly narrowed. Posterior dorsal margin straight near apex, anterior almost straight, with only a slight suggestion of concavity close to the apex. Surface ornaments as in *C. Lyelli* Sowerby, but the ridges more crowded and a little less developed. Ventral margins without crenulations. Length, 17^{mm}; height, 10^{mm}, but growing larger.

Santa Cruz; Lake Pueyrredon.

12. *Glycimeris regularis* sp. nov. Shell elongate, convex, with concentric lines of growth and undulations. Apex at ⅓ of the length, incurved. Anterior end rounded, posterior subtruncated, not narrower than anterior. Ventral margin straight in the middle. Long., 78; height, 45.

This form does not agree with any of the described Patagonian species, and accordingly I think it is new, although all the distinctive characters are taken only from the external form.

Santa Cruz; San Julian; Lake Pueyrredon.

13. *Corbula hatcheri* sp. nov. Shell small, solid and thick, subovate-triangular. Right valve very little larger than the left, both moderately convex. Anterior end rounded, posterior produced, subtruncated, an angular ridge running from apex to posterior angle. Ventral margin arcuate, posteriorly a little concave. Lower margin of right valve reflected toward the left valve. Surface with concentric ribs, which are rounded and rather crowded. Length, 11; height, 7·5; diameter (of right valve), 2·5^{mm}.

Santa Cruz; Las Salinas; Mt. of Observation; San Julian.

14. *Martesia pumila* sp. nov. This species resembles much *M. patagonica* Phil. but is much smaller, the callous plate of the anterior margin is very small, and the ribs of the anterior part of the shell form a very obtuse angle with the lines of

growth of the posterior part. The radiating furrow is narrower, and is more inclined posteriorly, so as to render the posterior part of the shell smaller in comparison with the anterior. Length, 9^{mm}; height, 4^{mm}.

This does not seem to be an immature stage of *M. patagonica.* Santa Cruz.

GASTROPODA.

15. *Liotia scotti* sp. nov. Shell small, rounded, flat above, with a large, open umbilicus below. Spire with four rounded whorls, increasing rapidly, suture deep. Last whorl with six revolving, equidistant keels, the keel nearest to the umbilicus the smallest, and disappearing within the umbilicus; the upper whorls show only the two uppermost keels. The keels are crossed by very fine striæ, and a number (15) of strong radial ribs; at the points of intersection of these ribs and the keels, there is a small conical tubercle. Last whorl a little deflected toward the mouth, which is circular and thickened. Height, 4^{mm}; diameter, 8^{mm}.

This species resembles much the recent *L. acrilla* of Dall. Santa Cruz.

16. *Calliostoma observationis* sp. nov. Shell low, conical, not umbilicated. Whorls flat, last whorl on the periphery bluntly angular. Above this angulation there are five distinct revolving ribs; near the mouth, between the second and third (counted from above), a sixth rib begins to appear. In the upper whorls the second and fourth ribs disappear, so that only three ribs remain, besides the peripheral angulation, which shows as a fourth rib immediately above the suture. All these ribs, when fully developed, are subequal, flattened, smooth, about as broad as the intervals between them. The base of the shell has 9–10 revolving ribs of the same character. The outermost of them is not separated from the peripheral angulation by a broader interval. Height, 10·5^{mm}; diameter, 12^{mm}.

Mt. of Observation.

17. *Calliostoma cossmanni* sp. nov. Shell conical, higher than broad, not umbilicated. Whorls flat, the last one angulated, with a keel on the periphery, which is wholly exposed on the upper whorls, being situated close to, but above the suture. Upper whorls with five revolving keels, the lowermost, formed by the peripheral keel just mentioned, is the strongest. It is smooth, with hardly any trace of granulations. The uppermost and the third keel are stronger than the second and the fourth; the first, second, and third are distinctly granulated, the fourth with finer granulations. Toward the apex of the shell, the second and fourth keels disappear, so that only three keels are present, the two upper

ones granulated, the lower one smooth. Base of shell hardly convex, with six revolving keels, which are subequal, smooth, and narrower than the intervals. Height, 8^{mm}; diameter, 6·5^{mm}.

Santa Cruz.

18. *Calliostoma garretti* sp. nov. Shell conical, as high as broad, not umbilicated. Eight whorls, which are very slightly convex, suture shallow. Last whorl very bluntly angulated at the periphery, without a distinct keel. Surface of whorls, above the periphery, covered with numerous fine, revolving threads: there are, on the third whorl, about 7 of them, increasing to about 17 on the last. The number of the threads increases by intercalation, the new keels being at first smooth, but soon they equal the others, and become, like the latter, finely, but distinctly granulated. These granulations, however, are developed only in the upper three quarters of the whorl, the lower four or five threads remain smooth. The threads continue over the periphery to the base of the shell, which is slightly convex; their number, on the base, is about 24, and they are smooth, resembling in all other respects those of the upper part of the whorls. Height and diameter 17^{mm}.

Santa Cruz.

19. *Calliostoma iheringi* sp. nov. Shell conical, broader than high, umbilicated. Six whorls, which are sharply angulated, one angulation being formed by a sharp revolving keel in the upper part of the whorls, a second one—exposed only on the last whorl—formed by a peripheral keel. Suture distinct. Upper part of whorls (above upper keel) oblique, flat, with 5–6 revolving threads, which are slightly granulated; lower part vertical, slightly concave on the last whorl, with 6–7 fine, smooth threads. Base of shell slightly convex, depressed toward the umbilicus, which is moderately large. About 18 revolving threads on the base, which are smooth, more crowded and finer toward the periphery, a little stronger near the umbilicus. Height, 9·5^{mm}; diameter, 12^{mm}.

Santa Cruz.

20. *Crucibulum dubium* spec. nov. Cast suborbicular, depressed-conical. Apex central. On one side is the impression of the internal cup-shaped lamina, which was attached to the inner wall of the shell. No further characteristics can be given, since only a single cast is represented in our collection.

Arroyo Gio.

21. *Sigapatella americana* sp. nov. Shell suborbicular or subelliptic, depressed. Apex distinctly excentric. Surface with irregular, concentric, slightly lamellate stiræ, crossed by very fine radial rugosities. Internal diaphragma spiral, colu-

mella excentric, margin of diaphragma slightly concave and slightly reflexed at the columella. Height, 16ᵐᵐ; diameter, 49ᵐᵐ.

Santa Cruz; Punta Arenas.

22. *Dolium ovulum* sp. nov.   Shell ovato-globular, spire short, conical, acute, last whorl large.   Surface with fine and crowded revolving striæ, which are sub-equal, only in the lower part finer ones are intercalated.   Mouth large, elongated-oval, canal very short, truncated, straight and comparatively narrow.   Inner lip without callous tubercles or folds.   Outer lip slightly thickened.   Height, 34ᵐᵐ; diameter, 25ᵐᵐ.

Santa Cruz.

23. *Tritonium morgani* sp. nov.   Shell subfusiform, elongated, with three varices.   Whorls with fine, unequal, spiral striæ and large tubercles, the latter, on the last whorl, in three spiral rows, those of the upper row large, ca. seven between two varices; those of the middle row (5–6) small, and those of the lower row (3–4) very indistinct.   Columella smooth, with a few indistinct crenulations in the lower part.   Canal comparatively long.   Outer lip distinctly crenulated, with an indistinct canaliform emargination in the upper part, opposite which is a fold on the inner lip.   Height, 63ᵐᵐ; diameter, 28ᵐᵐ.

Santa Cruz.

24. *Buccinum annæ* sp. nov.   Shell subfusiform, elongate-oval.   Spire long.   Whorls 7–8, angulated, the angulation with a series of tubercles, 12–14 of them on the last whorl, which are continued downward as irregular longitudinal ribs.   Upper part of whorls slightly concave, appressed toward the suture.   Exposed part of upper whorls, below angulation, subcylindrical.   Last whorl large.   Mouth ovate, elongated, upper end subcanaliculate, lower end truncated, and with a short reflexed canal, forming a varix on. the columella.   Outer lip thin, smooth within.   Height, 66ᵐᵐ: diameter, 30ᵐᵐ.   This species belongs into the subgenus *Cominella*.

Santa Cruz.

25. *Fusus archimedis* sp. nov.   Shell fusiform, spire shorter than the last whorl, scalariform.   Whorls very prominently angulated, suture very deep.   Upper part of whorls, above angulation, flat, obliquely descending from the suture, lower part, below angulation, very slightly convex, obliquely receding downward to the suture.   Angulation blunt, with a number (10–13) of blunt, often indistinct tubercles.   Surface of shell with fine revolving ribs on the lower part of the whorls and upon the angulation, but these ribs are absent on the upper part, above the angulation.   Whole surface with distinct lines of growth, which have a squamiform appearance, where they

cross the revolving ribs. Last whorl large. Mouth triangular, continued into a long and straight canal. Height, 50ᵐᵐ, but defective on upper end ; diameter, 25ᵐᵐ.

San Julian.

26. *Fusus torosus* sp. nov. Shell subturbinate or subfusiform. Spire short, rather depressed. Whorls four, last one very large. Surface with numerous fine spiral ribs, which are rather crowded and somewhat unequal, crossed by very fine, squamiform lines of growth. Whorls strongly convex, swollen, with ca. seven strong, variciform longitudinal ribs, which begin at the suture and become thick and swollen in the middle of the last whorl, attenuating again toward the lower end of the shell. Mouth ovate, continued into a canal of moderate length, which is slightly curved. Height, 31 ; diameter, 20ᵐᵐ. This species resembles somewhat *F. pyruliformis* Sow., from Navidad, and I would not hesitate to identify it with this species. But Sowerby's figure seems to be poor, and the account given by Philippi and Moericke of *F. pyruliformis* shows clearly that it is different.

Santa Cruz.

27. *Fusus cancellatus* sp. nov. Shell small, fusiform, elongated. Spire a little shorter than the last whorl. Whorls convex, surface ornamented by revolving and longitudinal ribs, cancellated. Spiral ribs, in the upper whorls, 4–5, 12–18 on the last whorl; they are sharp, but flat, equidistant, narrower in the intervening spaces between the longitudinal ribs, and on the points of intersection with them slightly broadened, giving the appearance of low tubercles. Longitudinal ribs 12–13 on one whorl, rounded, but distinct, running from suture to suture, but disappearing on the canal. Mouth elliptical, canal comparatively short. Outer lip crenulated within. Height, 16ᵐᵐ; diameter, 6·5ᵐᵐ.

Santa Cruz.

28. *Fusus pilsbryi* sp. nov. Shell thick, elongated, fusiform ; spire a little shorter than the last whorl. Whorls 7–8, convex, slightly appressed in the upper part, ornamented with 8–9 strong, rounded longitudinal ribs, which are slightly oblique and curved. On the upper whorls these ribs reach from suture to suture, on the last whorl they disappear below the middle. All of the surface of the shell is covered by very fine, numerous, distinct and subequal spiral striae. Mouth comparatively small, continued into a short canal. Outer lip thick. Height, 36·5ᵐᵐ (not quite complete); diameter, 12ᵐᵐ.

Santa Cruz.

29. *Murex hatcheri* sp. nov. Shell ovato-subfusiform. Whorls 5–6, rapidly increasing. Spire short, conical. Upper whorls angulated by a prominent, but blunt carina, which is

situated below the middle of the whorls; this carina forms an angulation on the last whorl, and below it there are 4–5 other carinae, decreasing in size. Upper part of whorls flat and obliquely descending from the suture, with a few revolving striae. Varices 5–6, lamelliform, strong and thick, at the points of crossing with the spinal carina produced into short leaf- or ear-like lobes, strongest on the uppermost carina. On the upper whorls only the uppermost row of lobes is visible. Mouth large, oval, with an open canal of medium length. Outer lip ornamented with 5–6 lobes, corresponding to those of the varices Height, 63^{mm}; diameter, 44^{mm}.

San Julian.

30. *Urosalpinx elegans* sp. nov. Shell ovato-fusiform; whorls 5–6, convex, with spiral striae and 7–8 longitudinal, variciform costae, which are rounded. Mouth oval, elongated into an open, but narrow canal, which is about as long as the mouth. Outer lip distinctly crenulated within. Height, 16·5^{mm}; diameter, 8^{mm}.

Santa Cruz.

31. *Marginella oliviformis* sp. nov. Shell elongated, sub-cylindrically-fusiform. Spire conical. Surface of shell smooth and shining. Suture quite indistinct. Mouth long and narrow, canal very short, represented only by a rounded sinus. Columella with four subequal folds. Outer lip thickened, smooth within. Height, 11^{mm}; diameter, 5^{mm}; length of mouth, 6·5^{mm}.

Santa Cruz.

32. *Voluta petersoni* sp. nov. Shell elongated, fusiform. Surface beautifully cancellated by spiral and longitudinal ribs. Spiral ribs strongly developed, equidistant, sharp; longitudinal ribs a little stronger than the spiral ribs, sharp, running from suture to suture, ca. 30 on the last whorl. Cancellations rectangular, about twice as broad as high on the last whorl, and about three or four times as broad on the upper whorls. Spire slender, conical, mouth not much longer than half of the shell. Upper whorls quite high. Whorls almost evenly convex, only slightly appressed and concave near the suture. Mouth elongated. Columellar folds at least two, indistinct. Height, 148^{mm} (not complete); diameter, 65^{mm}.

Santa Cruz.

33. *Drillia santacruzensis* sp. nov. Shell turrite, subfusiform. Whorls 8, last whorl hardly half as long as the shell. Whorls convex, but depressed and slightly concave in the upper part near the suture. This depression forms a shallow furrow, following the suture, and is sharply separated from the rest of the whorl, which is ornamented by oblique longitudinal ribs, which end abruptly at the sutural depression. These ribs number 12–15 in one whorl. Besides, there are

very fine lines of growth, but no trace of spiral sculpture. Mouth elongated, canal short. Sinus of outer lip semicircular, situated in the sutural depression, close to the suture; at the point of junction of the outer lip with the columella there is a distinct nodulose, callous swelling. Height, 13^{mm}; diameter, 4·5^{mm}.

Santa Cruz

34. *Borsonia patagonica* sp. nov. Shell subfusiform, biconical; whorls ca. 6, last whorl a little larger than half of the shell. Whorls convex, depressed in the upper part, with a slight swelling just below the suture. Depressed part smooth, the rest ornamented by 10–12 longitudinal rib-like swellings, which are slightly tuberculiform on the upper whorls; on the last whorl they are rib-like, but less distinct. Besides the ribs, there are spiral cords on the lower part of the whorls; they are wanting on the depressed part, but continue, on the last whorl, upon the canal. Mouth elongated, canal of medium length. Outer lip with a moderately developed sinus, which is situated in the sutural depression. Columella with two plaits, the lower one sometimes quite indistinct. Height, 19^{mm} (not complete); diameter, 9^{mm}.

Santa Cruz.

35. *Actæon semilævis* sp. nov. Shell elongated-ovate, rather slender, spire short, conical, about one-quarter of the length of the shell. Whorls four, convex. Suture distinct, a slight carina running close to the suture and parallel to it. Below this carina there is an indistinct spiral groove. Below the latter the surface of the shell is smooth; but in the lower third of the last whorl there are 5–7 spiral furrows, which are rather broad, almost as broad as the flat intervals. Mouth elongated, wider below, columella with a distinct fold below. Height, 7^{mm}; diameter, 3·5^{mm}.

Mt. of Observation.

CRUSTACEA.

36. *Scalpellum juliense* sp. nov. Only the carina known. Carina narrow, elongated, strong and solid, curved; basal margin bluntly pointed; surface smooth, only with lines of growth. Tectum strongly arched in its upper part, only slightly so in its lower; upper part solid, its cross section almost quadrangular, with a prominent ridge on the concave side, formed by the junction of the inflected parietes. Parietes very narrow, separated from the tectum by a distinct but blunt ridge. The carina of *S. solidulum* Steenstr. (See Darwin, Monogr. foss. Lepad., 1851, p. 42, pl. 1, f. 8) resembles so much the present fossil, that I have no doubt, we have to deal here with a closely allied species.

San Julian.

Species new for the Patagonian formation.

BRYOZOA.

1. *Cellaria fistulosa* (L.). (Hincks, Hist. Brit. mar. Polyzoa, 1880, p. 106.) This species is a living, almost cosmopolitan form, and has been found fossil from the Oligocene beds upward in Europe and New Zealand.
Shell Gap (Rio Chico)
2. *Aspidostoma giganteum* (Busk). (Busk, Rep. Challenger, vol. x, 1884, p. 161.) Known so far only living from southern Patagonia.
Santa Cruz and San Julian.
3. *Heteropora pelliculata* Waters (see Nicholson, Ann. Nat. Hist., ser. 5, v. vi, 1880). Known living from Japan and New Zealand, and fossil from New Zealand.
San Julian ; Arroyo Gio.

BRACHIOPODA.

4. *Rhynchonella squamosa* Hutton (Cat. Tert. Moll. New Zealand, 1873). Fossil from New Zealand and Australia, living (*pixydata* Dav.) from Kerguelen Islands.
Lake Pueyrredon.
5. *Terebratella dorsata* (Gmel.). (Davidson, Trans Linn. Soc., 1887, p. 75.) Fossil in New Zealand, and living on the Patagonian coast.
Santa Cruz ; Shell Gap ; Lake Pueyrredon.

PELECYPODA.

6. *Mytilus magellanicus* Chem. (Reeve, Conch. Icon., vol. x, 1858.) Living on the Patagonian coast.
San Julian.
7. *Leda oxyrrhyncha* (Philippi), (Tert. und Quart. Verst. Chiles, 1887, p. 197). Navidad beds of Chile.
Santa Cruz and Arroyo Gio.
8. *Leda errazurizi* (Philippi), (Ibid. p. 196). Navidad beds of Chile.
Santa Cruz ; Sierra Oveja (Rio Chico); Arroyo Gio ; Lake Pueyrredon.
9. *Cardita elegantoides* Ortm. (Amer. Journ. Sci., 1899, p. 428), described from the Magellanian beds of Punta Arenas.
Santa Cruz and Mt. of Observation.
10. *Cardita volckmanni* Phil. (l. c., p. 173). Navidad beds of Chile.
Lake Pueyrredon.
11. *Venus chiloënsis* Phil. (l. c., p. 121). Chile.
Punta Arenas. (Known previously from this locality, but stratigraphical position not ascertained.)

GASTROPODA.

12. *Vermetus cf. intortus* (Lmck.). (Moerch, Proc. Zool. Soc. London, 1861.) Known from Oligocene to Pliocene deposits in Europe. The identification is not beyond doubt.
Shell Gap; Lake Pueyrredon.

13. *Galerus araucanus* (Phil.), (l. c., p. 92). Navidad beds of Chile.
Shell Gap; Lake Pueyrredon.

14. *Aporhais araucana* (Phil.), (l. c., p. 35). Navidad beds of Chile.
Santa Cruz.

15. *Buccinum obesum* (Phil.), (l. c., p. 48). Navidad beds of Chile.
Santa Cruz.

16. *Cancellaria cf. medinae* (Phil.), (l. c., p. 68). Navidad beds of Chile. Identification not quite certain.
Santa Cruz and Mt. of Observation.

CRUSTACEA.

17. *Verruca laevigatæ* Sow. (Darwin, Monogr. Balan., 1854, p. 520). Living on the coast of S. America.
Upper Rio Chalia.

Remarks on Synonymy, etc.

1. *Magellania lenticularis* (Desh.). What v. Ihering mentions as *M. globosa* I take for *M. lenticularis.*

2. *Ostrea ingens* Zitt. The large Patagonian oyster is absolutely identical with the New Zealandian species described by Zittel as *O. ingens*, and differs from *O. patagonica* of d'Orbigny. The latter is not found at all in the Patagonian formation. There is only one species in the Patagonian beds.

3. *Pecten proximus* v. Ihering. This species has been called by v. Ihering in his text (Rev. Mus. S. Paulo, 1897, p. 229) by the name of *P. centralis* Sow., but is different; I accept for it the name given by v. Ihering on the plate.

4. *Pecten geminatus* Sow. Synonyms of this species are: *P. quemadensis* v. Ih. and *P. fissicostalis* v. Ih.

5. *Cucullæa alta* Sow. I cannot distinguish v. Ihering's *C. dalli* from this species.

6. *Cucullæa darwini* (Phil.). There is not the slightest doubt, that *Cucullaria tridentata* of v. Ihering is this species.

7. *Pectunculus ibari* (Phil.). Synonyms of this species are: *P. magellanicus* Phil. and *P. pulvinatus cuevensis* v. Ih.

8. *Nucula patagonica* (Phil.). *N. tricesima* v. Ih. is only a variety of this species.

9. *Cardita inæqualis* Phil. This species is the most abundant form of the genus at Santa Cruz. Large specimens of it have been sent by v. Ihering to the Princeton Museum under the name of *C. patagonica* Sow., and thus it seems apparent that *C. patagonica* of v. Ihering is identical with *C. inæqualis* of Philippi.

10. *Cardita patagonica* Sow. We possess only the small variety, called by v. Ihering in 1899 (Neues Jahrb. Miner., etc.) *C. pseudopatagonica.* I believe that this is really the true *patagonica* of Sowerby.

11. *Venus navidadis* Phil. Already v. Ihering suggests that his *V. striatolamellata* may be identical with this species. I think that is right.

12. *Dentalium sulcosum* Sow. Synonyms of this species are : *D. majus* Sow., and *D. patagonicum* Rochebrune and Mabille.

13. *Gibbula dalli* v. Iher. *G. fracta* v. Ih. seems to be nothing else than the young of this species.

14. *Infundibulum clypeolum* (Reeve). Called by v. Ihering *Trochita magellanica* Gray. But the specific name of Reeve has the priority.

15. *Natica ovoidea* Phil. I take *N. famula* Phil. for the young stage of this species.

16. *Natica secunda* Rochebr. and Mab. The specific name *secunda* of Rochebrune and Mabille was published in 1885 (Bull. Soc. Philom. Paris, ser. 7, vol. ix) and has the priority over *N. obtecta* Philippi, 1887.

17. *Natica darwini* v. Ihering. According to v. Ihering the specific name of *N. solida* Sow. has been preoccupied by Blainville. He attributes the name *darwini* to Hutton, but I cannot find it in any of the publications of Hutton.

18. *Odontostomia suturalis* v. Ihering. *O. synarthrota* Cossmann (Journal de Conchyliology, 1899) is indistinguishable from this species.

19. *Turbonilla cuevensis* v. Iher. *T. iheringi* Cossmann (ibid.) is indistinguishable from this species.

20. *Struthiolaria chilensis*·Phil. The extensive material at hand leaves it beyond doubt that *S. ameghinoi* v. Iher. is a synonym of this species.

21. *Fusus domeykoanus* Phil. *Siphonalia dilatata* var. *subrecta* of v. Ihering (1899) is this species.

22. *Trophon patagonicus* (Sow.). Abundant material of this species enables me to pronounce *T. laciniatus santacruzensis* v. Ihering as a form of this species, which is connected with it by numerous transitions.

23. *Urosalpinx cossmanni* nom. nov. For *U. cf. leucostomoides* Cossmann, non Sowerby.

24. *Voluta gracilior* v. Ihering. The specific name *gracilior* was introduced by v. Ihering in 1896 (Nachrichtsblatt Deutsch. Malakozool. Ges.) for *V. gracilis* Phil. (non Lea). It is impossible for me to distinguish from this species *V. quemadensis* v. Iher. The living *V. philippiana* Dall (1890) is different.

25. *Voluta domeykoana* Phil. *V. pilsbryi* v. Iher. (1899) does not seem to be different from this species.

26. *Pleurotoma subæqualis* Sow. What v. Ihering calls, in 1899, by the name of *Pl. discors* Sow. seems to belong to this species.

27. *Pleurotoma unifascialis* v. Ihering. This species has been regarded by v. Ihering (1897) as a variety of *P. discors*, but I think it is a good species.

28. *Geryon* (?) *peruvianus* (d'Orb.). *Cancer patagonicus* Philippi is apparently the same species as *Carcinus peruvianus* A. Milne Edwards (Ann. Sci. Nat., ser. 4, vol. xiv, 1860, p. 269). I do not think, however, that it is a *Carcinus.* It may belong to *Geryon.*

Princeton University, May, 1900.

Art. XXXVIII.—*The Cathode Stream and X-Light;*
by William Rollins.

The Cathode Stream.

There are two opinions about cathode rays.

1. The rays are some phenomenon in the ether. Lenard considered them smaller transverse waves than those of light. Michelson thought they were ether vortices.

2. They are flights of material particles. Varley, who published his results in 1871, considered the cathode stream composed of molecules of the residual gas in the vacuum tube, charged with negative electricity. He deflected the stream by a magnet: showed the force of its impact on a pivoted mica vane. Crookes, who illustrated the earlier work of Varley and Hittorf by beautiful experiments, agreed with Varley as to the nature of the stream. He said the cathode stream particles left the cathode normal to its surface, moved in straight lines, coming to a focus in the center of its curvature. My experiments showed that neither theory of the cathode stream could explain all the facts. If it was possible to remove all the ether from an X-light tube, there would be no X-light, for no cathode stream could form, because a strain in the ether is essential.

The Ether Theory.—One characteristic of the Varley or cathode stream has not been explained by this theory: the stream can be deflected by a magnet, X-light arising where the deflected stream strikes. As there are also other objections, the pure ether theory will not be further considered. The experiments supposed to support it apply as well to the material particle theory.

The Material Particle Theory.—Since Varley's experiments the opinion has slowly grown that the particles in the cathode stream are not as large as molecules. The facts of physics and chemistry appear to prove that electricity breaks molecules into ions. Schuster therefore said the particles were the same Faraday ions as appear in electrolysis. Other physicists have further reduced the size, Weichert giving it as 1/3000 of a hydrogen atom, and the speed as one-third that of light. In regard to this speed, Rowland, in a remarkable paper before the American Physical Society in 1899, said there was no way of producing this velocity in a body though it fell from infinite distance on the largest aggregation of matter in the universe.

Weichert made the first determination of the relation between the charge and the mass of a cathode stream particle.

[e gave this as $20 \times 10^\circ$ to $40 \times 10^\circ$ c.g.s. units. As this ratio about three thousand times as large for a cathode stream article as for a hydrogen ion in electrolysis, this physicist lought he must either assume a charge 3000 times as large, or mass 1/3000 of a hydrogen ion. He chose the latter and ther physicists have followed, though differing, some being illing to allow a cathode stream particle to be as large as /500 of a hydrogen ion. There is no proof of these suppo- tions. Weichert might have had the same ratio mean some ther size for the particles.

Kaufmann found that this ratio was not affected by the gas r the terminals in a vacuum tube. I shall not quote author- ies further, but group their statements to show what is called sufficient explanation of the cathode stream by those who old the material particle theory.

1. The cause of the cathode stream is a repulsion between 1e cathode and the charged particles of gas in the tube. This 2pulsion is due to the particles coming in contact with the athode receiving charges of the same nature.

2. The particles are given off from the cathode perpendicu- 1rly to its surface, and move in straight lines, coming to a 2cus at the center of curvature of the cathode.

3. The direction of the particles in high vacua is independent f the position of the anode.

4. All the cathode stream particles move with equal speed, 7hich is of the same order as that of light; they have the 1me charge; are of the same size.

5. In the cathode stream the charges are always carried by he same particles in all tubes, with any gas, with every cathode.

6. The cathode stream particles are the same in every atom 1 the universe. They are the ultimate particles of electricity.

7. In the cathode stream only one particle is detached from n atom. The remaining part has a positive charge.

8. The particles in the cathode stream are those liberated by ltra-violet light from a charged body; move with the same peed; have the same charge.

9. The particles in the cathode stream are those producing 3ecquerel light.

10. The particles in the cathode stream are the ultimate 1nits from which the elements are formed. Each has a mass f 1/3000 of a hydrogen atom.

I shall now mention some of the experiments with X-light ubes which I have described in the Electrical Review during he last three years, grouping them under the headings given :

1. The cathode stream is a purely material particle phe- 1omenon, with which the ether has no part. The sole cause of

the cathode stream is a repulsion between the particles of residual gas in the vacuum tube and the cathode. These two assertions are mentioned together, because one experiment applies to both.

An X-light tube was made with a biconcave aluminum cathode in the middle of its bulb. At each end of the tube was a platinum target. If the cathode stream was not affected by the ether, and depended entirely upon the repulsion between the cathode and the particles of residual gas, two cathode streams of equal intensity should have arisen from the cathode, producing the same amount of X-light at each target. All the X-light arising from either came from whichever target was made an anode. The experiment proved (1) that a space of strained ether was necessary, X-light arising with greatest intensity where the ether was most strained; (2) that the attraction of an anode was a cause of the cathode stream as well as a repulsion by a cathode.

2. The particles of the cathode stream are always given off from the cathode perpendicularly to its surface; move in straight lines, coming to a focus in the center of curvature of the cathode. Any one with experience can see the focus of the cathode stream when well developed in an X-light tube. By means of a pin-hole camera it may be photographed. Both methods prove that the distance varies not only with the degree of exhaustion in the tube, but also with the potential of the driving current.

I made a tube with a movable anode, whose distance from the cathode could be changed by a magnet. Under the conditions required for the economical production of X-light it was found necessary to have the target at twice the theoretical distance for it to be in the focus of the cathode stream. This experiment showed that the particles did not move in straight lines provided they left the cathode normal to its surface, nor did they come to a focus at the center of curvature. The explanation I offered was as follows: The particles of the cathode stream have the same kind of electricity, therefore they repel each other, coming to a focus beyond the place required by the accepted theory.

3. The direction of the particles in the cathode stream is independent of the position of the anode at high vacua. The experiment described in paragraph 1 appeared to disprove this, but another is also mentioned.

An X-light tube was made with connecting bulbs, one containing the usual cathode and anode. In the other, encircling the stem of the cathode already mentioned, as it passed through this second bulb, was a ring that could be made an anode. When the tube was connected in the usual way with the generator, the cathode stream arose from the concave side of the cathode, came to a focus on the normal anode, giving good X-light. This anode was then disconnected from the current. The supplementary ring terminal was made the anode. Under these conditions no cathode stream arose from the usual side of the cathode, consequently no X-light was produced at the normal anode. X·light arose in the other bulb from the spreading streams of cathode particles given off from the back or convex side of the cathode.

4. The cathode stream particles are always of the same size, move with the same speed, and carry the same charge.

If these assumptions were true we should always get the same effect when the particles struck the target in an X-light tube. We do not. Every one long familiar with the construction of such tubes knows that the quality of the light varies with the resistance of the tube. In the determinations of the speed no sufficient account was taken of the retarding effect of the gases in circulation in the tube. Suppose we try to keep this constant, even then the effect produced by the impact of these particles on the target in an X-light tube varies from other causes. To show this I made a tube in which the ordinary cathode could be covered by mercury. With the usual cathode the tube gave good X·light. With the mercury cathode there was not enough X-light to see the bones of the hand, but the whole tube was filled with a brilliant white light. I explained this by saying that the particles in the cathode stream were heavier when the cathode was mercury than when it was aluminum, and the stream was composed of some light gas. In consequence the particles did not strike the target at so high a velocity, and when stopped were therefore not heated to so high a degree; hence were not such efficient centers of radiation for the short ether movements we call X-light, most of the energy appearing as ordinary light instead. Some of the experiments on which I based this heat theory will be mentioned later. If the particles in the cathode stream were always of the same mass, all cathodes should lose equally in weight in the same time, with the same amount of current. They do not. I also found that when the gas had been considerably removed from the terminals, cathodes of heavy metals lost in weight more rapidly than those of light metals, like magnesium and aluminum.

5. In the cathode stream the charges are always carried by the same particles, in all tubes, with any gas, with every cathode.

An X-light tube was exhausted while using heat and heavy surges until it yielded no X-light. Had the particles been removed? Surely such a wonderful result was never before accomplished so easily.

My explanation was simpler. We depend upon gas amalgamated with the terminals to make an efficient cathode stream in an X-light tube. When the cathode stream no longer forms we have removed too much of this gas. Consider here the experiment with the mercury cathode already mentioned.

6. The cathode stream particles are the same in every atom. They are the ultimate units of electricity. If true, how explain the experiment mentioned in paragraph 5? Why should X-light have stopped? The terminals of the tube were constantly connected with a generator supplying electricity. They were good conductors. They were kept charged to a high degree. Is it not simpler to accept the explanation given in paragraph 5?

Consider another experiment. An X-light tube was exhausted by heat and pumping until no gas appeared in the pump. A current was then sent through it, producing X-light. Much gas appeared in the pump. The bubbles continued so long as X-light was produced. When, by continuing the current and pumping to remove the gas thus driven out of the terminals, the X-light died out, the bubbles stopped. While there were bubbles there was a normal cathode stream and X-light. Were the bubbles composed of the ultimate particles of electricity? Is electricity a gas with a familiar spectrum, or were these bubbles simply a common gas coming out of the terminals, thus forming a cathode stream, as stated in paragraph 5?

7. In the cathode stream only one corpuscle can be detached from an atom, the remainder of the atom being positive.

According to this theory, when the tube arrived at the condition mentioned in paragraphs 5 and 6, that is, when the normal cathode stream no longer formed, the cathode should have been left with a positive charge. The charge was negative.

8. The particles in the cathode stream are those liberated by ultra-violet light from a charged body, move with the same velocity, have the same charges. If true, we should get X-light from a cathode stream formed by allowing ultra-violet light to fall upon a cathode in an X-light tube connected with a capacity. I have not yet produced sufficient X-light in this easy way to use in my profession.

9. The particles in the cathode stream are those producing Becquerel light, have the same charges, move with the same velocity.

If true we could produce X-light from a vacuum tube with a radio active cathode, without an electric generator. It would nly be necessary to exhaust the tube to the usual X-light acuum. The particles in the cathode stream from the radiotive cathode would then meet with no more obstruction on 1eir way to the target to produce X-light, than would be ncountered by those of a cathode stream formed in the ordiary way.

A distinguished American physicist has expressed the pinion that the radio-active substances would be so intensified to act as substitutes for the complicated and troublesome pparatus now required for producing X-light for medical puroses. Having a high regard for his opinion I abandoned xperiments with X-light to work with these substances.

Experiments showed that the light from them was different 1 character from X-light, suffering diffusion in the tissues like rdinary light to such an extent that the bones even of the and were not visible. The experiments indicated a differnce, perhaps in velocity, between the units in the cathode .ream and those of the radio-active substances, because the haracter of the two resulting radiations toward human tissues as not the same when produced in the same vacuum. Experients of this nature require time and money. I was, therere, disappointed with the results, for my interest in the 1bjects mentioned in this paper was a desire to find the most fficient radiation, to aid in the relief of human suffering.

10. A cathode stream particle has a mass only 1/3000 of a ydrogen atom. The particles are the ultimate units of which all 1e elements are composed. That atoms are compound is probbly true, though not yet proved by experiment. The theory xplains phenomena better than that of the indivisible atom. 'hat the particles in the cathode stream are the ultimate units absurd.

I shall make a few suggestions in regard to each theory :

On the theory of indivisible atoms, how shall we explain ifferences in atomic weight, except by saying that such is the ature of atoms? This is not a satisfactory answer to an ctive mind.

On the theory of a compound atom it may be explained by 1ying that a light atom has fewer ultimate particles than a eavy one, for as each particle must have the same weight here would be more in a heavy atom than a light one, to ccount for differences in atomic weights.

That the particles in the cathode stream, however small, are ot the ultimate particles of which the universe is composed, is hown by their having a familiar spectrum. If we could make ltimate units give a spectrum, it would be a new one. In his connection I mention a statement made in my notes lready referred to :

In working with X-light tubes I appeared to get the spectrum of hydrogen in the cathode stream though other gases were in the tube. Therefore, before we reach the ultimate units of which the seventy-five or more elements are composed, we must seek more powerfully disruptive forces than those in the cathode stream. The so-called ultimate, indivisible corpuscles of this stream cannot be the final units because they have a familiar spectrum ; vibrating in too many ways at approximately the same temperature.

The experiments published by Trowbridge in this Journal for September are the most important contribution to celestial physics since Kirchhoff proved the law of exchanges and told the nature of the Fraunhofer lines. Trowbridge always obtained the spectrum of water vapor in the cathode stream when a condenser was used. Therefore the cathode stream particles are not necessarily elementary. If such minute particles are compound, it shows that when elements combine, it is not their atoms which unite to form new molecules ; the combination is far more intimate, a union of the particles of which atoms are composed. Long before we reach the heart of nature the present ultimate corpuscles will look to us more complicated than a wilderness of solar systems.

The X-Rays.

1. Röntgen considers them longitudinal vibrations in the ether.

2. Janmann believes they have also a transverse component.

3. Goldhammer stated they were short transverse vibrations differing from light only in size.

4. Stokes advanced a theory of irregular pulses in the ether, partly positive, partly negative.

5. J. J. Thomson has modified Stokes' theory. He believes when a cathode stream particle is stopped, its charge performs a single oscillation, giving rise to a pulse in the ether.

6. Michelson has suggested that X-rays may be ether vortices.

7. Several physicists believe they are flights of material particles.

This list of great names might easily be made longer. Yet what do we know of the true nature of X-light? Nothing.

I shall make a few suggestions and describe one or two experiments :

J. J. Thomson has accepted most of Weichert's views of the cathode rays, and Stokes' ideas of the X-rays. He believes when a corpuscle is stopped, there arises in the ether a pulse whose thickness is equal to the diameter of a corpuscle. These pulses are the X-rays. According to this theory the thickness of the radiant area on the target from which X-light arises

: be greater than the diameter of one of his corpuscles,
; a mass 1/1000 of an atom of hydrogen. Estimate the
ess on this basis. As a preliminary it is necessary to
the diameter of a hydrogen atom. This requires a
use of the imagination, as proved by the different esti-
Kelvin said an atom, or a molecule (he did not attempt
inguish them) was from 1/100000000 to 1/1000000 of a
leter in diameter. Meyer considered them smaller
\ sphere 1/1000000 of a millimeter in diameter, or as
as 0.2×10^{-7} centimeter. The physical chemists require
believe that a molecule is not a solid. Its atoms
ranged with ether between them. The diameter of
)m, therefore, is not equal to half the diameter of
nolecule with more than one atom, while in complex
iles its proportion is smaller. Atomic diameters are
)ly not greater than 1/3000000 of a millimeter. The
cle theory of the cathode stream gives the mass of a
cle as not more than 1/1000 of a hydrogen atom. The
ter of a corpuscle would not exceed 1/30000000 of a
ieter. This then would be the depth of the radiant area
: target from which X-light could arise.
sider the following experiment: An X-light tube had
get placed at an angle of 90° with the axis of the cathode
i. As close as possible to the area struck by the cathode
i was a narrow passage in a block of platinum, with
igs raised one millimeter above the surface of the target,
ngth of the passage parallel therewith. Under these cir-
ances, if X-light arose only from a radiant area on the
, whose depth was measured by 1/30000000 of a milli-
, direct rays could but to a small extent have illuminated
assage, for this was on the top of a cliff thirty million
as high as the depth of the radiant area at its base, in
hadow, except the opening nearest the radiant area. If
sual theory had been true, no bright image should have
ormed on the fluorescent screen, by light coming through
assage. The light was so bright Fomms' bands were
graphed with a short exposure. The experiment also
red to show that X-light was not composed of the
ed ether vortices of the cathode stream or of minute
ial particles. As these are supposed to move in straight
they would have had difficulty in going into the passage
e platinum block to brightly illuminate the screen.
, how could material particles shot from a charged anode
\ having a charge and being deflected by a magnet?
\se they did escape without a charge. When they went
3h charged aluminum they ought to have received a
; and been deflected by a magnet. They were not
ted. Was not the theory of X-light advanced in my

notes a better one? This was the theory: When the particles of the cathode stream strike the target, they are heated sufficiently to cause them to be radiant centers, from which the short ether waves we call X-light arise. As time is required for heat to decline, the particles are sufficiently hot, when they have rebounded opposite the opening of the passage, to be radiant centers. The wave fronts, therefore, go directly through the passage, to the fluorescent screen, brightly illuminating it. I have used the word heat not to distinguish this from electromagnetic phenomena, but to express a more persistent state, in the radiant particles of the cathode stream, after they have struck the target, than that represented by the single pulse of the usual theory of X-light. I mention an experiment bearing on this heat theory. An X-light tube was made with a hollow target. The area struck by the cathode stream could be cooled to a low temperature. When so cooled there was less X-light. This tube under the title of the A-W-L tube was made commercially, proving capable of converting a large amount of electricity into X-light, because with any cathode stream it was impossible to melt the target. I also designed a simpler tube for general use. The target rotated on an axis. Fresh metal could be brought to receive the force of the cathode stream when a hole had been melted.

Before closing this paper I shall consider the structure of the atom as seen by the cathode stream and X-light. After Röntgen discovered that some of Lenard's rays would show the bones of the hand, S. P. Thompson found the heaviest metals made the best targets in X-light reflecting focus tubes. The reason usually given is that light metals like aluminum allow X-light to pass through them, this part of the light being lost. Why should aluminum be transparent to X light if this is an electromagnetic phenomenon like ordinary light? Maxwell's theory requires conductors to be opaque. Aluminum is a good conductor and should be opaque to X-light. It is transparent. If X-light is an electromagnetic phenomenon, we must find some way of explaining the transparency of aluminum. Consider first the solid atom theory. As the nature of a substance depends on its atoms, conductors must have opaque atoms if these are solid. If X light is an electromagnetic phenomenon it cannot travel through such solid conducting atoms. If it passes through a metal with solid atoms it must pass in the ether between them. If one metal is more transparent than another, the more transparent must contain more ether. On the solid atom theory how can we get more ether in aluminum than in platinum? The solid atoms are of one size. There are about as many of them in aluminum as in platinum, for the ratios of the atomic weights are about the same as the ratios of the densities. There is not enough difference to account for aluminum being forty times as transparent as platinum. As we do not find enough difference in the amount of interatomic ether to account for the difference in transparency, we must give up the solid atom theory, for it fails here as it does in attempting to explain the cathode stream. Both phenomena require a compound atom. But the cathode stream theory, that the particles into which the atoms are broken are the ultimate units, cannot be true, for they have a familiar spectrum. A theory of the atom must meet these difficulties. The following is suggested. The atom is made of sub-atoms. The sub-atoms are the cathode stream particles. There are as many kinds of sub-atoms as there are elements. We should expect the cathode stream particles, therefore, to have familiar spectra. The sub-atoms are compounds. They are made of the ultimate units of which the elements are composed. These ultimate units have the same size and weight. As the atoms have the same size and different weights, light atoms must contain fewer ultimate units than heavy ones. There would be more space filled with ether in a light than in a heavy atom. When an electromagnetic phenomenon passes through an atom of a conductor, it can only travel in the contained ether, for this is the only transparent part. The atom with the most ether would be most transparent. Aluminum having more ether and fewer opaque particles, would be more transparent than platinum. Apply this theory to the following experiment, intended to show that the transparency of aluminum is not the only reason why aluminum is less efficient for a target in an X-light tube than platinum. I made a tube with the usual platinum target, covered with a thin veneer of aluminum on the surface struck by the cathode stream. Had the transparency theory been the true explanation, such a target should have acted toward X-light about as a glass mirror, silvered on the back, would have to ordinary light. As the X-light passing through the aluminum struck the platinum, only a little more could have been lost than with a target entirely of platinum. I never made an efficient X-light tube in this way. The following explanation was given. When a cathode stream particle strikes an aluminum target it is not so abruptly stopped as by a heavier metal. The heat of impact is not so high, the energy conversion extending over a longer time. The particles must be less efficient sources of X-light, if this is due to a high temperature. But why should a heavy metal target stop a flying particle more abruptly than a lighter one? I made this answer. Aluminum on the compound atom theory given is a more open structure than platinum: on this account it would stop a flying particle less abruptly.

September 28th.

SCIENTIFIC INTELLIGENCE.

I. CHEMISTRY AND PHYSICS.

1. *Ueber die physikalisch-chemishen Beziehungen zwischen Aragonit und Calcit.*—The object of this investigation by H. W. FOOTE was to show by purely physiochemical methods: 1. which of the two minerals under the existing conditions of temperature and pressure is the more stable; 2. whether the temperature of transformation is higher or lower than ordinary temperature. The method of investigation consisted in determining the relative solubility of the two minerals at different temperatures, as it is known on theoretical grounds that the least soluble must be the more stable. The best results were obtained by determining the electrical conductivity of aqueous solutions saturated with calcite or aragonite and carbon dioxide at atmospheric pressure and varying temperatures. It was found that calcite is more stable than aragonite at the temperatures experimented with, and that the solubility curves approach each other as the temperature rises. The author believes, however, from his own results and those of others, that at atmospheric pressure, calcite, below its melting point, can never become the less stable of the two minerals. From this it is concluded that the paramorphism of calcite after aragonite is theoretically possible, and the opposite paramorphism impossible. It is the author's opinion that rapidity of crystallization is one of the causes of the formation of aragonite.—*Zeitschr. physikal. Chem.*, xxxiii, 740.   H. L. W.

2. *Atomic Weight of Radio-Active Barium.*—It is well known that M. and Mdme. CURIE have discovered a substance "Radium" which is found with the barium of pitch-blende. By systematic fractional crystallization of the barium chloride obtained from the mineral, more concentrated products have been obtained. Finally by partial precipitation of solutions of the latter by means of alcohol or hydrochloric acid the concentration has been carried still further. The last product made in this way was examined spectroscopically by Demarçay and found to contain apparently only a trace of barium. This result indicates the isolation of radium chloride, but the quantity of the product was too small for use in an atomic weight determination. For the latter purpose Mdme. Curie was therefore obliged to use a less pure material, in which, from the aspect of its spectrum Demarçay thought that there was more radium than barium. Her results in two determinations gave the numbers 174·1 and 173·6 for the atomic weight of the metal in this chloride. Since the atomic weight of barium is 137·5, radium should have an atomic weight much higher than 174. The valency of radium is not discussed, but it is evident that if this is greater than that of barium the indicated atomic weight would be much larger.—*Comptes Rendus*, cxxxi, No. 6.   H. L. W.

3. *Artificial Radio-Active Barium.*—A. DEBIERNE, who, it will be remembered, is the discoverer of "Actinium," the radio-active substance that is found with the titanium of pitchblende, has apparently produced active barium chloride from the inact ve salt by induction due to contact with actinium in solution. The effect is made more intense by precipitating barium sulphate in a solution containing actinium. The latter is carried down by the barium sulphate, the sulphates are then changed to chlorides and the actinium is precipitated by ammonia in the form of hydrate. If barium and actinium are in contact in solution for a short time, the induced activity in the barium is insignificant; but this activity increases with the time of contact, at least for ten days or so, and a product was thus obtained which had several hundred times as much activity as ordinary uranium. The author decides that the radio-active barium chloride prepared in this way is different from the radiferous barium extracted from pitchblende (Curie's radium). In both cases the activity persists in all chemical transformations to which the barium may be subjected. The rays emitted seem to be similar; they ionize gases, cause phosphorescence in barium platinocyanide, act on photographic plates, a portion of the rays is deviated in the magnetic field, and the anhydrous chloride is spontaneously luminous in each case. Moreover, the "artificial" radio-active barium chloride may be concentrated by fractional crystallization from water or hydrochloric acid solution, when the crystallized part increases in activity in the same way as with radium. However, "artificial" radio-active barium differs from the radiferous kind in showing no spectrum. Demarçay examined a specimen of it which was about a thousand times as active as ordinary uranium and was unable to detect the radium lines, while with a product only ten times more active than uranium, extracted from pitchblende, the radium spectrum was very clearly visible. The author has observed a second difference, in that the activity of the artificial product diminishes with time. In three weeks the activity of some of it diminished to one-third, while the activity of radiferous barium chloride and of salts containing actinium increases at first and then remains constant. The author believes that the induced activity which he has studied is not due to traces of actinium or radium, on account of the methods of separation used, and moreover it is difficult to understand how contamination by active substances should take place only when the solutions had been mixed for a long time.—*Comptes Rendus*, cxxxi, p. 333.
H. L. W.

4. *The Relative Values of the Mitscherlich and Hydrofluoric Acid Methods for the determination of Ferrous Iron.*—It has been known for a long time that concordant results were not always obtained by the two methods mentioned above, in the analysis of rocks and minerals. In Mitscherlisch's method the substance is decomposed at a high temperature by means of a mixture of three parts of sulphuric acid and one part of water,

by weight, in a sealed tube from which air has been expelled. According to the other method the decomposition is accomplished by means of a mixture of hydrofluoric and dilute sulphuric acids in a platinum crucible kept full of steam, or protected from the air by being surrounded by carbon dioxide. In each case the ferrous iron is determined by titration with potassium permanganate solution. HILLEBRAND and SIOKES have now explained why, in many cases, the Mitscherlich method gives higher results than the other. It is due to the presence in many rocks of small quantities of pyrite or pyrrhotite, minerals which are readily oxidized by ferric sulphate under the conditions used in the method under consideration. The authors find that not only is the metal of the sulphide oxidized, but the sulphur is converted into sulphuric acid as well. The effect of sulphur is therefore very great, amounting in the maximum to an error corresponding to $13\frac{1}{2}$ times its weight of ferrous oxide, and the method should not be used for rocks and minerals which contain even a trace of free sulphur or sulphides. The authors have also shown that pure pyrite causes very little error in the hydrofluoric acid method.—*Jour. Am. Chem. Soc.*, xxii, p. 625.   H. L. W.

5. *Grundlinien der anorganischen Chemie*, von WILHELM OSTWALD, 12mo, pp. xx, 795. Leipsic, 1900 (Wilhelm Engelmann).—The appearance of each new work by Ostwald causes astonishment as to his prolificness, as well as enthusiasm on account of the excellence of his productions. Perhaps no book that he has written is as important from an educational standpoint as the elementary inorganic chemistry under consideration. The author himself considers it in a certain sense the keystone of a long and zealous activity in connection with the introduction and propagation of the new foundations of chemistry brought forward by van't Hoff and Arrhenius. Notwithstanding this aspect of the work, a careful examination shows that undue prominence has not been given to the new physical chemistry. Moreover, the treatment of the latter as well as other theoretical topics is admirably simple and philosophical. The descriptive arrangement, according to elements and their compounds, is used as the basis of the book, the general laws being introduced as opportunity and convenience permit. No attempt is made to arrange the elements according to the periodic system, in fact this system is only briefly dealt with at the very end of the book. The work contains comparatively few illustrations, and not many experiments are described in detail. A few inaccuracies are to be noticed in the descriptive matter, particularly in the paragraphs relating to metallurgy. It is to be hoped that these will be rectified in the next edition.

The author deplores the tendency that has often been shown in text-books of elementary chemistry to descend to a lower intellectual plane than is the case with works on physics and mathematics used by students at the same period of study, and he is convinced that this need not be the case now that the subject is

ess purely descriptive than formerly. It seems certain that this
ook will be extensively used, particularly by teachers and by stu-
ents who have had some preliminary training, and it is to be
oped that an English translation will soon appear.   H. L. W.

6. *The Elements of Inorganic Chemistry for Use in Schools
and Colleges;* by W. A. SHENSTONE, 12mo, pp. xii, 506. London,
900 (Edward Arnold).—The aim of this book is to provide a
graduated course which may be used at a period during which
pupils pass from childhood to adolescence. It begins with a
ourse of experimental work for quite young students and devel-
ps into a text-book for those who are older. A striking feature
f the work is the large number of experiments devised to enable
he student to verify quantitative laws and relations. These
xperiments are well chosen and very clearly explained, and they
vill doubtless furnish many useful suggestions to teachers of
hemistry. The sections on carbon include much that is usually
mitted from books on inorganic chemistry. The periodic system
s made the basis of classification after the non-metallic elements
ave been separated from the metallic. Electrolytic dissociation
nd ionization are only briefly discussed.   H. L. W.

7. *Die Kohlenoxydvergiftung,* von W. SACHS. 8vo, pp. x,
86, Braunschweig, 1900 (Friedrich Vieweg und Sohn).—This
monograph on poisoning by carbon monoxide treats the subject
ery fully in its clinical, hygienic and medico-legal aspects. An
laborate index of the literature is included in the work.

H. L. W.

8. *The Hall Effect in Flames.*—The theory of the wandering
f ions in metals gives renewed interest to the phenomenon dis-
overed by Professor Hall and known by his name. E. MARX
iscusses analytically the point whether the different velocities of
he positive and negative ions in a magnetic field can account for
he Hall effect, and reaches a value for the rotation coefficient.
Ie concludes that the only possible way of reaching quantitative
esults in regard to the Hall effect is in the direction of flame
onduction in a magnetic field. Arrhenius believes that the
onveyance of electricity in flames is accomplished through
lectrolytically dissociated ions. The velocity of the ions with
a fall of potential of unity is 10^6 times greater than in the case
of electrolytes and very different in the cases of the positive
and negative ions. So that we are led to expect a Hall effect a
million times greater than in the case of electrolytes. A full
account of the author's apparatus and his experiments is given
and he reaches the conclusion that the Hall effect in flame gases
has the dimensions which would be expected on the hypothesis of
of wandering electrical ions in a magnetic field. The author's
experiments, however, do not decide the question whether the
Hall effect in metals can be explained by pondero-motive effects
on wandering ions.—*Ann. der Physik,* No.8, pp. 798–834, 1900.

J. T.

9. *Wireless Telegraphy.*—It is well known that an earth connection is used in sending messages by wireless telegraphy, both in the sending and receiving apparatus. J. VALLOT and J. and L. LECARME maintain that this earth connection can be dispensed with for long distance transmission, and certain questions in regard to the part the capacity of the earth plays are thus opened. Their receiving apparatus was placed in a balloon, which rose to a height of 800 meters and was about 6 kilometers from the sending station.—*Comptes Rendus*, cxxx, pp. 1305–1307, May 14, 1900. J. T.

10. *The Spectrum of Radium.*—According to the investigations of Demarçay (Compt. Rend., cxxix, p. 717, 1899) radium is a new element and he gives a table of its wave lengths. C. RUNGE has examined the spectrum and finds most of the lines which are given by Demarçay in chloride of barium. The spectrum was investigated in air and the preparation was also heated in vacuum tubes and the spectrum showed no new lines.—*Ann. der Physik*, No. 8, pp. 742–745, 1900. J. T.

11. *Magnetic effect of moving electrical charges.*—M. CREMIEU has conducted an important experiment on this subject. He has endeavored to measure the inductive effect when the convective current studied by Rowland is started or stopped. He could not observe any such inductive effect, and therefore concludes that there is no magnetic effect such as Rowland observed and that the effect observed by him was not due to a moving electric charge. In a criticism of Lamor's treatise on the relations between ether and matter (Cambridge University Press, 1900) Professor Fitzgerald remarks, that the questions raised by this experiment is one of the most fundamental ones in the connection between ether and matter. If M. Cremieu's experiments are substantiated, they will overthrow existing theories of electro-magnetism.—*Nature*, July 19, 1900. J. T.

12. *What electric pressure is dangerous?*—In order to obtain light on this much mooted question Professor H. F. WEBER of the Zurich Polytechnic has conducted a number of experiments and reaches the following conclusions:

A simultaneous touching of both the poles of an alternating current circuit is dangerous as soon as the pressure exceeds 100 volts; and since it is impossible to free oneself the case must be regarded as fatal whenever immediate help is not at hand. With steady currents he found that all pressures between 100 and 1000 volts must be regarded as equally dangerous, and consequently there is no reason for not using higher pressures between 500 and 1000 volts since they lead to more economical working of traction lines. There is little danger of the public coming in contact with such lines.—*Nature*, Aug. 23, 1900. J. T.

II. Geology and Mineralogy.

1. *Recent Publications of the U. S. Geological Survey.*—The following volumes have been recently issued by the U. S. Geological Survey, Parts II, III, IV, V, VII of the Twentieth Annual Report. (See also this Journal, ix, p. 447.)

Part II. *General Geology and Paleontology.* Pp. 1–953, plates ɪ to cxcɪɪɪ. This volume includes six papers, as follows: G. F. Becker, Brief memorandum on the Geology of the Philippine Islands, pp. 1–8; T. N. Dale, A study of Bird Mountain, Vermont, pp. 9–24, pl. ɪ, ɪɪ; G. H. Girty, The fauna of the Ouray formation, pp. 25–82, pl. ɪɪɪ–vɪɪ; I. C. Russell, preliminary paper on the geology of the Cascade Mountains in northern Washington, pp. 83–210, pl. vɪɪɪ–xx; L. F. Ward, Status of the Mesozoic floras of the United States, pp. 211–748, pl. xxɪ–clxxɪx; David White, Stratigraphic succession of the fossil floras of the Pottsville formation in the southern anthracite coal field, Pennsylvania, pp. 749–918, pl. clxxx–cxcɪɪɪ.

Of these papers that by Prof. Ward was noticed at length in the last number of this Journal (p. 320). The paper by Prof. Russell is largely devoted to the glacial and post-glacial geology of the northern and eastern parts of the State of Washington. It is shown that during these times the valleys were filled with gravel and sand to a depth in general of several hundred feet. Subsequently the streams excavated channels through these deposits and in many instances removed them entirely. The portions of the gravel deposits which remain form terraces along the streams, some of which are a mile or more broad and have a down-stream gradient. The terraces occur even in mountain gorges and along torrential streams and in such cases have a conspicuous down-stream slope—not due to a tilting of the land—which is less steep than the gradient of the present streams. No evidence is found of a modern depression of the land of such a nature as to admit of the flooding of the valleys by the waters of the ocean. The prevalence of glaciers at this time shows a much colder climate and a greater amount of precipitation than at present.

Part III. *Precious-metal Mining Districts.* Pp. 1–595, plates ɪ–lxxvɪɪ. This volume will be noticed later. It contains three papers: J. S. Diller and F. H. Knowlton, On the Bohemia mining region of western Oregon; W. Lindgren, On the gold and silver veins of Silver City, De Lamar and other mining districts in Idaho; W. H. Weed and L. V. Pirsson, On the Geology of the Little Belt Mountains, Montana.

Part IV. *Hydrography.* Pp. 1–660, plates ɪ–lxɪɪɪ. This volume contains the report of progress on stream measurements for the calendar year 1898 by F. H. Newall, pp. 1–562; also an account of the hydrography of Nicaragua by A. P. Davis, pp. 563–638. The very rapid development of this new department of the Survey and its promise of great economic usefulness for the future are alike noteworthy. In the present volume we find not only the results of drainage measurements for prominent rivers over the whole extent of the country, but also the survey of reservoir sites and the discussion of conditions governing underground waters as reached for example by artesian wells, both the latter subjects of peculiar interest in the West. The discharge data are given with much fullness for a large number of rivers from Maine to California, and the results are presented in graphical form for the successive months of the year. In addition to the statistical matter, plans are presented whereby the limited water supply of certain arid districts and Indian Reservations may be more economically distributed. The closing portion of the volume gives the results of the investigations on the hydrography of Nicaragua made by Mr. Davis in 1898 ; the work was carried on in connection with the Nicaragua Canal Commission.

Part V. *Forest Reserves ;* by Henry Gannett. Pp. 1–478, plates I–CLIX, in part folded maps. The forest reserves of the United States numbered 37 on July 1, 1899, and are distributed over eleven States. The general condition of these reserves is given in this report, and there are also special papers on the following districts: Pikes Peak, Plum Creek and South Platte Reserves, by John G. Jack *;* White River Plateau and Battlement Mesa Reserves, by George B. Sudworth; Flathead Reserve, by H. B. Ayres ; Bitterroot, San Gabriel, San Bernardino and San Jacinto Reserves, by John B. Leiberg.

Part VII. *Explorations in Alaska* 1898. Pp. 494, with maps 1–25 and plates I–XXXVIII. In this volume are grouped the reports of G. H. Eldridge, J. E. Spurr, W. C. Mendenhall, F. C. Schrader and A. H. Brooks, on reconnaissance surveys in Western and Central Alaska. Besides the sketches of theoretical and economic geology much information is given regarding products, climate and feasible routes. Maps and routes of previous explorers are discussed and the general geographic knowledge of the region is brought down to date. Taken in connection with the "Map of Alaska" (U. S. G. S. 1898) and Schrader and Brooks, "Preliminary Report on the Cape Nome Gold Region" (U. S. G. S. 1900), this volume stands as the reference book for recent reliable information regarding Alaskan matters.

2. *Department of Geology and Natural Resources of Indiana, 24th Annual Report, 1899 ;* by W. S. BLATCHLEY, State Geologist. Indianapolis, 1900. Pp. 1–1078, 89 plates.—This report contains more than the ordinary Geological Annual Report. Six of the papers are concerning the Geology of the State. The first, by the Director, refers to the Natural Resources, and three others contain the reports of the State Inspector of Mines, the State Supervisor of Natural Gas, and the State Supervisor of Oil Inspection, respectively. Mr. Foerste contributes a discussion on the Middle Silurian rocks of the Cincinnati anticlinal region, the particular interest of which lies in his interpretation of the

synonymy of the several beds. The paper by J. A. Price on the Waldron Shale and its horizon gives details which will be of value to the general student of Structural Geology. In this paper a large number of detailed sections are given, running from the Silurian through the Devonian. The remainder of the book is devoted to four exhaustive treatises on the Natural History products of the State. Prof. E. B. Williamson furnishes a monograph on the Dragonflies of Indiana. Dr. R. E. Call gives an exhaustive and fully illustrated catalogue of the Mollusca of Indiana in which diagnostic descriptions of species are inserted. A catalogue of the Flowering Plants and the Ferns and their allies indigenous to Indiana is furnished by Prof. Stanley Coulter. And Mr. Blatchley contributes a short paper entitled "Notes on the Batrachians and Reptiles of Vigo County, Indiana." The volume is published in the usual style of the Indiana Reports; the paper being thin and the printing rather poor for reports of so great scientific value.   w.

3. *Geological Report on Monroe County, Michigan,* by W. H. SHERZER. *Geological Survey of Michigan,* ALFRED C. LANE, *State Geologist.* Vol. vii, Part I. Pp. 1–240, pl. i–xvii.—The nomenclature adopted in the stratigraphical part of this report is as follows:—*St. Clare Shale* is essentially the equivalent of the Genesee Shale of New York: the *Traverse Group* is the name for the equivalent of the Hamilton of New York; *Dundee Limestone* corresponds closely with the beds from the Oriskany Sandstone to the Onondaga limestone inclusive; the *Monroe beds* are correlated with the Salina, Rondout, and Manlius beds of New York State.   w.

4. *The Elements of the Geology of Tennessee,* prepared for the use of Schools of Tennessee, by J. M. SAFFORD and J. B. KILLEBREW, pp. 1–264, figs. 1–46.—Prof. Safford and Dr. Killebrew, the Commissioner of Agriculture, Statistics and Mines, have compiled a useful book for the use of Primary and Common Schools of the State. A glance through the book reveals nothing new; it appears to follow closely Dana's Manual of. Geology, from which many of .the illustrations are derived, and Safford's Geology of Tennessee, but the matter is brought down to the understanding of students of the lower' grades, for whom it will prove a useful guide.   w.

5. *Geological Survey of Canada,* G. M. DAWSON, Director.— The following papers of the Eleventh Annual Report have been received:

Report of the Section of Chemistry and Mineralogy, Part R, Annual Report, vol. xi, by G. C. Hoffmann. No. 695, pp. 1–55.

Section of Mineral Statistics and Mines, Annual Report for 1898, Part S, Annual Report, vol. xi, by E. D. Ingall. No. 689, pp. 1–103.

Catalogue of Canadian Birds. Part I. Water Birds, Gallinaceous Birds, and Pigeons, by John Macoun. No. 692, pp. 1–218.

Preliminary Report of the Klondike Gold Fields, Yukon District, Canada, by R. G. McConnell. No. 687, pp. 1–44.   w.

6. *Correlation between Tertiary Mammal Horizons of Europe and America*, by HENRY FAIRFIELD OSBORN. (Annals N. Y. Acad. Sci., vol. xiii, No. 1, pp. 1 to 72, July 21, 1900.)—In the two papers here published together Prof. Osborn sets forth his mature views upon the Zoögeography of Tertiary Mammals. In the introduction he makes an urgent plea for the establishment of uniform divisions of the Tertiary and for the international usage of common terms both as to life stages and life forms, and expresses confidence that approximate synchronisms can be established between the European and American Tertiary formations. For several years he has been attempting to correct these correlations and three trial sheets have been prepared and submitted to experts in Europe and this country for their criticism; as a result he publishes the following as a preliminary expression of the approximate correlations of formations:

I. STRATIGRAPHICAL CORRELATION: PRELIMINARY.

Lyell's System.			Approximate American Parallels.
Pleistocene	Upper	Post Glacial	
	Middle	Glacial and Interglacial	
	Lower	Preglacial	? Equus Beds
Pliocene	Upper	Sicilien	? Blanco
	Middle	Astien	
		Plaisancien	
	Lower	Messinien	Upper Loup Fork
Miocene	Upper	Tortonien	Loup Fork
	Middle	Helvetien	Lower Loup Fork
	Lower	Langhien	and Upper John Day
Oligocene	Upper	Aquitanien	Lower John Day (Diceratherium Layer)
	Lower	Stampien	White River
		Infra Tongrien	
Eocene	Upper	Ligurien	Bridger and Uinta
	Middle	Bartonien	Lower Bridger
		Lutetien	Wind River
	Lower	Suessonien	Wasatch
	Basal	Thanetien	Torrejon
		Montien	Puerco

All the levels above the Stampien are regarded by the author as imperfectly established.

In discussing the available evidences of parallelism, the following are cited as some of the more important tests : I. Common Genera; II. Similar Stages of Evolution; and he refers particularly, under this head, to the pattern of the molar teeth, the transformation of the pre-molar teeth, the complication of molar eth, and reduction of digits; III. Simultaneous introduction of ew forms, or immigration from outside regions; IV. The Preominance of Certain Types; V. The Convergence and Diverence of the Palearctic and Nearctic Faunae. In discussing the eographical distribution of Tertiary forms the original assumpoon is made that the animals of various families and orders have ther originated in, or migrated into their present habitat in past me, so that the geological record as to their order of appearance of first importance. And in investigating the distribution of le present time he finds it necessary to have an absolutely relible correlation time scale for the fossil faunac, and the problem e then sets for investigation is to connect living distribution ith distribution in past time and to propose a system which will e in harmony with both sets of facts. The author adopts Salr's classification of geographical areas, and the terms Arctogaea, otogaea, Neogaea, and makes an interesting distinction between ich large tracts of land as have, from their long separation, ecome the centers of adaptive radiation for orders of mammals, om those smaller regions which have been isolated from each her for shorter periods by climatic or physical barriers. The rst group of divisions of the earth's surface are called *realms*, id these are the main centers of adaptive radiation for the ders of mammals. The smaller divisions he calls *regions* and ley have served as the centers for the radiation of families of ammals. The author closes his paper with a discussion of the urces of migration of mammals to and from the three great alms of the earth as above described.   w.

7. *On the Flow of Marble under Pressure.*—A series of exper-aents has been carried on by F. D. ADAMS and J. T. NICOLSON ith pure Carrara marble, having as their object the investiga-on of the conditions under which the flow of rocks may take lace when subjected to differential pressure. Their experiments e to be extended later to other limestones, as also to granite and cks of different types.

The method of experiment adopted was as follows: heavy rought iron tubes were constructed of thin strips of Low Moor on about one-fourth of an inch in thickness; these were so made at the fibres of the iron ran around the tube. Polished columns the marble, about an inch or somewhat less in diameter, and 1 inch and a half in length, were inserted in the iron tubes, nder such conditions as to give perfect contact, and to leave an. 1 inch and a quarter of the tube free at each end. Pressures up 13,000 atmospheres were then exerted upon them by means of 1 accurately fitting steel plug. The marble was submitted to ressure under four conditions: first, dry, at ordinary tempera-ires, and again at 300° C. and at 400° C.; also in the presence f moisture at 300° C. The rate at which the pressure was pplied was also varied, the time of experiment extending in the fferent cases between the limits of ten minutes to sixty-four ys.

When the experiment was completed, the tube was slit longitudinally along two lines opposite to each other and the marble within exposed. It was uniformly found to be so firm and compact that a steel wedge was needed to split it through the middle, the half column adhering firmly to the sides of the tube. The deformed marble, while firm and compact, differs in appearance from the original rock in possessing a dead white color, somewhat like chalk, the glistening cleavage surfaces of the calcite being no longer visible. The difference was well brought out in certain cases where a certain portion of the original marble remained unaltered by the pressure. This had the form of two blunt cones of obtuse angle whose bases were the original ends of the columns resting against the faces of the steel plugs, while the apices extended toward one another into the mass of the deformed marble.

Special experiments on certain samples showed that, making all due allowance for the difference in shape of the specimens tested, the marble after deformation, while in some cases still possessing considerable strength, was much weaker than the original rock. It also appeared that when the deformation was carried on slowly the resulting rock was stronger than when the deformation was rapid.

By the study under the microscope of thin sections of the deformed marble, passing vertically through the unaltered cone and the deformed portion of the rock, the nature of the change that had taken place, could be seen clearly. The deformed portion of the rock could be at once distinguished by its turbid appearance, differing in a marked manner from the clear transparent mosaic of the unaltered cone. This turbid appearance was most marked along a series of reticulating lines running through the sections, which when highly magnified were seen to consist of lines or bands of minute calcite granules. These were lines along which shearing had taken place. The calcite individuals along these lines had broken down, and the fragments so produced had moved over and past one another, and remained as a compact mass after the movement ceased. In this granulated material were enclosed great numbers of irregular fragments and shreds of calcite crystals, bent and twisted, which had been carried along in the moving mass of granulated calcite as the shearing progressed. This structure was therefore cataclastic, identical with that seen in the feldspars of many gneisses.

Between these lines of granulated material the marble showed movements of another sort. Most of the calcite individuals in these positions could be seen to have been squeezed against one another and in many cases a distinct flattening of the grains had resulted, with marked strain shadows, indicating that they had been bent or twisted. They showed, moreover, a finely fibrous structure in most cases, which, when highly magnified, was seen to be due to an extremely minute polysynthetic twinning. The chalky aspect of the deformed rock was in fact due chiefly to the

struction by this repeated twinning of the continuity of the
cavage surfaces of the calcite individuals, thus making the
flecting surfaces smaller. By this twinning, the calcite individ-
ls were enabled under the pressure to alter their shape some-
hat, while the flattening of the grains was evidently due to
ovements along the gliding planes of the crystals. In these
rts, therefore, the rock presented a continuous mosaic of some-
hat flattened grains. From a study of the thin sections it
emed probable that very rapid deformation tends to increase
e relative abundance of the granulated material, and in this
ay to make the rock weaker than when the deformation is slow.
When the deformation was carried out at 300° C., or better at
0° C., the cataclastic structure described above, was not
veloped, and the whole movement was found to be due to
anges in the shape of the component calcite crystals by twin-
ng and gliding.

In the experiments where moisture was present, the deforma-
on of the marble taking place at 300°, no change in the charac-
r of the deformation was detected. In other words the presence
water appeared to exert no influence, the marble was deformed
when dry at 300° by twinning and gliding alone without cata-
astic action.

In order to ascertain whether the structures exhibited by the
formed marble were those possessed by the limestones and mar-
es of contorted districts of the earth's crust, a series of forty-
ro specimens of limestones and marbles from such districts in
rious parts of the world were selected and carefully studied.
f these, sixteen were found to exhibit the structures seen in the
tificially-deformed marble. In these cases the movements had
en identical with those developed in the Carrara marble. In
other cases the structures bore certain analogies to those in
e deformed rock but were of doubtful origin, while in the
maining twenty the structure was different.

It appears, therefore, that by submitting limestone or marble to
fferential pressures exceeding the elastic limit of the rock and
der the conditions described in this paper, permanent deforma-
on can be produced. This deformation, when carried out at
dinary temperatures, is due in part to a cataclastic structure
d in part to twinning and gliding movements in the individual
ystals comprising the rock. Both of these structures are seen
contorted limestones and marbles in nature.

At elevated temperatures only the changes of the second kind,
e. by twinning and gliding, take place. This latter movement
identical with that produced in metals by squeezing or ham-
ering, a movement which in metals, as a general rule, as in
arble, is facilitated by increase of temperature. There is therere a flow of marble just as there is a flow of metals, under
itable conditions of pressure. The movement is also identical
ith that seen in glacial ice, although in the latter case the move-
ent may not be entirely of this character.

The further results promised by the authors, of experiments on rocks of different kinds, will be looked for with interest.—*Proc. Roy. Soc.*, read June, 21, 1900.

8. *Report of the Section of Chemistry and Mineralogy;* by G. CHR. HOFFMANN, Geol. Survey of Canada. Part R, vol. xi, 1900. Among the various points of general interest in this report we note the following:

Celestite occurs in radiating columnar masses in the township of Bagot, Renfrew Co.; an analysis by Johnston showed the presence of 14 p. c. of barium sulphate. *Hübnerite* occurs in narrow seams and in irregular masses with coarsely laminated structure at Emerald, nine miles from Margaree Forks, Inverness Co., Nova Scotia. An analysis by Johnston showed it to be nearly pure manganese tungstate (0·47 FeO); sp. gravity = 6·975. *Hydromagnesite* occurs in considerable abundance on the Cariboo road, 93 miles north of Ashcroft, Lillooet district, British Columbia. Other extensive deposits have been found on the east side of Atlin Lake, Cassiar district. *Natron* has been identified in Goodenough Lake, 28 miles north of Clinton, Lillooet district, B. C. When examined, at the close of the dry season, it formed a deposit over nearly the entire bottom of the lake (20 acres) and with a thickness of about 8 inches. It was estimated that the deposit contained 20,000 tons. Natural soda also occurs in Last Chance Lake, 8 miles distant. *Polycrase* has been found in the township of Calvin, Nipissing district, Ontario. It forms crystalline masses, one of them weighing more than 700 grams; color pitch-black; luster resinous; sp. gravity 4·842. It is associated with xenotime, magnetite, etc. Only a qualitative analysis has been made thus far.

9. *Florencite, a new mineral.*—Florencite is a new hydrated phosphate of aluminum and the cerium earths recently described by HUSSAK and PRIOR. It was first found very sparingly in the cinnabar-bearing sands of Tripuhy, near Ouro Preto, Minas Geraes, Brazil, where it is associated with monazite, xenotime and the titano-antimonates, lewisite and derbylite. It also occurs more abundantly in diamond-bearing sand from Matta dos Creoulos, near Diamantina and with the well-known yellow topaz at Morro do Caixambú. It occurs in rhombohedral crystals with f $(02\bar{2}1)$ as the prominent form, also c (0001) and r $(10\bar{1}1)$ and m $(10\bar{1}0)$ both rare. The angle $ff' = 108°$ 26' is not far from that of hamlinite (108° 2'); the vertical axes are 1·1901 and 1·1353 respectively. Cleavage basal, fairly perfect; fracture splintery to subconchoidal; hardness about 5; specific gravity 3·586; luster greasy to resinous; color clear pale yellow.

The mean of two analyses by Prior gave:

P_2O_5	Al_2O_3	Ce_2O_3 (etc.)	Fe_2O_3	CaO	H_2O	SiO_2
25·61	32·28	28·00	0·76	1·31	10·87	0·48 = 99·31

For this the formula is deduced $3Al_2O_3.Ce_2O_3.2P_2O_5.6H_2O$, not far from that of hamlinite.—*Min. Mag.*, xii, 244, 1900.

10. *Elements of Mineralogy, Crystallography and Blowpipe Analysis from a practical Standpoint;* by ALFRED J. MOSES and CHARLES LATHROP PARSONS. New enlarged edition, pp. vii, 44. New York, 1900 (D. Van Nostrand Company).—This new edition of the Mineralogy by Professors Moses and Parsons appears in an enlarged and much improved form. The special features of the original work which made it particularly suitable for the use of the technical student, or the worker in the field,— as the classification of species and their clear and concise descriptions—have been retained, while the crystallographic portion has been rewritten and developed so as to conform to the modern method of treatment and classification. Other parts have also been revised and improved, so that the work as a whole may well commend itself to a wider range of students and readers than heretofore.

III. MISCELLANEOUS SCIENTIFIC INTELLIGENCE.

1. *Latitude-variation, Earth-magnetism and Solar activity.*— The following conclusions are reached by J. HALM after a discussion of the above subject.

(1) The changes in the motion of the pole of rotation round the pole of figure are in an intimate connection with the variations of the earth-magnetic forces.

(2) Inasmuch as the latter phenomena are in a close relation with the state of solar activity, the motion of the pole is also indirectly dependent on the dynamical changes taking place at the sun's surface.

(3) The distance between the instantaneous and mean poles decreases with increasing intensity of earth-magnetic disturbance.

(4) The length of the period of latitude-variation increases with increasing intensity of earth-magnetic disturbance.

(5) In strict analogy with the phenomena of auroræ and of magnetic disturbance, the influence of the eleven-year period of sunspots, as well as of the "great" period, is clearly exhibited in the phenomenon of latitude-variation; and the same deviations from the solar curve as are manifested by the auroræ are also evident in the motion of the pole.

(6) The half-yearly period of the earth-magnetic phenomena influences the motion of the pole of rotation in such a way that its path, instead of being circular, assumes the form of an ellipse, having the mean pole at its center.

(7) The half-yearly period also explains the conspicuous fact of a rotation of the axes of the ellipse in a direction opposite to that of the motion of the pole.—*Nature*, No. 1610.

2. *The location of the South Magnetic Pole.*—The October number of the Geographical Journal contains an interesting account by C. E. BORCHGREVINK on the "Southern Cross" Expedition to the Antarctic in 1898–1900. The party wintered at Cape Adare on South Victoria Land, in latitude 71° S. and car-

ried on extensive exploration along the coast line, the farthest southern point reached being about latitude 78° 45′ S. As the result of magnetic observations made by the Expedition, the present position of the southern magnetic pole is located at 73° 20′ S., and 146° E. At Cape Adare the dip was found to be −86° 34′ and the declination 56° 2′ E. Observations of the dip were taken at seven other geographical positions, the maximum dip, −88° 1½′, being taken at the foot of Mt. Melbourne; on the west side of Franklin Island a dip of −86° 52′ was found. As Sir James Ross, in 1841, observed a dip of −88° 24′ some twelve miles north of Franklin Island, there seems to have been a decrease in fifty-nine years of 1° 32′; it is concluded that there is very little doubt that the magnetic pole is very much farther north and west than in 1841.

3. *Ostwald's Klassiker der Exakten Wissenschaften* (Wilhelm Engelmann : Leipzig, 1900).—The following are recent additions to this valuable series:

Nr. 110. Die Gesetze des chemischen Gleichgewichtes für den verdünnten, gasförmigen oder gelösten Zustand, von J. H. Van't Hoff. Pp. 105.

Nr. 111. Abhandlung über eine besondere Klasse Algebraisch auflösbarer Gleichungen, von N. H. Abel (1829). Pp. 50.

Nr. 112. Abhandlung über bestimmte Integrale zwischen imaginären Grenzen, von Augustin-Louis Cauchy (1825). Pp. 80.

Nr. 113. Zwei Abhandlungen zur Theorie der partiellen Differentialgleichungen erster Ordnung, von Lagrange (1772) und Cauchy (1819). Pp. 54.

4. *Scientia.*—The inauguration of a series of small volumes on scientific subjects by G. Carré and C. Naud as publishers (Paris) was announced a year since (this Journal, July, 1899, p. 86). The seventh and eighth volumes of the physical-mathematical series have now been issued. The former is by H. LAURENT and is entitled *"L'Élimination.* The author remarks that the only monograph on the theory of elimination is that of Fàa de Bruno published in 1859; hence the discussion given by him has peculiar value. The other volume is on the subject *" Tonométrie "* by F. M. RAOULT.

5. *Introduction to Science;* by ALEX. HILL, M.D. 140 pp. New York and London (The Macmillan Co.).—This little volume contains a series of interesting chapters on some of the scientific problems now before the public, as on the age of the earth, the origin of species, the constitution of matter, etc.

The Recent Solar Eclipse.—The report of the Expeditions organized by the British Astronomical Association to observe the total Solar Eclipse of May 28th, 1900, will be contained in a volume shortly to be issued from the Office of "Knowledge." The work will be edited by Mr. E. Walter Maunder, and will contain many fine photographs of the various stages of the Eclipse.

BLACK HILLS CYCADS.

Black Hills Cycads.

722

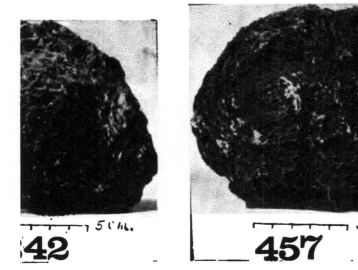

42   457

BLACK HILLS CYCADS.

ᴛT. XXXIX.—*Torsional Magnetostriction in Strong Transverse Fields and Allied Phenomena ;* by C. BARUS.

ᴌ. IN 1885 Kirchhoff* published his remarkable research on effect of stress on magnetization. This is the most comhensive treatment which the problem has received and ees in its applications with the deductions of Lorberg† and Korteweg‡ for the analogous cases of electrostriction. It is l known that Maxwell's§ original stresses refer to a medium hout structure. A subsequent generalization of von Helmtz‖ makes the stresses in the electric or the magnetic field ∙endent on changes of density in the medium while Kirch-f's stresses allow for a dependence both on bulk and on ᴉgation. These stresses thus contain three constants, the ᴌ of which is either the susceptibility or the permeability of medium (with the corresponding constants for dielectrics) ᴐrding as in the original statement the magnetization or the ɡnetic induction is to be expressed in terms of the field. ᴉ second and third constants correspond to the bulk expan-ᴉ and elongation in question. If the third constant is ∙ulled, Kirchhoff's stresses coincide with those of von Helm-tz ; if the two new constants are annulled Maxwell's stresses y be reduced. Other great authors¶ (J. J. Thomson, rtz) have contributed to the subject in similar ways.

Kirchhoff: Wied. Ann., xxiv, p. 52, 1885.
Lorberg: Wied. Ann., xxi, p. 300, 1884.
Korteweg: Wied. Ann., ix, p. 48, 1880.
Maxwell: Electricity; § 105, p. 146, vol i ; § 644. p. 255, et seq., vol. ii.
Helmholtz: Wied. Ann., xiii, p. 400, 1881.
The present meager account will suffice, since an excellent digest of the sub-has been given in a paper by Nagaoka and E. Taylor Jones in Phil. Mag., (5), p. 454, 1896.

Kirchhoff's theory has recently been examined in an elaborate experimental research due to Nagaoka and Honda,* with results showing an imperfect agreement with facts both for iron and for nickel. They not only include known reciprocal relations of stress and magnetization, but investigate new data particularly referring to changes of bulk. They find that Kirchhoff's new constants are complicated functions of strain, or that the expressions are applicable only to infinitely small strains.

As the phenomena, therefore, remain almost equally troublesome in terms of the constants of the theory, very little assistance has been gained from it; and in the entire absence of a better theory it is still permissible to look at magnetostriction from tentative points of view.

2. I have long been of the opinion that a statistical treatment of the subject, such as was suggested for viscosity by Maxwell, might have much in its favor as compared with the purely elastic treatment independent of the mechanism, sketched above. The reciprocal relations of stress and magnetization are, as it were, incidents in a much more varied phenomenon. Indeed the original explanation of the Wiedemann effect, as given by Wiedemann himself in terms of Weber's theory of revoluble molecular magnets (a theory which in Ewing's hands has been shown to include hysteresis) seems to have been too consistently ignored. The conception of a magnetic configuration which breaks down under stress but is restored when the stress has sufficiently vanished or been reversed seems to be a reasonable one, supposing the breakdown involves no chemical change, as it does for instance in tempering. In Maxwell's hypothesis *any deformation* due to molecular instability is a viscous deformation. Now when the breakdown is gradual in character, as it must be when depending on temporary local intensities in the distribution of heat motion, the deformation will be gradual as actually observed in the ordinary phenomena of viscosity. If however breakdown is instantaneous (kaleidoscopic as it were), due for instance to the molecular shake-up accompanying magnetization or its withdrawal, then viscosity is *instantaneously a minimum*, and the deformation correspondingly sudden or "static." There seems to be no theoretical gap here whatever, and I hope to show (more directly in the subsequent paper than in the present, which has an introductory character) that

* Nagaoka and Honda: Journ. of the College of Science, Tokyo, Japan, vol. ix, part iii, pp. 353–391, 1898.

with the interpretation I have ventured to give of Maxwell's conception of viscosity, the phenomena of magnetostriction become more closely akin, following the same course in all metals.

In discussing experiments like the present it is always desirable to devise a model of the simplest kind, which shall suggestively reproduce the phenomena in question as fully as possible. The model need not be looked upon as an actual occurrence. I will therefore insert a simple torsional mechanism around which the present experiments may be conveniently grouped. If we regard the rod to be twisted as a bundle of longitudinal fibers elastically bound together, then the effect of twisting is in the main an elongation of the fibers increasing from the axis to the circumference. Inasmuch as the external fibers are now helical in form, the stretch in question has a horizontal component acting along the circumference of any section, tending to restore the fiber to its original straight form. In a solid rod fibers inclined at 45° to the axis are subject to traction (with the accompanying compression at right angles thereto) only. Other fibers are both stressed and rotated. For the present purposes, however, a description in terms of the principle stresses is unnecessary.

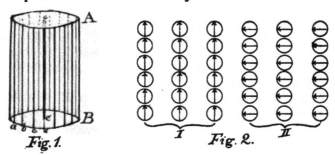

Fig. 1.   *Fig. 2.*

The case may be illustrated by the annexed model, in which the two parallel discs A and B are kept apart by the rod C, pivoted in a depression in each. Rubber fibers, $a, b, c, \ldots$ symmetrically grouped, hold the two discs together and secure the rod C in place. If torque be applied to B relative to A the rotation is forcibly resisted.

The bands $a, b, c, \ldots$ correspond to lines of stress which bind the non-magnetic molecules, longitudinally. In the simplest case one would assume that in a direction along the lines of stress molecules are closer together than are the successive lines themselves in a direction at right angles thereto; in other words, the distance apart of lines of stress is large as compared with the distance apart of the molecules which conduct stress.

If this condition were dropped the model, though still available, would lack simplicity.

The effect of longitudinal magnetization as inferred from the axial setting of the revoluble magnets, is increased tension along the fibers and therefore a corresponding increase of the horizontal component in question. Rigidity is apparently increased under all conditions, as is inferred from the distributions given in figure 2, type I. My own measurements[*] in addition to the more recent experiments of Mr. H. Day,[†] in which strong fields were applied, show an invariable increment of rigidity, increasing towards a limit as the field strength increases, and greater as the diameter of the rod is smaller. All this would be suggested by the model, remembering that slip takes place on twisting when the obliquity[‡] of the external fibers has reached a fixed quantity.

If the increment of rigidity due to longitudinal magnetization were to be associated with the Joule effect of magnetic elongation, one would expect a change of sign corresponding to the march of the latter in an increasing field. Rigidity however is a regularly increasing quantity and shows no change of sign of the kind in question; and this is just what the model indicates. Again for the case of nickel in which the Joule effect is at the outset opposite in sign to that of iron, the effect of longitudinal magnetization is nevertheless increased rigidity. An example of these results will be found below, §6. Figure 2, type I, is an extreme case. One may note that the poles of the magnets are closer together than the molecular centers, either longitudinally or laterally.

The effect of transverse magnetization, if the hypothesis of lines of stress relatively far apart be retained, is given as one extreme case by type II of figure 2. The molecular magnets tend to lie across the lines of stress and rigidity is therefore diminished. From the geometry of the figure, however, it appears that the poles of the molecular magnets now lie much farther apart than in case I. Hence the transverse effect should be a decrement of rigidity, under like circumstances of much smaller numerical magnitude than the increment of rigidity due to the longitudinal field. Thus transverse magnetization is adapted to test a variety of phenomena, and it seemed to me that an examination of these in very intense fields (much more intense than are available for longitudinal magnetization) might throw new light on the nature of magnetostriction.

The case of circular magnetization I had hoped to omit, as it was studied at considerable length by the original inves-

[*] Barus; this Jour., (3), vol. xxxiv, p. 175, 1887.
[†] Howard Day: this Jour., (3), vol. iii, p. 449, 1897.
[‡] Barus: l. c, pp. 182–183.

tigators, and recently in the exhaustive researches of C. G. Knott,[*] and others. In Knott's experiments longitudinal and circular fields are superposed and the data are made tributary to Maxwell's[†] and Chrystal's[‡] theories, in which the stresses of the Joule effect are sufficient to account for the torsional phenomena. Drude[§] uses similar reasoning to compute Kirchhoff's third constant.

The immediate effect of circular magnetism is a decrease of the rigidity of the wire (iron, nickel) through which the current flows. The case therefore corresponds at once to type II, figure 2, or to lines of stress relatively far apart as compared with molecular distances.

The presence of an electric current, however, is accompanied with an accession of heat in the wire through which the current flows, and the diminution of viscosity resulting manifests itself in my apparatus and perhaps inevitably, as a diminution of rigidity. Hence in circular magnetization the heat effect is superimposed on the magnetic effect of the same sign and in thin wires it is difficult to separate them satisfactorily, if at all ; for even if the wire carrying current is submerged in water there is still grave room for doubt, as will appear below.

For these and similar reasons I endeavored to investigate results with thin wires subjected to the strong transverse fields in the air gap of an electromagnet, for which conditions no data have as yet been forthcoming. In spite of the toil spent upon the work, however, my endeavors have not brought out sharp results, except in so far as they furnish a superior limit for the change of rigidity sought ; and even this is a small and uncertain residuum raising a doubt as to whether rigidity is at all influenced by strong transverse fields.

Having completed this work I thus found myself under the necessity of taking up the question of the effect of circular magnetization on rigidity for comparison. The results were again such as to place doubt on the occurrence of such an effect.

3. *Method of Experiment.*—The apparatus used was virtually the same as that of my earlier experiments on the subject, except that the helix formerly surrounding the upper wire is replaced by the fissure-like air gap of the strong tubular electromagnet, A, figure 3. The upper of two identical soft iron wires, ab and cd, is placed in this gap to be strongly magnetized when the current flows through A. Any twist may be applied to the system ab, cd, by the torsion heads D and E,

[*] Knott: Trans. Roy. Soc. Edinb., (2), xxxvi, pp. 485-535, 1891.
[†] Maxwell: Electr. and Mag., ii, p. 109.
[‡] Chrystal: Encyclop. Britanica, Art. Magnetism, p. 270.
[§] Drude: Wied. Ann., lxiii, p. 9, 1897.

and both wires are thus subject to the same torque. The change of rigidity due to the transverse magnetization is observed by the mirror, m, reflecting a beam of sunlight upon a wall 5 meters distant. Hence 1^{cm} of deflection corresponds to about ·001 radian or ·057°. The mirror is provided with vanes, vv, dipping in an annular trough (not shown) to deaden vibration, as well as to carry a current from E to v for the circular fields. The rod, e, is suitably provided with a pin and slot arrangement, so that the wires may be kept under the definite tension of the weight, W.

The core of the electromagnet, A, was of thick gas pipe about 37^{cm} long, 5·5^{cm} in diameter and ·3^{cm} thick, the longitudinal (vertical) air gap being about ·3^{cm} broad. It was wound from end to end with about 120 turns of wire. A fine vertical glass tube (not shown) about ·1^{cm} in bore was symmetrically secured within the gap running from end to end. Through this tube the wire, ab, was threaded to prevent adhesion to the sides of the jaws of the electromagnet. This occurrence in fact constitutes the chief difficulty in the experiments. Deflection of the wire toward either jaw is liable to be accompanied with spurious rotation. On the other hand too fine a tube interferes with the motion frictionally. In later experiments I discarded the glass tube in favor of a more carefully centered free wire.

During the tests with circular magnetization the lower wire carrying the current was submerged in a wide glass tube of flowing water.

The field H, within the air gap of the helix, measured ballistically with the aid of an earth inductor, had approximately the following c. g. s. values for different currents, C, in amperes. An attempt to secure sharp absolute values was needless.

$C=$	1	2	3	4	5	6	7	8	9	10	11	12	13
$H=$	1200	2700	4300	5700	7150	8500	10000	11100	12000	12650	13200	13700	13950

Between 2 and 7 amperes in the helix the variation of H with C is very nearly linear. Below this it is slightly accelerated, above rapidly retarded. Since all fields were made from zero, the effect of hysteresis could be disregarded as was specially tested. The field here referred to is that within the air gap and found by differential experiments. If B_1 be the induction within the gap and B_2 the total induction, including the

---

* Barus: this Jour., (3), xxxiv, p. 176, 1887.

stray field outside of the gap, the galvanometer for a loop within the gap will respond to $B_1-(B_2-B_1)$. Two measurements were therefore made to compute B_1.

To prevent excessive heating currents between 5 and 6 amperes were usually employed, though these were not sufficient to saturate the core of the helix as the above data and the results below show. I may add that the ordinary effects of viscosity and slip are eliminated by the method of experiment, since the field was alternately made and broken several times.

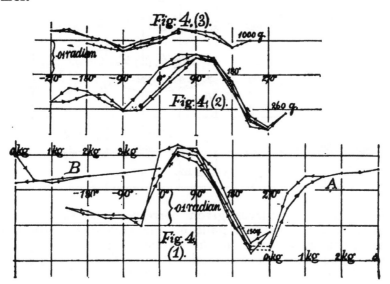

The change δn of rigidity n, is at once given in terms of the deflections $\delta\theta$ of the mirror; for $\delta n/n = \delta\theta/\theta$, very nearly, if for the same section and length, θ is the twist simultaneously imparted to each wire. In fact the above method was devised to secure this convenience. If $\delta\theta/\theta$ is positive, i. e. if in the above apparatus the twist is imparted clockwise looking down the wire and the deflection due to magnetization is also clockwise from the same point of view, then the magnetic change of rigidity is an increment for upper or magnetized wire; and vice versa.

4. *Results.*—The chief difficulty encountered has already been suggested. If the wire is carefully straightened and adjusted in the middle line of the vertical air gap, the wire nevertheless becomes slightly sinuous or wavy when the magnetic field is excited. Different parts of the length of the wire are unequally attracted. If these parts lie across the field such attraction is accompanied by rotation and the observed

phenomenon, otherwise quite consistant throughout, is spurious. As an example I may give the series of results, figure 4 (1, 2, 3,) and figure 5 (1, 2, 3), for annealed iron wires ·044^{cm} and ·024^{cm} in diameter, respectively, and 41^{cm} long each. The twist, θ, imparted to either wire is given in degrees by the abscissas, positive or negative as specified. The ordinates show the deflections $\delta\theta$, in radians, obtained when the field is made, so that ·057 $\delta\theta/\theta = \delta n/n$ is the increment of rigidity.

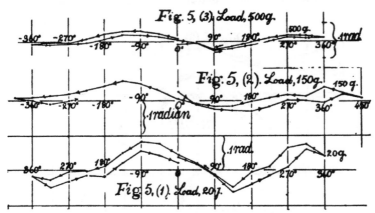

The consecutive curves given correspond to successively increasing loads or vertical pulls on the system, as indicated. Observations were made in the cyclical way shown by the arrows. The same zero was retained throughout. In general for positive twists the return curve is positive relatively to the outgoing curve; for negative twists the return curve is negative relatively to the outgoing curve. Definite consideration will be given to this interesting result in my next paper so that the present reference will suffice for this and succeeding cases. The effect of loading (among many results of a similar kind) is specially shown in figure 4 at A, B, giving evidence even of sign reversal. The deflections are large for the thin wires as compared with the thicker wires (different scales are needed in the two cases). Special experiments showed that if a series is repeated the curves obtained, however irregular, are identical.

Notwithstanding the trustworthy character of the results obtained, I became convinced that the excessive effects produced by loading, and hence the whole phenomena, were somehow in error.

5. In the experiments of §4, the magnetized wire was enclosed in the fine glass tube mentioned above. In the present experiments it is free (the tube being withdrawn) and visibly suspended in the air gap. The latter was adjusted after

each twist so that on making the field the transverse motion of the wire was the smallest possible. Rarely did I succeed in quite eliminating it. Moreover, to guard against errors of the

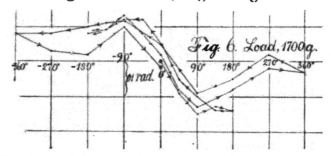

preceding kind the wire was as heavily loaded as its section permitted, without passing the limits of elasticity. An example of the results so obtained is given in figure 6. The load sus-pended from the system of thin wires was about 1700 grams. The character of the observations has remained essentially the same as in the foregoing experiments, and the additional care bestowed has not reduced the complications.

To test the trustworthiness of these observations and to interpret them, I now *rotated both wires* of the system around their common axis and examined them for each 90° of displace-ment. The data are given in figure 7 (1–5), which is an example of many similar results with other wires. To facil-itate the work the twists applied lay between +90° and −90°. The slopes of the lines obtained are about the same, but they are alternately positive and negative. In general when the deflection for no twist (fiducial zero) is positive, the slope is positive and vice versa: though neither this nor the alternation of sign seemed to be the invariable rule. One may note that the data for the initial (0°) and final (360°) positions agree very well as to slope.

The curves of figure 7 may be interpreted in two ways: either the wire is aeolotropically different in the four radial directions examined, which would indicate remarkably complex structure, or the wire is geometrically dissimilar, being either elliptical in section or not quite straight, or both. In both cases the curves obtained would be adventitious and the altera-tions observed easily explained: for if the wire is virtually ribbon-shaped with its plane oblique to the lines of magnetic force, the tendency of the ribbon will be to set "axially" when the field is made. For positions 180° apart the torques will be identical; for positions 90° apart they will vary as the sine and cosine, respectively of the obliquity, and will be numeri-cally opposite in sign. If the obliquity be 45° or nearly so, the torques will be equal and opposite.

Assuming that in the given curves, figure 7, the obliquity of 45° has been reached, then from two consecutive positions of the wires relatively to the field (0° and 90°, 90° and 180°, etc.)

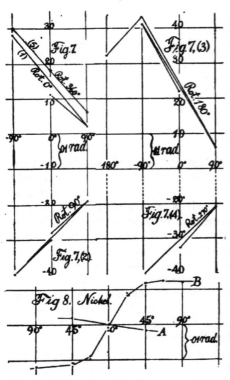

the spurious effect may be eliminated. The slope so found, however, lies within the errors of hysteresis, set, and the errors of orientation and can not be considered more than an estimate. For the mean slope indicates an effect of but ·11° per 180° of twist, or 2ᶜᵐ of deflection for a scale distance of over 5 meters. Hence even in the strongest transverse magnetic fields the persistent increment of rigidity remains small and is negligible in comparison with the spurious effects encountered.

6. After reaching this unexpected result it seemed necessary to verify it in other directions. In the first place the question occurred whether compatibly with figure 2, type I, the effect of longitudinal magnetization was an increment of rigidity in all metals. I therefore tested a nickel wire (length 41ᶜᵐ, diam. ·048ᶜᵐ) in the longitudinal field by replacing the transverse helix A, figure 3, by an ordinary helix. An example of the uniform results is given by figure 8, B, and bears this out (current 4 amperes, load 150 grams, field 400 c.g.s. units).

On the same diagram will be found the apparent effect, B, of circular magnetization produced by a current of about 2 amperes passed through the wire, freely suspended in air. The obvious decrement of rigidity resulting seemed suspicious, however, inasmuch as the deflection on making the circuit was rapid and the return to zero on breaking, prolonged. This is nearly what would occur if the observed decrement of rigidity was due to heat alone.

7. Final experiments were therefore directed to a comparison of the effect of a circular field in iron and the effect of a similar field in brass, yielding it would appear a straightforward method of decision. In figure 9, I give a preliminary

example of the decrement of rigidity due to a current of one
ampere in an iron wire suspended in air, compared with the
effect when the wire is submerged in flowing water, other
things being equal. It will be seen how relatively large the
heat effect is, the decrements for the submerged wire being

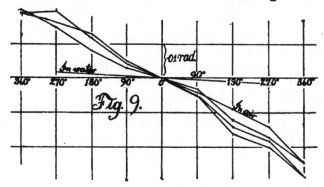

less than 1ᶜᵐ at 5 meters of distance and 90° of twist lying
within the errors of measurement. Apart from this the question is open whether even the submerged wire is not heated
to an extent appreciable by the method of observation.

Hence in my concluding experiments I tested an iron and a
brass wire of about the same dimensions, in the same apparatus
and under like condition consecutively. Both were submerged
in running water. The results for iron are given in figure 10,
those for brass in figure 11. The wires were sufficiently thin

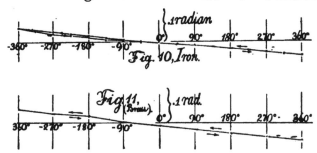

and the current strong enough to give large deflections (up to
50ᶜᵐ). The current and diameter in brass (6 amp., ·020ᶜᵐ) are
smaller, in iron (7·5 amp., ·044ᶜᵐ) larger. In both cases,
however, there is a definite decrement of rigidity due to the
circular field of about the same order, and in both cases so far
as can be made out it is a mere heat effect.

I am obliged therefore to conclude, that in so far as these
experiments have weight the effect of longitudinal magnetiza-

tion is an increment of rigidity in all paramagnetic metals; whereas the permanent effect of a transverse or a circular field is relatively inappreciable so far as rigidity is concerned. At least its certain detection would have to be left to researches of a higher order of refinement than was reached in the method pursued. This conclusion is at variance with much of the earlier work on the subject, but I do not see how the step for step march toward this result which the above simple experiments contain can be evaded.

Finally, the suggestions of the model (figures 1 and 2) are in keeping with the data found. Cf. § 2.

The quantity which I have here considered is the *permanent* effect of magnetization or rigidity, i. e., that which persists after making and breaking the field many times. Besides this there is a very striking temporary effect of marked relative value, in the interest of which the present experiments have largely been made. The relation of this temporary effect to viscosity will be considered in the paper soon to appear.

Brown University, Providence, R. I.

ART. XL.— *Notes on Tellurides from Colorado;* by CHARLES PALACHE.

1. *Sylvanite from Cripple Creek.*

THE presence of sylvanite among the telluride ores of Cripple Creek was first made known by the analysis of Pearce,[*] and the considerable silver content of the ores was credited by him, in large part at least, to that mineral. His analysis however was made on massive material, and the lack of crystallographic evidence of the presence of sylvanite, together with the failure of other investigators to find the mineral, led to some mistrust of his results and the silver content of the ores was doubtfully attributed to the common ore, calaverite, which was known to carry a small percentage of silver.

Of recent years, however, crystallized sylvanite appears to have been found in considerable amount in several of the Cripple Creek mines and it is the purpose of this paper to describe a series of sylvanite crystals which offer most satisfactory proof of the correctness of Pearce's original conclusion.

The crystals to be described were placed at the writer's disposal for examination by Professor Hobbs of the University of Wisconsin, to whom they had been presented by Mr. F. M. Woods of Victor, Colorado, the collector. The crystals were in two lots labelled respectively: "Sylvanite," Mabel M. Property, Beacon Hill, Victor; and "Calaverite," Victor, Col.

All proved to be alike sylvanite and consisted of isolated crystals and crystal fragments varying in size from 1^{mm} to 8^{mm} in greatest dimension. The color is pure silver-white and many of the fragments display the perfect cleavage parallel to the clinopinacoid which is characteristic of this mineral and serves well to distinguish it from the more common calaverite. The specific gravity, determined on three isolated crystals on the hydrostatic balance, was 8·161.

The dominant habit of the crystals is thin tabular parallel to *b*, 010, the edges of the tables being bounded by planes of the orthodome zone. The tables are frequently as thin as paper and the edge planes become too small to be measurable. A second well-marked habit is prismatic, determined by the pronounced development of the zone of the positive unit pyramid and orthodome, (111) and (101), and other pyramids of this zone. As will be seen from the figures, crystals of this habit are frequently rich in forms, some of which proved to be new to the mineral.

[*] Proc. Colo. Sci. Soc., 1894.

A skeletal development, well known on sylvanite from other localities, is common on these crystals; but it did not affect the character of the crystal planes which were in general brilliant, giving good reflections.

Measurements were made on the two-circle goniometer and the adjustment of the crystals was rendered very accurate by the use of the three pinacoids, nearly always all present.

The following twenty-nine forms were observed on the five crystals measured:

c,	001	m,	101	r,	111	u,*	723
a,	100	n,	201	w,*	343	II,	341
b,	010	N,	$\bar{2}$01	s,	121	y,	123
R,	120	d,	011	o,	131	ρ,	$\bar{1}$11
e,	110	v,*	525	q,	141	σ,	$\bar{1}$21
f,	210	γ,	212	i,	321	J,	$\bar{3}$21
g,	310	t,	323	j,*	521	κ,	$\bar{5}$21
						Y,	$\bar{1}$23

In the following table are given the computed and measured angles for the four new forms. The forms o, 131, q, 141 and II, 341 while not new, have not been observed since Miller's studies, and are here added, with their computed angles, in order to complete the tables of Goldschmidt by whom they were omitted as uncertain. The new form u, 723, is the most interesting, occurring in nearly all the measured crystals with good development and being apparently characteristic for this locality.

Letter.	Miller symbol.	Computed angle.		Average measured angle.		Variations.				Number of observations.	Quality of faces.
		φ	ρ	φ	ρ	φ +	φ −	ρ +	ρ −		
v	525	57°·06′	39°·41′	57°·03′	39°·36′	12′	13′	7′	3′	5	fair
w	343	24·53	58·52	24·49	58·57	9	8	7′	6	2	poor
u	723	65·04	60·42	65·01	60·36	14	5	12	15	6	good
j	521	56·53	76·22	56·50	76·25					1	excellent.
II	341	24·44	78·36								
o	131	11·39	73·50								
q	141	8·47	77·38								

The figures illustrate the typical prismatic habit, as shown in various combinations of planes. In fig. 1 the pyramids are developed at the expense of orthodome and clinopinacoid; in fig. 2 the latter forms predominate, giving a tabular-prismatic character to the crystal. This zone of pyramids is generally

---

* Forms marked thus are new.

rich in forms as shown in fig. 2 and is sometimes deeply striated. It forms a satisfactory means of distinguishing the positive from the negative octants, which in its absence, owing to the nearness of the angle β to 90° in sylvanite, is not always easy to do.

Of the other pyramidal forms, σ, $\bar{1}21$, is the most common, being rarely absent and generally relatively large in size. The other pyramids, both positive and negative, and the prisms are

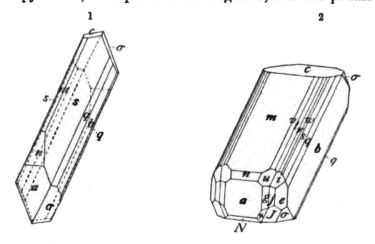

of very subordinate importance in defining the outlines of the crystals.

Two of the measured crystals were found to be twinned according to the common law for sylvanite, twinning plane the

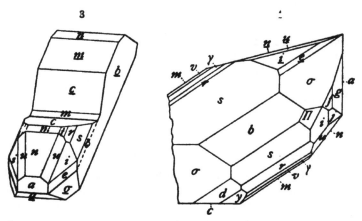

orthodome, 101. Figs. 3 and 4 reproduce these twinned crystals in about the proportions of the originals. Fig. 3 will be seen to be a simple contact twin, in which, however, the two individuals, in contact along the plane 101, are very unlike in development. The crystal figured in normal position is bounded by numerous positive pyramids and domes while the twinned individual is much larger and a much simpler combination. The reëntrant angle between the two basal planes with the intervening narrow portion of the twinning plane 101 made the twinned character of the crystal quite evident. In a subordinate development this mode of twinning is quite frequent, the twinned individual appearing as a narrow transverse ridge on the surface of the *m* plane of a larger simple crystal such as shown in fig. 2.

A less simple phase of the twinning is represented in fig. 4 in orthographic projection on 010, where apparently no plane boundary exists between the two individuals. The planes $\underline{n}$, $\underline{u}$, $\underline{z}$, and $\underline{e}$, constituting the upper part of that figure were only recognized as in twin position by the complex symbols which they yielded in the projection, whose study, in the light of the simple twin already observed, showed their true nature. It is possible, however, that this is a contact twin like No. 3; for the remaining forms shown in the upper part of fig. 4, *s*, *r*, *v*, etc., are not affected in their position by the twinning; but in the absence of the planes bounding the opposite end of the crystal the course of the twinning boundary cannot be accurately defined.

A chemical analysis was made upon about ·5 grams of the sylvanite crystals, which were carefully picked over under the lens until apparently entirely freed from adhering gangue. The analysis shows, however, that considerable of the siliceous gangue was still present. The composition is that of a normal sylvanite, $Au,AgTe_4$.

	Sylvanite.	Cal. to 100%.	Mol. ratio.
Insol.	1·02		
Au	26·09	26·25	·1334 } = 1
Ag	12·49	12·57	·1164 }
Te	60·82	61·18	·4894 = 1·96
Fe	1·19		
Total	101·61	100·00	

2. *Crystallographic Identity of Goldschmidtite with Sylvanite.*

In 1899, Hobbs* described goldschmidtite, a new silver-gold telluride, intermediate in composition between sylvanite and calaverite. The analysis was made on a very small amount of material and seemed therefore open to question, but the crystallographic results appeared to place the mineral distinctly

* This Journal, vii, 357, 1899.

apart from others. Through the kindness of Professor Hobbs the writer was entrusted with two of the five type crystals of the new mineral for examination and thus familiar with its appearance, recognized it again in a specimen in the Harvard collection secured some time before from Cripple Creek and labelled as from the Little May Mine. Several measurable crystals were obtained from the latter specimen and they were found to agree in general with the two type crystals of goldschmidtite examined. The author's measurements of Hobbs' two type crystals confirmed in general his results as shown by the table of angles given below but the use of the two-circle goniometer made possible the measurement of a number of minute pyramid planes which Hobbs had been compelled to leave undetermined. These same forms and some additional ones were also found on the new material, thus confirming the identity of the two specimens. These pyramid forms, projected on the goldschmidtite axes, gave symbols far from simple, and unsatisfactory relations with the other forms. They were studied in gnomonic projection, and a comparison of such a projection of the goldschmidtite forms on the clinopinacoid with a similar projection of sylvanite forms showed a striking analogy between the two. By making the orthopinacoid 100 of goldschmidtite equivalent to the orthodome 101 of sylvanite many forms of the first become identical with known forms of the second, and the remainder, while apparently new to sylvanite, receive comparatively simple symbols on the sylvanite axes. In the following table the agreement of angles of the common forms in the two principal zones is well shown.

Symbols Goldschm.	Symbol Sylvanite.	Obs. (Hobbs).	Obs. (Palache).	Calc. (Sylvanite).
100 : 110	=101 : 121	61° 41′	61° 41′	61° 35½′
100 : 210	=101 : 111	42 43	42 56	42 45
100 : 230	=101 : 131	....	70 11	70 10
100 : 310	=101 : 323	31 55	31 40	31 38½
100 : $\bar{1}$01	=101 : 100	55 35	55 15	55 08
100 : 201	=101 : 001	34 13	34 16	34 27
$\bar{1}$01 : 201	=100 : 001	89 48	89 31	89 35
100 : 001	=101 : $\bar{2}$01	89 25	88 38	88 48
100 : $\bar{4}$01	=101 : 201	19 18	19 30	19 20

Much had been learned concerning the crystallographic character of goldschmidtite when the sylvanite crystals above described came to hand, and their study helped to clear up and make certain the relation between the two minerals. The general similarity between the habit and forms of the two series of crystals led to a surmise that a concealed twinning like that found in the sylvanite crystals was also present in goldschmidtite, giving rise to the many apparently new forms referred to above. And a comparison of the measurements and projections finally proved that this was the case for most of these forms. The following table shows the sylvanite forms equivalent to those observed on goldschmidtite. It will be observed that it is divided into three parts containing (1) forms given by Hobbs and reobserved by the author; (2) forms observed by the author only; and (3) forms given by Hobbs which were not observed on any of the author's material and which lead to new forms for sylvanite.

(1)
Forms observed by Hobbs and Palache.

Goldschmidtite.		Sylvanite.	
Letter.	Symbol.	Letter.	Symbol.
c	001	N	$\bar{2}01$
b	010	b	010
a	100	m	101
g	310	t	323
f	210	r	111
m	110	s	121
k	032	θ	$\bar{2}31$
w	401	n	201 twinned
n	201	c	001
s	101	a	100 twinned
S	$\bar{1}01$	a	100
N	$\bar{2}01$	c	001 twinned
W	$\bar{4}01$	n	201

(2)
Additional Forms observed by Palache.

Goldschmidtite.	Sylvanite.	
Symbol.	Letter.	Symbol.
$\bar{2}32$	e	110
232	e	110 twinned
$\bar{4}34$	f	210
434	f	210 twinned
$\bar{2}12$	g	310
212	g	310 twinned
$\bar{1}31$	R	120
$\bar{2}34$	ρ	$\bar{1}11$ twinned
132	σ	$\bar{1}21$
$\bar{1}32$	σ	$\bar{1}21$ twinned

(2) (cont'd)
Additional Forms observed by Palache

Goldschmidtite.	Sylvanite,	
Symbol.	Letter.	Symbol.
$\bar{1}34$	J	321
$\bar{1}12$	κ	$\bar{5}21$
230	o	131
120	q	141
832	I	211 twinned
$\bar{5}32$	i	321
532	i	321 twinned
734	j	521 twinned
102	M	$\bar{1}01$

(3)
Forms observed by Hobbs not confirmed by Palache and yielding new sylvanite forms.

Goldschmidtite.		Sylvanite.
Letter.	Symbol.	Symbol.
t	370	292
l	130	161
v	35·0·1	9·0·10 ?
x	10·0·1	403 twinned
q	801	203
r	703	2·0·25 or / 301 twinned
y	508	708 or / $\bar{8}01$ twinned
X	$\bar{1}0·0·1$	403
Z	$\bar{1}4·0·1$	504

The forms in (3) appear to need confirmation before being added to the sylvanite series as the following considerations will show. None of them was observed by the author on the twelve or more supposed goldschmidtite crystals from the Little May Mine measured by him nor on the complex sylvanite crystals from the same region. In the absence of definite statements concerning the quality or frequency of occurrence of particular forms in the paper by Hobbs it is difficult to know what weight to assign to these forms. *t*, 292 and *l*, 161, are pyramids of the principal zone of sylvanite and may well be good forms. The seven remaining forms are orthodomes. Of these, two, *q*, 203 and *X*, 403 with its twin, *x*, 403 are probably good forms as they have simple symbols and agree closely in angle with calculated position:

			Meas.	Calcul.
q,	203	⋀ 001	24° 37′	24° 20′
X,	403	⋀ 001	42 34	42 36
x,	403 twin	⋀ 001	26 25	26 24

v, 9·0·10, stands only 2° 20′ from the dominant form 101 and may be vicinal to it; *r*, 2·0·25, is inclined but 3° 10′ to 001, of which it is very likely a vicinal; *y*, 708, is inclined 3° 27′ to 101 and may be vicinal to it; lastly *Z*, 504 is inclined 2° 16′ to 403 and might be considered vicinal to that form were it confirmed.

It seems evident in the light of the above facts that so far as crystallographic character is concerned goldschmidtite can not be distinguished from sylvanite but represents a peculiar habit of that mineral, common, as shown in the preceding paper, to the sylvanite of Cripple Creek. Fig. 1 of the preceding paper on sylvanite illustrates fairly well the prevailing habit of goldschmidtite when in the sylvanite position, except that *σ*, 121, should be very minute, and most crystals are simple contact twins.

The results of this study were submitted to Prof. Hobbs in the hope that he might be able to complete them by a new analysis of the type material on a larger amount of substance. This proves unfortunately to be impossible, since the specimen from which the type crystals were obtained was only temporarily in his hands and is not now available.

In default of this analysis it was thought that it might be worth while to determine the composition of the supposed goldschmidtite from the Little May Mine, and by sacrificing the whole specimen and carefully picking the crushed gangue material enough was obtained (about 0·4 gram) for an analysis. This purpose was, however, defeated by an accident early in the work, so that only the gold content could be determined. For the sake of comparison this is given below together with that of goldschmidtite and sylvanite.

Little May Mine.	Goldschmidtite.	Sylvanite $AuAgTe_4$.	Sylvanite Cripple Creek.
Au 28·89	31·44	24·45	26·09

While little weight can be attached to a comparison based thus on a defective analysis, still it seems clear that the material studied differs but little from the typical goldschmidtite in composition. On the other hand, while the gold content is high when compared with that of the theoretical $AuAgTe_4$, it should be clearly remembered that no sylvanite with exactly that composition has been as yet analyzed, the gold content actually found varying from 25·87 per cent to 29·35 per cent with proportionately varying silver content. These results seem to show conclusively that the ratio of gold to silver in sylvanite may vary considerably from the theoretic proportion of 1 : 1, without affecting the physical characteristics materially, and it is a question whether it is advisable to attempt to establish species based upon these variations of composition.

Note by Professor W. H. Hobbs.—In view of the results of the investigation by Dr. Palache, above detailed, it seems proper for me to say that the name goldschmidtite should be withdrawn from mineralogical literature as representing a distinct mineral species. Dr. Palache's study shows that goldschmidtite can be referred to the same set of axes as sylvanite, of which it represents a peculiar type. This being true my analysis, which had to be made on an extremely small amount of material, must contain a large error. The method used was the oxidation of the tellurium on charcoal, weighing the button of combined silver and gold, and, after solution of the silver, weighing the gold in the form of powder. The danger of this method lies in the possibility that tellurium will not be completely eliminated and that some silver will oxidize. The button obtained was, however, bright and apparently freed from tellurium. I regret that material is not now available for a second analysis, yet in view of Dr. Palache's crystallographic study of better material, checked as it has been by analysis, no course is open to me but to discredit the results of my analysis.

3. *Hessite Crystals from Colorado.*

A specimen of well-crystallized hessite from Boulder Co., Colorado, has been recently acquired by the Harvard Mineral Cabinet from Mr. G. B. Frazer. A description of this specimen is here offered because the crystals present certain interesting peculiarities of habit and because so far as the author has been able to discover no crystals of this mineral from the United States have yet been figured.

The specimen consists of a small fragment of bluish vein quartz in one side of which is a drusy, quartz-lined cavity. On the quartz walls of the cavity are eight or ten brilliant hessite crystals, a millimeter or less in height, and two small tufts of wire gold. Two of the steel-gray hessite crystals and a small irregular fragment were detached, the latter yielding before the blowpipe the characteristic reactions of hessite.

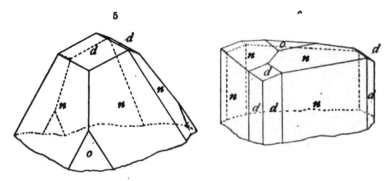

The two crystals (figs. 5, 6) which are shown in about their natural proportions in the figures, presented two entirely different habits, both hexagonal in appearance; but measurements proved them to be isometric and combinations of the three forms:

o, (111); *d*, (101); and *n*, (211).

The hexagonal appearance is due to the development about a trigonal axis, an axis, that is, normal to an octahedral face, and in order to bring out the peculiar symmetry of the distortion the drawings have been made with this trigonal axis vertical, as though the crystals were truly hexagonal.

The difference in habit is due to the fact that different faces ·of the above forms are developed on the two crystals. In the one, six faces of the trapezohedron equally inclined to the trigonal axis give the effect of a scalenohedron whose summit is modified by a rhombohedron composed of three faces of the dodecahedron, while three octahedron faces form a steeper rhombohedron of the same sign. In the other crystal the habit is prismatic; three only of the six trapezohedron faces parallel to the trigonal axis are developed, yielding a trigonal prism whose edges are beveled by the prism of second order, that is by six planes of the dodecahedron; the termination consists of positive and negative rhombohedrons consisting of dodecahedron and trapezohedron faces respectively, and a single face of the octahedron forming a basal pinacoid.

The two habits were represented about equally on the other crystals in the cavity as far as could be judged without their removal. It should be said that the faces present were sharp and clear and no trace of the missing faces of any form could be detected.

Harvard Mineralogical Laboratory, August, 1900.

ART. XLI. — *New Species of Merycochœrus in Montana.*
Part I.   By EARL DOUGLASS.

Description of the Skull of Merycochœrus laticeps n. sp.

WHILE making a collection of vertebrate fossils from the
Loup Fork beds of the Lower Madison Valley in Montana,
I obtained several parts of mandibles, representing three or
four different species, which were doubtfully referred to the
genus Merycochœrus. Three of the species were remarkable
for the depth of the rami. One, which was more slender
than the others, possessed characters which made it probable
that it belonged to the same genus; though, measuring
beneath the middle lobes of the last molars, the deepest jaw
was double the depth of the narrower one. I could not identify any of these with species that had been described.

In the summer of 1899 in some clay bluffs near the village
of New Chicago in Granite County, Montana, a nearly complete skull and lower jaw with some other bone fragments
were found. The jaw was at once recognized as similar to one
of those found in the Madison bluffs. The greater part of the
skull was enclosed in a clay nodule. From some of the teeth
which were exposed it was supposed to be one of the Oregon species of Merycochœrus; but when cleaned from the
matrix it was seen to be very different, possessing many
peculiar characters which separate it from those forms. It is
more nearly related to the type represented by *Merycochœrus
proprius* and *M. rusticus*, between which it is intermediate in
size.

Its most striking characteristics are the following:

Skull low, broad behind the orbits, narrowing rapidly toward
the front and back. Brain case short, the length behind the
post-frontal process being about one-half the distance in front
of it. Premaxillaries united in front forming a trough-shaped
depression, evidently for the accommodation of a proboscis.
Maxillaries deeply concave on the sides of the face—this, with
the malo-maxillary ridge which widens outward rapidly toward
the zygomatic arch, forming a broad nearly horizontal shelf
above the posterior premolars and anterior molars. Larger
part of external narial opening nearly between the orbits, but
continuing forward in a horizontal slit between the maxillaries.
Nasal bones short and ascending, placed far back, the anterior
tips being about midway between the inion and incisive border.
Foramen infraorbitale placed farther back than in any other
species. Bending down of the face upon the basicranial
axis carried to the extreme, the shape of the posterior basal

part of the skull indicating that the head was carried in a nearly vertical or hanging down position. Back part of zygomatic arch small and simple. Length of the alveolar border twice the distance from the last molar to the occipital condyle. Otic bullæ not inflated. Mandible heavy and very deep in the region of the angle. First lower incisor absent. Jaw nearly as deep as skull exclusive of nasals, and nearly as long as the skull at base.

Merycochœrus laticeps.
Skull and Mandible, side view, × ⅓.

Description of Skull.—The premaxillaries are coössified with the maxillaries. Anteriorly the maxillo-premaxillaries are united for a distance of 5·5^{cm}. Here they form a trough-like depression which becomes broader as it curves gently upward and backward. This widening of the trough is caused by the converging of the ridges from the upper borders of the maxillaries which bound laterally the horizontal wedge-shaped forward projection of the narial opening. and by the conical shape—as seen from the front—of the large convexities in the region of the canines. These convexities have their bases close together in the middle of the incisive alveolar border, make a broad lateral sweep and die out on the face in front of the anterior projection of the malo-maxillary ridge. The first mentioned ridges die out on the anterior inner sides of these convexities a little distance above the alveolar border.

The anterior part of the narial opening begins anteriorly in a rather blunt point, widens gradually as it extends backward and slightly upward for a distance of 3^{cm}, then the sides are nearly parallel and horizontal about the same distance, then it expands and ascends to form the inferior part of the posterior oval opening. This posterior part of the external nares looks forward and slightly upward, thus forming a high angle with the anterior part. Above it is arched over by the short backward sloping nasals. Its lower border is in advance of the orbits, but in front of the bases of the nasals a notch extends back of the orbital border. The vertical height is about 5^{cm}. Between the narial opening and the orbit the tongue of the maxillary curves upward and then backward, apparently ending in a wedge-shaped process between the nasals and the frontals in a plane with the posterior parts of the orbits. The suture separating the frontal and maxillary extends from this point outward and forward and then downward, passing close to the anterior border of the orbit. The skull is a little injured just at the border of the orbit and the orbit is still filled with the matrix so that the lachrymal bone cannot be made out, but it occupies an extremely small space if any in front of the orbit. The malo-maxillary suture continues downward from the orbital border to a line 2·2^{cm} lower than the lower border of the orbit, where it curves forward and downward to near the lower anterior border of the overhanging malo-maxillary ridge. From this place it cannot be traced. The suture where seen is well defined, is complex but forms a quite regular band 3 to 4^{mm} broad. On the roof of the mouth the maxillo-palatine suture appears near the root of the second molar, extends forward and slightly inward to opposite the posterior lobe of *m1*, then transversely across the palate in nearly a straight line. The roof of the mouth is broad and concave. The incisive foramina are confluent, but this is apparently due to the breaking away of the thin median partition, part of the superior portion of which still remains. It now appears as an oval opening, the smaller end being directed backward. Its length is 3^{cm}, and its greatest width 1·5^{cm}.

On the side of the face is a large depression difficult to ɜfine. It is nearly triangular in shape. Above its boundary the median upper border of the maxillary; behind it is the ɪtward expansion of the posterior part of the maxillary in front ᷣ the orbit. Beneath is a broad shelf above the malo-maxillary dge. The anterior angle of the depression is occupied by a ɜeper elliptical one, the deepest part of which is above the ɜt premolar and the first molar. These two concavities ɪtending inward on opposite sides of the maxillaries make the ɜe quite thin transversely in this region, being in fact only 4^{cm} thick, while the skull at its widest place is 20^{cm}. The malo-axillary ridge dies out on the anterior border of this depres-ɔn but expands rapidly posteriorly, thus forming the broad ɪelf above mentioned which is broadest in front but extends ɪtward and backward toward the zygomatic arch. The infra-·bital foramen is large, is near the inferior posterior border of ɪe oval depression above *m2* and looks forward and outward, ᷣening on the horizontal maxillary platform or shelf.

The nasals are nearly triangular if I make out their pos-rior borders correctly. They are short, extending upward ɪd forward to form the roof of the external nareil opening. hey are convex transversely and longitudinally and are ᷣinted in front. These points are about midway between the cisor and the inion.

The orbits are oval with the larger end upward. The pos-rior inferior border is nearly straight or a trifle convex, but ɪis may be due to a slight displacement of the post-orbital ·ocess of the malar. Only one orbit is preserved.

What I take to be the boundary between the nasals and fron-ls is a line where the bone is broken. It passes from the ᷣsterior angle of the maxillary transversely and somewhat ɪckward to the suture between the nasals.

The frontal appears to meet the malar posterior to the orbit ᷣ the median line, where there is a roughening on the narrow thmus of bone. From the upper posterior border of the ·bit the supra-orbital ridges converge rapidly backward and ɪen less rapidly, uniting to form a prominent narrow sagittal ·est. The form of this part of the skull back of the nasals ɪd including the upper part of the brain case is almost like that ɡured in Bettany's paper, "On the Genus Merycochœrus," ᷣ *M. temporalis;* but the supra-orbital foramina occupy a ɪfferent position. In the present species they are above a ɪe uniting the posterior borders of the orbits, are far apart, ɪve no grooves leading into them, and are a little nearer the edian line of the skull than the outer border. The tempero-ɪrietal suture extends upward and backward until, beneath ɪe anterior part of the sagittal crest it curves downward and then upward again, passing just beneath the foramina at the base of the inion. These foramina are not the same on opposite sides of the skull. They are farther forward and farther apart on the right side of the skull and are more uniform in size. On the left side the anterior one is circular and several times larger than the one which is a little behind and beneath it. Just posterior to this smaller foramen the parieto-occipital suture extends upward to the broken sagittal crest. The temporo-occipital extends backward and downward but is lost on the broken border. The ridge near the parieto-squamosal suture extends from the larger foramen above mentioned, downward and forward, at first coinciding with the suture and then running beneath and nearly parallel to it. The brain case on each side of this is broadly concave. The brain case is short antero-posteriorly and slopes downward quite rapidly behind the forehead, though the sagittal crest evidently continued nearly on a level.

The crest of the inion is broken but was considerably posterior to the occipital condyles. The wing-like expanses extending to the zygomatic arch were broad and thin. Just above the foramen magnum the occiput is broadly convex; but a short distance above a depression begins on the median line, becomes broader and deeper and then shallower to near the crest of the inion. The ridges that bound this concavity separate it from larger lateral concavities situated farther down. These concavities are bounded above by the wing-like expansions above mentioned and outwardly by the broad convex post-tympanics which appear as large swellings on the sides of the occiput. This post-tympanic with the horizontally expanded portion of the squamosal above and the long posterior flat surface of the post-glenoid encloses a quite large triangular cavity, the post-tympanic and post-glenoid nearly coming in contact below. The meatus auditorius externus evidently did not fill all this space, but the shape cannot be definitely made out on account of the matrix and injury of the bone; but it passes inward and forward a distance of 4^{cm} to connect with the otic bullæ.

As seen from below, the foramen magnum is lenticular in section and its transverse is double its antero-posterior diameter, being respectively $2 \cdot 5^{cm}$ and $1 \cdot 2^{cm}$. With a line joining the incisive border, the lower extremities of the pterygoids and the occipital condyles, this opening forms an angle of about 65°. The occipital condyles are also narrow antero-posteriorly, the diameter being only $1 \cdot 7^{cm}$ in this direction while the transverse diameter is $6 \cdot 1^{cm}$. The anterior articular faces are much deeper and broader than the posterior ones and they are partially

eparated by an angle. The anterior faces are separated by an
xcavation 1·5ᶜᵐ wide while the posterior ones are 3ᶜᵐ apart.

The basi-occipital and basi-sphenoid ascend at a steep angle.
This with the upper contour of the skull makes the brain cavity
ery small. Beneath the forehead the basi-sphenoid bends forward and becomes nearly parallel with it. The basi-sphenoid
s angulate for a short distance between the glenoid surfaces.

The paroccipital processes are broken off at about a level
with the lower points of the occipital condyles, but they appear
o have been much longer. They are broad transversely but
ather narrow antero-posteriorly. At the bases they are nearly
rescent-shaped in cross-section. The more convex portion
aces inward and backward, while the concave area faces forward and outward. The outer portion is a wing-like expansion
f the more robust inner part, which slants forward and outward
rom the occipital condyles, forming with them an angle of
5°. The anterior inner horns or lobes are in a line with the
ost-glenoids. Just in front of these horns and closely in con-
act with them at the bases but with their anterior bases higher
re the prismatic tympanic bullæ. These bullæ are not inflated
ut quite long vertically, especially on the anterior inner side,
nd do not extend much below the posterior bases of the par-
ccipitals. The shape of these bullæ is nearly that of a quarter
f a cylinder terminated by a cone, the angle joining the nearly
lane or slightly concave faces being directed outward and
ackward.

The post-glenoid processes are of moderate length and trans-
erse breadth, are flat behind and moderately convex in front.
The outer border slopes outward and upward, dying out on the
nferior posterior surface of the zygomatic arch, which slopes
pward and backward from the glenoid surface. This surface
s slightly concave, being bounded exteriorly by the ridge on
he lower outer border of the arch. The glenoid surface is
road transversely (6ᶜᵐ) and uniformly convex antero-poste-
iorly.

The posterior angle of the zygomatic arch is on a line with
he anterior borders of the paroccipital process, and extends
pward to the line of the lower border of the orbit.

The malar has four approximately equal sides if the upper
s measured along the border of the orbit; though the lower is
onger in a straight line. The vertical width below the orbit
nd the antero-posterior length just below it are nearly the
ame. The upward extension in front of the orbit is narrower
han the post-orbital process.

The palatines and pterygoids extend about 2·9ᶜᵐ back of the
ast molars. It is only 2ᶜᵐ from the posterior edges of the
terygoids to the anterior faces of the otic bullæ. This

approximation is due to the extreme shortening of the posterior basal elements of the skull so that the post-glenoid and par-occipital processes, the occipital condyles and otic bullæ occupy a comparatively narrow transverse zone.

Between the posterior lobes of the last molars the palato-pterygoid lobes extend downward and backward until they reach a point about on a level with the incisive border and the lower extremities of the occipital condyles. They formed a median trough extending downward and backward from the palate.

2

Merycochœrus laticeps.
Skull, view of under side, nearly × ⅘.

Lower Jaw.—The rami were not coössified at the symphysis, and in the present specimen they are slightly spread apart. The anterior border is concave downward and nearly uniformly convex transversely. The *foramen mentale* is beneath *pm 3*, as is also the angle of the chin. At this angle a small process projects downward. Back of this the lower border of the ramus curves slightly upward and then downward to a point beneath the posterior part of *m 3*. At this point it is twice the depth of the shallower part under *m 1*. Back of this the posterior border ascends with a steep curve to a point 4·5^{cm} below the top of the condyle, where it ascends nearly vertically to the condyle, forming a low angle (about 25°) with the continuation f the posterior border. The posterior border is slightly raised aterally. The condyle is not quite so broad transversely as le glenoid surface with which it articulated and is more arrowly convex antero-posteriorly. The masseteric fossa escends only a short distance—to a line connecting the base f the lobe of *m 3* and the posterior angle above mentioned. 'he bone is injured in the region of the coronoid process so its rm cannot be made out. A broad convexity begins at the osterior part of *m 3*, extends forward and downward to the iterior lower border of the ramus, branches at the mental ramen and continues to the alveolar border at *pm 2*, thus aving a depression under *pms 3 and 4* amd *ms 1 and 2*. 'he length of the dental series is two-thirds the length of the aw.

3

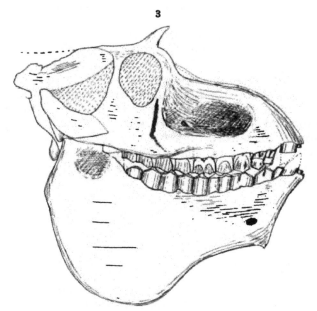

Merycochœrus laticeps.
Skull with jaw, × ¼.

The mandible is too far ahead on account of slight forward displacement of st-glenoid process.

Dentition.—Superior: The crowns of the incisors and mines are not preserved. The upper incisors were small, aite close together and in a transverse row. The roots of the rst and second are laterally compressed. They were nearly le same size, but the third was a little larger. In cross-section the roots of the canines are three-sided with rounded agles, the posterior side being the broadest: *Pm 1* is compressed longitudinally, is inserted diagonally by two roots, and the posterior part overlaps *pm 2* outwardly. *Pm 2* is considerably worn, but there is still a shallow oblique enamel lake near the posterior margin. *Pm 3* has a longitudinal lake with two pits, near the inner border of the tooth. *Pm 4* is like the corresponding tooth in *Merychyus elegans.* The posterior lobe on the last molar is not developed as in the *Merycochœrus proprius;* though the posterior horn of the posterior external crescent is convex behind, it is narrower and extends much more outward than in that species. The premolars and molars are all longer than in *M. proprius.* This is due in part, but I think not wholly, to less amount of wear.

Inferior: There is no trace of a first incisor. The second and third are laterally compressed at the roots, and are 2^{mm} apart. They were about the size of upper incisors one and two. *I 3* is set obliquely close to the canine on its anterior inner side and a little more anteriorly than *I 2*. The canine is small, but a cross-section at the root is nearly the same in form as the upper canine. *Pm 1* is much larger than the lower canine. It has a longitudinal groove on the outside of the root. The *crowns* of the above teeth are not preserved. *Pm 2* is more oblique than in *M. proprius* or *M. rusticus*, being closely crowded between *pms 1 and 3*. *Pm 3* is proportionally narrower posteriorly than in the former species, judging by Leidy's figure. In *pm 4* the posterior fossa is more oblique, opening on the posterior inner angle instead of on the posterior border.

Measurements.

SKULL.

M.

Length of skull from incisive border to posterior of occipital condyles... ·245
Length from incisive border to back of pterygoids......... ·185
Length from incisive border to front of otic bullæ......... ·203
Width of skull at glenoid articulation..................... ·200
Width at middle of last molars........................... ·180
Width at third premolars................................. ·080
Width at canines... ·057
Width of palate between last molars...................... ·063
Distance from anterior tips of nasals to incisive border, measured in a straight line............................ ·166
From tips of nasals to crest of inion, about ....·150 to ·160
Distance from front of orbit to incisive border............ ·135
From front of orbit to back of occiput.................... ·150
Diameter of orbit, antero-posterior....................... ·038
Diameter of orbit, vertical............................... ·045

M.

Width of top of skull above orbits .... ·118
External nares, greatest width............................... ·044
External nares, greatest length .... ·112
Length of three upper incisors............................... ·015
Length of dental series from front of canine............. ·153
Length of molar-premolar series............................. ·136
Length of premolar series .... ·056
Length of molar series .... ·080
Width of last molar.. ·030

LOWER JAW.

M.

Length of jaw, greatest...................................... ·235
Depth at chin.. ·057
Depth under m 1... ·053
Depth under back part of m 2 .... ·063
Depth under last lobe of m 3............................... ·109
Length of inferior dental series .... ·157
Length of molar-premolar series .... ·150
Length of premolar series .... .058
Length of molar series...................................... ·092
Space occupied by i 2 and 3 .... ·008
Width of canine at alveolus, transverse.................... ·009
Length of canine antero-posterior .... ·0065
Width of pm 1 .... ·011
Length of pm 1 antero-posterior .... ·014
Length of pm 2.. ·014
Width of pm 2 .... ·006
Length of pm 3.. ·017
Width of pm 3 .... ·0095
Length of pm 4.. ·0185
Width of pm 4 .... ·012
Length of m 3.. ·044
Width of m 3 .... ·0165
Length of foramen mentale................................. ·008
Width of foramen mentale.................................. ·005

From the parts of *Merycochœrus proprius* that have been described *M. laticeps* differs in the lateral aspect of the maxillaries; the smaller size and, apparently, the more transverse position of the incisors and incisive border; the absence of a space between *pms 1* and *2;* the more hypsodont character of the teeth; the more posterior position of the infra-orbital foramen; the smaller size and different shape of the small posterior lobe on the last upper molar; and the different shape of the mandible, especially its greater posterior depth.

M. rusticus approaches more nearly to the present species in the facial concavities, in the size and position of the incisors, and the trough-like shape of the maxillo-premaxillaries above; the crowding together of *pms 1* and *2;* and possibly in the posterior deepening of the mandible. But, as will be seen by comparing the present descriptions with Leidy's figures and descriptions, the face and anterior border are quite different in the two species. In *M. rusticus* the anterior part of the face as viewed from the side rises more abruptly for a short distance and then more gradually backward. The form of the concavity on the side of the face is different; the infra-orbital foramen and the anterior inferior root of the zygomatic arch are farther forward; the latter also is higher as is also the lower border of the malar. The space between the upper canine and first pre-molar is greater and *pm 1* does not overlap *pm 2.* The mental foramen is longer and the chin is less concave, the pro-tuberance at the chin is larger and longer, and the anterior inferior border of the jaw is more nearly straight. *M. rusticus* is smaller than *M. laticeps.*

RT. XLII.—*On Mohawkite, Stibio-domeykite, Domeykite, Algodonite and some artificial copper-arsenides;* by GEORGE A. KOENIG.

THE knowledge of the existence of copper arsenides in the Leweenaw copper formation is nearly as old as the mining perations in this region themselves. In the matter of occurnce there are two points to be noted : (1) The arsenides are ot found in the bedded deposits of native copper, but always i fissures, intersecting the beds. (2) These veins have thus ir only been observed in the lower beds, near the foot of the rmation to the southeast. Arsenic, however, is found in the nelted and refined copper of all the mines. This element is a iinimal amount in the copper from the Great Conglomerate f the Calumet and Hecla mine and becomes a maximum in ie copper from the amygdaloid beds of the lower measures, a which the Mohawk, Wolverine, Arcadian, Sheldon-Columbia, le-Royale, Atlantic, Baltic, Champion, etc., are located. The heldon-Columbia location near the shore of Portage Lake in ie village of Houghton was the first mine which furnished pper arsenides, notably domeykite. The larger part of the ecimens in collections come probably from this mine. Whiteyite was found in a quartz vein in dark melaphyr, not far om the present Mohawk, but the existence of any copper rock as not suspected at that time at that point. Several masses f domeykite have been found in the drift on the Hancock iore of Portage Lake, very much decomposed, cuprite and senates being the chief products. In the spring of 1898 the pening of a new street in East Houghton on the old Sheldonolumbia location, disclosed a quartz vein containing some folied domeykite. Many good specimens were gathered and ive come into collections. A similar vein has been known or years on the old Huron location, now the southern end of ie Isle-Royale property. Algodonite was found in the ewabic mine, located on the Quincy amygdaloid bed and hich lies about 5000 feet higher than the Isle-Royale amygdaid. In developing the Mohawk property a cross vein at ght angles to the copper-bearing amygdaloid was met in ecember, 1899. This cross vein was from 12 to 15 inches ide when struck, but has since shown very varying dimenons. In a gangue of quartz and calcite the copper arsenides ive been found in this vein more abundantly than in any of ie locations mentioned above. Early in January Mr. Fred. mith, Superintendent of the Mohawk mine, sent me a solid iece weighing from 4 to 5 pounds, with the request to make a thorough investigation and report. On January 18th I reported to Mr. Smith that the mineral substance which he had sent me was a new mineral species which I would call *Mohawkite.* Later on I received other material from Mr. Smith in which I identified an antimonial domeykite for which I propose the name *Stibio-domeykite,* and also some very peculiar intimate mixtures of Mohawkite with Whitneyite. It appears that the Stibio-domeykite is the more prevalent of the arsenides, and not the Mohawkite as was thought at the start. The vein has since been traced to the outcrop where it was found to form a ridge owing to the quartz in the gangue.

1. *Mohawkite.*

Physical properties.—Form is massive; no crystallization of any sort has been observed. The structure is mostly fine granular, sometimes compact. The color on fresh surface is gray with a faint tinge of yellow. Tarnishes very easily, and the tarnish is apt to be ultimately dull purple. A yellow brassy tarnish is brought about by boiling, though in cold water the original color lasts for two days and more. The color is, however, no sure guide for identification. The mineral is very brittle, and owing to the granular structure is not possible to fix the degree of hardness; it is approximately 3·5. Spec. gr. at 21° C. (in boiled water but without corrections) = 8·07, mean of three closely agreeing trials and with perfect material, of which there was plenty,—4.6294 grams were taken.

Chemical properties.—In closed tube gives no sublimate of arsenic; only a slight sublimate of As^2O^3, owing to the air in the tube. The substance melts in the tube at cherry heat, and colors the glass blue if the heating be kept up for a few minutes (Cobalt). In the open tube the reactions are similar but more pronounced owing to the vigorous oxidation. On charcoal in O. F. copious vapors of As^2O^3, odor of arsenic and ultimately a globule of metallic copper. If a fragment of the mineral be placed in a shallow cavity on charcoal, along side of a borax bead of equal size or somewhat larger, and both fused together in the point of the blue flame, so that the metallic globule be exposed to the air, then the borax bead will assume the pure blue color of cobalt. If this treatment be kept up for sometime and a fresh borax bead be taken every minute, then a brown nickel bead will be obtained and finally a blue or red copper bead; thus proving all the metals present except the trace of iron. This test (Plattner's) should be used always in examining metallic arsenides. Thus the mineral contains copper, nickel, cobalt, iron (trace) and arsenic. Boil-

concentr. nitric acid dissolves the mineral without leaving
sidue, forming first a green, then a murky blue solution
ng to the interference between nickel-nitrate green and
lt-nitrate red). HCl does not act upon the mineral. The
ititative analysis was made by several methods:

) Nitric solution made ammoniacal, alcohol added and
nesia mixture. This is not a good method, because part of
nickel and cobalt enter into the magnesium-ammonium
iate.

) Nitric solution made ammoniacal, diluted properly and
  passed into hot solution to saturation. Filtrate evapo-
l to dryness, residue oxidized with HNO' and magnesium-
ionium arsenate precipitate. Separation of copper from
el and cobalt by H'S; separation of nickel and cobalt by
te method.

) Powder fused with nitrate and carbonate of sodium.
 gives very good results for arsenic; but for the fine pul-
lent condition of the oxides these latter must be carefully
lled to avoid loss.

find that the magnes. amm. arsenate can be heated on the
stos pad of a Gooch crucible without loss into pyro-arse-
. From four analyses of perfect material I obtain

$$
\begin{array}{ll}
\text{Cu} & 61\cdot67 \\
\text{Ni} & 7\cdot03 \\
\text{CoO} & 2\cdot20 \\
\text{Fe} & \text{trace} \\
\text{As} & 28\cdot85 \\
\hline
& 99\cdot75
\end{array}
$$

e the atomic ratio

Cu	0·9803		
Ni	0·1200		
Co	0·0373		

	1·1376	2·958	$(Cu, Ni, Co)'$
As	0·3847	1·000	As

i is an exact ratio of 3 : 1, the ratio of domeykite. I
upon this as a case of isomorphic replacement. We may
ict to find all gradations of replacement within the ratio of
. Synthetic experiments lead me to consider this ratio as
esenting an especially strong one, as of bonds well satisfied;
use I have been enabled to obtain it artificially in well
ied crystals, of which more hereafter. The molecule
As is the strongest of all, the other ratios seem to partake
e of the nature of alloys, of unsatisfied bonds. It may be
id that a mechanical mixture of the copper arsenide with

nickel-cobalt arsenide is here presented. Against such a view speaks the physical condition of the substance in the first place and also that a ratio of (Ni, Co)'As has not been observed so far. The question presented itself at once as to whether any of the domeykite occurrences in Houghton Co. contain nickel and cobalt. Winkler gives this metal in the Zwickau occurrence at 0·44 per cent. F. A. Genth does not mention it, and even from the Michipicoten Island location where niccolite accompanies it, the nickel is absent in the domeykite. I have examined every accessible specimen qualitatively. Neither nickel nor cobalt could be found by Plattner's test; but nevertheless it is very probable that all domeykite contains both metals in traces, as my careful analysis of the mineral from the Sheldon-Columbia location, both old and recent, demonstrates. From 0·5 gram of substance just sufficient cobalt oxide was obtained to give color to 50 mg. of borax glass and the nickel was proved by reducing the bead with tin on charcoal. Another matter engaged my attention at once, namely the discrepancy in the specific gravities given by the several authors and recorded on page 44, Dana Min., 6th edit. These densities vary between 6·70 and 7·547 against my determination of 8·07 for the Mohawkite. The collection of the Michigan College of Mines contains one fine specimen of domeykite from the Sheldon-Columbia mine. The material is very uniform, a few specks of calcite the only visible gangue. Of this 3·8559 grams were selected. The specific gravity at 21° C. was found = 7·9486. The whole material was then dissolved, no residue was left and in a part of the liquid calcium was looked for, but only a trace was found. Hence we have every reason to take this specific gravity as that belonging to domeykite; the figures in the literature must be wrong. It is true that the analysis of this substance does not agree with the theoretical composition exactly, namely:

Cu	74·00	Spec. gr. found	7·9486
(Fe, Ni, Co)	0·06	" calculated	8·1020
As	26·14		

100·20

Now if we assume that the elements in this combination possess the same specific gravity as they do in the free state, we may calculate the specific gravity. If specific gravity of Cu = 8·96 and of As = 5·63 then the above composition must weigh $8·96 \times 74·00 + 5·63 \times 26·14 = 8·1020$. The specific gravity of the theoretical Cu'As will be = 8·017. For Mohawkite the calculated specific gravity will be, taking Co = 8·96 and Ni = 9·5 (the highest number on record):

$$
\begin{array}{rclcl}
Cu &=& 61\cdot67 \times 8\cdot96 &=& 5\cdot5256 \\
Ni &=& 7\,03 \times 9\cdot5 &=& 0\cdot6678 \\
Co &=& 2\cdot02 \times 8\cdot96 &=& 0\cdot1809 \\
As &=& 23\,85 \times 5\cdot63 &=& 1\cdot6242
\end{array}
$$

$$7\cdot9985$$
$$8\cdot0650 \text{ found}$$

$$+0\cdot0665 \text{ difference}$$

This is quite a satisfactory agreement. Yet it occurred to e at this stage of the investigation to verify the relations tween specific gravity and composition by artificial com- unds and also to see in how far domeykite might be obtained crystals.

A combustion tube was closed at one end over the lamp. esublimed arsenic, roughly powdered, was placed into the ttom to the amount of 7·9 grams. On top of this was ured 20·0 grs. of copper filings. The latter had been the d product in a determination of oxygen in refined copper. otal non-copper in this material 0·07 per cent, consisting of senic and iron. An asbestos plug was put over the copper d the tube placed in horizontal position into an Erlenmeyer mbustion furnace. The copper was first heated to a dull erry redness and then the arsenic was heated to the sublima- on temperature. The heating was kept up for two hours, the en end of the tube being provided with a mercury valve to event any air currents from entering the tube. The copper sorbed the arsenic with avidity. One could see that only e part nearest the vapors changed in color and 30 minutes om the beginning this front portion began to melt off. After l the arsenic had become volatilized and a considerable part of e liquid compound had been formed the temperature of this uid was raised to bright redness and kept so until the end of e experiment; it being 6 o'clock P. M. The tube was clined so that the liquid covered the solid portion and left to ol slowly. In the morning three distinct substances were und. A dark colored fused mass, then a porous gray part d then the apparently unaltered copper filings, merely ightly caked together. (*a*) The dark, fused portion. On the acture blue-gray color; strongly developed radial structure d, in fact, so closely resembling chalcocite that one could not ll by the eye one from the other. The weight of all three rts was 27·27 grams. The weight of the unaltered (appar- tly) part was 9·65 grams; hence in the altered part there ust be approximately 10·35 of copper and 7·2 grams of senic, giving an atomic ratio of 9·6 : 16·4. Now if the quanti- es taken of copper and arsenic were in the ratio of 3 : 1 it is evident that at red heat the two elements form the compound Cu^2As, leaving much copper unaffected.   The specific gravity of this fused arsenide was found at 7·71, (3·8837 grams being taken).   The analysis gave

		Calculated for Cu^2As.
Cu	63·30	Cu 62·69
As	37·00	As 37·31
	100·30	100·00

The calculated specific gravity is = 7·754, giving a minus difference of 0·044 against the actual specific gravity.

(b) The porous zone between a and the unaltered copper. An ordinary lens showed that this mass consisted of groups of minute crystals, exhibiting very brilliant faces.   A power of 50 diameters differentiated the crystals easily.   They appear to be combinations similar to those of arsenopyrite.   Believing that I had found the method by which these crystals could be prepared readily I destroyed them before making an attempt at measurements, leaving only one group, and this one was lost clumsily.   There are still crystals on the copper cylinder show-ing the forms and building up of the grape-like bunches but I do not consider them worth the trouble connected with micro-goniometric work.   253 milligrams of these groups, which separate easily from the base, were taken for analysis and only the copper was determined, as there could not be anything but it and arsenic in the material.   Found

$$Cu = 71·39; \text{ theoretical for } Cu^2As = 71·6 \text{ Cu}$$

No doubt can exist as to the identity of these crystals with domeykite.   I have made several attempts since to get the crystals but evidently did not hit the right temperature again. I obtained domeykite as crystalline mass.   In one experiment I melted together 28·886 grams of the Cu^2As with 7·8 grams of copper in a glass tube.   Liquidity maintained for one hour. After cooling the column was found of varying composition from bottom up.   The bottom portion showed a fracture sim-ilar in color and structure to white iron, with a faint yellow, in fact it looks exactly like the domeykite of Houghton Co. Specific gravity = 8·05; percentage of copper = 75·4, consider-ably above the ratio Cu^2As; calculated specific gravity=8·085.

Above this was found a laminated or scaly zone, resembling some of the recently found domeykite from Sheldon location. It gives copper = 71·14, very close to Cu^2As.

The top portion resembles in color and structure (very fine grain) the Cu^2As.   It gives copper = 64·8.

2. *Stibio-domeykite.*

With this name I propose to designate the domeykite from the Mohawk mine, Keweenaw Co., and any other locality where mineral of similar composition will be found. In physical properties it is undistinguishable from the domeykite of the Houghton Co. mines. It is remarkable for the massiveness of its occurrence, the freedom from admixtures. It has a decided subconchoidal fracture and takes on a brass tarnish very soon, which, however, ultimately turns into bluish purple. It is very brittle but not as friable as the Mohawkite. Its hardness is very near 4, a little below.

Specific gravity at 21° C. = 7·902 (made with 4·5595 grams) B. B. In the open tube after heating the molten globule for five minutes one observes the forming of yellow spots near the globule and a slight ring or patches farther on, which turn yellow on applying higher heat and volatilize in part. Farther on there is a copious sublimate of arsenic trioxide. If now one washes out the tube in a jet of water and passes H_2S into the tube one observes the above ring and patches turn red, whilst the arsenic oxide turns gold-yellow. I was astonished myself to find that one could demonstrate thus the presence of 0·1 per cent antimony in the mineral. On charcoal blow the point of the oxidizing flame upon the melting mineral for two minutes, then drive off the white arsenic with a gentle flame, the antimony trioxide remains. The open tube reaction is preferable and certain. Concentrated nitric acid does not dissolve the mineral completely; a white cloudiness or a white sediment will be left. When fused on charcoal alongside of a borax bead, the latter colors greenish after several minutes' action (cobalt and nickel cannot be thus found, though present in small amount).

The analysis gave (type specimen):

Cu	72·48
(Fe, Ni, Co)	0·24
As	26·45
Sb	0·78
	99·95

This is evidently a typical domeykite, and one would expect likewise that the percentage of antimony is variable. In order to ascertain this I took twelve pieces, all over one pound each, and treated one gram of each with 10ᶜᶜ of concentrated nitric acid, boiling until the color was blue, then added 15ᶜᶜ of water to each and stood the beaker glasses in a row. (The type specimen with its 0·78 antimony was among the lot.) It appeared that all contained antimony, but no two an equal amount. Some showed more, some less than the type specimen. I selected the one with the heaviest sediment and determined the antimony as trisulphide. It gave Sb = 1·29. This must be taken as the maximum until other tests show greater amounts. The Houghton domeykites show no trace of antimony.

3. *Mohawk Whitneyite.*

With this name I shall designate not a species, nor even a variety but simply a most intimate mixture of the two species Whitneyite and Mohawkite, of which considerable masses are at present encountered in the Mohawk mine. This material is distinguishable at once to the layman even. It is very tough and approaches Whitneyite in that respect. Ordinary blows with the hammer fail to break the masses, the chisel is often required and the hammer leaves a dent as in soft metal. The fresh fracture is gray, fine granular, even hackly. These fractures assume a dull brown or olive-green tarnish, much like algodonite and Whitneyite. There are, of course, all gradations, from nearly pure Mohawkite to nearly pure Whitneyite. The material looks homogenous, but is not, as the following analyses show. The samples were broken with the chisel in close proximity to one another.

	No. 1.		No. 2.	No. 3.		No. 4.	
Cu	85·36	⎫	83·61	Cu	79·36	Cu	84·86
(NiCo)	0·32	⎬		Co	0·82	(Ni+Co)	0·64
(by diff.) As	13·59	⎭		Ni	0·61	As	13·06
CaCO³	0·73			Fe	0·36	(CaMg)Co³	0·71
				As	15·07		
	100·00			CaCO³	2·41		99·27
				MgCO³	0·60		
					99·23		

A notable fact is that in No. 3 the cobalt exceeds the nickel. Calcite is the gangue of this material. All manner of atomic ratios can be calculated from these analyses.

No. 1 has the ratio Cu $^{7 \cdot 1}$ As
No. 2   "   Cu $^{6 \cdot 7}$ As
No. 3   "   Cu $^{6 \cdot 46}$ As
No. 4   Cu $^{7 \cdot 7}$: As

In connection with these results, attention may be called to a communication made by Mr. J. Stanton, Secretary of the Mohawk Mg. Co., to the Engineering and Mining Journal of April 7, 1900. Mr. Stanton here gives Dr. Ledoux's report upon some of the Mohawk mineral. Dr. Ledoux figures from

analysis the ratio Cu_4As (including $Ni+Co$) and says that
ι this formula Mohawkite will probably be accepted by
ιce. If this ratio does exist then it is *not* Mohawkite ; for
name was given by me to Cu^3As (including $Ni+Co$). But
following sentence regarding the material induces me to
ιve that it was a mixture of Mohawkite and Whitneyite.
report says : "The sample was pulverized and concentrated
νashing until under a powerful glass no other minerals were
ent. Thus purified the mineral was analyzed with the
ιwing result : $Cu = 68·6$; $As = 22·67$; $Ni = 6·55$; $Co =$
$Fe = 0·23$; $S = 0·53$. The sulphur and iron are impuri-
[why ? if the powerful glass showed no other minerals
ent ? Koenig] and it is also reasonably probable that the
el and cobalt are mechanically mixed in the sample,
ough it is possible that some of the copper has been replaced
nickel and cobalt." How can Dr. Ledoux calculate a
ιula at all if he holds such views as here expressed ? He
s the specific gravity at $7·8$, but as I have shown above, a
position such as he finds must have a higher specific
'ity even than I find for the Mohawkite proper, namely
0, leaving out both iron and sulphur. The atomic quo-
ts of Dr. Ledoux's percentages are :

Cu........	1·089	As........ 0·302
Ni.........	0·111	
Co.........	0·020	

$$1·220$$

ng the ratio of $4:1$, but if nickel and cobalt are thrown
the ratio is nearer $3:1$ than $4:1$.
'nder the existing conditions I permit myself to doubt the
tence of a molecule $(Cu, Ni, Co)^4As$ and merely present
facts to the mineralogical public.

4. *Algodonite.*

his species has not yet been observed at the Mohawk mine,
ιgh some of the Mohawk-Whitneyite resembles it very
:h. I was induced to bring it within the present investiga-
merely by the specific gravity, which is given by Dr. F. A.
ιth as the only authority at $7·62$. To Dr. Lucius Hubbard
n indebted for the material. This gentleman possesses a
lsome specimen from the find at the Pewabic mine, many
's ago. The specimen looks like the segment of a nodular
:e. It showed the chocolate-brown dull tarnish. But the
h fracture is beautiful. It has a color and texture exactly

like razor-steel. Although very tough, it does not dent like Whitneyite. The nodule was covered with Whitneyite and this mineral shows on Dr. Hubbard's specimen like a thin fringe around the algodonite. I could break off some 5 grams of faultless material. The specific gravity was found at 21° C. = 8·383 (using 3·8418 grams). The analysis made with 0·5 gram. Of silver I could find no trace. The analysis gave

Cu	83·72	for Cu⁶As	
(Fe, Ni, Co)	0·08	Cu	83·5
As	16·08	As	16·5
	99·88		

Calculated from these figures, the specific gravity is 8·406. This gives a minus difference of 0·023 from my experimental value of 8·383.

Michigan College of Mines, June, 1900.

RT. XLIII.—*Heat of Solution of Resorcinol in Ethyl Alcohol;* by C. L. SPEYERS and C. R. ROSELL.

WHEN a system undergoes a change depending upon an xsorption of some particular energy, then as the intensity of ιat particular energy is increased on the outside of the chang- g system, the amount of change inside the system is ιcreased; when the system undergoes a change depending pon a rejection of some particular energy, then as the intensity that particular energy is increased on the outside of the ιanging system, the amount of change inside the system is minished. This statement is to be considered as universally ue though now and then a seeming exception comes up.

The latest seeming exception that we have come across is sorcinol with ethyl alcohol. Resorcinol dissolves in a large ccess of ethyl alcohol with rejection of heat* and yet the lubility of resorcinol increases with the temperature. That , a change depending upon a rejection of heat energy seems be favored instead of being hindered, by increasing the tensity of the heat energy on the outside of the system.

But the mistake lies in joining a heat of solution in a rge excess of solvent, giving a very dilute solution, to a solu- lity which implies a saturated solution and one more or less con- ntrated. The two solutions must be of equal concentration. ᵀherefore the heat of solution in a nearly saturated solution is ιe quantity that must be used, and its sign alone is enough, ɜ value is not needed, to test the validity of the above state- ent.

The sign can be found by diluting a saturated solution with small quantity of pure solvent. The method by supersatu- tion fails for resorcinol, does not crystallize fast enough from saturated solution even to get the sign; at least it did not do in the calorimeter we used.

About 200^{cc} of saturated solution at about $27°$ were diluted ith about 10^{cc} of pure solvent. The temperature rose more ιan $0·2°$.

The act of solution may be separated into two parts; namely, ιe in which the solute liquefies, which always involves a gain energy to the system and is +, the other in which the lique- ᵈd solute mixes with the rest of the liquid to form the sired solution. The second part may be + or − ; when it + then obviously the whole heat of solution must be +. nd this is the case for resorcinol, because concentrating a

* Speyers, Journ. Am. Chem. Soc., xviii, 146, 1896.

nearly saturated solution absorbs heat, and the heat of fusion is +.

Consequently the above statement does express the behavior of resorcinol with ethyl alcohol so far as heat of solution and solubility are concerned.

Since heat is rejected when resorcinol dissolves in a large excess of ethyl alcohol and since heat is absorbed when it dissolves in a small quantity, the temperature should not change when these substances are mixed in some certain proportion. This proportion was found to be about 6 grams of resorcinol with about 100 grams of ethyl alcohol.

The solubilities in all the following 37 cases increase with the temperature and so the heat of solution of all the solids in a nearly saturated solution should be +. This demands that the heat of fusion should be larger than the heat of mixing the liquid solute and solvent whenever the sign of this latter heat is −.

The signs refer to the heat of dilution of a saturated solution with about 5 per cent or 10 per cent of pure solvent.

	H_2O	CH_3OH	C_2H_6OH	C_3H_7OH	$CHCl_3$	C_7H_8
Urea	+	....	±	....	....	....
Urethane	+	+	+	+	−	+
Chloral hydrate...	,−	....	−	....	+	+
Succinimid	+	....	±	....	....	....
Acetamid	−	....	+	....	....	....
Mannitol.........	±	....	....	....	....	....
Resorcinol	+	....	−	....	....	....
Benzamid	....	....	±	....	....	....
p-Toluidin	....	...	+	....	....	....
Acetanilid	....	−	±	....	+	....
Acenaphthene	....	±:	±	±	−	−
Naphthalene	....	−	±	−	−	−
Phenanthrene	....	....	±	....	....	−
Cane sugar.......	−	....	....	....	....	....

Rutgers College, Sept. 1, 1900.

ART. XLIV.—*The Sulphocyanides of Copper and Silver in Gravimetric Analysis;* by R. G. VAN NAME.

[Contributions from the Kent Chemical Laboratory of Yale University—XCIX.]

Cuprous Sulphocyanide.

As early as 1854 attention was drawn by Rivot* to the possibility of estimating copper gravimetrically by weighing as cuprous sulphocyanide, and to the advantages which the process afforded in separating copper from other metals. Rivot's method of procedure consisted in dissolving the substance to be analyzed in hydrochloric acid, reducing the copper with hypophosphorous or sulphurous acid, and precipitating with potassium sulphocyanide. The precipitate dried at a moderate temperature was weighed as cuprous sulphocyanide and then as a control converted by ignition with sulphur into cuprous sulphide and weighed in that condition.

In his well known work upon quantitative analysis Fresenius in one place† denies the practicability of the direct weighing of copper as cuprous sulphocyanide on account of the tendency of the latter to hold water even when heated to the temperature of incipient decomposition. As authority for this statement he cites Claus,‡ who found 3 per cent of water in the precipitate after drying at 115°, and Meitzendorff, who gave the percentage of water under the same conditions as 1·54.

On a later page of the same volume,§ however, Fresenius, after a trial of the process which gave 99·66 per cent of the theory for copper, concludes that the method is practicable although apt to give low results, particularly in the presence of free acid.

The process was again recommended in 1878 by Busse,‖ who had employed it for the estimation of copper, both alone and in the presence of iron, nickel, zinc and arsenic, obtaining results very near the theory and plainly comparable with the figures obtained by afterwards igniting the cuprous sulphocyanide with sulphur in hydrogen.

In spite of the evident advantages for certain purposes of Rivot's method over other modes of determining copper, it has never come into general use. The chief reason for this has apparently been the difficulty and inaccuracy attendant upon the weighing of the precipitate upon dried paper filters, a process which can hardly be depended upon unless managed with extreme care.

* Comptes Rendus, xxxviii, 868.
‡ L. Gmelin Handb., iv, 472.
‖ Zeitschr. Anal. Chem., xvii, 53.
† Fresenius, 6th Aufl., i, 187.
§ 6th Aufl., i, 335.

In the experiments to be described this difficulty was avoided by performing the filtering and weighing upon asbestos in a perforated platinum crucible. The method of conducting a determination was as follows: A suitable quantity of a standard copper sulphate solution was run from a burette, diluted to a convenient volume, a few cubic centimeters of a saturated solution of ammonium bisulphite added, and the copper precipitated by an excess of ammonium sulphocyanide. The precipitate was allowed to settle, collected upon asbestos in a weighed crucible, washed with cold water and dried at 110° until no further loss of weight took place.

In Table I are given the results of a number of determinations made in this way. The copper sulphate solution was made up exactly decinormal and the standard confirmed electrolytically. As the ammonium sulphocyanide solution was slightly above decinormal, 13^{cm^3} represent a small excess (about one cubic centimeter) above the amount theoretically required to precipitate 25^{cm^3} of the copper sulphate solution. The ammonium bisulphite, which had been recently prepared by saturating aqueous ammonia with sulphur dioxide, was always used in sufficient quantity to give the liquid a strong and permanent odor of the latter.

TABLE I.

25^{cm^3} of N/10 $CuSO_4$ solution, equivalent to 0·0795 grm. Cu, taken for each experiment.

	H_2SO_4 concentrated. cm^3.	HNH_4SO_3 sat sol. cm^3.	NH_4SCN approx. N/10. cm^3.	Final volume. cm^3.	Time of standing. hours.	Cu found. grm.	Error grm.
1.	none	5	13	68	¼	0·0795	0·0000
2.	"	3	13	66	48	0·0793	−0·0002
3.	"	3	25	78	½	0·0796	+0·0001
4.	"	3	25	78	12	0·0796	+0·0001
5.	1·5	10	*13	85	12	0·0792	−0·0003
6.	1·5	8	13	105	48	0·0785	−0·0010
7.	1·5	3	25	85	4	0·0783	−0·0012
8.	1·5	5	25	85	21	0·0795	0·0000
9.	5	5	25	85	3	0·0797	+0·0002
10.	15	10	25	115	21	0·0793	−0·0002
	HCl concentrated cm^3.						
11.	10	5	25	100	20	0·0795	0·0000
12.	25	10	25	100	28	0·0784	−0·0011

When there is no free acid present the time of standing before filtration and the amount of the excess of ammonium sulphocyanide are practically without effect, as experiments 1 to 4 of the table show.

)eriments 5 to 10 were carried out in the presence of
ıs amounts of free sulphuric acid up to 12 per cent of
)tal volume of liquid. The acid, at least within this
does not exert a sufficient solvent effect upon the cuprous
)cyanide to interfere materially with the accuracy of the
ıs, but it retards the precipitation, making it necessary to
se the time of standing before filtering in proportion to
aount of acid present. In several of these determina-
:he precipitation was visibly incomplete even after sev-
ours standing. This effect of the acid, however, hardly
iu the results of the table because the standing was
s prolonged until the copper appeared to be all down
) filtering.

: low results of No. 7 was probably due chiefly to incom-
precipitation, although No. 9 shows that even with a
larger amount of acid precipitation may be complete
ı three hours. In general, however, it is safer to allow
time (twelve hours or more) for the precipitation when
is much free acid present.

nparison of Nos. 5 and 6, for which only a bare excess
monium sulphocyanide was used, with Nos. 7 to 12
an apparent advantage in the larger excess in the pres-
)f acid. Hydrochloric acid, judging from the results of
1 and 12, has no greater disturbing influence than sul-
: acid, although in No. 12, where the concentrated acid
tuted one-fourth of the entire volume, there was appar-
a slight solvent action. The filtrate from this determi-
ı when concentrated to about 25$^{cm^3}$ and treated with potas-
ferrocyanide gave a strong test for copper, as did also the
ə from No. 6. Several of the other filtrates were tested
same way, but none showed more than an insignificant
)f copper. No. 7, however, was not tested.

ıle II contains the results of a series of experiments con-
ı as before, except that larger amounts of copper were

TABLE II.

)u ɩen. m.	H$_2$SO$_4$ conc. cm^3.	NH$_4$SCN approx. N/10. cm^3.	Final volume. cm^3.	Cu$_2$S$_2$(CN)$_2$ found calc. as Cu. grm.	Error. grm.	Cu in filtrate.
ɩ75	none	60	500	·3176	+0·0001	none
ɩ75	"	60	500	·3177	+0·0002	"
ɩ75	"	60	500	·3176	+0·0001	"
ɩ75	10 HCl conc. cm^3.	100	500	·3175	0·0000	"
ɩ75	20	100	500	·3165	−0·0010	distinct

employed. The copper sulphate solution was approximately N/5 and standardized by the battery. The solution of ammonium sulphocyanide was the same previously used and a considerable excess was employed in every determination. More than twice the amount theoretically required was used in every case where free acid was present, and at least twenty hours were allowed for the precipitation, which was made in cold, and as the table shows, rather dilute solutions. If the solution is too concentrated the copper is apt to be thrown down in a finely divided condition, making it hard to filter.

The time required to dry the cuprous sulphocyanide at 110° is in general from two to three hours. Heating much longer than this is not to be recommended, as a gradual increase in weight begins to take place, as is shown by the following example, which gives a series of weights of the same precipitate at different stages.

	$Cu_2S_2(ON)_2$. grm.	Calculated as Cu. grm.
After 2 hours at 110°	·6060	·3167
" 4 " " "	·6059	·3167
" 19 " " "	·6067	·3171
" 23 " " "	·6069	·3172

This tendency to increase in weight is, however, usually less marked than in the above example, and in any case need not interfere materially with the accuracy of the process unless the drying is prolonged far beyond the necessary length of time.

The method is easily handled and as the results of Tables I and II show is capable of considerable accuracy. From the nature of the process it is evident that it is much less likely to be interfered with by the presence of other metals than the other gravimetric methods for copper, and may therefore be directly applied with good results in many cases where the use of the electrolytic or the oxide method would involve a previous separation.

Silver Sulphocyanide.

The sulphocyanide of silver, unlike that of copper, is soluble in an excess of ammonium or alkali sulphocyanides and this fact prevents the use of the latter to precipitate silver for gravimetric estimation. The reverse process, however, the precipitation of a soluble sulphocyanide by an excess of silver nitrate, as will be shown by the experiments to be described, furnishes a convenient means of standardizing sulphocyanide solutions and in general for estimating thiocyanic acid.

When freshly precipitated the sulphocyanide of silver resembles the chloride in appearance, but when allowed to

ıd a few hours becomes finely granular and is very easily
ıred and washed. It may be safely dried to a constant
ght upon an asbestos filter at 110°–120° but at a somewhat
her temperature is decomposed, leaving a residue of silver
ıhide.

'he determinations which are tabulated below were made as
ows. Portions of 25ᶜᵐˢ of an approximately decinormal
ıtion of ammonium sulphocyanide were measured from a
ette, diluted with 100ᶜᵐˢ of water and silver nitrate added
ıxcess. The precipitate was collected upon asbestos in a
inum crucible, washed with cold water and dried to a con-
ıt weight at 115° the drying requiring usually between two
three hours.

'he filtering is facilitated by allowing a few hours for the
:ipitate to settle; but this is by no means essential, as it is
, with a little care to obtain a clear filtrate even when the
ring is performed at once.

'he solution of ammonium sulphocyanide was prepared
ı a pure salt, especially tested and found free from choride.
s point is of importance, as chlorine is a common impurity
its presence in any considerable quantity will vitiate the
ılts.

TABLE III.

Final volume of liquid 150ᶜᵐ³

25ᶜᵐ³ of NH₄SCN sol. equivalent to 25·15ᶜᵐ³ of AgNO₃ sol.

	NH₄SCN cm³	AgNO₃ cm³	Excess of AgNO₃ cm³	AgSCN found. grm.
1.	25	25·3	0·15	.4372
2.	25	25·3	0·15	·4376
3.	25	25·4	0·25	·4373
4.	25	25·4	0·25	·4375
5.	25	30·4	5·25	·4382
6.	25	Rough excess.		·4366
7.	25	"		·4381
8.	25	"		·4373
9.	25			·4372
10.	25			·4369

ı order that the effect of varying the excess of silver
ht be investigated, an approximately decinormal solution
ıilver nitrate was titrated against the ammonium sulpho-
ıide and the ratio between the two solutions determined.
s silver nitrate solution was used for the first five determi-
ons of Table III. For the rest the quantity of silver nitrate
not measured but regulated by the eye alone, thus making

the conditions the same as would be the case in practical use of the method.

These results are as uniform as could be expected considering the variations which would be produced by even very small errors in measuring out 25^{cm^3} of decinormal sulphocyanide solution. It is moreover clearly shown that there is no difference in the results whether a bare excess or a moderately large excess of the silver nitrate is used.

The mean of the values in the last column is ·4374, which is equivalent to ·2006 grm. of ammonium sulphocyanide for every 25^{cm^3} of the solution.

The standard of the sulphocyanide solution was also determined volumetrically by Volhard's process. The mean of four titrations carried out with great care against a standard silver nitrate solution gave as the standard ·2003 grm of ammonium sulphocyanide for 25^{cm^3} of solution. This difference between the standards as determined by the two methods (one part in 670) is much less than the variations which frequently appear between successive determinations by Volhard's method, under like conditions as to strength of solutions and amounts used. It is about equal to the error that would be produced in a single volumetric determination by a mistake of one drop in measuring one of the solutions, or of one half drop in the same direction on each.

It is therefore evident that the standard of a sulphocyanide solution obtained in the above way may be applied directly to the estimation of unknown amounts of silver by Volhard's method without sensible error.

To remove a possible doubt as to whether the silver sulphocyanide dried at 115° was entirely free from water, a number of electrolytic determinations of the silver contained in the previously weighed precipitates of Table III were made in the following way.

The perforated platinum crucible containing the silver sulphocyanide and asbestos was hung in a loop of heavy platinum wire and served as the anode. For the cathode a deep platinum dish of about 200^{cm^3} capacity was used. An ammonical solution of potassium cyanide was employed as the electrolyte and gave the best results when made up by dissolving 2 grams of potassium cyanide in 15^{cm^3} of strong ammonia and 15^{cm^3} of water. The crucible which served as the anode was filled with this solution in full strength, and the remainder was put into the platinum dish and diluted to the required volume with water. In this medium the silver sulphocyanide is slowly dissolved and diffuses through the asbestos felt into the space between the electrodes where the silver is deposited in the usual way. This diffusion is, however, aided but little if at all by the current, and there is a tendency for traces of the silver to remain behind in the crucible. The current density employed was about ·0012 ampére per square centimeter of cathode surface and the time about twelve hours. After weighing the silver deposited, it was dissolved in nitric acid, precipitated by hydrochloric acid and weighed again as the chloride, giving a check upon the results.

Seven of the ten determinations of Table III were thus treated, but owing to the imperfections of the process the results were all slightly low, the worst showing a deficiency of ·0025 grm. of silver, an error of less than 0·9 per cent. The results of the two best of these determinations given below are, however, sufficient to prove the point in question, namely that the silver sulphocyanide dried at 115° has the theoretical constitution and contains no water. The numbers are those under which the determinations appear in Table III.

	AgSCN taken grm.	Calc. as Ag. grm.	Ag found by battery. grm.	Error. grm.	Weighed as AgCl. grm.	Calc. as Ag. grm.	Error. grm.
4.	·4375	·2844	·2839	−0·0005	·3765	·2834	−0·0010
10.	·4369	·2840	·2838	−0·0002	·3761	·2831	−0·0009

It is clear therefore that the estimation of sulphocyanides by precipitation with silver nitrate and direct weighing of the precipitate is wholly permissible. The method is extremely simple and, as has been shown, the results are quite accurate.

In conclusion I wish to thank Prof. F. A. Gooch for many valuable suggestions given during the course of this investigation.

SCIENTIFIC INTELLIGENCE.

I. CHEMISTRY AND PHYSICS.

1. *On the action of Permanganate upon Hydrogen Peroxide.*—Several views have been advanced in explanation of this well known decomposition, in which the permanganate is decolorized and oxygen is evolved. Berthelot believes that a trioxide of hydrogen is produced which undergoes spontaneous decomposition into water and oxygen. This idea is based upon the fact that he produced the decomposition at $-12°$ without effervescence taking place. Schönbein's theory is that the two substances simultaneously give up atomic oxygen which condenses to the molecular condition. It is the view of Weltzien and Traube that permanganic acid oxidizes the hydrogen of hydrogen peroxide to water, thus liberating oxygen. According to the last explanation all of the evolved oxygen originates in the hydrogen peroxide.

BAEYER and VILLIGER have recently shown that the phenomenon observed by Berthelot was due to the supersaturation of the liquid by oxygen, and that if the liquid be gently agitated the evolution of gas takes place regularly. They observe that the hypothetical formation of hydrogen trioxide is an assumption which is supported by the formation of potassium polysulphide when sulphur and potassium sulphide are brought together, and also by the formation of potassium tri- or tetra-oxide, as observed by Schöne, when a solution of hydrogen peroxide is evaporated with potash; but the assumption now lacks foundation in fact. According to Schönbein's theory both substances act as oxidizing agents. The authors believe, however, that hydrogen peroxide is hardly to be considered as such an agent, since in the pure condition it decomposes dilute hydriodic acid only very slowly. Moreover, if the theory is true, the more strongly oxidizing derivatives of hydrogen peroxide should oxidize permanganate still more easily and evolve oxygen, while in fact exactly the opposite is the case, for example, with " Caro's acid " which does not attack permanganate. This view, therefore, is not supported by facts. The authors strongly favor the view of Weltzien and Traube that it is the hydrogen of the hydrogen peroxide that is oxidized to water. Hydrogen peroxide thus shows a behavior analogous to that of hydrogen sulphide, with which it shows a striking similarity in many other ways. They urge the adoption of this theory, agreeing as it does with the present state of our knowledge, by the authors of elementary text-books.—*Berichte,* xxxiii, 2488.   H. L. W.

2. *An Instance of Trivalent Carbon: Triphenylmethyl.*—In a preliminary paper GOMBERG describes a body produced by the action of zinc upon triphenylchlormethane, $(C_6H_5)_3C\,Cl$, which is an unsaturated compound apparently containing but one carbon atom and corresponding to the formula $(C_6H_5)_3C$. The substance

combines with atmospheric oxygen with great readiness, forming di-triphenylmethylperoxide $(C_6H_5)_3C.O_2.C(C_6H_5)_3$. This property renders the preparation of the unsaturated hydrocarbon in the pure state extremely difficult, and although the author has succeeded in obtaining large crystals of it, he has not yet been able to wash them without the formation of the peroxide. If subsequent investigation confirms the preliminary conclusions, the compound will possess great theoretical interest as being the first unsaturated compound of the kind, and furnishing the first definite evidence of the existence of the trivalent carbon atom.—*Jour. Am. Chem. Soc.*, xxii, 757. H. L. W.

3. *On the Phosphorescence of Inorganic Chemical Preparations.*—This subject has been studied by E. GOLDSTEIN. The method of investigation consisted in placing the substance under examination in a properly shaped tube and waving the latter back and forth in a cone of cathode rays. The nature of the phosphorescence could be determined from the length of the observed trail of light. The examination of various pure samples of a substance showed that the color of light attributed to it by previous investigators becomes weaker as the substance becomes purer. On the contrary, the light which appears at the point of contact with the cathode rays, and which lasts only during the illumination, becomes continually brighter as the purity increases. The substances investigated may be divided into two groups in respect to their phosphorescent light. In the case of the sulphates, carbonates, phosphates, borates, silicates, chlorides, bromides, fluorides, oxides, and hydroxides of Li, Na, K, Rb, Cs, Ca, Sr, Ba, Al, Zr, Mg, Be, Zn, and Cd the color is blue, violet-blue, or in a few instances violet. Salts of Cu, Cr. Mn, U, Ni, Co, Pb, Ce, La, Y, Er, Pr, and Ne belong to the other group. The latter, when admixed in very small amounts with compounds of the first group, give a light which in most cases exceeds the blue light in intensity and duration, but corresponding phenomena could not be observed when reciprocal additions from the first group were made. There is a maximum in the strength of the colored light, for it decreases again when the small amount of substance from group 2 is increased beyond a certain limit. The presence of moisture may possibly be assigned as the reason for the similar luminosity of the substances of the first group. The effect of additions from the second group may possibly be due to greater power of absorption or else to the occurrence of decomposition.— *Chem. Central-Blatt*, 1900, ii, 756. H. L. W.

4. *Weight of Hydrogen Desiccated by Liquid Air.*—LORD RAYLEIGH has found that the density of hydrogen dried by cooling with liquid air is the same as that obtained when the gas is dried with phosphorus pentoxide. This was to be expected, for it was shown by Morley in an article published in this Journal in September, 1887, that this oxide is a practically perfect dryer for gases, but it is satisfactory to have a confirmation of this important fact.—*Proc. Roy. Soc.*, lxvi, 334. H. L. W.

5. *The Separation of Tungsten Trioxide from Molybdenum Trioxide.*—M. J. RUEGENBERG and EDGAR F. SMITH describe a method of separation for the oxides mentioned in the title. It is based upon the fact that tungstic acid is insoluble in concentrated or dilute sulphuric acid, hot or cold, whereas molybdenum trioxide is very easily and rapidly dissolved. Upon trial it was found that sulphuric acid of specific gravity 1·378 was well suited for the purpose. Test analyses were made by mixing about 1 or 2 grams of molybdic anhydride with varying quantities of tungstic anhydride and digesting with 25^{cc} of warm sulphuric acid of the strength mentioned above for a few minutes, then filtering, washing with water containing sulphuric acid, and weighing the tungstic acid. The results of six analyses were extremely sharp. The authors found also that tungstic acid may be separated from ferric hydroxide in precisely the same way.—*Jour. Am. Chem. Soc.*, xxii, 772. H. L. W.

6. *On Radio-active Barium and Polonium.*—When concentrated uranium nitrate solution is mixed with a little sulphuric acid, then with a solution of a barium salt, care being taken that sufficient barium salt is not added to cause a precipitate, and finally diluted with water, a strongly radio-active precipitate of barium sulphate is obtained. When this sulphate is converted into a soluble barium salt and this is treated with ammonium hydroxide, a small amount of precipitate is obtained which is even more strongly radio-active than the original sulphate; but when the barium is subsequently converted into carbonate, it has completely lost its activity. The radio-activity is probably due to small quantities of radium, or more probably of actinium. Polonium preparations obtained from lead chloride from uranium residues have proved to consist of bismuth hydroxide, which, either in this form or as oxychlorides exhibits strong activity.— *Berichte*, xxxiii, 1665. *Jour. Chem. Soc.*, lxxviii, 480.

7. *Lecture Experiments Illustrating the Electrolytic Dissociation Theory and the Laws of the Velocity and Equilibrium of Chemical Change.*—A. A. NOYES and A. A. BLANCHARD have published a description of seventeen experiments which were originally devised as an accompaniment to an extended course of lectures on theoretical chemistry. Most of the principles illustrated are of such fundamental importance that they should soon be generally introduced into elementary lecture courses on inorganic and analytical chemistry. The experiments are devised with much skill and will furnish very striking illustrations to the audience. They are described in great detail, so that they may be readily performed by the lecturer, and the principles illustrated are clearly stated in each case. It is certain that this series of experiments will be of great assistance to teachers of theoretical chemistry, and it will doubtless serve to hasten the introduction of instruction in the principles of chemical equilibrium and of electrolytic dissociation into elementary courses.—*Jour. Am. Chem. Soc.*, xxii, 726. H. L. W.

8. *Physikalisch-chemische Propädeutik ;* von H. Griesbach, zweite Hälfte, 3 Lieferung, Band I, pp. 945–992 and Band II, pp. 1–352. Leipsic, 1900 (Wilhelm Engelmann). The first two installments of this work were noticed in this Journal, December, 1896. The present, fourth installment gives a most favorable impression of the author's success in continuing the excellence of his work. The size of the book as well as its scope have been enlarged beyond the original intention, so that a second volume is now begun. In giving a general view of modern scientific knowledge the work will be valuable to many classes of readers.

H. L. W.

9. *A School Chemistry ;* by John Waddell. 12mo, pp. xiii, 278. New York, 1900 (The Macmillan Company).—This text-book on elementary chemistry possesses many excellent features. The author has shown much originality in his presentation of the subject, good judgment in regard to the scope of the book, and accuracy in his statements of facts and theories. The interrogatory method is largely employed with the object of stimulating the pupil's thought. It is the aim of the book to help the student in the discovery of facts, to enable him to see their connections, and to show how facts lead to theory. H. L. W.

10. *Leçons de Chimie Physique ;* by J. H. van't Hoff. Volume III. Relations between the properties and the composition. Translated from the German edition by M. Corvisy. 8vo, pp. 170, Paris, 1900 (A. Hermann).—This book is a very literal translation of the concluding volume of van't Hoff's "Vorlesungen über Chemie" which appeared early in the year. The subject treated is, however, practically independent of the two earlier volumes. The relations between the physical properties and the composition are first taken up. Under this heading, come the relations and behavior of matter in regard to space, pressure, temperature, heat, surface tension and optical properties. The second half of the book treats of the relations between the chemical properties and the composition.

Van't Hoff has made a wide departure from other books on physical chemistry which have thus far appeared. He aims not to treat the subject exhaustively but to take up the more important or more interesting parts, and these are treated in his usual brilliant manner. H. W. F.

11. *On the Viscosity of Gases as affected by Temperature ;* by Lord Rayleigh.—A former paper* describes the apparatus by which I examined the influence of temperature upon the viscosity of argon and other gases. I have recently had the opportunity of testing, in the same way, an interesting sample of gas prepared by Professor Dewar, being the residue, uncondensed by *liquid hydrogen,* from a large quantity collected at the Bath springs. As was to be expected, it consists mainly of helium, as is evidenced by its spectrum when rendered luminous in a vacuum tube. A line, not visible from another helium tube, approximately in the position of D, (Neon) is also apparent. The result

* This Journal, ix, 365.

of the comparison of viscosities at about 100° C. and at the temperature of the room was to show that the temperature effect was the same as for *hydrogen.*

In the former paper the results were reduced so as to show to what power (*n*) of the absolute temperature the viscosity was proportional.

	n	*c*
Air	0·754	111·3
Oxygen	0·782	128·2
Hydrogen } Helium }	0·681	72·2
Argon	0·815	150·2

Since practically only two points on the temperature curve were examined, the numbers obtained were of course of no avail to determine whether or no any power of the temperature was adequate to represent the complete curve. The question of the dependence of viscosity upon temperature has been studied by Sutherland, on the basis of a theoretical argument which, if not absolutely rigorous, is still entitled to considerable weight. He deduces from a special form of the kinetic theory as the function of temperature to which the viscosity is proportional

$$\frac{\theta^{\frac{1}{2}}}{1+c/\theta}$$

c being some constant proper to the particular gas. The simple law $\theta^{\frac{1}{2}}$, appropriate to "hard spheres," here appears as the limiting form when θ is very great. In this case, the collisions are sensibly uninfluenced by the molecular forces which may act at distances exceeding that of impact. When, on the other hand, the temperature and the molecular velocities are lower, the mutual attraction of molecules which pass near one another increases the number of collisions, much as if the diameter of the spheres was increased. Sutherland finds a very good agreement between his formula (1) and the observations of Holman and others upon various gases.

If the law be assumed, my observations suffice to determine the values of *c*. They are shown in the table, and they agree well with the numbers for air and oxygen calculated by Sutherland from observations of Obermayer.—*Proc. Roy. Soc.*, lxxvii, 137.

12. *Absorption of gases by glass powder.*—Early investigators claimed that the absorption of gases ceased in a few days; that it increased with increasing pressure; diminished with increasing temperature; and that the presence of moisture had little influence. Bunsen on the other hand found that the absorption continued even for years; increased with increasing temperature, and was largely independent of variations of pressure. The cause of the absorption he claimed to be in the capillary coating of moisture. P. MÜLFARTH (Bonn dissertation) gives an exhaus-

tive list of references of work upon this subject. His conclusions vary from those of the previous investigators.

Fully dried glass powder absorbs a large amount of CO_2. The amount increases with diminishing temperature and rises with increasing temperature: it is accomplished in a few hours. The presence of moisture hinders this absorption. The absorption of SO_2 is similar to that of CO_2: it is accomplished in a couple of hours—increases with increasing pressure, diminishes with increasing temperature. The following absorb in the order written: C_2H_2, N_2O, CO_2, SO_2, NH_3. The investigation bears upon the preparation of X-ray tubes, and also upon the question of the influence of the glass walls of vessels in spectrum tubes.— *Ann. der Physik*, No. 10, 1900, pp. 328-352.   J. T.

13. *Normal lines of iron.*—Rowland has published a list of iron lines which serve for normals in spectra comparisons. KAYSER finds that this list is not sufficient for certain regions of the solar spectrum, and, therefore, has given a large additional number between wave lengths 2300 and 4500. The author claims a higher accuracy than that reached by Rowland since every wave length in his table is the mean of from 6–15 measures, taken with gratings of different orders. The probable error lies between 0·001 and 0·003.—*Ann. der Physik*, No. 10, 1900, pp. 195-203.
   J. T.

14. *Arc spectra of some metals in an atmosphere of hydrogen.*— HENRY CREW finds that a hydrogen atmosphere profoundly modifies certain arc spectra of metals, and that all the lines in the arc spectra which are affected by hydrogen belong to the spark spectrum also. Moreover that the lines which belong to Kayser and Runge's series are unaffected by the change from air to hydrogen. —*Phil. Mag.*, Nov., 1900, pp. 497-505.   J. T.

15. *Recognition of the Solar Corona independent of the total eclipse.*—H. DESLANDRES has made some observations at Meudon which apparently show the possibility of observing the corona under ordinary conditions by the aid of a thermopile.— *Comptes Rendus*, Oct. 22, 1900.   J. T.

16. *Loss of charge by evaporation.*—Various physicists have arrived at different conclusions on this subject. W. CRAIG HENDERSON, working in the Cavendish laboratory, has investigated the subject anew and comes to the conclusion that there is no appreciable loss of charge by the evaporation of a charged liquid.—*Phil. Mag.*, Nov., 1900, pp. 489-497.   J. T.

17. *On the action of the coherer.*—Previous investigations seem to show that the action of the coherer is due to the welding action of minute sparks in the powders which are employed. T. MIZUNO of Kyoto University has studied the behavior of a large number of substances. He finds that the action of the iron powder coherer is conspicuously irregular and concludes that the action of the coherer can plausibly be explained on the hypothesis of the welding action of minute sparks.—*Phil. Mag.*, Nov., 1900, pp. 445-459.   J. T.

18. *Loss of electrical charges in air which is traversed by ultra-violet rays.*—P. LENARD published a paper on the effect of ultra-violet rays on gas in the Annalen der Physik, No. 1, 1900, p. 486. He now continues his investigation and shows that the effect of ultra-violet light is fourfold. The air is divided into bearers of negative electricity which appear to be charged atoms or molecules; bearers of positive electricity of larger dimensions; nuclei of vapor which are non-electrified; and ozone.—*Ann. der Physik*, No. 10, 1900, pp. 298–319.   J. T.

19. *Handbuch der Spectroscopie* von H. KAYSER. Erster Band. Pp. xxiv, 781; large 8vo; 781 figures in the text. Leipzig, 1900 (S. Hirzel).—The preparation of a comprehensive work on a subject which has had so remarkable a development as that of spectroscopy in its different branches, is a labor of such magnitude that we may well honor the energy of the author who ventures to undertake it. But here we have not simply the promise but also its fulfilment in the completion of a considerable part of the undertaking. The present volume of nearly eight hundred pages by Professor Kayser, is so thorough and so satisfactory a treatment of the portions of the subject here included, that the reader feels justified in looking forward with confidence to the completion of the remaining four volumes promised as a well-rounded whole.

The volume opens with an admirable summary, scientific and yet thoroughly readable, of the history of spectroscopy from the time of Newton down to the present. Although an exhaustive treatment of this subject might well extend the chapter to the limits of a volume by itself, the author presents the salient points in the successive steps of progress, and the contributions of each worker so clearly, that the chapter as a whole leaves little to be desired. In the treatment of the delicate question of the claims of earlier writers to the discoveries which are usually connected with the name of some physicist who has come later, the author shows much discrimination and critical care; for example in regard to the relative amount of credit to be given to Wollaston as contrasted with Fraunhofer, or to Talbot, Miller and Stokes as compared with Kirchhoff. The same critical method appears throughout. The five other chapters of the volume deal with the practical working of spectroscopy. The first of these discusses the production of incandescent vapors in the flame, arc-light, or vacuum tubes. This subject is treated both historically with a multitude of references to original authorities and descriptively as needed by a present worker. The two chapters following discuss prisms and diffraction gratings, giving the theory, construction and use in detail. In the former subject the author has been aided by Dr. H. Konen, while Prof. C. Runge has furnished for the volume his theory of concave gratings. The concluding chapters are devoted to spectroscopic apparatus and measurements, the use of photography and other methods. Here, as throughout the volume, we are struck with the simplicity and

'ectness of statement, and the thoroughness with which the
rk of the many physicists who have contributed is recognized.
'en where only a very brief allusion is possible to some matters
minor importance, they are not ignored, but the literature ref-
:nces at the bottom of each page embrace all that has been
ne on the subject. The great steps of progress that have been
ide, as the Rowland grating, the methods for the investigation
the infra-red spectrum by Abney, the echelon spectroscope of
chelson, are treated with all necessary fullness.

The second volume is to be devoted to Kirchhoff's law and its
isequences, emission spectra of different kinds and as obtained
der varying conditions of pressure, temperature, magnetic
'ce, etc. The third volume will include the absorption phe-
mena with also fluorescence and phosphorescence; the fourth
lume will give all spectroscopic data available in regard to
:h individual element, and a fifth is also planned to include the
:ctroscopy of the heavenly bodies including the Sun.

20. *A Brief Course in General Physics, experimental and*
plied ; by GEORGE A. HOADLEY, Swarthmore College. Pp.
3, New York, 1900 (The American Book Company).—This is a
ar, systematic statement of the fundamental principles and laws
physics, such as is needed by the high school student ; it is
:ompanied by good illustrations and a well-chosen series of
periments. The subjects in general are well treated, although
occasional points open to criticism ; thus we find in one article
brief statement of surface tension and in one following, an
planation of capillary attraction, but there is no suggestion
to any connection between the two subjects.

II. GEOLOGY AND MINERALOGY.

1. *U. S. Geological Survey, 20th Annual Report 1898–99.*
rt III. Precious Metal Mining Districts. (Washington,
)0, pp. 595, pls. 77.)—This volume consists of three papers, two
them having appended reports by specialists.

[, *A preliminary report on the Bohemia mining region of*
stern Oregon, with notes on the Blue River mining region and
the structure and age of the Cascade range by J. S. DILLER,
iich is accompanied by a report on the fossil plants associated
tb the lavas of the Cascade range by F. H. KNOWLTON (pp.
34, pls. 1–6). The brief sketch of the district shows that the
:ks of Calapooya Mountain, upon which the district lies, are
iolly of igneous origin. They consist of dacite porphyry,
lesites and basalts with volcanic tuffs. It is in these rocks
it the mines are situated which are described in the report.
e veins are gold-bearing. The fossil plants found in the tuffs
)w that the latter were deposited in lakes of Miocene age.
e various localities from which fossil leaves have been col-
ted are mentioned and in his special report Knowlton describes

these leaves, showing them to represent a flora of Miocene age made up of conifers, oaks, etc., and including several new species.

II, *Gold and silver veins of Silver City, De Lamar and other mining districts in Idaho*, by W. LINDGREN (pp. 65–256, pls. 7–35). This begins with a sketch of the general geology of western central Idaho, accompanied by a geologic map, based in part on actual surveys made by the author and in part on reconnaissance work. The map shows a central area of post-Carboniferous granite of great size, bordered on the east by Carboniferous beds in which it has produced contact metamorphism. The Snake River runs through the southern and, turning to the north, the western portions of the area and along its course is bordered by immense outflows of basalts, some of Miocene, others of Pliocene age. The Silver City and De Lamar district is in this southern part on a granite ridge which rises above the flood of basalt and rhyolite. The gold and silver veins of this district, their mode of occurrence, the associated minerals, the individual mines, etc., are carefully and fully described. In the same manner the author gives an account of the Wood River district, where the silver-lead ores occur in veins cutting the Carboniferous beds. At Florence and Warren, situated in the great granite area, the veins are gold quartz and some placer mining is done. In the Seven Devils district the ore is copper in the contact zone of the granite mass with the Carboniferous beds on its western side. The ores are bornite and chalcopyrite and are supposed to have been deposited by pneumatolytic processes. The paper contains a wealth of careful observations on a large number of ore deposits and is a most important contribution to the study of ore deposits in general.

III, *Geology of the Little Belt Mountains of Montana*, with notes on the mineral deposits of the Neihart, Barker, Yogo and other districts by W. H. WEED, with a report on the petrography of the igneous rocks of the district by L. V. PIRSSON.

The Little Belt Mountains are an anticlinal uplift of a rather broad, even character. In the central portion erosion has disclosed the fundamental Archean gneisses and schists upon which the later sediments lie. These range through the Belt formation (here having a considerable local development and supposed to be of Algonkian age), the Cambrian, Siluro-Devonian, Carboniferous, up to the Cretaceous forming the outer foot hills, slopes and boundary plains of the area. The streams rising in basins of Belt or Cambrian shales pass through deep and narrow canyons or gateways of Carboniferous limestone. Chiefly on the outer margin of the range, along the spring of the arch, great laccolithic intrusions of igneous material have taken place, which now present either wholly or partly uncovered masses of porphyries of several types. Accompanying them are numbers of intruded sheets and dikes.

The basins cut in shale, the flat-topped plateaux of limestone, the deep gorges cut on the outer flanks of the range and the reat mountains of igneous rock form the dominating features of opographic relief and express the geologic relations. The ormations are described and also the local geology of the various istricts is given in detail and several features of general interest a structural and dynamical geology, particularly with reference o the intrusion of laccoliths are brought out.

The ores are found chiefly in two districts; at Barker where hey are of the type of silver-lead ores occurring in limestones at he contact of intrusive igneous masses, and at Neihart where atrusions of diorite and other rocks have occurred in the archean complex. At this place the ores occur in a well defined ein system and are gold and silver, the latter occurring in the alphides, ruby silver, polybasite, etc. The important mines, heir ores and workings are described.

The petrographical report contains a full description of the gneous rocks which are chiefly porphyries of several kinds, ranite, syenite and diorite porphyries with intermediate types. At Yogo peak is a differentiated mass changing from granite orphyry through monzonites to shonkinite. The intruded heets and dikes present several rock varieties, the most interest- ag of which are the minettes and analcite basalts. Of all these ypes detailed analyses by W. F. Hillebrand and H. N. Stokes re given and in the theoretical discussion of the rocks of the istrict are used to show that all of the main types are so related o one another in a rock series that any one of them may be xpressed in terms of the others; these relations being shown by diagram in graphic form. L. V. P.

2. *Iowa Geological Survey, Vol. X. Annual Report, 1899, ith accompanying papers.* Samuel Calvin, State Geologist, I. F. Bain, Assistant State Geologist. Pp. 1–166, plates i–xi, gures 1–102, and 10 folded maps, 1900.—The Geological Board f the State of Iowa is to be congratulated both upon the dmirable organization of its geological workers and upon the qually admirable results it is annually producing. The board is onstituted of the governor, the auditor of State, two college residents and the president of the Iowa Academy of Sciences, hus combining for effective service the forces of politics, educa- ion and science. The geological staff for the year, besides the eologist and assistant geologist, is drawn from the officers of six f the colleges of the State and one high school and one from the hicago University. The papers of the present volume are hiefly connected with the detailed mapping of the geology of ounties not previously so mapped. Forty-two counties in all ave been mapped, eight of them having been completed within ae year. The great value of the topographical sheets prepared y the U. S. Geological Survey is acknowledged, particularly in espect to the Dubuque region, the important lead mines of rhich are scientifically described in the chapter on Dubuque ounty, prepared by the geologist and assistant geologist.

Mr. Frank Wilder brings to notice, the fact "that the Hawarden beds, heretofore referred to the Ft. Pierre, really belong to the Benton shales and accordingly the limestones and chalk of the Inoceramus beds, as developed in Iowa, are not to be correlated with the Niobrara farther west. They represent instead that portion of the Fort Benton sub-stage exposed in the 'Oyster Shell Rim' of the Black Hills." The principle of coöperation expressed in the organization of the survey, is further carried out in the formation of a joint commission to investigate the clay resources of the State. This commission is composed of the Iowa Brick and Tile Makers' Association, Iowa Engineering Society, the State College, and the State University. The results of their work during the year is given in Dr. Beyer's appended report on the Mineral Production of Iowa, in 1899.   H. S. W.

3. *Cleopatra's Emerald Mines.*—An account is given in the November number of the Geological Journal of an expedition led by Mr. DONALD A. MAC ALISTER to the so-called Cleopatra's Emerald Mines in the neighborhood of Jebel Sikait in Northern Etbai. The locality is situated nearly due east of Edfu, and some fifteen miles in a direct line from the Red Sea, between 25° and 24° 30′ N. Lat. The place was reached by the expedition starting from Daraw. The mountain, Jebel Sikait, has an altitude of 1800 feet above sea-level and commands an extensive view over the wild and desolate surrounding country.

The emeralds occur in mica schist and talc schist, which rocks occur over a large area. The mica schist is in some cases highly micaceous, looking like a finely bedded, contorted sandstone, and again contains very little quartz. In some cases the rock is much hydrated, passing into a soft, powdery rock. Tourmaline, garnet, actinolite, chrysolite are common minerals of the locality. The schists of Jebel Sikait are inclined at an angle of about 45°, and overlie gneiss on one side, while on the other they are inclosed by a tough, green serpentine, of which the upper part of the mountain is composed. The search for emeralds was carried on two thousand years ago very extensively, a large number of mines being opened and extensive workings being carried forward. The author remarks in regard to this subject:

"The mining is of a most primitive character. The ancients simply excavated, in the likely emerald-bearing schist, a network of long and very tortuous passages just large enough to allow of the body being dragged through, and only in a very few cases was any attempt made at stopping (or excavating) the entire seam. It has been suggested that the passages were made small on account of the absence of timber, suitable to be hewn into supports, which would be necessary to prevent collapses in the case of larger excavations. So small are the openings, that a casual observer would not notice the existence of a mine at all but for the gray *débris* thrown out at the mouth. Along the schists at Sikait alone we visited considerably over a hundred mines, some of which took more than an hour to crawl through. That these mines have been worked at widely distant periods, is evident from the different styles of work. There are seven or eight groups of mines in different places within a couple of hours or so of Jebel Sikait."

4. *Handbuch der Mineralogie*, von Dr. CARL HINTZE. Erster Band, fünfte Lieferung; pp. 641–800. Leipzig, 1900 (Veit & Company).—The fifth part of the first volume of Hintze's Mineralogy (No. 17 of the entire series) has recently been issued. It contains the closing pages in regard to the pyrrhotite group, descriptions of the species of the cinnabar group, and the opening portion of the pyrite group. The treatment is as thorough and exhaustive as that which has characterized the parts already published.

5. *Contributions to Chemistry and Mineralogy from the Laboratory of the U. S. Geological Survey*, F. W. CLARKE, Chief Chemist. Pp. 166, Washington, 1900. (Bulletin 167, U. S. G. Surv.)—This Bulletin, like others which have preceded it from the same source, contains a series of chemical and mineralogical researches carried on in the laboratory under the charge of Prof. Clarke. Several of the papers included, as that on the constitution of tourmaline, and others, have already been published in part or entire in this Journal. Among others may be mentioned a series of papers by H. N. Stokes on the chloronitrides of phosphorus and the metaphosphimic acids.

6. *Johnstonotite, a supposed new Garnet.*—W. A. MACLEOD and O. E. WHITE have given the name johnstonotite (after R. M. Johnston) to a brownish-yellow garnet occurring in trapezohedral crystals in trachyte at Port Cygnet, Tasmania. An analysis gave:

SiO_2	Al_2O_3	FeO	MnO	MgO	CaO	Ign.
36·87	7·28	17·12	13·68	12·49	11 98	0·29 = 99·71

It would seem to be analogous to the spessartite of Colorado, which is similar in occurrence; however, although regarded as differing from other garnets in the relative proportions of the bases, the analysis cannot be accepted as accurate since (as noted by the reviewer, L. J. Spencer) the formula only approximates to that of garnet, even if all the iron is made Fe_2O_3.—*Proc. R. Soc. Tasmania*, 1898–99, 1900, 74, noticed in *J. Chem. Soc.*, lxxviii, 663.

III. MISCELLANEOUS SCIENTIFIC INTELLIGENCE.

1. *Annual Report of the Board of Regents of the Smithsonian Institution showing the operations, expenditures and condition of the Institution for the Year ending June 30, 1898.* Pp. lv, 713. Washington, 1899.—The report of the Secretary of the Smithsonian Institution, Prof. S. P. LANGLEY, which has just been issued, contains the usual interesting summary of the recent progress of the Institution in its various departments of research and exploration, collections, library and publications, and foreign

exchange. The growth in the *National Zoological Park* is noted in particular and also the results obtained in the *Astrophysical Observatory* by Mr. C. G. Abbott, the Aid in charge. Professor Langley states that the work of the last year "has resulted in the discovery and determination of position of 500 new absorption lines, so that we have now over 700 new lines of well-determined positions, and we may be said now to know, by the aid of the bolometer and the labors of the observatory, more lines in this invisible spectrum than were known in the visible one up to the great research of Kirchhoff and Bunsen, which opened the era of modern spectrum analysis. Moreover, these lines, the exactness of whose determination has now reached a surprising degree of perfection through the recent improvements in the delicacy as well as the precision of both bolometer and galvanometer, and through other improvements in the apparatus (improvements due principally to the present Aid acting in charge), depend not only on the instruments, but on the labors of those who have used them, the comparator measurements alone having included, as stated in the body of the report, about 44,000 separate observations.

"A great deal of other work has been done at the observatory, but nothing which in importance and present and prospective interest compares with this main research in the infra-red spectrum, which is now known throughout nearly the whole of the invisible portion of the solar energy, and extends through a range of wave lengths considerably over twelve times that known to Sir Isaac Newton, the present exact knowledge of this region being due not exclusively, but it may properly be said principally, to the labors of this observatory.

"I call attention in this connection to the interesting remarks made in the report to the effect that very marked changes of absorption have been observed at various parts of the infra-red spectrum. In one part of the invisible region a decrease in absorption, amounting to nearly half the total, took place in February, and this abnormal state continued until May, when the usual condition gradually returned. As this change is found to occur yearly at about the same period, the idea presents itself that the growth of vegetation, so rapid in these months, may abstract from the air large quantities of vapor active in absorption at this point in the spectrum, but this interesting possibility has not yet, it will be understood, been fully verified."

The second part of the volume, pp. 101–696, contains a series of well-selected papers on important topics, many of them not readily accessible in the original, so that their republication here will be a great convenience to many readers.

2. *Report of the U. S. National Museum under the direction of the Smithsonian Institution for the year ending June 30, 1898.* Pp. xviii, 1270; plates 1–36. Washington, 1900.—This volume contains the report by the Director, Mr. Charles D. Walcott, on the condition and progress of the United States National Museum

during the year ending June 30th, 1898; also the reports of the head curators, W. H. Holmes in Anthropology, F. W. True in Biology, and G. P. Merrill in Geology; further lists of accessions to the collections and library (pp. 1–152). Part II, following, is given to an extended paper by the late Professor E. D. Cope, on the Crocodilians, Lizards, and Snakes; this covers pages 153 to 1270, with thirty-six plates and numerous text figures.

The Director, in his brief statement, shows the remarkable growth which the Museum has made and emphasizes the imperative need of a larger, more dignified and more convenient building, not simply for exhibition purposes, but also to include laboratories and a lecture hall. He shows further the directions in which it is to be desired that the National Museum should develop in the immediate future, and the great service which it may thus do to the country at large. It is greatly to be hoped that his appeal may be regarded and adequate provision made, not only for the present needs of the Museum but also for its future growth.

3. *Publications of the Earthquake Investigation Committee in Foreign Languages.* Tokyo, 1900. No. 3, pp. 1–103. No. 4, pp. 1–141.—These publications of the Japan Earthquake Committee contain a series of papers of much interest to seismologists, chiefly in English, in part also in French. Among these are accounts, with seismical records, of several destructive earthquakes (1891–1894). A paper of general interest is by H. NAGAOKA, on *the elastic constants of rocks and the velocity of seismic waves.* This gives a long series of observations on specimens of rocks from Archean chlorite schist, through the Paleozoic to quartz-sandstones and andesites of the Tertiary. The modulus of elasticity, and modulus of rigidity were determined in each case, as also the density, and from these data the corresponding velocities for elastic waves were then calculated. The author finds a distinct gradation in the elastic constants from the higher values characteristic of the oldest rocks down to those of the youngest; the velocity of wave-propagation is not, however, proportionally larger, since the density also enters into the problem. Thus for an Archean chlorite schist a velocity of 6·4 to 7 kilometers per second is obtained for longitudinal waves; for a Paleozoic clay slate a velocity of 4·5 to 6; while for a Tertiary sandstone the value is only 2·2. The experiments made seem to show that the elastic constants increase more rapidly than the density as the rocks become more dense, and consequently elastic waves travel with greater velocity in the interior than on the surface of the earth's crust. Eruptive rocks are more isotropic than those of non-igneous origin and have inferior elasticity, but there is with them also the same distinction as to age. The schistose structure of the deeper rocks gives a greater value of the elastic constants in a certain particular direction, coinciding with that of swiftest propagation of elastic disturbances. It seems to be shown that the ratio of elastic constants to density, and hence

the velocity of elastic waves, increases from the surface downward to *a stratum of maximum velocity*, probably at a considerable depth, and then diminishes again. The conclusion of Professor Omori from records in Italy and Japan, that the velocity of the first tremor is almost always equal to 13 kilometers per second, while longitudinal waves in steel have only a velocity of 6·2, is difficult to account for; it may be explained by the probable supposition that the rocks deeper in the earth's crust have elastic constants several times greater than those near the surface. The velocity named would roughly correspond to the values $E = 6 \times 10^{12}$ and $\rho = 3·5$. The high value of the elastic constant relative to the density is not impossible, since, in passing from Tertiary to Archean rocks an increase of density from 2 to 3 is here shown to be accompanied by an increase of the modulus of elasticity more than ten times in some cases.

4. *The Graduate Bulletin of the University of Nebraska.* Volume I, No. 1, pp. 1–68. Lincoln, Nebraska, 1900.—Among the papers here published is one by E. J. Rendtorff, on acromatic polarization with crystalline plates; another by C. C. Engberg on the Cartesian oval, and one by C. A. Fisher on the geology of Lincoln and environs.

5. *National Academy of Sciences.*—The autumn meeting of the National Academy was held at Providence, Nov. 13–15th. The following is a list of the papers read (several by title):

J. TROWBRIDGE: Investigations of light and electricity with the aid of a battery of twenty thousand cells.

ALPHEUS HYATT: Progressive evolution of characters in the young stages of Cephalopods. Descriptive method of presenting the phenomena of the cycle of evolution among Cephalopods.

T. W. RICHARDS: The porous cup voltameter. An account of the study of growing crystals by instantaneous microphotography.

S. L PENFIELD: Stereographic projection and some of its possibilities from a graphical standpoint.

C. S. MINOT: On the development of the pig. Normal plates illustrating the development of the rabbit and the dogfish.

T. C. MENDENHALL: Note on the energy of recent earthquakes.

A. A. MICHELSON: Spectrum of sodium in a magnetic field. Progress in the Echelon spectroscope.

H. A. ROWLAND: A report of the spectrum work carried on with the aid of a grant from the Bache Fund. On the explanation of inertia and gravitation by means of electrical phenomena.

A S PACKARD: Distribution of philogeny of Limulus. Male preponderance (androrhopy) in Lepidopterous insects.

A. MICHAEL: The synthesis and reactions of sodium acetate ester, and their relation to a new interpretation of chemical metathesis On the genesis of matter

C. BARUS: Demonstration of the projection of one grating by another. Exhibition of certain novel apparatus: A wave machine; an expansion lens; a recording system of two degrees of freedom; A tube showing colored cloudy condensation On stability of vibration and on vanishing resonance Hysteresis-like phenomena in torsional magnetostriction and their relation to viscosity.

CHARLES D. WALCOTT: Folding and faulting of strata in the Cordilleran Area.

J. M CRAFTS: Report on the Meeting of the Committee of the International Association of Academies held at Paris.

INDEX TO VOLUME X.*

A

, C. G., observations at Wash-
n Astrophysical Laboratory,

)tion of gases by glass powder,
arth, 462.

my, National, meeting at Prov-
:e, 472.

ty, Optical, and Chemical Com-
ion, Landolt, 76.

, F. D., flow of marble under
mre, 401.

ating Currents, joint trans-
ion with direct, Bedell, 163.

surements with, Rowland-Potts,

ator, string, Honda and
uzu, 64.

ws, C. W., geology, etc., of
stmas Island, 79.

iation, American, meeting at
York, 169.

:ish, meeting at Bradford, 169.

pheric electricity, variation in,
tveau, 161.

B

Land, Lower Silurian fauna,
chert, 81.

, C., torsional magnetostric-
407.

, F., joint transmission of di-
and alternating currents, 163.

er, C. E., restoration of Stylo-
s lacoanus, 145.

of N. America, Cory, 89.

NY.

x of oriental Asia, Franchet,
.

es from Alaska, Holm, 266.

raceæ, studies in, No. xiii,
lm, 88; No. xiv, Holm, 266.

tyledones, anatomy, Solereder,

enhair tree (Gingko), Seward,
8.

Brazil, schists of gold regions, Derby,
207.
British Museum, catalogues, 170.
Browning, P. E., separation of
nickel from cobalt, 316.

C

Cady, W. G., energy of cathode
rays, 1.
Canada, Geol. Survey, Summary
report for 1899, 165; vol. xi,
399.
— report on minerals, Hoffmann,
404.
Cathode Rays, effect on conductivity
of gases, McLennan, 319.
— energy of, Cady, 1.
— and X-light, Rollins, 382.

CHEMICAL WORKS:
Elementary Inorganic Chemistry,
Ostwald, 394.
— Inorganic Chemistry, Shenstone,
395.
Leçons de Chimie Physique, Van't
Hoff, 461.
Physikalisch-chemische Propädeu-
tik, Griesbach, 461.
School Chemistry, Waddell, 461.

CHEMISTRY.
Acetylene, products of explosion,
Mixter, 299.
Arsenic acid, iodometric estimation,
Gooch and Morris, 151.
Barium, radio-active, Lengyel, 74;
artificial, Debierne, 393; atomic
weight, Curie, 392.
— and polonium, 460.
Cæsium, properties of, Eckhardt
and Graefe, 75.
Carbon monoxide poisoning, Sachs,
395.
— trivalent, Gomberg, 458.
Chromite, method of decomposing,
Fieber, 76.

Index contains the general heads, BOTANY, CHEMISTRY (incl. chem. physics), GEOLOGY,
LS, OBITUARY, ROCKS, and under each the titles of Articles referring thereto are
ed.

CHEMISTRY.

Copper arsenides, artificial, Koenig, 439.

— estimation as oxalate, etc., Peters, 359.

— sulphocyanides, Van Name, 451.

Electrolytic dissociation theory, Noyes and Blanchard, 460.

Hydrogen, desiccated by liquid air, weight of, Rayleigh, 459.

Iron, atomic weight, Richards and Baxter, 75.

— determination of ferrous, Hillebrand and Stokes, 893.

Mercuric antimonide, 75.

— titration by sodium thiosulphate, Norton, 48.

Nickel, separation from cobalt, Browning and Hartwell, 316.

Permanganate, action upon hydrogen peroxide, Baeyer and Villiger, 458.

Radinm, spectrum, Runge, 396.

Resorcinol, solution in ethyl alcohol, Speyers and Rosell, 449.

Selenium interference rings, Longden, 55.

Silver sulphocyanides, Van Name, 451.

Stibonium compounds, 75.

Sulphur, forms of, 161.

Telluric acid, compounds with iodates, Weinland and Prause, 76.

Triphenylmethyl, Gomberg, 458.

Tungsten trioxide, separation from molybdenum trioxide, Ruegenberg and Smith, 460.

Uranium, radio-activity, Crookes, 318.

Christmas Island, geology, Andrews, 79.

Clarke, F. W., rock analyses of U. S. Geol. Survey, 250.

— Contributions to Chemistry and Mineralogy, 469.

Coherer, action of, Mizuno, 463.

Colorado, carnotite, etc., from, Hillebrand and Ransome, 120.

— tellurides from, Palache, 419.

Colorado Canyon, geology, Davis, 251.

Corona, see solar.

Cory, Birds of N. America, 89.

Crookes, Sir Wm., radio-activity of uranium, 318.

D

Davis, B., stationary sound-waves, 281.

Davis, W. M., notes on the Colorado Canyon, 251.

Day, A. L., gas thermometer at high temperatures, 171.

Derby, O. A., schists of the gold and diamond regions, Brazil, 207 ; notes on monazite, 217.

Devil's Lake, Wis., geology, Salisbury and Attwood, 248.

Diller, J. S., Bohemia Mining region of Oregon, 465.

Douglass, E., new species of Merycochærus in Montana, 428.

E

Earthquake investigation committee, publications, 471.

— waves, velocity, 471.

Electric charge, loss by evaporation, Craig, 468 ; loss in air from ultraviolet radiation, Lenard, 464.

— charges, magnetic effect of moving, Cremieu, 896.

— conductivity of gases traversed by cathode rays, McLennan, 219.

— discharge in rarefied gases, Skinner, 76.

— measurements, Rowland's methods, Potts. 91.

— pressure, when dangerous, Weber, 896.

— resistance, experiments on high, Part I, Rood, 285.

Electricity, atmospheric, variation of, Chauveau, 161.

Electrolytic Dissociation, Jones, 76; theory of, Noyes and Blanchard, 460.

Emerald Mines of Cleopatra, MacAlister, 468.

Emerson, B. K., Carboniferous bowlders from India, 57 ; bivalve from Connecticut river Trias, 58.

Expansion, co-efficient of, for platinum, porcelain, etc., Holborn and Day, 172.

F

Farrington, O. C., mineralogical notes, 83.

Finland, Geol. Survey Bulletins, 249.

— Bjurböle meteorite, 250.

Flight, possibilities of, Rayleigh, 77.

Flink, minerals from Greenland, 323.

Ford, W. E., Calcite crystals from Union Springs, N. Y., etc., 237.

G

Gas thermometer at high temperatures, Holborn and Day, 171.

Gases, absorption by glass powder, Mulfarth, 462.

GEOLOGICAL REPORTS.
Alaska, 398.
Canada, Summary report for 1899, 165; 11th annual report, 399; reporton minerals, Hoffmann, 404.
Indiana, geology and natural resources, Blatchley, 398.
Iowa, vol. x, 467.
Michigan, vol. vii, 399.
United States. 20th annual report, 320, 397, 465.
— Bulletin No. 157, 168; No. 158, 249; No. 167, 469.
—Monographs, xxxiii, 168; xxxiv, 247; xxxvii, 166; xxxviii, 247.

GEOLOGY.
Bohemia Mining region of Oregon, Diller and Knowlton, 465.
Carboniferous bowlders from India, Emerson, 57.
Christmas Island, geology, Andrews, 79.
Colorado Canyon, geology, Davis, 251.
Copper-bearing rocks of Douglas Co.. Wis., Grant, 249.
Cycads of Wyoming, Ward, 321; fossil, in the Yale Museum, Ward, 327.
Face de la Terre, la, Suess-Margerie, 167.
Flora of Lower Coal Measures of Missouri, D. White, 166.
Fossils, invertebrate, from So. Patagonia, Ortmann, 368.
Glacial gravels of Maine, Stone, 274.
— lobe, Illinois, Leverett, 247.
Gold and silver veins in Idaho, Lindgren, 466.
Jurassic plants, Seward, 328.
— vertebrates, new, Knight, 115.
Little Belt Mountains, Mont. geology, Weed and Pirsson, 466.
Louisiana, geology, Harris and Veatch, 81.
Lower Silurian fauna of Baffin Land, Schuchert, 81.
Marble, flow of, under pressure, Adams and Nicolson, 401.
Merycochœrus in Montana, Douglass, 428.
Mesozoic floras of United States, Ward, 320.
Moraines of South Dakota, Todd, 249.
Narragansett Basin, geology, Shaler, Woodworth and Foerste, 163; Carboniferous fauna, Packard, 164.

Paleobotany, Elements, Zeiller, 88.
Panama geology, Bertrand, 82.
Pflanzenpalæontologie, Potonié, 88.
Preglacial drainage in Michigan, Mudge, 158.
Schists of gold and diamond regions, Brazil, Derby, 207.
Stylonurus lacoanus, restoration, Beecher, 145.
Tertiary Mammal horizons of Europe and America, Osborn, 400.
Trias of Connecticut R., bivalve from, Emerson, 58.
Wealden flora of Bernissart, Seward, 322.
Gooch, F. A., iodometric estimation of arsenic acid, 151.
Grammar of Science, Pearson, 170.
Grant, U. S., Copper-bearing rocks of Wisconsin, 249.
Greenland, new minerals from, Flink, 328.
Gregory, H. E., volcanic rocks of Temiscouata Lake, Quebec, 14.
Groth, Zeitschrift für Krystallographie, Index, 84.

H
Hall, C. W., rocks of Southwestern Minnesota, 168.
Hall effect in flames, Marx, 395.
Hart, J. H., action of light on magnetism, 66.
Hartwell, J. B., separation of nickel from cobalt, 316.
Hill, A., Introduction to Science, 406.
Hillebrand, W. F., carnotite, etc., of Colorado, 120; rock analyses, 250.
Hintze, C., Mineralogie, 469.
Hoadley, G. A., Brief Course in General Physics, 465.
Hoffmann, G. C., Canadian minerals, 404.
Holborn, L., gas thermometer at high temperatures, 171.
Holm, T., studies in the Cyperaceæ, No. xiii, Carex willdenowii, 83; No. xiv, 266.
Honda, K., string alternator, 64.
Huxley, T. H., Elementary Physiology. 90.
Hydrogen, spectrum, Trowbridge, 222.

I
Idaho, gold and silver veins, Lindgren, 466.

India, Carboniferous bowlders from, Emerson, 57.

Indiana, geology and natural resources, Blatchley, 398.

Interference rings, selenium, Longden, 55.

Iowa Geological Survey, vol. x, 467.

J

Jones, H. C., Electrolytic dissociation, 76.

K

Kayser, H., Handbuch der Spectroscopie, 464.

Knight, W. C., new Jurassic vertebrates, 115.

Knowlton, Bohemian mining region of Oregon, 465.

Koenig, G. A., on mohawkite, stibiodomeykite, domeykite, algodonite, etc., 439.

Kohlrausch, F., Practical Physics, 320.

L

Landolt, H., Optical Activity and Chemical Composition, 76.

Langley, S. P., Smithsonian Report for 1898, 469.

Latitude, variation, earth magnetism, etc., Halm, 405.

Leverett, F., Illinois glacial lobe, 249.

Light, action on magnetism, Hart, 66.

Lindgren, W., gold and silver veins in Idaho, 466.

Longden, A. C., selenium interference rings, 55.

Louisiana, geology, Harris and Veatch, 81.

M

Magnetic effect of moving electric charges, Cremieu, 396.

— screening, DuBois and Wills, 77.

Magnetic Pole, South, location, Borchgrevink. 405.

Magnetism, action of light on, Hart, 66.

Magnetostriction, torsional, Barus, 407.

Maine, glacial gravels, Stone, 247.

Margerie, E. de, translation of Suess Antlitz der Erde, 167.

Melting points of the metals, Holborn and Day, 187.

Meteoric stone, Bjurböle, Finland, 250.

Michigan Geol. survey, vol. vii, 399.

— preglacial drainage in, Mudge, 158.

Mineral analyses, interpretation, Penfield, 19.

— Tables, Weisbach, 5th ed., 84.

— veins, enrichment, Weed, 82.

Mineralogie, Handbuch, Hintze, 469.

Mineralogy, Elements of, Moses and Parsons, 405.

Minerals, determination by maximum birefringence, Pirsson and Robinson, 260.

MINERALS.

Algodonite, 447. Ancylite, Greenland, 324. Aragonite and calcite, relative stability, 392.

Beryl (emerald) mines in Egypt, 468. Britholite, Greenland, 325.

Calcite crystals, Joplin, 84 ; Union Springs, N. Y., 237 ; Cumberland, England, 240, 241 ; Lancashire, 242 ; Montreal, 243. Caledonite, New Mexico, 84. Carnotite, Colorado, 120. Celestite, Canada, 404. Chalcolamprite, Greenland, 324. Chromite, method of decomposing, 76. Cordylite, Greenland, 324. Corundum, No. Carolina, 295. Cubosilicite, 168. Cupro-goslarite, Kansas, 168.

Domeykite, 439.

Endeiolite, Greenland, 325. Epistolite, Greenland, 325.

Florencite, Brazil, 404.

Goldschmidtite, Colorado, 422.

Hessite crystals, Colorado, 426. Hübnerite, Nova Scotia, 404. Hydromagnesite, British Columbia, 404.

Inesite, Mexico, 83.

Johnstonotite, Tasmania, 469.

Leucosphenite, Greenland, 324. Lorenzenite, Greenland, 324.

Melite, 168. Mohawkite, 440. Monazite, Brazil, Derby, 217. Müllerite, 168.

Narsarsukite, Greenland, 324. Natron, British Columbia, 404.

Polycrase, Canada, 404.

Robellazite, 168. Roscoelite, Colorado, 128, 130.

Schizolite, Greenland, 325. Spodiophyllite, Greenland, 324. Stibiodomeykite, 445. Sylvanite, Colorado, 419.

Tainolite, Greenland, 324. Tellurides, Colorado, 419. Tourmaline, constitution, Penfield, 19. Turquois, chemical composition, 346.

Whitneyite, 446.

Missouri, flora of Lower Coal Measures, D. White, 166.

Mixter, W. G., products of explosion of acetylene, 299.

Montana, Geology of Little Belt Mountains, Weed and Pirsson, 466.

— vertebrate fossils from, E. Douglass, 428.

Morris, J. C., iodometric estimation of arsenic acid, 151.

Moses, A. J., Mineralogy, 405.

Mudge, G. H., preglacial drainage in Michigan, 158.

N

Nagaoka, H., velocity of seismic waves, 471.

Narragansett Basin, geology of, Shaler, Woodworth and Foerste, 163 ; Carboniferous fauna, Packard, 164.

Nebraska, Bulletin of the University of, 472.

Nicolson, J. T., flow of marble under pressure, 401.

Nipher, F. E., properties of light-struck photographic plates, 78.

Norton, J. T., Jr., titration of mercury by sodium thiosulphate, 48.

Norwegian North Atlantic Expedition, 170.

O

OBITUARY.
Keeler, J. E., 325.
Ulrich, G. H. F., 250.

Ortmann, A. E., invertebrate fossils from Patagonia, 368.

Osborn, H. F., Tertiary Mammal horizons of Europe and America, 400.

Ostwald, Elementary Inorganic Chemistry, 394.

Ostwald's Klassiker der exakten Wissenschaften, 90, 406.

P

Packard, A. S., Carboniferous fauna of Narragansett Basin, 164.

Palache, C., tellurides from Colorado, 419.

Paleobotany, Zeiller, 88 ; Potonié, 88.

Panama, geology of, Bertrand, 82.

Parsons, C. L., Mineralogy, 405.

Patagonia, invertebrate fossils, Ortmann, 368.

Penfield, S. L., interpretation of mineral analyses and constitution of tourmaline, 19.

— calcite crystals from Union Springs, N. Y., etc., 237.

— composition of turquois, 346.

Peters, C. A., volumetric estimation of copper, etc., 359.

Phosphorescence of inorganic chemical preparations, Goldstein, 459.

Photochromie, Lehrbuch, Zenker, 162.

Photographic plates, properties of light-struck, Nipher, 78.

Photometry, Manual of, Stine, 320.

Physics, Brief Course in General, Hoadley, 465.

— Practical, Kohlrausch, 320.

Physiology, Elementary. Huxley, 90.

Pirsson, L. V., determination of minerals by maximum birefringence, 260.

— Little Belt Mountains, Montana, 466.

Potonié, Pflanzenpalaeontologie, 88.

Potts, L. M., Rowland's methods of electric measurements with alternating currents, 91.

Pratt, J. H., two new occurrences of corundum in No. Carolina, 295.

R

Radio-active barium, Lengyel, 74.

— barium and polonium, 460.

— uranium, Crookes, 318.

Ransome, F. L., carnotite, etc., of Colorado, 120.

Rayleigh, possibilities of flight, 77.

— viscosity of gases as affected by temperature, 461.

— weight of hydrogen desiccated by liquid air, 459.

Robinson, H. H., determination of minerals by birefringence, 260.

ROCKS.

Analyses by Hillebrand, Stokes, etc., collated by Clarke, 250.

— statement of, Washington, 59.

Gneisses, etc., of S. W. Minnesota, Hall, 168.

Igneous rocks of Temiscouata Lake, Quebec, Gregory, 14

Ijolite, Kunsamo, Finland, 249.

Marble, flow, under pressure, Adams and Nicholson, 401.

Quartz-muscovite rock, Belmont, Nevada, Spurr, 351.

Scapolite rocks from Alaska, Spurr, 310.

Schists of gold and diamond regions, Brazil, Derby, 207.

Rollins, W., cathode stream and X-light, 382.

Röntgen rays, nature of, Rollins, 382.

Rood, O. N., experiments on high electrical resistance. Part I, 285.

Rosell, C. R., solution of resorcinol in ethyl alcohol, 449.

Rowland, H. A., electric measurements with alternating currents, 91.

Safford, J. M., Geology of Tennessee, 399.

Salisbury, R. D., surface geology of Wisconsin, 248.

Schuchert, C., Lower Silurian fauna of Baffin Land, 81.

Science, Introduction to, Hill, 406.

Scientia, 406.

Seward, A. C., Jurassic plants, 323; maidenhair tree (*Gingko*), 322; Wealden flora of Bernissart, 322.

Shaler, N. S., geology of Narragansett Basin, 163.

Shenstone, W. A., Inorganic Chemistry, 395.

Shimizu, S., string alternator, 64.

Smithsonian Institution, Report for 1898, 469.

Solar corona, nature of, Pringsheim, 77; recognition, independent of total eclipse, Deslandres, 463.

— eclipse, 89.

Sound waves, stationary, Davis, 231.

Spectra of some metals in an atmosphere of hydrogen, Crew, 463.

Spectroscopie, Handbuch der, Kayser, 464.

Spectrum, normal lines of iron, Kayser, 463.

— of hydrogen and aqueous vapor, Trowbridge, 222.

Speyers, C. L., solution of resorcinol in ethyl alcohol, 449.

Spurr, J. E., scapolite rocks from Alaska, 310; quartz-muscovite rock, Belmont, Nevada, 351.

Stevens, J. S., method of measuring surface tension, 245.

Stine, W. M., Photometry Manual, 320.

Stone, G. H., glacial gravels of Maine, 247.

Suess, Antlitz der Erde, translation by Margerie, 167.

Surface tension, measurement of, Stevens, 245.

T

Telegraphy, wireless, Vallot and Lecarme, 396.

Temiscouata Lake, Quebec, volcanic rocks of, Gregory, 14.

Temperatures, measurement of high, by gas thermometer, Holborn and Day, 171.

Tennessee, geology, Safford and Killebrew, 399.

Tierreich, das, Schulze, 89.

Todd, J. E., moraines of South Dakota, 249.

Trowbridge, J., spectra of hydrogen, spectrum of aqueous vapor, 222.

U

United States Geol. Survey, see GEOL. REPORTS.

— National Museum, Report, 470.

V

Van Name, R. G., sulphocyanides of copper and silver in gravimetric analysis, 451.

Van't Hoff, J. H., Leçons de Chimie Physique, 461.

Viscosity of gases, as affected by temperature, Rayleigh, 461.

W

Waddell, J., School Chemistry, 461.

Ward, L. F., Mesozoic floras of the U. S., 320.

— fossil cycads in the Yale Museum, 327.

Washington, H. S., statement of rock analyses, 59.

Weed, W. H., enrichment of mineral veins, 82.

— Little Belt Mountains, Montana, 466.

Weisbach, Determination Tables, 84.

White, D., Flora of Lower Coal Measures of Missouri, 166.

Wisconsin, copper-bearing rocks, Grant, 249.

— Dalles of the, Salisbury and Attwood, 248.

X

X-rays, see Röntgen rays.

Z

Zeiller, Paleobotany, 88.

Zenker, Lehrbuch d. Photochromie, 162.

Zoological results from New Britain, etc., Willey, 89.

GENERAL INDEX

OF

VOLUMES I-X OF THE FOURTH SERIES.

☞ In the references, **heavy-faced type** is used for the numbers of the volumes.

NOTE—The names of minerals are inserted under the head of MINERALS; all obituary notices are referred to under OBITUARY. Under the heads BOTANY and BOTAN. WORKS, CHEMISTRY and CHEM. WORKS, GEOLOGY, ROCKS, the references to the topics in these departments are grouped together; in many cases the same references appear also elsewhere.

A

Abbot, C. G., longitudinal aberration of prisms, **2**, 255.
— observations at the Washington Astrophysical Observatory, **9**, 214; **10**, 470.
Abney, W. de W., color sensations in terms of luminosity, **1**, 464.
Absorption of gases by glass powder, Mülfarth, **10**, 452.
— infra-red waves by rock salt, etc., Rubens and Trowbridge, **5**, 33.
— of light in a magnetic field, Rhigi, **7**, 239.
Academy of Sciences, National, meeting at Boston, 1897, **4**, 479; New Haven, 1898, **6**, 512; New York, 1896, **2**, 462; 1899, **8**, 474; Providence, 1900, **10**, 478, Washington, 1896, **1**, 493; 1897, **3**, 426; 1898, **5**, 396; 1899, **7**, 401; 1900, **9**, 393.
— — Report for 1897, **5**, 397.
— Washington, Proceedings, **9**, 394.
Acetylene, see **CHEMISTRY**.
Acoustics, researches in, Mayer, **1**, 81.
— See **Sound**.
Actinometers, electro-chemical, use of, Rigollot, **5**, 228.
Action at a distance, Drude, **4**, 390
Activity, Optical, and Chemical Composition, Landolt, **10**, 76.
Adams, F. D., new alkali hornblende and titaniferous andradite, Ontario, **1**, 210.
— Grenville and Hastings series of Canadian Laurentian, **3**, 173.
— flow of marble under pressure, **10**, 401.

Adams, G. I., extinct Felidæ of North America, **1**, 419; **4**, 145.
Adams, J. S., Law of Mines and Mining in the U. S, **6**, 436.
Aerolite, see **Meteorite**.
Africa, Origin of Culture in, Frobenius, **7**, 80.
Agassiz, A., Great Barrier Reef of Australia, **2**, 240.
— coral reefs of the Fiji group, **5**, 113; **8**, 80.
— explorations of the "Albatross" in the Pacific, **9**, 33, 109, 193, 369.
— Tertiary limestone reefs of Fiji, **6**, 165.
Agricultural Experiment Stations of the Univ. of California, report, **1**, 158.
Air, composition of expired, Billings, Mitchell and Bergey, **1**, 154.
— diffusion through water, Barus, **9**, 397.
— liquid, density, Ladenburg and Krugel, **9**, 64.
— stratified brush discharges in, Toepler, **7**, 67.
— See **CHEMISTRY**.
Air-thermometer bulb, measurement, Cady, **2**, 341.
Air-waves, observation of, Emden, **5**, 70.
Alabama, Geological Survey, publications, **4**, 393; **9**, 69.
— iron-making in, Phillips, **7**, 398.
— report of the valley regions, McCalley, **3**, 350.
Alaska, Klondike gold fields, **5**, 305; Cape Nome gold region, **9**, 455; explorations in, **10**, 398.
— Caribes in, Holm, **10**, 266.

Albatross, explorations in the Pacific, Agassiz, **9,** 33, 109, 193, 369, 390 ; report on fishes by S. Garman, **9,** 390.

Aldous, J. C. P., Physics, **6,** 100

Algebra, Hall and Knight, **1,** 328 ; of Quantics, Elliott, **1,** 328.

Allen, E. T., native iron in coal measures of Missouri, **4,** 99.

Alternating Currents, D. C. Jackson and J. P. Jackson, **2,** 455.

— change into direct, Graetz, **5,** 70.

— joint transmission with direct, Bedell. **10,** 160.

— measurements with. Rowland, **4,** 429 ; Rowland and Penniman, **6,** 97 ; **8,** 35 ; Rowland and Potts, **10,** 91.

Alternator, string, Honda and Shimizu, **10,** 64.

Aluminum, acoustic properties, Mayer, **1,** 103.

— See **CHEMISTRY.**

Ameghino, Fl., publications of, Roth, **9,** 261.

American Association for the Advancement of Science, see **Association.**

— Geological Society, see **Geological.**

— Microscopical Society, transactions, **4,** 256 ; **5,** 321 : **8,** 399.

— Museum of Natural History, catalogue of types, **9,** 69.

Ames, J. S., Theory of Physics, **3,** 420.

— Manual of Experiments in Physics, **5,** 302.

— Zeeman effect, **6,** 99 ; Free Expansion of Gases, **6,** 504.

— Prismatic and diffraction Spectra, **7,** 69.

Andrews, C. W., geology, etc., of Christmas Is., **10,** 76.

Aneroid, coil, Barus, **1,** 115.

Antillean Valleys, submarine, Spencer, **6,** 272.

Appalachians, extension across Mississippi, etc., Branner, **4,** 357.

Arctic Discoveries, Greeley, **1,** 403.

— sea ice, geological action, Tarr, **3,** 223.

Argentine, Paleozoic faunas, Kayser, **5,** 72.

Argon, spectra of, Trowbridge and Richards, **3,** 15.

— See **CHEMISTRY.**

Arizona, Governor's report, **7,** 169 ; report on petrified forest, L. F. Ward, **9,** 461.

Arkansas, Devonian interval in northern, Williams, **8,** 139.

Arkansas, thickness of Paleozoic in, Branner, **2,** 229.

Artesian wells in the Dakotas, Darton, **5,** 161.

Association, American, meeting at Buffalo, 1896, **2,** 90, 307 ; at Detroit, 1897, **4,** 160, 250 ; at Boston, 1898, **5,** 397 ; **6,** 199, 363 ; at Columbus, 1899, **8,** 86, 238, 311 ; at New York, 1900, **9,** 461 ; **10,** 169.

— Australasian, **2,** 397.

— British, meeting at Liverpool, 1896, **2,** 90, 397 ; at Toronto, 1897, **4,** 324 ; at Bristol, 1898, **6,** 372 ; at Dover, 1899, **8,** 238 ; at Bradford, 1900, **10,** 169.

Astronomical Observatory, Harvard College, Publications, **1,** 75 ; **3,** 252, 492 ; **5,** 80, 320 ; **8,** 87 ; **9,** 311.

— — University of Chicago, Publications, **9,** 311.

Astronomy, Elements of, Campbell, **8,** 88.

— Herschels and Modern, Clerke, **1,** 76

Astrophysical Observatory at Washington, observations, Abbot, **9,** 214 ; **10,** 470.

Atkinson, E, Electricity and Magnetism, **2,** 83.

Atmospheric air, see **Air.**

— electricity, variation in, Chauveau, **10,** 161.

Atom, relation to the charge of electricity carried, Thomson, **1,** 140.

Atoms, existence of masses smaller than, Thomson, **8,** 463.

— in Space, Arrangement of, Van't Hoff, **5,** 388.

— ions and molecules, color relations, Lea, **1,** 405.

— masses of, Berthelot, **9,** 62.

Audition, limits of, Rayleigh, **4,** 69.

Auroral period, 27-day, Clayton, **5,** 81.

Austin, M., oxidation of manganese, **5,** 260 ; estimation of manganese, **5,** 209, 382.

— manganese as pyrophosphate, **6,** 233.

— ammonium-magnesium phosphate, **7,** 187.

— double ammonium phosphates of beryllium, zinc, and cadmium, **8,** 206.

— ammonium-magnesium arseniate, **9,** 55.

Australia, Great Barrier Reef of, Agassiz, **2,** 240.

— meteorites, Ward, **5,** 135.

Avery, E. M., School Physics, **1,** 57.

B

Bacteria and decomposition of rocks, Brauner, **3**, 438.

Baffinland, observations, Bell, **4**, 476; Lower Silurian fauna of, Schuchert, **10**, 81.

Bailey, G. H., Tutorial Chemistry, **3**, 357; **5**, 390.

Bailey, L. H., Plantbreeding, **1**, 151; Survival of the Unlike, **3**, 77; Lessons with Plants, **5**, 318; Evolution of our Native Fruits, **7**, 78.

Bailey, W. W., Botanizing, **9**, 80.

Bain, H. F., limestone at Bethany, Missouri, **5**, 433.

Baldwin, DeF., action of acetylene on the oxides of copper, **8**, 354.

Bancroft, W. D., The Phase Rule, **4**, 67.

Barbour, E. H., on Dæmonelix, **4**, 77.

Barlow, A. E., Grenville and Hastings series of Canadian Laurentian, **3**, 173.

Barnes, C. R., Analytic keys to N. American Mosses, **3**, 354; Plant Life, **7**, 244.

Barringer, D. M., Law of Mines and Mining in the U. S, **6**, 436; Minerals of Commercial Value, **5**, 155.

Barrington-Brown, rubies of Burma, **1**, 64.

Barus, C., counter-twisted curl aneroid, **1**. 115.

— liquid carbon dioxide, **2**, 1.

— interferential induction balance, **3**, 107; excursions of a telephone diaphragm, **3**, 219.

— combination tones of siren and organ pipe, **5**, 88; electro-magnetic theory of light, **5**, 343.

— colloidal glass, **6**, 270; compressibility of colloids, **6**, 285.

— thermodynamic relations of hydrated glass, **7**, 1.

— motion of a submerged index thread of mercury, **9**, 139; thermodynamic relations of waterglass, **9**, 161; diffusion of air through water, **9**, 397.

— torsional magnetostriction, **10**, 407.

Bascom, F., volcanic rocks of South Mt., Pennsylvania, **3**, 160.

Bashore, H. B., glacial gravels in Lower Susquehanna Valley, **1**, 281,

Battelli, H., physical papers from the University of Pisa, **3**, 493.

Battery, large storage, John Trowbridge, **3**. 246.

Bauer, Survey of Maryland, **6**, 513.

Baumhauer, Crystal-Symmetry, **9**, 73.

Bayley, W. S., rocks of Crystal Falls iron-bearing district, **9**, 451.

Beach, F. E., General Physics, **7**, 314.

Bearpaw Mountains, Montana, geology, Weed and Pirsson, **1**, 283, 351; **2**, 136, 188.

Becker, G. F., gold fields of the S. Appalachians, **1**, 57.

— rock differentiation, **3**, 21; method of computing diffusion, **3**, 280.

— fractional crystallization of rocks, **4**, 257.

— Kant as a natural philosopher, **5**, 97; auriferous conglomerate of the Transvaal, **5**, 193; determination of plagioclase feldspars, **5**, 349.

Becquerel rays, Elster and Geitel, **8**, 463; **9**, 378; chemical effects of, Curie, **9**, 145, 144; deviation in magnetic field, Giesel, **9**, 147.

— and Röntgen rays in a magnetic field, Strutt, **9**, 376.

— See also **Uranium-radiation** and **Radio-active**.

Bedell, F., Principles of the Transformer, **2**, 453; joint transmission of the alternating and direct currents, **10**, 163.

Beecher, C. E., morphology of Triarthrus, **1**, 251; comparative morphology of the Galeodidæ, by Bernard, **1**, 491.

— natural classification of the trilobites, **3**, 89, 181.

— origin of spines, **6**, 1, 125, 249, 329.

— notice of Othniel Charles Marsh, **7**, 403.

— Conrad's types of Syrian fossils, **9**, 176; Uintacrinus from Kansas, **9**, 267.

— restoration of Stylonurus lacoanus, **10**, 145.

Bell, R., rising of land around Hudson Bay, **1**, 219; observations on Baffinland, **4**, 476.

Bergen, J. Y., Elements of Botany, **3**, 490.

Berlin Academy of Sciences, 200th anniversary, **9**, 312.

Bermudas, geology of, Verrill, **9**, 313.

Bernard, H. M., comparative morphology of the Galeodidæ, **1**, 491.

Bibliotheca Zoologica II, **1**, 77; **6**, 103.

Bigelow, F. H., solar and terrestrial magnetism, **5**, 455; international cloud work for the U. S., **8**, 433.

Billings, composition of expired air, **1,** 154.

Binary star, 70 Ophiuchi, **1,** 159.

Biological Lectures, Wood's Holl, 1895, **2,** 457.

— Variation, Davenport, **8,** 399.

Birds of Colorado, Cooke. **4,** 326.

— of Eastern N. America, Cory, **8,** 398 ; **10,** 89.

— Hand-list, Sharpe, **8,** 398.

— Our Native, Lange, **9,** 81.

— North American, Nehrling, **1,** 404 ; **3,** 358.

Bitumen, Trinidad, Peckham and Linton, **1,** 193.

Black Hills, geology of northern, Irving. **9,** 384.

Blair, A. A., Chemical Analysis of Iron, **2,** 450.

Bliss, W. J. A., Manual of Experiments in Physics. **5,** 302.

Block Island geology, Marsh, **2,** 295, 375 ; prehistoric fauna of, Eaton. **6,** 137.

Blowpipe Analysis, Getman, **9,** 82.

— Penfield-Brush, **2,** 459 ; **6,** 436.

Boiling points, in Crookes vacuum, **3,** 67 ; curves for, Speyers, **9,** 341.

Bolometer, Langley, **5,** 241 ; Abbot, **9,** 217, **10,** 470.

Bolton, H C., Catalogue of Scientific Periodicals 1665-1895, **6,** 513 ; Bibliography of Chemistry, 1492–1897, **7,** 322.

Bonney, T. G., Charles Lyell and Modern Geology, **1,** 322.

— origin of diamonds of South Africa, **5,** 76.

BOTANICAL WORKS—

Analecta Algologica, iii, Agardh, **3,** 78 ; **9,** 78.

Australian Fungi, McAlpine, **1,** 324.

— Gazette, **2,** 396.

Botanizing, Bailey, **9,** 80.

Elements of Botany, Bergen, **3,** 490 ; Darwin, **3,** 490 ; Van Tieghem, **7,** 77.

Flora of Franz Joseph Archipelago, Fisher. **5,** 236.

— of Newfoundland, Robinson and von Schrenk, **3,** 77.

— of North America, Synoptical, Robinson, vol. i, pt. I, **4,** 249.

— of Northern United States, Canada, etc., Britton and Brown, **3,** 76 ; **4,** 250 ; **6,** 277.

— of Pribiloff Islands, Macoun, **9,** 232.

— of Southern United States, Chapman. **3,** 425.

— of the West Indies, Urban, **7,** 244.

BOTANICAL WORKS—

Flowering Plants and Ferns, Willis, **3,** 353.

Forestry, Primer of, Pinchot, **8,** 399.

Fossil Plants, Seward, **5,** 472.

— See also **GEOLOGY.**

Index Desmidiacearum, Nordstedt. **3,** 354.

Kryptogamen-Flora von Deutschland, etc., Rehm, **1,** 492.

Laboratory Practice for Beginners, Setchell, **3,** 490.

Lessons in Elementary Botany, MacBride, **3,** 490.

Marine Algæ of Greenland, Rosenvinge, **7,** 400.

Missouri Botanical Garden, 7th Report, **2,** 89 ; 8th Report. **5,** 78 ; 11th Report, **9,** 233.

Morphology, Experimental, Part I, Davenport, **4,** 397 ; Part II, **7,** 474.

Mosses, keys to N. American, Barnes and Heald, **3,** 354.

— and Ferns, Campbell, **1,** 72.

Mycetozoa, British, Lister, **1,** 153.

Native Fruits, Evolution of, Bailey, **7,** 78.

Paleobotany, see **GEOLOGY.**

Pflanzenfamilien, die natürlichen, Engler, **7,** 169.

Pflanzengeographie, Warming, **2,** 89.

Pflanzenphysiologie, Pfeffer, **5,** 317.

Phycotheca Boreali-Americana, Collins, Holden and Setchell, **1,** 73, 493 ; **3,** 78, 354.

Phytogeography of Nebraska, Pound and Clements, **5,** 471.

Plantbreeding, Bailey, **1,** 151.

Plant Life, Barnes, **7,** 244.

Plants, Catalogue of useful fiber, Dodge, **4,** 478.

— Guide to the Study of Common, Spalding, **3,** 490.

— Lessons with, Bailey, **5,** 318.

— Nature and work of, MacDougal, **9,** 391.

Practical Botany for Beginners, Bower, **3,** 490.

Rhodora, **7,** 245.

Study of Seaweeds, Murray, **1,** 74.

Survival of the Unlike, Bailey, **3,** 77.

The Teaching Botanist, Ganong, **9,** 79.

Textbook of Botany. Strasburger, Schimper, Noll and Schenck, **5,** 471.

Zellen u. Befruchtungslehre, Häcker, **9,** 77.

BOTANY—

Anemone apennina, Hildebrand, **9,** 231.

Ascomyceten, Früchtentwickelung, Harper, **3,** 78.

Berberitze, Hexenbesenrost der, Eriksson, **3,** 353.

Bush-fruits, Card, **7,** 78.

Carex of oriental Asia, Franchet, **10,** 84.

Carex-vegetation des aussertropischen Südamerika, **9,** 231.

Carices in Alaska, Holm, **10,** 266.

Caulerpæ, monograph of the, Bosse, **7,** 400.

Cell-wall, modifications, Van Wisselingh, **1,** 152; structure of, Gardiner, **5,** 470.

Characeæ of America, Allen, **1,** 492; of Japan, Allen, **5,** 472.

Cycads, see **GEOLOGY.**

Cyperaceæ, studies on, Holm, No. i, **1,** 348: No. ii, **2,** 214; No. iii, **3,** 121; No. iv, **3,** 429; No. v, **4,** 13; No. vi, **4,** 298; No vii, **5,** 47; No. viii, **7,** 5; No. ix, **7,** 171; No. x, **7,** 435; No. xi, **8,** 105; No. xii, **9,** 355; No. xiii, **10,** 33; No. xiv, **10,** 266.

Delesseria, Agardh, **7,** 401.

Dicotyledones, anatomy, Solereder, **10,** 85.

Dualistic theory of descent, Sachs, **2,** 396.

Embryophyta siphonogama, Engler, **5,** 156.

Flowers, colors of, Hervey, **8,** 471.

Hepaticæ and Anthocerates of California, Howe, **8,** 309.

Hickory nuts, abnormal, Herrick, **2,** 258.

Hypogæous fungi, Californian, Harkness, **8,** 310.

Lithothamnion, Norwegian forms, Foslie, **1,** 153.

Lupinus albus, toxic action of acids on, True, **9,** 183.

Magnolias, formation of pollengrains, Guignard, **7,** 77.

Maiden-hair tree (*Gingko*), Seward, **10,** 323.

Mosses, reproduction of, Correns, **9,** 78.

Nitrogen, appropriation of, by plants, Lutz, **8,** 85.

Permeability of the bark of tree trunks to atmospheric gases, Devaux, **5,** 318.

Phenols, effect on living plants, True, **7,** 76.

BOTANY—

Phænogamia, new system of classification, van Tieghem, **4,** 79.

Plant diseases, caused by cryptogamic parasites, Marsee, **8,** 471.

Pleiotaxy in the androecium of *Epidendrum cochleatum*, Mead, **1,** 72.

Pogonia ophioglossoides, Holm, **9,** 13.

Protoplasma, A. Fischer, **9,** 77.

Protrophie, Minks, **3,** 355.

Reis-Brand und der Setaria-Brand, Brefeld, **1,** 325.

Respiration after injury, increase of activity, Richards, **2,** 464.

Root-pressure, cause of, Leavitt, **7,** 301.

Seeds, effect on, of temperature of —250° C., Thiselton-Dyer, **9,** 74.

Soja bean, tubercles on roots, Kirchner, **1,** 151.

Sphærotheca Castagnei, die Entwickelung des Perithecinms bei, **1,** 326.

Sphagna Boreali-Americana exsiccata, Eaton and Faxon, **3,** 77.

Sugar in root of beet, Maquenne, **1,** 152.

Sugar-cane, propagation, Wakker, **1,** 324.

Timber pines of the Southern United States, Mohr, **2,** 463.

Toxic action of acids on *Lupinus albus*, True, **9,** 183.

Transpiration of plants, psychrometer for, Leavitt, **5,** 440.

Trees, winter conditions of reserve food substances, Wilcox, **6,** 69.

Variation under grafting, Daniel, **8,** 84.

Würzeln, Beiträge zur Physiologie der, Rimbach, **9,** 230.

See also **GEOLOGY.**

Boulenger, G. A, Catalogue of Perciform Fishes, **1,** 397.

Branner, J. C., thickness of the Paleozoic in Arkansas, **2,** 229; phosphate deposits of Arkansas, **3,** 159; bacteria and decomposition of rocks, **3,** 438; extension of the Appalachians across Mississippi, etc., **4,** 357.

Brazil, argillaceous rocks and quartz veins in, Derby, **7,** 343.

— Bendegó Meteorite, **4,** 159.

— schists of gold regions, Derby, **10,** 207.

Breckenridge, J. E., separation of potassium and sodium, **2,** 263.

Briggs, L. R., Röntgen rays, **1,** 247.

British Museum Catalogues. 1, 396, 397, 398; 5, 75, 319; 10, 170.
Britton, N. L., Flora of the Northern United States, Canada, etc., 3, 76; 4, 250; 6, 277
Broadhead, G. C., Devonian of North Missouri, 2, 237.
Brögger, W. C., Igneous rocks of Predazzo, 1, 399; of Christiania, 6, 273.
— Williams Memorial Lectures on Geology, 9, 456.
Brown, A., Flora of the Northern United States, Canada, etc., 3, 76; 4, 250; 6, 277.
Browning, P. E., interaction of chromic and arsenious acids, 1, 35; estimation of cadmium, 2, 269; of vanadium, 2, 355; detection of sulphides, etc., 6, 317; volumetric estimation of cerium, 8, 451; estimation of thallium as the chromate, 8, 460; estimation of thallium, 9, 137; separation of nickel from cobalt, 10, 316.
Brush, C. F., a new gas, 6, 431.
Brush, G. J., Manual of Determinative Mineralogy, revised by Penfield, 2, 459; 6, 436.
Brush discharges in air, Toepler, 7, 67.
Burbank, J. E., X-rays and mineral phosphorescence, 5, 53; phosphorescence by electrification, 5, 55; source of X-rays, 5, 129
Bush, K. J., Ledidæ and Nuculidæ of N. Atlantic coast, 3, 51.
Byerly, W. E., Differential Calculus, 1, 328.

C

Cadmium normal element, Jaeger and Wachsmuth, 3, 71.
Cady, W. G., measurement of an air thermometer bulb, 2, 341; energy of cathode rays, 10, 1
Cajori, F., History of Elementary Mathematics, 3, 79.
— History of Physics, 7, 394.
— search for solar X-rays on Pike's Peak, 2, 289.
Calculus, Lambert, 6, 513; for Engineers, Perry, 4, 398.
California, Mesozoic plants from, Fontaine, 2, 273
— tin deposits of Temescal, Fairbanks, 4, 39.
Call, R. E., Rafinesque's Ichthyologica Ohiensis, 7, 473.
Calvin, S., geology of Johnson Co., Iowa, 5, 149.

Campbell, D. H., Structure, etc., of Mosses and Ferns, 1, 72.
— Evolution of Plants, 9, 79.
Campbell, H. D., Specimens from Chichan-Kanab, Yucatan, 2, 413.
Campbell, W. W., Elements of Practical Astronomy, 8, 88.
Canada Geological Survey, publications, 1, 150, 490; Hoffmann's report, 2, 88; vol. vii. 1894, 3, 72, 421, 488; Hoffmann's report, 4, 78; vol. viii. 1895, 4, 394; vol. ix, 1896, 5, 232; 1897, 6, 434, 510; vol. ix, 1896, 7, 71; 1898, 8, 232; 9, 68; vol. x, 1897, 9, 156, 302, 456; 1899, 10, 165; vol. xi, 10, 399; Hoffmann's report, 10, 404.
— Laurentian of, Adams and Barlow, 3, 173.
— minerals of, Hoffmann, 2, 88; 4, 78; 6, 437; 10, 404.
Canadian Paleontology, Whiteaves, 7, 71.
— Paleozoic corals, revision of, Lambe, 9, 155.
Cape Breton, Etcheminian fauna of, Matthew, 9, 158.
Cape Cod, Geology, Shaler, 8, 76.
Cape of Good Hope, geological commission, 1896, 4, 395.
Cape Nome gold region, 9, 455.
Carbon, see **CHEMISTRY**.
Card, F. W., Bush-fruits, 7, 78.
Carmichael, N. R., Röntgen rays, 1, 247.
Carpenter, G. H., Insects, their Structure and Life, 8, 473.
Case, E. C., foramina in cranium of a Permian reptile, 3, 321.
Cathetometer, Wadsworth, 1, 41.
Cathode light and the nature of the Lenard rays, 4, 475.
— photography, Trowbridge, 1, 245.
— ray phenomena, Elster and Geitel, 1, 139
— rays and their effects, Wright, 1, 235.
— — change of energy into light rays, Wiedemann, 6, 433.
— — dispersion by magnetic force. Birkeland, 8, 463.
— — effect on air, Lenard, 5, 149.
— — effect on conductivity of gases, McLennan, 10, 319.
— — energy of, Cady, 10, 1.
— — nature of, Thomson, 4, 71, 391.
— — reflection, Starke, 6, 433.
— — velocity and susceptibility to magnetic action, Wiechert, 9, 148.
— — and X-light, Rollins, 10, 382.
— See **RONTGEN RAYS**.

Cell in Development and Inheritance, Wilson, **3**, 161.

Chalmers, R., Pleistocene marine shore lines, **1**, 302; glaciation in New Brunswick, etc., **3**, 72; preglacial decay of rocks in East Canada, **5**, 273.

Chapman, A. W., Flora of the Southern United States, **3**, 425.

CHEMICAL WORKS—
Analytische Chemie, Ostwald, **5**, 221.

Chemical Experiments, Williams, **1**, 317.

Chemistry. Bibliography, 1492–1897, Bolton, **7**, 322.

— in Daily Life, Lassar-Cohn and Muir, **2**, 449.

— of Fatty Compounds, Whiteley, **1**, 53.

Chimie Physique, Van't Hoff, **7**, 157; **10**, 461.

Copper Smelting, modern, E. D. Peters, Jr., **1**, 54.

Dictionary of Inorganic Solubilities, Comey, **1**, 484.

Fermentations, Schützenberger, **1**, 484.

Grundlinien der anorganischen Chemie, Ostwald, **10**, 394.

Grundriss d. allgemeinen Chemie, Ostwald, **9**, 65.

Industrial Chemistry, Thorp, **7**, 157.

Inorganic Chemical Preparations, Thorp, **3**, 357; **5**, 222.

Inorganic Chemistry, Shenstone, **10**, 395.

Laboratory Manual, Hillyer, **9**, 65; Williams, **3**, 357.

Lehrbuch der allgemeinen Chemie, Ostwald, **3**, 357; **5**, 222; **8**, 74; **9**, 64.

Lessons in Physical Chemistry, Van't Hoff, **7**, 157.

Lexikon der Kohlenstoff-Verbindungen, Richter, **9**, 445.

Manual of Chemical Analysis, Fleurent, **5**, 147; Newth, **7**, 67.

— of Qualitative Analysis, Fresenius and Wells, **4**, 474.

Méchanique chimique, Duhem, **3**, 419.

Molekulargewichtsbestimmung, Fuchs, **1**, 53.

Organic Chemistry, Cohn, **1**, 53; Noyes, **5**, 147.

Physical Chemistry, Lehfeldt, **9**, 445; Speyers, **5**, 390; Van't Hoff, **7**, 157, **10**, 461.

CHEMICAL WORKS—
Physikalisch-chemische Propädeutik, Griesbach, **2**, 450; **5**, 321; **10**, 461.

Practical Inorganic Chemistry, Turpin, **1**, 317.

Methods of Organic Chemistry, Gattermann, **2**, 450.

Progress of Scientific Chemistry, Tilden, **8**, 385.

Proofs of Chemical Laws, Cornish, **1**, 52.

Qualitative Chemical Analysis, H. L. Wells, **6**, 269; J. S. C. Wells, **6**, 269.

School Chemistry, Waddell, **10**, 461.

Spirit of Organic Chemistry, Lachman, **8**, 73.

Sucres, et leurs principaux Derivés, Maquenne, **9**, 445.

Tutorial Chemistry, Bailey, **3**, 357; **5**, 390.

CHEMISTRY—
Acetic series, fractional distillation of acids of, Sorel, **3**, 70.

Acetylene, action of, on the oxides of copper, Gooch and Baldwin, **8**, 354.

— certain derivatives of, Erdman and Kothner, **7**, 469.

— explosive properties, Berthelot and Vieille, **3**, 483.

— products of the explosion of, Mixter, **9**, 1; **10**, 299.

Actinium, new radio-active substance, Debierne, **9**, 444.

Ætherion, Crookes, **7**, 64.

Air, combustion in rarefied, Benedicenti, **6**, 95.

— influence of the silent discharge on, Shenstone and Evans, **5**, 464.

— new constituents of, Ramsay and Travers, **6**, 192, 360; Brush, **6**, 431.

Alkali metals, spectra of fused salts, De Gramont, **3**, 150.

— nitrates, preparation, Divers, **7**, 311.

Alumina, separation from molten magmas, Pratt, **8**, 227.

Aluminum, separation from beryllium, Havens, **4**, 111; from iron, Gooch and Havens, **2**, 416; from zinc, etc., by hydrochloric acid, **6**, 45.

— nitride, Francke, **6**, 501.

Ammonia, compounds of metallic salts with, Wiede and Hoffmann, **1**, 393.

Ammonium chloride, action upon minerals, Clarke and Steiger, **9**, 117, 345.

CHEMISTRY—

Ammonium cyanide, manufacture of, Lance, **5,** 69.
— magnesium arseniate, Austin, **9,** 55.
— magnesium phosphate, Gooch and Austin, **7,** 187.
— peroxide, Melikoff and Pissar-jewski, **6,** 195.
Argon and its combinations, Berthe-lot, **8,** 383; determination in air, Schloesing. **1,** 49; passage through thin films of India-rubber, Ray-leigh, **9,** 292; refractivity, Ram-say and Travers, **5,** 227; spectra of, **3,** 15; viscosity, Rayleigh, **9,** 375.
— and helium, Lockyer, **3,** 152; electric discharge in, Collie and Ramsay, **3,** 241; expansion of, **3,** 241; electric discharge through, Strutt, **9,** 294; homogeneity of, Ramsav and Collie. **2,** 300; from a natural spring, Moureu, **2,** 301; properties, Rayleigh, **1,** 315.
— — and other gases, relative rates of effusion, Donnan. **9,** 443
Arsenic acid. iodometric estimation, Gooch and Morris. **10,** 151.
— sulphide, new. Scott, **9,** 442.
Atomic masses, calculation, Berthe-lot. **9,** 62.
Barium, radio-active. Lengyel, **10,** 74; artificial, Debierne, **10,** 393; atomic weight of, Curie. **10,** 392.
Barium chloride, radio-active, Curie, **9,** 114.
— and polonium. **10,** 460.
Beryllium, Retgers. **2,** 448; electro-lytic preparation, Lebeau. **7,** 155.
— zinc and cadmium. double am-monium phosphates, Austin, **8,** 206.
Boric acid, estimation of. Gooch and Jones, **7,** 34; Jones, **7,** 147, **8,** 127.
Borides of calcium, barium and strontium, Moissan and Williams. **5,** 388.
Cadmium. estimation of, Browning and Jones, **2,** 269.
Cæsium. properties of. Eckardt and Graefe. **10,** 75
— and rubidium, double halogen salts. Wells and Foote. **3,** 461.
— and zirconium, double fluorides, Wells and Foote, **1,** 18
Caffeine, synthesis, Fischer and Ach. **1,** 316.
Calcium, metallic, properties, Mois-san. **6,** 428; **7,** 393; preparation of, von Lengyel. **9,** 63.

CHEMISTRY—

Calcium group, eutropic series, Eppler. **7,** 470.
— hydride, Moissan, **6,** 428.
— nitride, Moissan, **6,** 500.
Carbon. conductivity for heat and electricity, Cellier, **5,** 223.
— dioxide, action on soluble borates, Jones, **5,** 442; iodome-tric method for determination. Phelps, **2,** 70; experiments with liquid, Barus, **2,** 1.
— electro-chemical equivalent. Coehn, **5,** 218.
— monoxide, direct elimination, Engler and Grimm, **6,** 193; in-fluence of water on the combus-tion of. Martin, **9,** 293; poison-ing, Sachs, **10,** 395.
— trivalent, Gomberg, **10,** 458.
— and hydrogen, direct union, Bone and Jordan, **3,** 481.
— and oxygen, presence in sun, Trowbridge, **1,** 329.
— varieties of. Moissan. **5,** 220.
Cerium, volumetric estimation of, Browning. **8,** 451.
Charcoal in purification of spirit, Glasenapp. **8,** 161.
Chlorine, bromine. etc., separation, Bennett and Placeway, **2,** 300.
— and hydrogen, action of light on, Gautier and Hélier, **5,** 144.
— peroxide, explosion with carbon monoxide, Dixon, **4.** 472.
Chromic and arsenious acids, inter-action of, Browning, **1,** 35.
Chromite. method of decomposing, Fieber, **10,** 76.
Chromium tetroxide and perchro-mates, Wiede, **5,** 299.
Combustion of organic substance in the wet way, Phelps, **4,** 372.
Copper arsenides, artificial, Koenig. **10,** 439.
— estimation as oxalate, Peters, **10,** 359.
— sulphocyanides, Van Name, **10,** 451.
— voltameter. Foerster, **5,** 219.
Crookes vacuum. boiling points, Krafft and Weilandt, **3,** 67.
Cuprous chloride and acetylene, Chavastelon, **7,** 237.
Cyanide process, chemistry of, Bod-laender. **2,** 448.
Cyanogen. volumetric determina-tion, Denigès, **1,** 50; spark spec-trum, Hartley. **4,** 388.
Dalton's law for solutions, Wilder-mann, **4,** 387.

CHEMISTRY—

Double salts, Van't Hoff, **4**, 68.

Electric discharge, synthetic action of dark, Losanitsch and Jovitschitsch, **4**, 66.

— oscillations, chemical action, Hemptinne, **4**, 471.

Electro-chemical method for changing currents, Graetz, **5**, 218.

Electrolysis of water, Sokoloff, **3**, 149.

Electrolytes, conductivity of, **3**, 391.

Electrolytic decomposition, **5**, 66; dissociation and osmotic pressure, **5**, 65, 463; dissociation theory, Noyes and Blanchard, **10**, 460; production of hypochlorides, etc., **3**, 149; of iodoform, **5**, 466; solution and deposition of carbon, **4**, 389.

Electrosynthesis, Mixter, **4**, 51; **6**, 217.

Elements, relations between atomic weights of, Lea, **1**, 386; classification, Lea, **1**, 405.

Enantiomorphism, Kipping and Pope, **6**, 502.

Ferric chloride, action on metallic gold, McIlhiney, **2**, 293.

Fertilizers, use of heavy solutions in the examination of, Bryant, **2**, 82.

Fluorescence and chemical composition, Meyer, **5**, 387.

Fluorine, liquefaction, Moissan and Dewar, **4**, 318.

Gallium in the clay-ironstone of Yorkshire, Hartley and Ramage, **2**, 378.

Gas, action of heating on detonating, V. Meyer and Raum, **1**, 138.

Gaseous elements, specific heat of, Berthelot, **4**, 65.

Gases, molecular masses, Berthelot, **7**, 154; new methods for measuring, Bleier, **5**, 385.

Glycogen, formative property of, Creighton, **3**, 426.

Gold, aqueous solutions of metallic, Zsigmondy, **7**, 236.

— experiment with, Lea, **3**, 64.

— iodometric determination, Gooch and Morley, **8**, 261.

— and silver in sea-water, Liversidge, **2**, 304.

Graphitic acid, Staudenmaier, **7**, 65.

Hæmochromogen, Von Zeynek, **8**, 162.

Helium, action of silent electric discharge on, Berthelot, **4**, 152; density, Ramsay, **3**, 241; experi-

CHEMISTRY—

ments with, **6**, 499; homogeneity of, Ramsay and Travers **7**, 810; liquefaction, Olszewski, **2**, 801, 379; occlusion by palladium, **5**, 224.

— argon and krypton, position in scheme of elements, Crookes, **6**, 189.

— See *Argon* (p. 486).

Hydrazine, free, de Bruyn, **1**, 816; **3**, 479.

Hydrazoic acid, Curtius and Rissom, **8**, 382.

Hydrocarbon, new, Schickler, **3**, 70.

Hydrochloric acid in titrations by sodium thiosulphate, Norton, **7**, 287.

Hydrogen, action on sulphuric acid, **5**, 465; boiling point, Dewar, **6**, 361; desiccated by liquid air, weight, Rayleigh, **10**, 459; liquefaction, Dewar, **6**, 96; liquid, Dewar, **8**, 160; solidification, Dewar, **8**, 382, viscosity of, Rayleigh, **9**, 375; **10**, 461.

— peroxide, Jannasch, **2**, 81; Traube, **1**, 136.

— and oxygen, combination, Berthelot, **5**, 220.

Hyponitrous acid, Kirschner, **6**, 499.

Inorganic compounds, structural isomerism, Sabanéeff, **4**, 66.

— molecular mass of, Werner, **6**, 195.

Iodic acid in analysis of iodides, Gooch and Walker, **3**, 293.

Iodine, in the analysis of alkalies, etc, Walker, **6**, 455; method of preparing pure, **5**, 387; spectra of, Konen, **4**, 67.

— and bromine solutions, absorption spectra, Wood, **3**, 67.

Ions, see **Ions**.

Iron, atomic weight, Richards and Baxter, **10**, 75.

— determination of ferrous, Hillebrand and Stokes, **10**, 393; of ferric, Norton, **8**, 25.

— separation of, Havens and Way, **8**, 217.

— carbide, direct production, Moissan, **5**, 67.

— silicide, preparation, Lebeau, **8**, 72.

Krypton, Ramsay, **6**, 192; **9**, 62, 442; Crookes, **6**, 189.

Lead and bismuth in zinc, solubility, Spring and Romanoff, **3**, 418.

Liquids, solubilities, Aignan and Dugas, **5**, 297.

CHEMISTRY—

Lithium, preparation, Warren, **3**, 243.

— and beryllium, Borchers, **3**, 151.

— ammonium, etc.. Moissan, **8**, 384.

— and calcium with ammonium, Moissan, **7**, 65.

Lucium, Barrière, **2**, 378.

Manganese, estimation of, Gooch and Austin, **5**, 209; Austin, **5**, 382; oxidation of, Gooch and Austin, **5**, 260; determination as pyrophosphate, Gooch and Austin, **6**, 233.

— carbide, Moissan, **1**, 392.

Melting point and critical temperature, relation, Clarke, **2**, 299.

Mercuric antimonide, **10**, 75.

Mercury in the colloidal condition, **8**, 884.

— determination as mercurous oxalate, Peters, **9**, 401.

— titration by sodium thiosulphate, Norton, **10**, 48.

Metallic hydroxides, preparation by electrolysis, Lorenz, **3**, 244.

Metals, diffusion, Roberts-Austen, **3**, 147; "excited" Wislicenus, **3**, 244; preparation by means of aluminum, Goldschmidt, **7**, 154.

Metargon, Dewar, **6**, 360, 361; **9**, 62.

Methane and air, explosion by electric current, Couriot and Meunier, **7**, 236.

Molecular mass of solid substances, Traube, **6**, 95; determined by the boiling point, Walker and Lumsden, **6**, 429.

Molybdenum, estimation iodometrically, Gooch, **3**, 237; **6**, 168.

Molybdic acid, iodometric estimation, Gooch and Fairbanks, **2**, 156.

Neodymium, Boudouard, **7**, 157.

Neon and Metargon, Ramsay and Travers, **6**, 360; Ramsay, **9**, 62.

Nickel, extraction by the Mond process, Roberts-Austen, **7**, 64.

— and cobalt, separation by hydrochloric acid, Havens, **6**, 396; separation, Browning and Hartwell, **10**, 316.

Nitrates, presence in the air, Defren, **3**, 418; conversion into cyanides, of, Kerp, **4**, 390.

Nitric acid, action upon potassium cobalti-cyanide, Jackson and Comey, **2**, 82.

Nitrogen, absorption by carbon compounds, Berthelot, **4**, 473; absorption by lithium, Deslandres, **1**, 138.

CHEMISTRY—

Nitrogen gas, oxidation, Rayleigh, **3**, 416.

— pentasulphide, Muthmann and Clever, **3**, 480.

Nitrous acid, action in a Grove cell, Ihle, **3**, 150.

Oceanic salt deposits, formation of, Van't Hoff and Meyerhoffer, **8**, 73.

Orthophthalic acids, supposed two forms, Howe, **1**, 485; non-existence of two, Wheeler, **2**, 449.

Oxalic acid, titration, Gooch and Peters, **7**, 461.

Oxygen, determination in air and in aqueous solution, Kreider, **2**, 361; evolution during reduction, **5**, 298.

— and hydrogen, atomic weights. Thomsen, **1**, 316; occlusion, Mond, Ramsay and Shields, **7**, 468.

— sulphur and selenium, series spectra, **5**, 145.

Ozomolybdates, Muthmann and Nagel, **8**, 160.

Ozone, boiling point, Troost, **6**, 362; properties, Ladenburg, **7**, 310.

Palladium, reduction in presence of, Zelinsky, **7**, 393.

Permanganate, action on hydrogen peroxide, Baeyer and Villiger, **10**, 458.

Peroxides, production of, Bach, **5**, 68; Melikoff and Pissarjewsky, **8**, 72.

Persulphuric acid, formation, Elbs and Schönherr, **3**, 68.

Petroleum, composition of American, Young, **7**, 311; formation of natural, Engler, **5**, 300; normal and iso-pentane from America, Young and Thomas, **4**, 319.

Phosphorus in iron, determination, Fairbanks, **2**, 181.

Platinum, fusibility, V. Meyer, **2**, 81.

— and gold, solution in electrolytes, Margueles, **7**, 236.

— and potassium, double halides, Herty, **1**, 315.

Refraction, relation of, to density, Traube, **3**, 479.

Refractivities of air, oxygen, etc, Ramsay and Travers, **5**, 227.

Resorcinol, solution in ethyl alcohol, Speyers and Rosell, **10**, 449.

Rotation angle, new substance for increasing, Walden, **5**, 463.

Rubidium and its dioxide, Erdmann and Köthner, **3**, 482.

TRY—
hotoelectric properties of
l, Elster and Geitel, **6**, 95.
'r, gold and silver in, Liver-
2, 304.
s and selenic acid. iodo-
determination, Gooch and
, **1**, 31.
l, gravimetric determina-
Peirce, **1**, 416: separation
ellurium, Gooch and Peirce,

!erence rings, Longden, **10**,

xide, Peirce, **2**, 163.
, decomposition by boric
Jannasch and Heidenreich,
; experiments with, Clarke,
; **9**, 117, 345.
spectrum of, de Gramont,

:olloidal, Lottermoser and
'eyer, **7**, 156.
, reaction upon hydrogen
de, Riegler. **3**, 69.
ide and peroxynitrate,
r, **7**, 156.
ynitrate, Sulc, Mulder, and
ta, **3**, 69.
lphate, transition tempera-
lichards, **6**, 201.
:arbide. Matignon. **6**, 196.
xide and peroxide, For-
6, 501.
ilphate. titration with iodic
Valker, **4**, 235.
m compounds, **10**, 75.
n compounds, Sorensen, **1**,

ne, new bases from, Tafel,

of nitrogen. Cleves and
iann, **1**, 391.
s, etc., detection of, Brown-
l Howe, **6**, 317.
'anides of copper and sil-
an Name, **10**, 451.
forms of, **10**, 161.
mlar weight, Orndorff and
se, **1**, 483
le, combination with
i, Russell and Smith, **9**, 293.
aration and supercooling.
d, **4**, 151
acid. compounds with
i, Weinland and Prause,

a, determination, Gooch
organ, **2**, 271.
s acid in presence of haloid
Jooch and Peters, **8**, 122

CHEMISTRY—
Thallium, estimation of, Browning
and Hutchins, **8**, 460 ; Browning,
9, 137.
Thermochemical method for deter-
mining the equivalents of acids,
Berthelot, **4**, 151.
Thermodynamics of the swelling of
starch, Rodewald, **5**, 297.
Titanium, properties, etc., **1**, 52.
Triphenylmethyl, Gomberg, **10**, 458.
Tungsten trioxide, separation from
molybdenum trioxide, Ruegen-
berg and Smith, **10**, 460.
Uraninite and eliasite, gases ob-
tained from, Lockyer, **3**, 242.
Uranium, radio activity, Crookes,
10, 318.
Vanadium, distribution, **7**, 470;
estimation of, Browning and
Goodman, **2**, 355.
Vapor pressure of reciprocally sol-
uble liquids, Ostwald, **6**, 93.
Victorium, Crookes, **9**, 146.
Voltaic action, influence of proxim-
ity on, Gore, **5**, 144.
Water, constitution of, Brühl, **1**,
138.
Zirconium with lithium, etc., double
fluorides, Wells and Foote, **3**,
466.
Chester, A. H., Dictionary of Mineral
Names, **1**, 401 ; Catalogue of Min-
erals, **5**, 78 ; krennerite, from
Cripple Creek, Colo., **5**, 375.
Chili, Geology and Petrography of,
von Wolff, **9**, 228.
Chittenden, H. M., Yellowstone
National Park, **1**, 327.
Christiansen, C., Theoretical Physics,
3, 419.
Christmas Island, geology, Andrews,
10, 79.
Clark, W. B., Potomac river section
of Atlantic Coast Eocene, **1**, 365 ;
Eocene deposits of middle Atlantic
slope, **3**, 250.
— Geological Survey of Maryland,
7, 69 ; **9**, 223 ; Maryland Weather
Service, **9**, 234.
Clarke, F. W., Constants of Nature,
3, 245.
— hydromica from New Jersey, **7**,
365 ; composition of roscoelite, **7**,
451.
— constitution of tourmaline **8**, 111 ;
experiments with pectolite, etc., **8**,
245.
— action of ammonium chloride on
analcite and leucite, **9**, 117 ; on
natrolite, etc., **9**, 345.

Clarke, F. W., rock analyses of U. S. Geol. Survey, 10, 469.

Clarke, J. M., Dictyospongidæ, 9, 69.

Clayton, H. H., seven-day weather period, 2, 7; 27-day auroral period and the moon, 5, 81.

Clements, J. M., contact metamorphism, 7, 81; iron-bearing district of Crystal Falls, Michigan, 9, 451.

Clerke, A. M., the Herschels and Modern Astronomy, 1, 76.

Climate of Davis' and Baffin's Bay, Tarr, 3, 315.

— of Frankfurt a. M., Ziegler and Konig, 3, 358.

— of Geological Past, Dubois, 1, 62.

Cloud work for the U. S., international, Bigelow, 8, 433.

Coast Survey, U. S., Report, 8, 87.

Coherer, action of, Mizuno, 10, 463.

— quantitative investigation, A. Trowbridge, 8, 199.

— theory of, Van Gulik, 6, 433; Aschkinass, 6, 503.

— use in measuring electric waves, Behrendsen, 7, 158.

Cohn, L., Organic Chemistry, 1, 53.

Collins, W. D., surface travel on electrolytes, 5, 59.

Colloids, compressibility of, and jelly theory of the ether, Barus, 6, 285.

Color relations of atoms, ions and molecules, 1, 405.

— sensations in terms of luminosity, Abney, 8, 464.

— vision, Rood, 8, 258.

Colorado canyon, geology, Davis, 10, 251.

— carnotite, etc, of, Hillebrand and Ransome, 10, 120.

— Devonian strata in, Spencer, 9, 125.

— tellurides from, Palache, 10, 419.

Colors, analysis of contrast, Mayer, 1, 38.

Comet of 1843 I, Kreutz, 1, 75.

Comets, orbits of, Stichenroth, 3, 160.

Conductivity of carbon, 5, 223.

Connecticut, lava beds of Meriden, Davis, 1, 1

Conrad's types of Syrian fossils, Beecher, 9, 176

Constants of Nature, Clarke, 3, 245.

Cooke, W. W., Birds of Colorado, 4, 326.

Cooley, L. C., Physics, 4, 390.

Coolgardie gold field, geology, Blatchford, 8, 396.

Coral barrier reef of Australia, Agassiz, 2, 240.

— boring at Funafuti. 5, 75; 7, 317.

— islands of the Pacific, Agassiz, 9, 33, 109, 193, 369.

— reefs of the Fijis, Agassiz, 5, 113; 6, 165; 8, 80.

Cordoba Photographs, Gould, 4, 480.

Cornish, V., Proofs of Chemical Laws, 1, 52.

Corona, see Solar.

Corrigan, S. J., Constitution of Gases, etc., 1, 328.

Cory, C. B., Birds of N. America, 8, 398; 10, 89.

Crater Lake, Oregon, Diller, 3, 165; 7, 316.

Crew, H., Elements of Physics, 9, 146.

Crookes, W., position of helium, argon and krypton, in scheme of elements, 6, 189; radio-activity of uranium, 10, 318.

Crookes' tube, images in the field of, Oumoff and Samolloff, 2, 452.

— use for X-rays, Hutchins and Robinson, 1, 463.

Crosby, W. O., geology of Newport Neck and Conanicut Island, 3, 230.

Cross, W., igneous rocks in Wyoming, 4, 115.

Crystalline liquids, Schenck, 9, 63.

Crystallization, emission of light during, Bandrowski, 1, 51.

Crystallography, Dewalqne, 2, 307; Lewis, 9, 73; Linck, 2, 306; Panebianco, 2, 307; Patton, 2, 306.

Crystals, Characters of, Moses, 8, 84.

— detection of dextro- and lævo-rotating, Kreider, 8, 133.

— drawing of, Moses, 1, 462.

— measurement of, Palache, 2, 279.

— rotation of circular polarizing, Landolt, 3, 416.

Currents, rapid break for large, Webster, 3, 383.

— See Alternating.

Cycads, American fossil, Wieland, 7, 219, 305, 383; Ward, 9, 70, 384, 10, 327.

D

Dæmonelix, papers on, 4, 77,

Dale, T. N., structural details in the Green Mountain region, 2, 395.

Dall, W. H., Tertiary Fauna of Florida, 7, 71.

Daly, R. A., studies in the amphiboles and pyroxenes, 8, 82, 83.

Dana, E. S., Text-book of Mineralogy, new edition, 6, 275.

. S., First Appendix to Sys-Mineralogy, **8**, 236.
D., Life of, by D. C. Gilman,

)ook of Geology, new edition, N. Rice, **5**, 393.
N. H., catalogue of contribu-o North American geology, 391, **3**, 350; geothermal data rtesian wells in the Dakotas. ; dikes of felsophyre and in Virginia, **6**, 305; hy-a from New Jersey, **7**, 365.
F., Elements of Botany, **3**,

rt, **C. B.**, Experimental)logy, Part I. **4**, 397; Part II. Biological Variation. **8**, 399.
l., rotation caused by station-nd-waves, **10**, 231.
V. M., lava beds at Meriden. **1**, 1; origin of freshwater y of Rocky Mts., **9**, 387; n the Colorado Canyon, **10**,

cal Geography, **7**, 248.
, **C.**, diurnal periodicity of lakes. **1**, 402; Hereford lake of December 17, 1896, ; earthquake sounds, **9**, 307.
, **J. M.**, wardite, **2**, 154; m and iridium in meteoric , 4.

umphrey, Poet and Philos-Thorpe, **2**, 449.
, **J. W.**, genus Lepidophloios,

L., gas thermometer at high atures, **8**, 165, **10**, 171; -electricity in certain metals,

. **D.**, residual viscosity on l expansion, **2**, 342; magnetic ent of rigidity, **3**, 449.
osition, non-explosive, Hoit-), 63
it, **L.**, reflection of Hertzian **8**, 58
nay, **L.**, Gold Mines of the aal, **2**, 88.
ints du Cap. **5**, 77.
es thermo-minérales, **7**, 474 , relation to refractive power, ', **3**, 479.
). **A.**, Bendegó meteorite, **4**, :cessory elements of itacolum-:., **5**, 187; argillaceous rocks artz-veins in Brazil, **7**, 343.
s of the gold and diamond i of Brazil, **10**, 207; notes on te, **10**, 217.

Devil's Lake, Wis., geology, Salis-bury and Attwood, **10**, 248.
Dewar, boiling point of hydrogen, **6**, 361; liquefaction of hydrogen, **6**, 96. **8**, 160; solid hydrogen, **8**, 382: neon and metargon, **6**, 360, 361, **9**, 62; effect of extreme cold on seeds, **9**, 74.
Diamond, artificial production, **3**, 243; **5**, 469.
— of So. Africa, genesis and matrix of, Lewis and Bonney, **4**, 77; origin, Bonney, **5**, 76.
— — work on, DeLaunay, **5**, 77.
Dielectric constants, determination, Nernst, **1**, 317.
— resistance, Drude, **1**, 318.
Dielectrics, polarization and hys-teresis, Schaufelberger, **7**, 312.
Diffusion of rocks, Becker **3**, 21; method of computing, Becker, **3**, 280.
Diller, J. S., geological reconnais-sance of northwestern Oregon. **3**, 155; Crater Lake, Oregon, **3**, 165, **7**, 316; origin of Paleotrochis, **7**, 337; Bohemia Mining region of Oregon, **10**, 465.
Discharge rays, relation to cathode and Rontgen rays, Hoffmann, **3**, 246.
— See **Electric**.
Dodge, C. R., catalogue of useful fiber plants, **4**, 478.
Dolbear, A. E., Modes of Motion, **5**, 148; Natural Philosophy, **3**, 486.
Douglass, E., new species of Mery-cochærus in Montana, **10**, 428.
Drawing of crystal forms, Moses, **1**, 462.
Duane, W., electrical thermostat, **9**, 179.
Dubois, E., Climates of the Geologi-cal Past, **1**. 62.
— On Pithecanthropus erectus, **1**, 475.
Dufet, Optical physical data, **7**, 472.
Duff, A. W., seiches on the Bay of Fundy, **3**, 406.
Duhem, P., Méchanique Chimique, **3**, 419; Thermodynamics, **7**, 68.
Dunstan, A. St. C., broadening of the sodium lines, etc., **3**, 472.
Durward, A., temperature coefficients of hard steel magnets, **5**, 245.

E

Eakle, A. S., erionite, **6**, 66; biotite-tinguaite dike in Essex Co., Mass., **6**, 489.
Earth, age of, Kelvin, **7**, 160; Geikie, **8**, 387: Joly. **8**, 390.

Earth and ether, relative motion of Michelson, **3**, 475.
— form of, Putnam, **1**, 186.
— Sculpture, Geikie, **7**, 72.
Earthquake, of Dec. 17, 1896 at Hereford, Davison, **8**, 235.
— investigation committee of Japan, publications, **10**, 471.
— motion, propagation to great distances, Oldham, **9**, 305.
— sounds, Davison, **9**, 307.
— waves, velocity, **10**, 471.
Earthquakes, diurnal periodicity, Davison **1**, 402.
— in Japan, Omori, **9**, 305.
— of Pacific coast, Holden, **6**, 200.
Eastman, C. R., relations of certain body-plates in the Dinichthyids, **2**, 46 ; Ctenacanthus spines from the Keokuk limestone of Iowa, **4**, 10 ; Tamiobatis vetustus, **4**, 85 ; Devonian Ptyctodontidea, **7**, 314 ; von Zittel's text-book of Paleontology, translated, **9**, 388.
Eaton, G. F., prehistoric fauna of Block Island, **6**, 137.
Ebonite, transparency, Perrigot, **4**, 72.
Echelon spectroscope, Michelson, **5**, 215 ; and the Zeeman effect, Blythswood and Marchant, **9**, 380.
Eclipse, of the sun, total, **9**, 391 ; work on, Todd, **9**, 393.
— Party in Africa, Loomis, **3**, 80.
Edwards, A. M., bacillaria of the Occidental Sea, **8**, 445.
Elasticity, modulus determined, Mayer, **1**, 81.
Electric (Electrical) arc, acoustic phenomena in, Simon, **5**, 302.
— — between aluminum electrodes, Lang, **5**, 149.
— — temperature of carbons of, Wilson and Gray, **1**, 394.
— charge, loss by evaporation, Craig, **10**, 463 ; loss in air from ultra-violet radiation, Lenard, **10**, 464.
— — magnetic effect of moving, Cremieu, **10**, 396.
— condenser, new, Bradley, **9**, 220.
— conductivity of carbon, **5**, 223 ; of the ether, J. Trowbridge, **3**, 387 ; of gases traversed by cathode rays, McLennan, **10**, 219.
— convection of dissolved substances, Picton and Linder, **4**, 150.
— currents excited by Röntgen rays, Winkelmann, **6**, 432 ; break for, Webster, **3**, 383 ; electro-chemical method for changing, Graetz, **5**, 218.
— — See **Alternating.**

Electric discharge, chemical effects of, Berthelot, **6**, 430 ; Losanitsch and Jovitschitsch, **4**, 66.
— — in free air, Toepler, **5**, 149 ; in Geissler tubes, theory of stratification, Gill, **5**, 399 ; in rarefied gases, Skinner, **10**, 76 ; in vacuum tubes, J. E. Moore, **6**, 21.
— discharges in air, J. Trowbridge, **4**, 190.
— — explosive effect, Trowbridge McKay and Howe, **8**, 239.
— — rapid spark, Simon, **9**, 294.
— energy by atmospheric action, Warren, **6**, 93.
— excitation, Coehn, **5**, 302.
— fields, rotating bodies in, Heydweiller, **9**, 66.
— — rotation in constant, Quincke, **3**, 71.
— furnace, fusion, Oddo, **6**, 194.
— indices of refraction, Drude, **2**, 380.
— light in capillary tubes, Schott, **3**, 152.
— measurements by alternating currents, Rowland, **4**, 429 ; Rowland and Penniman, **8**, 35 ; by Rowland's methods, Potts, **10**, 91.
— oscillations, Rydberg, Kayser and Runge, **2**, 380 ; chemical effect, Hemptinne, **4**, 471 ; measurement of slow, König, **7**, 395 ; influence on vapors, Kauffman, **7**, 468.
— — for determining dielectric constants, Nernst, **3**, 485.
— pressure, when dangerous, Weber, **10**, 396.
— resistance, experiments on high, Part I, Rood, **10**, 285.
— — of thin films from cathode discharge, Longden, **9**, 407.
— — standards, **5**, 391.
— shadows, duration, **1**, 141.
— spectrum, dispersion in, Marx, **7**, 68.
— tension at the poles of an induction apparatus, Oberbeck, **4**, 391.
— thermostat, Duane and Lory, **9**, 179.
— waves, absorption of, Branly and Lebon, **7**, 471.
— — compact apparatus for the study of, Bose, **3**, 245.
— — detection of, Neugschwender, **8**, 75.
— — double refraction, Mack, **1**, 140.
— — interference, von Lang, **1**, 394.
— — interferential refractor, Wiedeburg, **3**, 71.
— — magnetic detector, Rutherford, **2**, 381.

aves, measured by radio-
er, Pierce, **9**, 252.
nethod for showing, Neug-
r, **7**, 312.
tion of, DeForest, **8**, 58.
ctive index and reflective
' water and alcohol for,
!18.
f the coherer in measur-
endsen, **7**, 158.
ity in air, Maclean, **8**, 1.
; wires. Sommerfield, **7**,
lidge, **7**, 396.
rrhenius, **5**, 148.
, atmospheric, variation
reau, **10**, 161.
by the ions produced by
rays. Thomson, **7**, 158.
ion through gases by
ons, Thomson, **7**, 312.
;e through gases, **1**, 140.
acuum conduct? J. Trow-
, 343.
ial, Graffigny-Elliot, **6**,

gnetism, Nipher, **6**, 432 ;
d Atkinson, **2**, 83; Perkins,
Webster, **4**, 72.
ion by chemical means,
5, 146.
of, from Central Stations,
Yeaman, **9**, 221.
n and Country Houses,
, **8**, 88.
body, discharge from, by
the Tesla spark, Smith, **2**,

iemical equivalents of cop-
silver, Richards. Collins,
nrod, **9**, 218.
mistry, ionic reactions,
5, 93.
is of water, **3**, 149.
e, theory of Hall effect in,

es, conductivity of, Rich-
Trowbridge, **3**, 391 ; sur-
rel on, Fiske and Collins,

ic decomposition of aque-
ions, Nernst, **5**, 66.
tion, Jones, **10**, 76 ; Noyes
nchard, **10**, 460 ; and os-
ressure, Compton, **5**, 65 ;
5, 463.
oter, Swinton, **7**, 395.
ion of hypochlorites and
i, Oettel, **3**, 149, of iodo-
ierster and Meves, **5**, 466.
1 and deposition of carbon,
l, 389.

Electro-magnetic theory of light,
Barus, **5**, 343.
Electromotive force, J. Trowbridge,
5, 57.
Electroscope, vacuum, Pflaum, **9**,
294.
Electrostatic and electromagnetic
units, ratio of, Hurmuzesca, **1**, 140.
Electrosynthesis, Mixter, **4**, 51 ; **6**,
217.
Elkin, W. L., November meteors, **4**,
480.
Elliott, E. B., Algebra of Quantics,
1, 328.
Ells, R. W., Grenville and Hastings
series of Canadian Laurentian, **3**,
173.
Emerald Mines of Cleopatra, Mac-
Alister, **10**, 468.
Emerson, B. K., Mineral Lexicon of
Franklin Co., etc., Mass., **2**, 306 ;
geology of old Hampshire Co., Mass.,
8, 393 ; Carboniferous bowlders of
India, **10**, 57 ; bivalve from Con-
necticut Trias, **10**, 58.
Endothermic gases, experiments
with, Mixter, **7**, 323.
Energy of cathode rays, Cady, **10**, 1.
— Doctrine of. **6**, 503.
— luminous and chemical, Berthelot,
7, 309.
Entomology, Text-book of, Packard,
6, 103.
Ether, electrical conductivity of,
Trowbridge, **3**, 387.
— jelly theory, Barus, **6**, 285.
— movements. Mil, **8**, 75.
— properties of. Rowland, **8**, 405.
— relative motion of earth and,
Michelson, **3**, 475.
Ethnology, 16th Ann. Report of
American Bureau of, **4**, 480.
Europe, Races of, Ripley, **8**, 474.
European Fauna. Scharff, **8**, 395.
Evolution, organic, Fairhurst, **7**, 80.
Ewell, A. W., rotatory polarization
of light produced by torsion, **8**, 89.
Expansion, coefficients of, for plati-
num, porcelain, etc., Holborn and
Day, **10**, 172.

F

Fairbanks, C., iodometric estimation
of molybdic acid, **2**, 156 ; phos-
phorus in iron, **2**, 181.
Fairbanks, H. W., contact meta-
morphism, **4**, 36 ; tin deposits at
Temescal, So. California, **4**, 39.
Fairchild, H. L., glacial lakes in
Central New York, **7**, 249.

Farrington, O. C., datolite from Guanajuato, 5, 285 ; mineralogical notes, 10, 83.

Fassig, O. L., March weather in the United States, 8, 319.

Feldspars, see MINERALS.

Ferrière, E., La Cause première, 3, 494.

Field Columbian Museum, publications, 4, 481 ; 6, 104 ; 7, 248.

Fiji Islands coral reefs, Agassiz, 5, 113, 8, 80 ; Tertiary limestone reefs of. Agassiz, 6, 165.

Finland, peat bogs of, Andersson. 8, 467 ; geological commission, 8, 467; 10, 249.

Fisher, H., flora of Franz Joseph Land, 5, 236.

Fishes, Perciform, in British Museum, Boulenger, 1, 397.
— fossil, in British Museum, Woodward, 1, 396.

Fiske, W. E., surface travel on electrolytes, 5, 59.

Flame temperatures, Waggener, 2, 879.

Flames, theory of singing, Gill, 4, 177.

Fleurent, E., Manual of Chemical Analysis, 5, 147.

Flight, possibilities of, Rayleigh, 10, 77.

Flink, G., minerals of Greenland, 10, 323.

Flint, J. M., recent Foraminifera, 9, 158.

Flora, see BOTANY, GEOLOGY.

Florida, Tertiary fauna of, Dall, 7, 71.

Flotation of disks and rings of metals, Mayer, 3, 253.

Fluids, transparency for long heat waves. Rubens and Aschkinass, 5, 891.
— See Liquids.

Fluorescence and actinic electricity, Schmidt, 5, 467 ; and chem. composition. Meyer, 5, 387.

Fluorine, liquefaction, 4, 318.

Fontaine, W. M., Mesozoic plants from California, 2, 273.

Foote, H. W., double fluorides of cæsium and zirconium, 1, 18; pollucite, mangano-columbite and microlite in Maine, 1, 457.
— rœblingite from Franklin Furnace. N. J., 3, 413 ; wellsite. a new mineral, 3, 443 ; double halogen salts of cæsium and rubidium, 3, 461 ; double fluorides of zirconium with lithium, etc., 3, 466.

Foote, H. W., bixbyite and topaz, 4, 105 ; composition of ilmenite, 4, 108 ; clinohedrite from Franklin, N. J., 5, 289.
— composition of tourmaline, 7, 97.

Foote, W. M., meteoric iron, Sacremento Mts., New Mexico, 3, 65; native lead with rœblingite, Franklin Furnace, N. J., 6, 187; new meteoric iron, Alabama, 8, 153: new meteoric iron, Texas, 8, 415.

Foraminifera, recent. Flint, 9. 158.

Forbes, E. H., chrysolite-fayalite group, 1, 129; epidote, Huntington, Mass., 1, 26.

Ford, W. E., siliceous calcites from So. Dakota, 9, 352 ; calcite crystals from Union Springs, N. Y., etc, 10, 237.

Fossils, see GEOLOGY.

Foster, G. C., Electricity and Magnetism, 2, 83.

Fowle, F. E., Jr., longitudinal aberration of prisms, 2, 255.

Fox Islands, Me.. geology, Smith, 3, 161.

Franklin, W. S., Elements of Physics, 1, 319 ; 2, 454 ; 4, 73.

Franz Joseph Land, geology, 5, 233; flora, Fisher. 5, 236.

Frazer, P., Weisbach's Tables for the determination of Minerals, 3, 162.

Freezing points, determination, Harker, 2, 390.

Frenzel, A., chalcostibite and guejarite, 4, 27.

Fresenius, C. R., Manual of Qualitative Analysis, 4, 473.

Frog's Egg, Development. Morgan, 4, 161.

Fuchs, G., Molekulargewichtsbestimmung, 1, 53.

Fulgurite, spiral, Wisconsin, Hobbs, 8, 17.

Funafuti, coral boring at, 5, 75 ; Sollas, 7, 317.

Fur seals of the North Pacific Ocean. Jordan. 9, 390.

Fusion, see melting point.

G

Gage, A., Physical Experiments, 5, 222.

Galvanometer, new method of reading. Rice, 2, 276.
— shunt box for, Stine, 5, 124.

Ganong, W. F., The Teaching Botanist, 9, 79.

Gas, a new, Brush, 6, 431.
See also CHEMISTRY.

anufacture of, text-book on,
y, **1**, 317.
1ometer at high temperatures,
ınd Holborn, **8**, 165; **10**,

absorption by glass powder,
-th, **10**, 462; absorption, in a
acuum, Hutchins, **7**, 61.
stitution of, etc., Corrigan,

ro-synthesis of, Mixter, **4**,
217.
hermic, Mixter, **7**, 323; ex-
1 of, Mixter, **7**, 827.
ed on heating mineral sub-
s. Travers, **7**, 244.
Expansion of, **6**, 504.
:ic Theory of, Meyer, **9**, 221.
sfaction of, Hardin, **9**, 221.
ple spectra of, Trowbridge
.chards, **3**, 117.
(neon, krypton, metargon),
5, **9**, 62.
ed, influence of temperature
3 potential-fall, Schmidt, **9**,

erature and ohmic resistance,
ridge and Richards, **3**, 827.
CHEMISTRY.
A., geological time, **8**, 387.
J., Earth Sculpture, **7**, 72.
tubes, electric discharge in,
Gill, **5**, 399.
erature in, Wood, **2**, 452.
hical Congress, Interna-
7, 316.
ibution of marine animals,
nn, **1**, 321.
ohie, La, **9**, 312.
:al (Geologic) Atlas of the
l States, **9**, 157, 387 · of
'stone National Park, **3**,

gy, H. S. Williams, **1**, 63.
ress, International, **3**, 351,
4, 477; **5**, 313; **7**, 316; **9**,

itute of Mexico, Bulletin,
ra, **3**, 422; **5**, 152.
res at Harvard University,
1, **4**, 397; at Baltimore,
r, **9**, 456.
aclature, **3**, 313.
ty of America, meeting at
elphia, **1**, 147; at Washington,
; Cordilleran section, **9**, 156.
OGICAL REPORTS—
aa, 1896, **3**, 350; **4**, 393;
making, Phillips, **7**, 398;
, **9**, 69.
, **10**, 398.
2

GEOLOGICAL REPORTS—
Canada, **1**, 150, 490; report of
Hoffmann, **2**, 88; vol. vii, 1894,
3, 72, 421, 488; report of
Hoffmann, **4**, 78; vol. viii,
1895, **4**, 394; vol. ix, 1896, **5**,
232; 1897, **6**, 434, 510; vol. ix,
1896, **7**, 71; 1898, **8**, 232; **9**, 68;
vol. x, 1897, **9**, 156, 302, 456;
1899, **10**, 165; vol. xi, **10**, 399;
report of Hoffmann, **10**, 404.
Cape of Good Hope, 1896, **4**, 395.
Georgia, corundum, **3**, 489; phosphates, etc., **5**, 394; clays, **8**, 469.
Great Britain, Memoirs, vol. i, **9**, 300.
India, **5**, 394.
Indiana, 21st Ann. Report, **4**, 394;
23d Report, **9**, 67; 24th Report,
10, 398.
Iowa, 1894, **1**, 149; 1895, **2**, 303;
vol. vii, 1896, **5**, 152; 1897, **7**,
168; vol. ix, 1898, **8**, 466; vol. x,
10, 467.
Kansas, vol. i, 1896, **1**, 489; mineral
resources, **8**, 396; gypsum de-
posits, **8**, 466.
Maryland, vols. i, ii, Clark, **7**, 69;
vol. iii, Clark, **9**, 223.
Michigan, vol. vi, **8**, 466; vol. viii,
10, 399.
Minnesota, vol. iii, part 2 **3**, 349;
vol. iv, **9**, 149; 24th Ann. Report,
9, 456.
Missouri, 1894, **1**, 149.
New Jersey, 1894, **1**, 60; vol. iv,
5, 468; 1898, **8**, 394.
New York, 1898, **1**, 61.
Norway, **1**, 399.
Nova Scotia, **6**, 510.
Pennsylvania, **1**, 488.
South Dakota, **7**, 316.
United States, 15th Ann. Report, **1**,
487; other publications, **1**, 144,
146; topographic maps, **1**, 488.
— 16th Ann. Report, **1**, 142; **2**, 84,
450; other publications, **2**, 84.
— 17th Ann. Report, **3**, 153, 249;
4, 155; other publications, **3**,
154, 155, 156, 250; **4**, 156, 157, 392.
— 18th Ann. Report, **5**, 308; **6**,
433; **7**, 166; **8**, 75; other publications, **5**, 304, 305, 469; **6**, 508,
509; topographic maps, **6**, 102.
— 19th Ann Report, **7**, 166; **8**, 76,
392; **9**, 447; other publications;
7, 167; **8**, 393, 394, 465.
— 20th Ann. Report, **9**, 447, 448;
10, 320, 397, 465; other publications, **9**, 297, 448, 451; **10**, 163,
166, 168, 247, 249, 469; geologic
atlas, **9**, 157, 387.

GEOLOGICAL REPORTS—

West Virginia, White, **7**, 398, 399.

Wisconsin, **9**, 69.

Witwatersrand, Southern Transvaal, Hatch, **5**, 393.

Géologiques, Catalogue des Bibliographies, de Margerie, **3**, 250.

Geologische Uebersichtskarte der Schweiz, Schmidt, **3**, 160.

Geology, Bibliography of N. American, Weeks, **2**, 303 ; **6**, 510 ; **8**, 393 ; **9**, 448.

— Catalogue of N. Amer. contributions, 1732–1891, Darton, **3**, 350.

— Dana's Text-book, revised by W. N. Rice, **5**, 393.

— Elementary, Tarr, **3**, 351.

— Elements of, Le Conte, **2**, 303.

— Experimental, Meunier, **8**, 468.

— Introduction to, Scott, **3**, 422.

— the Student's Lyell, Judd, **2**, 86.

— Williams Memorial Lectures at Baltimore, Brögger, **9**, 456.

GEOLOGY—

Alkali carbonate solution, geologic efficacy, Hilgard, **2**, 100.

Amphictis, skull of, Riggs, **5**, 257.

Anticlinorium and synclinorium, Rice, **2**, 168.

Apodidæ, revision, Schuchert, **4**, 325.

Appalachians, extension across Mississippi, etc., Branner, **4**, 357.

Archelon ischyros from S. Dakota, Wieland, **2**, 399.

Arctic Sea ice as a geological agent, Tarr, **3**, 223.

Arkansas Valley, Colorado, underground water, Gilbert, **3**, 156.

Aspen mining district, Colo., geology, Spurr, **8**, 465.

Auriferous conglomerate of the Transvaal, Becker, **5**, 193.

— deposits of Quebec, Chalmers, **8**, 394.

Bacillaria of the Occidental Sea, Edwards, **8**, 445.

Bacteria and decomposition of rocks, Branner, **3**, 438.

Bearpaw Mts., Mont., petrography, Weed and Pirsson, **1**, 283, 351 ; **2**, 136, 188.

Belodont reptile, new, Marsh, **2**, 59.

Bermudas, geology of, Verrill, **9**, 313.

Bethany limestone, Missouri, Bain, **5**, 433 ; Keyes, **2**, 221.

Black Hills, Dakota, geology, Irving, **9**, 384.

Bohemia Mining region of Oregon, Diller and Knowlton, **10**, 465.

GEOLOGY—

Bowlders of the Mattawa Valley, scoured, Taylor, **3**, 208.

Brachiopod fauna of Rhode Island, Walcott, **6**, 327.

Calamaria of the Dresden Museum, Geinitz, **6**, 198.

Cambrian faunas, Matthew, **9**, 69.

— rocks of Pennsylvania, Walcott, **2**, 84.

Camden chert of Tennessee, Safford and Schuchert, **7**, 429.

Cape Cod district, geology, Shaler, **8**, 76.

Cape Nome gold region, Schrader and Broom, **9**, 455.

Carboniferous invertebrates, index of North American, Weller, **7**, 70.

— bowlders from India, Emerson, **10**, 57.

Castle Mt. district, Montana, geology, Weed and Pirsson, **3**, 250.

Cephalonia, sea mills of, Crosby, **3**, 489.

Cephalopoda, fossil, of the British Museum, Crick, **6**, 198.

Ceratopsia, new species, Marsh, **6**, 92.

Cerrillos coalfield, Stevenson, **1**, 148.

Chautauqua grape belt, geology, Tarr, **1**, 399.

Christmas Is., geology, Andrews, **10**, 79.

Clays of Pennsylvania, Hopkins, **8**, 237.

Climates of Geological Past, Dubois, **1**, 62.

Coal deposits of Indiana, Blatchley, **9**, 67.

— of Missouri, age of the lower, White, **3**, 158.

Colorado canyon, geology, Davis, **10**, 251.

Comanche series in Oklahoma and Kansas, Vaughan, **4**, 43.

Copper-bearing rocks of Douglas Co., Wis., Grant, **10**, 249.

Coral boring at Funafuti, **5**, 75.

— islands of Pacific, Agassiz, **9**, 109, 193, 369.

— reefs and islands of Fiji, Agassiz, **5**, 113 ; **8**, 80.

— revision of Canadian Paleozoic, Lambe, **9**, 155.

Crangopsis vermiformis of Kentucky, Ortmann, **4**, 283.

Crater Lake, Oregon, Diller, **7**, 316 ; **3**, 165.

Cretaceous of Athabasca river, Tyrrell, **5**, 469.

GEOLOGY—

Cretaceous foraminifera of New Jersey, Bagg, **6**, 509.

— formations of the Black Hills, Ward, **9**, 70 ; of Nebraska, Gould, **9**, 429 ; in Kansas, Gould, **5**, 169.

— paleontology of Pacific Coast. Stanton, **1**, 320.

— section at El Paso, Texas, Stanton and Vaughan, **1**, 21.

— and Tertiary plants of North America, Knowlton, **7**, 168.

— turtles of South Dakota, Wieland, **9**, 237.

Ctenacanthus spines from the Keokuk limestone of Iowa, Eastman, **4**, 10.

Currituck Sound, Virginia and No. Carolina, Wieland, **4**, 76.

Cycad horizons in the Rocky Mountain region, Marsh, **6**, 197.

Cycadean monœcism, Wieland, **8**, 164.

Cycadofilices, Wieland, **8**, 309.

Cycads, American fossil, Wieland, **7**, 219, 305. 388 ; Ward, **9**, 70, 384 ; **10**, 327.

Deposits from borings in the Nile Delta, Judd, **4**, 74.

Devonian Amphibian footprints, Marsh, **2**, 374.

— fauna of black shale of Kentucky, Girty, **6**, 384.

— fishes, Eastman, **7**, 314.

— formation of Southern U. S., Williams, **3**, 393.

— interval in northern Arkansas, Williams, **8**, 139.

— of North Missouri, Broadhead, **2**, 237.

— strata in Colorado, Spencer, **9**, 125

— Upper, Key to, Harris. **9**, 156.

Dictyospongidæ, Hall and Clarke, **9**, 69, 224.

Dinichthyids, relations of body-plates, Eastman, **2**, 46.

Dinosaurs of North America, Marsh, **2**, 458 ; European, Marsh, **4**, 413.

Earth, age of, Kelvin, **7**, 160; Geikie, **8**, 387 ; Joly, **8**, 390.

— movement, Van Hise, **5**, 230 ; in the Great Lakes region, Gilbert, **7**, 239.

Edwards plateau. geology, Hill and Vaughan, **7**, 315.

Eocene of Atlantic Coast, Potomac river section of, Clark, **1**, 365 ;

— deposits of the Middle Atlantic slope. Clark, **3**, 250.

— carnivore, Marsh, **7**, 397.

GEOLOGY—

Eopaleozoic hot springs and siliceous oölite, Wieland, **4**, 262.

Etcheminian fauna of Cape Breton, Matthew, **9**, 158.

Face de la Terre, Suess–Margerie, **5**, 152 ; **10**, 167.

Fault at Jamesville, N. Y., Schneider, **3**, 458 ; in Meade Co., Kansas, Haworth, **2**, 368.

Fauna of the Ithaca group, relation of, Kindle, **3**, 159.

— prehistoric, of Block Island, Eaton, **6**, 187.

Felidæ of North America, extinct, Adams, **1**, 419 ; **4**, 145.

Flora of Lower Coal measures of Missouri, D. White, **10**, 166.

Floras of North America, Newberry, **8**, 394.

Flore des couches permiennes de Trienbach, Zeiller, **3**, 74.

Florencia formation, Hershey. **4**, 90 ; Pilsbry, **5**, 232.

Foramina in cranium of a Permian reptile, Case, **3**, 321.

Fossil fishes in British Museum, catalogue, Pt. III, Woodward, **1**, 396.

— insects of New Brunswick, Matthew, **4**, 394.

— invertebrates and plants in British Museum, **3**, 489.

— Medusæ, Walcott, **6**, 509.

— plants, Seward, **5**, 472.

Fossils from Canada, two new, Whiteaves, **6**, 198.

— invertebrate, from So. Patagonia. Ortmann, **10**, 368.

— of the Midway Stage, Harris, **2**, 86.

— for the National Museum, Marsh, **6**, 101.

— use of, in determining geological age, Marsh, **6**, 483.

— vertebrate, of the Denver basin, Marsh, **3**, 349.

Fox Islands, Me., geology, G. O. Smith, **3**, 161.

Franklin white limestone, New Jersey. Wolff and Brooks, **7**, 397.

Franz Joseph Land geology, **5**, 283.

Galeodidæ, comparative morphology, by Bernard, Beecher, **1**, 491.

Geothermal data from artesian wells in the Dakotas, Darton, **5**, 161.

— gradient in Michigan, Lane, **9**, 434.

Glacial, glaciers, see **Glacial**, etc.

Glyphioceras and phylogony of the Glyphioceratidæ, Smith, **5**, 315.

GEOLOGY—

Gold fields of S. Appalachians, Becker, 1, 57; of Alaska, 5, 305; 9, 455; of Georgia, 7, 168; of Transvaal, 5, 193.

Gold and silver veins in Idaho, Lindgren, 10, 466.

Grand River. Michigan, mouth of, Mudge, 8, 31.

Green Mountain region, structural details, Dale, 2, 395: monograph, Pumpelly, Wolff and Dale, 1, 146.

Hampshire Co., Mass., geology, Emerson, 8, 393.

Hesperornis, affinities of, Marsh, 3, 347.

Hudson Bay, elevation of land around, Bell, 1, 219·

Ichthyodectes, species of, Hay, 6, 225.

Igneous rocks, see **ROCKS.**

Iron-bearing district of Crystal Falls, Michigan, Clements and Smyth, 9, 451.

Johnson Co., Iowa, geology, Calvin, 5, 149.

Judith Mountains, geology, Weed and Pirsson, 6, 508.

Jura and Neocomian of Arkansas, etc., Marcou, 4, 197, 449.

Jurassic Dinosaurs, footprints, Marsh, 7, 229.

— formation on the Atlantic Coast, Marsh, 2, 295, 375, 433; 6, 105.

— plants, Seward, 10, 328.

— of Texas, alleged, Hill, 4, 449.

— times, climatic zones in, Ortmann, 1, 257.

— vertebrates from Wyoming, Knight, 5, 186, 378; 10, 115.

Lapiés, l'étude des, Chaix, 1, 321.

Laurentian. Canadian, Adams, Barlow and Ells, 3, 173.

Lava beds at Meriden, Conn., Davis, 1, 1.

— flows of the Sierra Nevada, Ransome, 5, 355.

Leitfossilien, Koken, 3, 160.

Lepidophloios, Dawson, 5, 394.

Lichenaria typa, W. and S., Sardeson, 8, 101.

Limestones, oblique bedding in, 1, 398.

Lingulepis, Walcott, 3, 404.

Linuparus atavus of Dakota, Ortmann, 4, 290.

Little Belt Mountains, Mont., geology, Weed and Pirsson, 10, 466.

Loess, eolian origin of, Keyes, 6, 299; origin discussed, Sardeson, 7, 58.

GEOLOGY—

Louisiana, geology, Harris and Veatch, 10, 81.

Lower Cambrian in Atlantic Province, Walcott, 9, 302.

— Cretaceous Gryphæas of Texas, Hill and Vaughan, 7, 70.

— Silurian fauna of Baffin land, Schuchert, 10, 81.

Lytoceras and Phylloceras, development, Smith, 7, 398.

Madreporarian corals in the British Museum, Bernard, 5, 319.

Magellanian beds of Chili, fauna, Ortmann, 8, 427.

Mammals, catalogue of, Trouessart, 3, 351; 7, 79; 8, 397.

— origin of, Marsh, 6, 406; Osborn, 7, 92.

Man, antiquity, in Britain, Abbott. 3, 158.

Manganese nodules at Onybygambah, New South Wales, 9, 72.

Marble, flow of, under pressure. Adams and Nicolson. 10, 401.

Massanutten Mountain, Virginia, geology. Spencer. 5, 231·

Mercer mining district, Spurr. 1, 395.

Merycochærus in Montana, Douglass, 10, 428.

Mesozoic floras of U. S., Ward, 10, 320.

— plants, catalogue, British Museum, Seward, 1, 397: from California. Fontaine, 2, 273.

Metamorphism, contact, Fairbanks, 4, 36; 7, 81.

— of rocks and rock flowage, Van Hise, 6, 75.

Miocene of N. Jersey, mollusca and crustacea. Whitfield, 1, 61.

Moraines of Minnesota, Todd, 6, 469; of So. Dakota, 10, 249.

Narragansett Basin. geology, Shaler, Woodworth and Foerste. 10, 163; Carboniferous fauna of, Packard, 10, 164.

Niagara Falls, duration of. etc., Spencer, 1, 398, episode in the history of, Spencer, 6, 439.

Paleobotany, Elements, Zeiller, 10, 88; Potonié, 10, 88.

Paleotrochis in Mexico, Williams, 7, 335; origin of. Diller, 7, 337.

Paleozoic faunas of the Argentine. Kayser, 5, 72.

— fossils from Baffinland, Kindle, 2, 455.

— terrane beneath the Cambrian, Matthew, 8, 79.

GEOLOGY—

Paleozoic, thickness of, in Arkansas, Branner, **2**, 229.

Panama, geology. Bertrand, **10**, 82.

— and Costa Rica, geology, Hill, **6**, 435, 505.

Patagonia, geology of, Hatcher, **4**, 246, 327 ; **9**, 85 ; invertebrate fossils from, Ortmann, **10**, 368.

Peat in the Dismal Swamp, depth of, Wieland, **4**, 76.

— bogs of Finland, Andersson, **8**, 467.

Petrography of the Boston Basin, White, **5**, 470.

Pflanzenpaleontologie, Potonié, **10**, 88.

Phosphate-deposits of Arkansas, Branner, **3**, 159 ; in Tennessee, Safford, **2**, 462.

Pithecanthropus erectus of Dubois, Marsh, **1**, 475; Manouvrier, **4**, 213.

Pleistocene deposits of Chicago area, Leverett, **4**, 157.

— glaciation ·in New Brunswick, etc., Chalmers, **3**, 72.

— marine shore lines, Chalmers, **1**, 302.

Pleurotomaria providencis, Broadhead, **2**, 287.

Popocatepetl and Ixtaccihuatl, observations, Farrington, **4**, 326.

Pre-Cambrian fossiliferous formations, Walcott, **8**, 78.

— geology, North America, Van Hise, **2**, 205.

— rocks and fossils, **3**, 157.

Pre-glacial decay of rocks in Eastern Canada, Chalmers, **5**, 273.

— drainage in Michigan, Mudge, **4**, 388 ; **10**, 158.

Protoceratidæ, principal characters, Marsh, **4**, 165.

Protostegan plastron, Wieland, **5**, 15.

Pseudoscorpion, a new fossil, Geinitz, **4**, 158.

Reconnaissance in Oregon, J. S. Diller, **3**, 155.

Road-building stones of Massachusetts, Shaler, **1**, 489.

Rock differentiation, Becker, **3**, 21.

Rocks, flow and fracture as related to structure, Hoskins, **2**, 213 ; fractional crystallization, Becker, **4**, 257.

— See further, ROCKS.

San Clemente Island, geology, W. S. T. Smith, **7**, 315.

Santa Catalina Island, geology, W. S. T. Smith, **3**, 351.

GEOLOGY—

Saurocephalus. Hay, **7**, 299.

Sauropodous Dinosauria, Marsh, **6**, 487.

Schists of gold and diamond regions of Brazil, Derby, **10**, 207.

Silurian rocks of Britain, Peach and Horne, **9**, 300.

Silurian-Devonian boundary in N. America, H. S. Williams, **9**, 203.

Silveria formation, Hershey, **2**, 324.

Slate Belt of Eastern New York and Western Vermont, Dale, **9**, 382.

Species, distribution and origin discussed, Ortmann, **2**, 63.

Spines, origin of, Beecher, **6**, 1, 125, 249, 329.

Stylinodontia, Marsh,˙**3**, 187.

Stylolites, Hopkins, **4**, 142

Stylonurus lacoanus, restoration, Beecher, **10**, 145.

Syrian fossils, Conrad's types of, Beecher, **9**, 176.

Tamiobatis vetustus, Eastman, **4**, 85.

Tapirs, recent and fossil, Hatcher, **1**, 161.

Tertiary fauna of Florida, Dall, **7**, 71.

— floras of Yellowstone Park, Knowlton, **2**, 51.

— formations of the Rocky Mts., Davis, **9**, 387.

— horizons, new marine, Ortmann, **6**, 478.

— limestone reefs of Fiji, Agassiz, **6**, 165.

— mammal horizons of Europe and America, Osborn, **10**, 400.

Testudinate humerus, evolution, Wieland, **9**, 413.

Trap Rock of Conn., Davis, **1**, 1 ; of the Palisades, Lyman, **1**, 149 ; of Rocky Hill, N. J., Phillips, **8**, 267.

Trenton rocks, original. White, **2**, 430 ; at Ungava, Whiteaves, **7**, 433.

Triarthrus, morphology of, Beecher, **1**, 251.

Triassic formation of Connecticut, Davis, **8**, 76; bivalve from, Emerson, **10**, 58.

Trilobites, classification of, Beecher, **3**, 89, 181.

Uintacrinus from Kansas, Beecher, **9**, 267.

Vertebral centra, terminology, Wieland, **8**, 163.

Washington, So. Western, geology, Russell, **3**, 246.

GEOLOGY—
  Wealden, age of, Marsh, 1, 234.
  — flora of Bernissart, Seward, 10, 322.
  Wind deposits, mechanical composition, Udden, 7, 74.
  Yellowstone National Park, atlas, 3, 246; geology, 9, 297.
Geometry, Beman and Smith, 1, 328.
Geophysik, Beiträge zur, Gerland, 1, 322.
Georgia, clays of, Ladd, 8, 469; Geol. survey, 5, 394; gold deposits, 7, 168.
Gesteinslehre, Rosenbusch, 7, 78.
Geyser eruption, conditions affecting, Jaggar, 5, 323.
Gibbs, J. W., Hubert Anson Newton, 3, 359.
Gilbert, G. K., underground water of Arkansas Valley, Col., 3, 156; earth movement in the Great Lakes region, 7, 239.
Gill, H. V., theory of singing flames, 4, 177; stratification of electric discharge in Geissler tubes, 5, 399.
Gillespie, D. H. M., iodine in the analysis of alkalies, etc., 6, 455.
Gilman, D. C., life of J. D. Dana, 9, 80.
Girty, G. H., fauna from Devonian black shale of Kentucky, 6, 384.
Glacial deposits in subalpine Switzerland, Du Riche Preller, 2, 301.
  — gravels in Lower Susquehanna Valley, Bashore, 1, 281; of Maine, Stone, 10, 247.
  — ice, "plasticity," Russell, 3, 344.
  — Lakes in Central New York, Fairchild, 7, 249.
  — lobe in Illinois, Leverett, 10, 249.
  — observations in Greenland, Barton, 4, 395.
Glaciation of Central Idaho, Stone, 9, 9; in Greenland, Tarr, 4, 325; in Wayne Co., Pennsylvania, Kummel, 1, 113.
  — Pleistocene, in New Brunswick, Chalmers, 3, 72.
Glaciers of Mt. Ranier, Russell, 8, 76; of N. America, Russell, 3, 423.
  — variation in length, Rabot, 4, 395; 8, 88; Richter, 9, 71.
Glarner Alpen, das geotektonische problem der, Rothpletz, 9, 303.
Glass, colloidal, Barus, 6, 270.
  — from Jena, properties and use, 9, 445.
  — thermodynamic relations of hydrated, Barus, 7, 1; 9, 161.

Glazebrook, R. T., Mechanics and Hydrostatics. 1, 327; James Clerk Maxwell and Modern Physics, 1, 404.
Gold fields of S. Appalachians, 1, 57; of Alaska, 5, 305; 9, 455; of Georgia, 7, 168; of Transvaal, 5, 193.
  — Mines of the Transvaal, De Launay, 2, 88.
  — and platinum layers, optical relations of, Breithaupt, 8, 74.
  — in sea-water, Liversidge, 2, 304.
  — See CHEMISTRY.
Goldschmidt, V., Krystallographische Winkeltabellen, 5, 153.
Gooch, F. A., separation of selenium from tellurium, 1, 181: iodometric determination of selenious and selenic acids, 1, 31.
  — iodometric estimation of molybdic acid, 2, 156; determination of tellurium, 2, 271; separation of aluminum from iron, 2, 416.
  — estimation of molybdenum iodometrically, 3, 237; iodic acid in the analysis of iodides, 3, 293.
  — estimation of manganese, 5, 209; oxidation of manganese, 5, 260.
  — iodometric determination of molybdenum, 6, 168; manganese as pyrophosphate, 6, 283.
  — estimation of boric acid, 7, 34; ammonium-magnesium phosphate, 187; volatilization of iron chlorides in analysis, 7, 370; titration of oxalic acid, 7, 461.
  — tellurous acid in presence of haloid salts, 8, 122; iodometric determination of gold, 8, 261; action of acetylene on the oxides of copper, 8, 354.
  — iodometric determination of arsenic acid, 10, 151.
Goode, G. B., History of the Smithsonian Institution, 5, 158.
Goodman, R. J., estimation of vanadium, 2, 355.
Gould, B. A., Cordoba Photographs, 4, 480; Fund established by, 7, 320.
Gould, C. N., transition beds from the Comanche to the Dakota Cretaceous, 5, 169; the Dakota Cretaceous of Nebraska, 9, 429.
Grammar of Science, Pearson, 10, 170.
Grant, U. S., copper-bearing rocks of Wisconsin, 10, 249.
Gravitation in gaseous nebulæ, Nipher, 7, 459.

Gravitation constant and mean density of the earth, 6, 503.
— determined by pendulum, Putnam, 1, 186.
Gray Herbarium of Harvard University, Contributions, No. xi, Greenman, 4, 249.
Greely, A. W., Arctic Discoveries, 1, 403.
Greenland, glaciation in, Tarr, 4, 325; Barton, 4, 395.
— marine algæ of, Rosenvinge, 7, 400.
— new minerals from, 10, 323.
Greenleaf, J. L., hydrology of the Mississippi, 2, 29.
Gregory, H. E., volcanic rocks of Temiscouata Lake, Quebec, 10, 14.
Groth, P., Mineralogical Tables, 5, 154; Zeitschrift für Krystallographie, Index, 9, 229; 10, 84.
Gulf stream water, origin, Cleve, 9, 310.
Guthe, K. E., measurement of self-inductance, 5, 141.

H

Hague, A., age of the igneous rocks of the Yellowstone, 1, 445.
Hall, C. W., rocks of S. W. Minnesota, 10, 168.
Hall, James, obituary, 6, 284, 487.
— Dictyospongidæ, 9, 69, 224.
Hall effect in an electrolyte, 6, 504; Poincaré, 7, 312; in flames, Marx, 10, 395.
Hallock, W., underground temperatures at great depths, 4, 76.
Hardin, W. L., Liquefaction of Gases, 9, 221.
Hardness of minerals, Auerbach, 2, 390; Jaggar, 4, 399.
Harkness, W., achromatic objectives of telescopes, 9, 287.
Harmonic analyzer, new, Michelson and Stratton, 5, 1.
Harper's Scientific Memoirs, 6, 199, 504; 7, 69, 402; 8, 400.
Harrington, B. J., new alkali hornblende and titaniferous andradite, Ontario, 1, 210.
Harris, G. D., the midway stage of the Eocene, 2, 86; key to the Upper Devonian, 9, 156.
Hart, J. H., action of light on magnetism, 10, 66.
Hartwell, J. B., separation of nickel from cobalt, 10, 316.
Harvard College Observatory, 1, 75; 3, 252, 492; 5, 80, 320; 8, 87; 9, 311.

Hastings, C. S., telescope objectiv e 7, 267; General Physics, 7, 314.
Haswell, W. A., Text book of Zoology, 5, 319; Manual of Zoology, 9, 390.
Hatcher, J. B., recent and fossil tapirs, 1, 161; geology of southern Patagonia, 4, 246, 327; sedimentary rocks of Southern Patagonia, 9, 85.
Havens, F. S., separation of aluminum from iron, 2, 416; of aluminum and beryllium, 4, 111; of aluminum by hydrochloric acid, 6, 45; of nickel and cobalt by hydrochloric acid, 6, 396; volatilization of iron chlorides in analysis, 7, 370; separation of iron, 8, 217.
Hawaiian soils, chemical composition, Lyons, 2, 421.
Haworth, E., deformation of strata in Kansas, 2, 368.
Hay, O. P., species of Ichthyodectes, 6, 225; unknown species of Saurocephalus, 7, 299.
Headden, W. P., tin compounds from an old furnace in Cornwall, 5, 93.
Heald, F. DeF., analytic keys of North American Mosses, 3, 354.
Hearing, limits of, Koenig, 9, 66, 148.
Heat of combustion, Mendeléeff, 4, 319.
— conductivity, relative, Voigt, 5, 223; of glass, Winkelmann, 7, 471.
— long waves of, separated by quartz prisms, Rubens and Aschkinass, 7, 312.
— insulators for, Hempel, 8, 74.
Heliostat, Mayer, 4, 306.
Helium, see CHEMISTRY.
Hereford earthquake of December 17, 1896, Davison, 8, 235.
Herrick, F. H., abnormal hickory nuts, 2, 258.
Herrmann, Steinbruchindustrie, 8, 81.
Herschels and Modern Astronomy, Clerke, 1, 76.
Hershey, O. H, Silveria formation, 2, 324; Florencia formation, 4, 90.
Hertzian waves, reflection, L. De-Forest, 8, 58. See Electric.
Hervey, Colors of Flowers, 8, 471.
Hidden, W. E., rhodolite, 5, 294; sperrylite in North Carolina, 6, 381; twinned crystals of zircon, 6, 323; associated minerals of rhodolite, 6, 463; ruby in North Carolina, 8, 370; iron meteorite, Hayden Creek, Idaho, 9, 367.
Hilgard, E. W., geologic efficacy of alkali carbonate solution, 2, 100.

Hill, A., Introduction to Science, 10, 406.

Hill, R. T., alleged Jurassic of Texas, reply to Marcou, 4, 449; geological history of Panama and Costa Rica, 6, 435, 505; Lower Cretaceous Gryphæas of Texas, 7, 70; geology and physical geography of Jamaica, 9, 222.

Hillebrand, W. F., phosphorescence in wollastonite, 1, 323; vanadium and molybdenum in rocks, 6, 209; mineralogical notes, 7, 51; analyses of biotites and amphiboles, 7, 294; roscoelite, 7, 451; mineralogical notes, 8, 295; methods of analysis, 9, 73; carnotite, etc., of Colorado, 10, 120; rock analyses, 10, 250.

Hintze, C., Handbuch der Mineralogie, 3, 251; 5, 316; 6, 435; 9, 229; 10, 469.

Hoadley, G. A., Brief Course in General Physics, 10, 465.

Hobbs, W. H., chloritoid from Michigan, 2, 87; goldschmidite, 7, 357; 10, 426; spiral fulgurite from Wisconsin, 8, 17.

Hoffmann, Canadian Minerals, 4, 78; 6, 437; 10, 404; xenotime in Canada, 5, 235; baddeckite, 6, 274; polycrase in Canada, 7, 243.

Holborn, L, gas thermometer at high temperatures, 8, 165; 10, 171; thermo-electricity in certain metals, 8, 303.

Holden, E. S., Earthquakes of the Pacific Coast, 6, 200.

Holm, T., Studies in Cyperaceæ, No. i, 1, 348; No. ii, 2, 214; No. iii, Carex Fraseri, 3, 121; No. iv, Dulichium spathaceum, 3, 429; No. v, Fuirena squarrosa and F. scirpoidea, 4, 13; No. vi, Dichromena leucocephala and D. latifolia, 4, 298; No. vii, Scleria, 5, 47; No. viii, Scleria, 7, 5; No. ix, Lipocarpha, 7, 171; No. x, Fimbristylis, 7, 435; No. xi, Carex stipata, 8, 105; No. xii, Carex filifolia, 9, 355; No. xiii, Carex wildenowii, 10, 33; No. xiv, Carices from Alaska, 10, 266.

— Pogonia ophioglossoides, 9, 13.

Holman, S. W., Computation Rules and Logarithms, 1, 328; Matter, Energy, Force and Work, 7, 237.

Holtz machine, influence of light on discharge, Elster and Geitel, 1, 393.

Honda, K., string alternator, 10, 64.

Hopkins, T. C., stylolites, 4, 142.

Hornby, J., Text-book of Gas Manufacture, 1, 317.

Horne, J, Silurian rocks of Britain, 9, 300.

Hoskins, L. M., flow and fracture of rocks as related to structure, 2, 213.

Hovestadt, H., Jenaer Glas, 9, 445.

Hovey, E. O., acid dike in the Connecticut Triassic, 3, 287; pseudomorphs after halite, Jamaica, 3, 425.

Hovey, H. C., Mammoth Cave of Kentucky, 4, 326.

Howe, E., detection of sulphides, etc., 6, 317.

Howe, J. C., explosive effect of electrical discharges, 8, 239.

Howe, J. L., specimens from Chichan-Kanab, Yucatan, 2, 413.

— Bibliography of Platinum Metals, 5, 322.

Howe, M. A., Hepaticæ and Anthocerates of California, 8, 309.

Hudson Bay, rising of land around, Tyrrell, 2, 200.

— region, geological report, Tyrrell, 7, 72

Hutchins, C. C., use of Crookes tubes for X-rays, 1, 463; irregular reflection, 6, 373; absorption of gases in a high vacuum, 7, 61.

Hutchins, G. P., estimation of thallium, 8, 460.

Huxley, T. H., Elementary Physiology, 10, 90.

— memorial, 1, 249.

Hydrogen, spectrum, Trowbridge. 10, 222.

— See CHEMISTRY.

Hydrology of the Mississippi, Greenleaf, 2, 29.

I

Ice, plasticity, Mügge, 1, 56; geological action of Arctic, Tarr, 3, 223.

Idaho, glaciation of Central, Stone, 9, 9; geological reconnaissance of, Eldridge, 1, 145; gold and silver veins, Lindgren, 10, 466.

Illinois State Museum, 1, 150, 398.

India Geological Survey, 5, 394; Carboniferous bowlders from, Emerson, 10, 57.

Indiana, annual report of department of geology, 1896, 4, 394.

— coal deposits of, Blatchley. 9, 67.

— geology and natural resources, Blatchley, 10, 398.

Induction balance, interferential, Barus, 3, 107.

— Coil, Wright, 4, 324.

Insects, Aquatic, Miall, 1, 249.
Interference rings, selenium, Longden, 10, 55.
Interpolation, H. L. Rice, 9, 394.
Ions, color of, Lea, 1, 405.
— and electric conduction, Thomson, 7, 312; also 7, 158.
— in gases at low pressures, masses of, Thomson, 9, 66.
Iowa, Geol. Survey, 1894. 1, 149; 1895. 2, 303; 7, 168; vol. ix, 8, 466; 10, 467.
— Johnson Co., geology, Calvin, 5, 149.
Iron, chemical analysis, Blair, 2, 450.
— See **CHEMISTRY.**
Irrigation and Drainage, King, 9, 394.
Isham, G. S., registering solar radiometer, 6, 160.
Italian volcanic rocks, analyses, Washington, 1, 375; 8, 286.
Iwasaki, C., orthoclase crystals from Japan, 8, 157.

J

Jackson, D. C., and J. P., Alternating Currents, 2, 455.
Jacoby, H., division errors of a straight scale, 1, 333.
Jaggar, T. A., Jr., instrument for inclining a preparation for the microscope, 3, 129; conditions affecting geyser eruption, 5, 323; microsclerometer for determining the hardness of minerals, 4, 399.
Jamaica, late formations and changes of level, Spencer, 6, 270.
— Geology and Physical Geography, Hill, 9, 222.
— Yellow limestone, Hill, 3, 251.
Japan, earthquakes in, Omori, 9, 305; 10, 471.
Jenaer Glas, Hovestadt, 9, 445.
Johanniskäfer light, Muruoka, 5, 224.
Joly, J., geological age of the earth, 8, 390.
Jones, H. C., Electrolytic Dissociation, 10, 76.
Jones, L. C., estimation of cadmium, 2, 269; action of carbon dioxide on soluble borates, 5, 442; estimation of boric acid, 7, 34, 147; 8, 127.
Judd, J. W., rubies of Burma. 1, 64; structure-planes of corundum, 1, 323; deposits from borings in the Nile Delta, 4, 74; petrology of Rockall Island, 7, 241; ruby in North Carolina, 8, 370.
— The Student's Lyell, 2, 86.
Judith Mts., Montana, geology, Weed and Pirsson, 6, 508.

K

Kansas, Geol. Survey, 1, 489.
— meteorites, 5, 447; 7, 283; 9, 410.
— mineral resources, 1899, Haworth, 8, 396; gypsum deposits of, 8, 466.
Kant as a natural philosopher, Becker, 5, 97.
Kaolins and fire clays of Europe, Ries, 7, 243.
Kathode, see **Cathode.**
Kayser, E., Paleozoic faunas of the Argentine, 5, 72.
Kayser, H., Handbuch der Spectroscopie, 10, 464.
Keith, A., dikes of felsophyre and basalt in Virginia, 6, 305.
Kemp, J. F., Handbook of Rocks, 3, 76.
— Ore Deposits of the United States and Canada, 9, 303.
Keyes, C. R., Bethany limestone, 2, 221; eolian origin of loess, 6, 299.
Kindle, E. M., Palæozoic fossils from Baffinland, 2, 455.
King, F. H., Irrigation and Drainage, 9, 394.
Kingsley, J. S., Text-book of Vertebrate Zoology, 8, 472.
Kite, Observations of, 1898, Frankenfield, 9, 394.
Kleinasiens Naturschätze. Kannenberg, 5, 79.
Klondike gold fields, 5, 305; 9, 456.
Knight, W. C., Jurassic vertebrates from Wyoming, 5, 186, 378; 10, 115.
Knipp, C. T., new form of make and break, 5, 283.
Knowlton, F. H., Tertiary floras of the Yellowstone Park, 2, 51.
— Bohemian mining region of Oregon, 10, 465.
Koenig, G. A., on mohawkite, stibiodomeykite, domeykite, etc., 10, 439.
Koenig, R., limit of hearing, 9, 66, 148.
Kohlenstoff-Verbindungen, Lexikon, Richter, 9, 445.
Kohlrausch, F., Practical Physics, 10, 320.
Koken, E., die Leitfossilien, 3, 160.
Kraus, C. A., broadening of the sodium lines, etc., 3, 472.
Kreider, D. A., potassium and sodium, separation of, 2, 263; oxygen in air and in aqueous solution, 2, 361; structural and magneto-optic rotation, 6, 416; detection of dextro- and lævo-rotating crystals, 8, 188.

Krypton, Ramsay, 6, 192. 360; 9, 62, 442; Crookes, 6, 189.
Kummel, H. B., glaciation in Pennsylvania, 1, 113.
Kunz, G. F., sapphires from Montana, 4, 417; native silver in No. Carolina, 7, 242.

L

Labrador Peninsula, traverse of, Low, 6, 510.
Lachman, A., Spirit of Organic Chemistry, 8, 78.
Lacroix, A., Minéralogie de la France et ses Colonies, 1, 401; 2, 461; 5, 155.
— granite of the Pyrenees, 6, 511.
Ladd, clays of Georgia, 8, 469.
Lambe, L. H., revision of Canadian Paleozoic corals, 9, 155.
Lambert, P. A., Differential and Integral Calculus, 6, 513.
Landolt, H., Optical Activity, 10, 76.
Lane, A. C., geothermal gradient in Michigan, 9, 434.
Lange, D., our Native Birds, 9, 81.
Langley, S. P., the bolometer, 5, 241.
— Smithsoniam Reports, 5, 289; 7, 80, 246, 402; 9, 233; 10, 469.
Lassar-Cohn, Chemistry in Daily Life, 2, 449.
Latitude, determinations in the Hawaiian Is., Preston, 1, 75.
— variation, Earth magnetism, etc., Halm, 10, 405.
Lea, M. C., Röntgen rays not present in sunlight, 1, 363; relations between atomic weights of the elements, 1, 386; color relations of atoms, ions and molecules, 1, 405; experiment with gold, 3, 64.
— obituary notice of, 3, 428.
Leavitt, R. G., psychrometer for study of transpiration, 5, 440; cause of root-pressure, 7, 381.
Le Conte, J., Elements of Geology, 2, 303.
Lehfeldt, R. A., Physical Chemistry, 9, 445.
Lenard rays, des Coudres, 4, 391; Goldstein, 4, 475.
— See Röntgen rays.
Leverett, F., Pleistocene features and deposits of the Chicago area, 4, 157; Illinois glacial lobe. 10, 249.
Lewis, H. Carvill, Genesis of the Diamond, 4, 77.
Lewis, W. J., Crystallography, 9, 73.
Lexikon der Kohlenstoff-Verbindungen, Richter, 9, 445.

Liebig, Justus von, Shenstone, 1, 76.
Light, absorption in a magnetic field, Righi, 6, 433.
— action on magnetism, Hart, 10, 66.
— electro-magnetic theory of, Barus, 5, 343.
— longitudinal, Lenard, 1, 249.
— nature of white, Carvallo, 9, 220.
— rotatory polarization of, Ewell, 8, 89; magneto-optic, Wright and Kreider. 6, 416; detection of, Kreider. 8, 133.
— and Sound, Nichols and Franklin, 4, 73.
— standards of, Petavel, 9, 295.
— Visible and Invisible, Thompson, 5, 71.
Light-emission power at high temperatures, St. John, 1, 54.
Lighting, Groves and Thorp, 1, 139.
Lightning, globular, Marsh, 1, 13; spectrum of, 7, 68,
Linck, G., relation between the geometric constants and the molecular weight of a crystal, 4, 321.
Lindgren, W., granitic rocks of Pyramid Peak, Cal., 3, 301: monazite from Idaho, 4, 63; orthoclase as gangue mineral, 5, 418; granodiorite and other intermediate rocks. 9, 269; gold and silver veins in Idaho, 10, 466.
Linebarger, C. E., surface tensions of liquids, 2, 108; tension of mixtures of normal liquids, 2, 226; viscosity of mixtures of liquids, 2, 331.
Linton, L. A., Trinidad pitch, 1, 193.
Liquid air, density and composition, Ladenburg and Krügel, 7, 234.
— carbon dioxide, Barus, 2, 1.
Liquids, solubilities, 5, 297.
— surface tension, Linebarger, 2, 108; Mayer, 3, 253; Stevens, 10, 245.
— tension of mixtures of normal, Linebarger, 2, 226.
— viscosity of mixtures, Linebarger, 2, 331; Thorpe and Rodger, 4, 65.
Lister, A., British Mycetozoa, 1, 153.
Lithology, E. H. Williams, 1, 150.
Littlehales, G. W., form of isolated submarine peaks, 1, 15; problem of finding isolated shoals in the open sea, 1, 106.
Liversidge, A., amount of gold and silver in sea-water, 2, 304.
Lockyer, N., Sun's Place in Nature, 5, 300.

Lodge, O., influence of a magnetic field on radiation-frequency, 4, 153.

Longden, A. C., electrical resistence of thin films from cathode discharge, 9, 407; selenium interference rings, 10, 55.

Loomis, E. J., Eclipse party in Africa, 3, 80.

Lory, C. A., electrical thermostat, 9, 179.

Loudon, W. J., Physics, 1, 141.

Louisiana, geology, Harris and Veatch, 10, 81.

Low, A. P., traverse of Labrador Peninsula, 6, 510.

Lucas, F. A., contributions to Paleontology, 6, 399.

Luedecke, O., die Minerale des Harzes, 2, 460.

Luminous night-clouds, height, 2, 89.

Luquer, Minerals in rock sections, 7, 319.

Lyell and Modern Geology, Bonney, 1, 322.

— Students Geology, Judd, 2, 86.

Lyman, B. S., trap rock of the Palisades, 1, 149.

Lyons, A. B., chemical composition of Hawaiian soils, 2, 421.

M

Mabilleau, Philosophie Atomistique, 1, 54.

McAlpine, D., Australian Fungi, 1, 324.

MacCurdy, G. G., translation of paper on Pithecanthropus erectus, 4, 213.

McIlhiney, P. C., action of ferric chloride on metallic gold, 2, 293.

McKay, T. C., explosive effect of electrical discharges, 8, 239.

Maclean, G. V., velocity of electric waves in air, 8, 1.

McLennan, J. C., Physics, 1, 141.

Magma, minerals formed from, 8, 80; separation of alumina from, 8, 227.

Magnetic declination in Alaska, 1895, Schott, 1, 75.

— field, influence on radiation-frequency, Lodge, 4, 153.

— increment of rigidity, H. D. Day, 3, 449.

— method of detecting metallic iron, Duane, 4, 475.

— Pole, South, location, Borchgrevink, 10, 405.

— screening, Dubois and Wills, 10, 77.

Magnetism, action of light on, Hart, 10, 66.

— of Earth, determination of, Meyer, 5, 467.

— influence on light. Zeeman, 3, 486; Dunstan, Rice and Kraus, 3, 472.

— solar and terrestrial, Bigelow, 5, 455.

— See Electricity.

Magnetismus der Planeten, Leyst, 1, 56.

Magnetizing constants of inorganic substances, Meyer, 8, 464.

Magnetostriction, torsional, Barus, 10, 407.

Magnets, induction coefficients of hard steel, Pierce, 2, 347; properties of seasoned, Pierce, 5, 334; temperature-coefficients of, Durward, 5, 245.

Maine, glacial gravels. Stone, 10, 247.

Make and break, new form, Knipp, 5, 283.

Mammals, Catalogue of, Trouessart, 3, 351; 7, 79; 8, 397.

— origin of, Marsh, 6, 406; Osborn, 7, 92.

Mammoth Cave of Kentucky, Hovey and Call, 4, 326.

Manouvrier, L., Pithecanthropus erectus, 4, 218.

Maquenne, L., les Sucres et leurs principaux Dérivés, 9, 445.

Marcou, J., Jura and Neocomian of Arkansas, etc.. 4, 197; reply to, by R. T. Hill, 4, 449.

Margerie, E. de, translation of, Suess, Antlitz der Erde, 5, 152; 10, 167.

Marine animals, geographical distribution, Ortmann, 1, 321.

Marsee, G., Plant diseases caused by Cryptogamic parasites, 8, 471.

Marsh, O. C., globular lightning, 1, 13; age of the Wealden, 1, 234; Pithecanthropus erectus, 1, 475.

— new Belodont reptile, 2, 59; geology of Block Island, 2, 295, 375: Amphibian footprints from the Devonian, 2, 374; Jurassic formation on the Atlantic Coast, 2, 433; Dinosaurs of North America, 2, 458.

— the Stylinodontia, 3, 187; affinities of Hesperornis, 3, 347; vertebrate fossils of Denver basin, 3, 349.

— principal characters of the Protoceratidæ, 4, 165; observations on European Dinosaurs. 4, 413.

— Cuvier prize awarded, 5, 79; collections presented to Yale University, 5, 156.

Marsh, O. C., Ceratopsia. 6, 92 ; vertebrate fossils for the National Museum, 6, 101 ; Jurassic formation on the Atlantic coast, 6, 105 ; cycad horizons in the Rocky Mountain region, 6, 197 ; value of type specimens, 6, 401 ; origin of mammals, 6, 406 ; fossils in determining geological age, 6, 483 ; families of Sauropodos Dinosauria, 6, 487.

— footprints of Jurassic Dinosaurs, 7, 229 ; note on a Bridger Eocene carnivore, 7, 397.

— obituary notice by C. E. Beecher, 7, 403 ; bibliography of, 7, 420.

Martin, G. C., dunite in Western Massachusetts, 6, 244.

Maryland, Geological Survey, Clark, vols. i, ii, 7, 69 ; vol. iii, 9, 223.

— Weather Service, Clark, 9, 234 ; Abbe, 9, 81.

Massachusetts, geology of Green Mts., 1, 146.

Mathematical congress, International, papers read, 2, 90.

Mathematics, History, Boyer, 9, 284 ; History of Elementary, Cajori. 3, 79.

Matter, Energy, Force and Work, Holman, 7, 237.

Matthew, G. F., Palæozoic terrane beneath the Cambrian, 8, 79 ; Etcheminian fauna of Cape Breton, 9, 158.

Mauna Loa. eruption, 8, 237.

Maxwell, James Clerk, and Modern Physics, Glazebrook, 1, 404.

Mayer, A. G., improved heliostat of A. M. Mayer, 4, 306.

Mayer, A. M., analysis of contrast-colors, 1, 38 ; researches in acoustics, 81 ; Röntgen rays, 1, 467 ; flotation of disks and rings of metal, 3, 253.

— obituary notice of, 4, 161.

Mazama, Crater Lake number, 5, 79.

Means, T. H., determination of soluble mineral matter in a soil, 7, 264.

Measurement of crystals, Palache, 2, 279.

Mechanics, Applied, Perry, 5, 80.

— and Hydrostatics, Glazebrook, 1, 327.

Melting points of metals, Holman, Lawrence and Barr, 1, 395; Holborn and Day, 10, 187 ; of platinum, Meyer, 2, 81.

— and critical temperatures, 2, 299.

Membranes, semi-permeable, Mijers, 7, 235.

Mercury, motion of a submerged index thread of, Barus, 9, 139.

— resistance, pressure co-efficient of, Palmer, 4, 1.

— See **CHEMISTRY..**

Merrill, G. P., free gold in granite, 1, 309 ; meteorite from Hamblen Co., Tenn., 2, 149.

— Treatise on Rocks, 3, 423.

Metals, capillary constants of molten, Siedentopf, 4, 320.

— melting points, 1, 395 ; 10, 187.

— reflective power of, Hagen and Rubens, 9, 294.

Metamorphism, see **GEOLOGY.**

Meteor of Dec. 4, 1896, 3, 81.

Meteoric iron, containing platinum and iridium, Davison, 7, 4.

— structure and origin, Preston, 5, 62.

Meteorite, iron, from Australia, Ward, 5, 135.

— Ballinoo, Australia, H. A. Ward, 5, 136.

— Bendegó. Brazil, Derby. 4, 159.

— Central Missouri, Preston, 9, 285.

— Forsyth Co., N. C., de Schweinitz, 1, 208.

— Hamblen Co., Tenn., mesosiderite, Merrill, 2, 149.

— Hayden Creek, Idaho, Hidden, 9, 367.

— Illinois Gulch, Montana, Preston, 9, 201.

— Iredell, Bosque Co., Texas, Foote, 8, 415.

— Luis Lopez, N. Mexico, Preston, 9, 283.

— Mooranoppin, Australia, H. A. Ward, 5, 140.

— Mungindi, Australia, H. A. Ward, 5, 138.

— Murphy, Cherokee Co., N. C., H. L. Ward, 8, 225.

— Roebourne, Australia, H. A. Ward, 5, 135.

— Sacramento Mts., N. Mexico, Foote, 3, 65.

— San Angelo, Texas, Preston, 5, 269.

— Thurlow, Hastings Co., Canada, 4, 325.

— Tombigbee River, Alabama, Foote, 8, 153.

Meteorite, stone, Allegan, Michigan, Ward, 8, 412.

— Bjurbole, Finland, 10, 250.

— Jerome, Gove Co., Kansas, Washington, 5, 447.

— Ness Co., Kansas, Ward, 7, 233.

— Oakley, Kansas, Preston, 9, 410.

Meteorites, collection in British Museum, Fletcher, **3,** 424; in Peabody Museum, Yale University, **3,** 83; at Vienna, Brezina, **2,** 461; Ward-Coonley collection, **9,** 304.
— gases yielded by, Travers, **7,** 244.
— worship of, Newton, **3,** 1.
Meteors, November 1897; **4,** 480; November 1899, **8,** 473; **9,** 80.
Meteorological Society, New England, **1,** 494.
Meteorology, Elementary, Waldo, **3,** 80.
Meunier, Experimental Geology, **8,** 468.
Mexico, Geological Institute, Bulletin, Aguilera, **3,** 422; **5,** 152.
Meyer, O. E., Kinetic theory of Gases, **9,** 221.
Miall, L. C., Aquatic Insects, **1,** 249.
Micas, crystal symmetry of, Walker, **7,** 199; percussion figures, **2,** 5.
— See **MINERALS.**
Michel-Lévy, feldspaths dans les plaques minces, **1,** 402.
Michelson, A. A., theory of the X-rays, **1,** 312; relative motion of earth and ether, **3,** 475; new harmonic analyzer, **5,** 1; spectroscope (echelon) without prisms or gratings, **5,** 215.
Michigan Geological Survey, vol. vi, **8,** 466; vol. vii, **10,** 399.
— geothermal gradient in, Lane, **9,** 434.
— iron-bearing district, Crystal Falls, Clements and Smyth, **9,** 451.
— mouth of Grand River, Mudge, **8,** 21.
— pre-glacial drainage, Mudge, **4,** 388; **10,** 158.
Microsclerometer for determining hardness of minerals, Jaggar, **4,** 399.
Microscopical Society, America, **4,** 256; **5,** 321; **8,** 399.
Mikroskopische Physiographie der massigen Gesteine, Rosenbusch, **1,** 63, 460.
Millikan, R. H., General Physics, **5,** 389.
Mineral analyses, interpretation of, Penfield, **10,** 19.
— Industry, vol. iv, Rothwell, **2,** 396.
— Names, Dictionary of, Chester, **1,** 401.
— Resources of the United States, Day, **1,** 145; **4,** 478; 1896, **5,** 469; **8,** 77, 392; **9,** 447.
— Tables, Weisbach, **10,** 84; Weisbach-Frazer, **3,** 162; Groth, **5,** 154.
— veins, enrichment Weed, **10,** 82.

Mineralchemie, Rammelsberg, **1,** 151.
Minerale des Harzes, die, Luedecke, **2,** 460.
Mineralogia, A. D'Achiardi, **9,** 160.
Mineralogica, Synopsis, Weisbach, **5,** 78.
Mineralogical Lexicon of Franklin Co., etc., Mass., Emerson, **2,** 306.
— Tables, Groth, **5,** 154; Weisbach-Frazer, **3,** 162; Weisbach, **10,** 84.
Minéralogie de la France et ses colonies, La Croix, **1,** 401; **2,** 461; **5,** 155.
Mineralogie, Elemente der, Naumann-Zirkel, **3,** 424.
— Handbuch der, Hintze, **3,** 251; **5,** 316; **6,** 435; **9,** 229; **10,** 469.
Mineralogy, First Appendix to Dana's System, **8,** 236.
— Elements of, Moses and Parsons, **10,** 405.
— Manual of Determinative, Brush and Penfield. **2,** 459; **6,** 436.
— Rutley, **1,** 401.
— Text Book, Dana, **6,** 275.
Minerals, Catalogue of, Chester, **5,** 78; Foote, **5,** 155.
— of Commercial Value, Barringer, **5,** 155.
— determination by maximum birefringence, Pirsson and Robinson, **10,** 260; determination by physical properties, Frazer, **3,** 162; determined by refractive indices, **9,** 229.
— formation in a magma, Morozewicz, **8,** 80.
— hardness, Auerbach, **2,** 390; Jaggar, **4,** 399.
— of Mexico, Aguilera, **8,** 236.
— in rock sections, Luquer, **7,** 319.
MINERALS—
Albite, etching-figures, **5,** 182. Algodonite, **10,** 447. Altaite, British Columbia, **4,** 78. Amphibole, alkali, Ontario, **1,** 210 : analyses, **7,** 297; etching figures, Daly, **8,** 82. Analcite, formula of, Lepierre, **2,** 81; analysis, Nova Scotia, **8,** 251. Ancylite, Greenland, **10,** 324. Andradite, titaniferous, Ontario, **8,** 210. Anorthite, No. Carolina, **5,** 128. Anthophyllite, No. Carolina, **5,** 429. Aragonite and calcite, relative stability, **10,** 392. Arzrunite, Chili, **8,** 468. Ascharite, **1,** 70. Asphalt, Indian territory, **8,** 219. Autunite, **6,** 42. Axinite, etching-figures, **5,** 180. Baddeckite, **6,** 274. Bastnäsite, Colorado, **7,** 51. Batavite, **7,** 76. Berthierite, California, **5,** 428.

MINERALS—

Bertrandite, Maine, 4, 316. Beryl, No. Carolina, 5, 432; Beryl (emerald) mines in Egypt, 10, 468. Biotite, 7, 202, 294. Bismutosmaltite, 4, 159. Bitumen, 1, 198. Bixbyite, Utah, 4, 105. Bliabergsite, Sweden, 2, 306. Britholite, Greenland, 10, 325. Bronzite, Jackson Co., N. C., 5, 431.

Calamine, New Jersey, 8, 248. Calcite crystals, Kansas, Rogers, 9, 365; Joplin, Mo., Farrington, 10, 84; siliceous from So. Dakota, Penfield and Ford, 9, 352; Union Springs, N. Y., England, Montana, Penfield and Ford, 10, 237. Caledonite, N. Mexico, 10, 84. Carnotite, Colorado, 9, 83; 10, 120. Cassiterite in California, 4, 39; etc., from tin furnace, 5, 93. Caswellite, New Jersey, 2, 305. Cedarite, 7, 76. Celestite, Lansdowne, Ontario, 2, 88; Renfrew Co., Canada, 10, 404. Chalcanthite, etching-figures, 5, 182. Chalcolamprite, Greenland, 10, 324. Chalcostibite, Bolivia, 4, 31. Chloritoid, Michigan, 2, 87. Chromite, North Carolina, 7, 281; method of decomposing, 10, 76. Chrysolite. 1, 132. Clinohedrite, Franklin, N. J., 5, 289. Coloradoite(?), California, 8, 297. Copper, native, Franklin Furnace, N. J., 6, 187. Cordylite, Greenland, 10, 324. Corundum deposits of Georgia, King, 3, 489; in Canada, 7, 318, 9, 389; in India, 7, 318; in Montana, 4, 417, 421, 424; origin of, 8, 227; origin in N. Carolina, 6, 49; new occurrences in N. Carolina, 10, 295; structure-planes, Judd, 1, 323. Covellite, Montana, 7, 56. Crocoite, Tasmania, 1, 389. Cubosilicite, 10, 168. Cuprogoslarite, 10, 168. Cupro-iodargyrite, Peru, 1, 70. Cyanite, No. Carolina, 5, 126; etching-figures, 5, 181.

Danaite, British Columbia, 4, 78. Datolite, Mexico, 5, 285. Diamond, artificial production, 3, 243, 5, 469; occurrence, Africa, Carvill-Lewis, Bonney, 4, 77; origin, Bonney, 5, 77; work on, De Launay, 5, 77. Diaphorite, Washington and Mexico, 6, 316. Dicksbergite, Sweden, 4, 158. Domeykite, 10, 439. Dundasite, Tasmania, 3, 352.

MINERALS—

Eliasite, gases from, 3, 242. Enargite, Montana, 7, 56. Endeiolite, Greenland, 10, 325. Enstatite, No. Carolina, 5, 430. Epidote, Huntington, Mass., 1, 26; Idaho, 8, 299. Epistolite, Greenland, 10, 325. Erionite, 6, 66. Fayalite, Rockport, Mass., 1, 129. Federovite, Italy, 8, 83. Feldspars in thin sections, Michel-Lévy, 1, 402; determination of plagioclase, 5, 349. Florencite, Brazil, 10, 404. Fluorite, photo-electric, 4, 474. Fuggerite, 4, 159.

Ganomalite, Sweden, 8, 348. Garnet, Idaho, 8, 299; Ontario, 8, 210. Gaylussite, Calif., 2, 130. Gersbyite, Sweden, 5, 316. Glaucocochroite, New Jersey, 8, 343. Gold, crystalline structure of nuggets, 5, 235; in granite, Merrill, 1, 309; of Georgia, 7, 168; Cape Nome, 7, 455; Klondike, 9, 456. Gold ores containing tellurium, selenium and nickel, 5, 427. Goldschmidtite, 7, 357, 10, 422. Graftonite, N. Hampshire, 10, 20. Graphite, graphitite and graphitoid, 5, 146, 220. Graphite in pegmatite, 1, 50. Grünlingite, 7, 76. Guejarite, 4, 27. Gypsum, Kansas, 8, 466; 9, 364.

Halite pseudomorphs, Hovey, 3, 425. Hamlinite, Maine, 4, 313. Hancockite, New Jersey, 8, 339. Hanksite, California, 2, 133. Hardystonite, New Jersey, 8, 82. Hastingsite, Hastings Co., Ont., 1, 212. Heazlewoodite, Tasmania, 3, 352. Hessite, Mexico, 8, 298; crystals, Colorado, 10, 426. Hoererite, Bohemia, 1, 70. Hortonolite, Orange Co., N. Y., 1, 131. Hübnerite, Nova Scotia, 10, 404. Hydromagnesite, 10, 404. Hydromica, New Jersey, 7, 365.

Ilmenite, composition, 4, 108. Inesite, Mexico, 10, 83. Iron, native, Missouri, 4, 99.

Jadeite, Thibet, 1, 401. Jeffersonite, 7, 55. Johnstonite, Tasmania, 10, 469.

Kalgoorlite, West Australia, 6, 199. Kamarezite, Greece, 1, 7. Kentrolite, New Mexico, 6, 116. Krennerite, Colorado, 5, 375. Ktypeite, 7, 320.

Lagoriolite, 7, 319. Langbeinite, 5, 316. Lawsonite, 3, 489. Lead, native, Franklin Furnace, N. J.,

MINERALS—

6, 187. Leonite, 4, 158; 5, 316. Lepidolite, 7, 202. Leucite, British Columbia, 2, 88; Wyoming. 4, 115. Leucophœnicite, 8, 351. Leucosphenite, Greenland, 10, 324. Lewisite, Brazil, 1, 71. Loranskite, Finland, 8, 469. Lorenzinite. Greenland, 10, 324. Lossenite, Greece, 1, 71. Maltesite, Finland, 4, 158. Mangan-andalusite, Sweden, 4, 158. Mangano-columbite, Maine, 1, 460. Mauzeliite, Sweden, 1, 71. Melanotekite, New Mexico, 6, 116. Melite, 10, 168. Melonite (?) California, 8, 295. Mica, percussion figures, Walker, 2, 5. crystallographic symmetry, 7, 199. Mica-pseudomorphs, 4, 309. Microlite, Maine, 1, 461. Miersite, New South Wales, 6, 199. Mitchellite, 7, 286. Mohawkite. 10, 440. Molybdenite. Calif., 5, 426. Monazite, Idaho, 4, 68; Brazil, 10, 217. Monticellite, Magnet Cove, Ark., 1, 134. Mossite, 7, 75. Müllerite, 10, 168. Munkforssite, Sweden, 4, 159. Muscovite, 7, 203.

Narsarsukite, Greenland, 10, 324. Nasonite, New Jersey, 8, 346. Natron, Br. Columbia, 10, 404. Nitre, 4, 118. Northupite, California, 2, 123; artificial production, Schulten, 3, 75.

Orthoclase, Japan. 8, 157; as gangue mineral, 5, 418.

Paralaurionite, 8, 469. Parisite, Montana, 8, 21. Pearceite, 2, 17. Pectolite, New Jersey, 8, 245. Petzite, California, 8, 297. Phenacite, pseudomorphs after, 6, 119. Philipstadite, Sweden, 8, 82, 83. Phlogopite, 7, 201. Pirssonite, Calif., 2, 120. Planoferrite, 7, 76. Pollucite, Maine. 1, 457. Polybasite, crystallization, 2, 23. Polycrase, Canada, 7, 243; 10, 404. Powellite crystals, 7, 367. Prosopite, Utah, 7, 53. Pyrophyllite, North Carolina, 8, 247. Pyroxene pseudomorph, 4, 309. Pyroxenes, etching figures, 8, 82. Quirogite, 4, 158.

Rafaëlite, Chili, 8, 468. Ransätite, Sweden, 2, 306. Raspite, 5, 315. Rathite, 2, 305. Rhodolite, 5, 294; associated minerals of, 6, 468. Rhodonite. etching-figures, 5, 182. Rhodophosphite, Sweden,

MINERALS—

2, 306. Robellazite, 10, 168. Rœblingite, Franklin Furnace, N. J., 3. 413; 6, 187. Roscoelite, 7, 451; Colorado, 10, 128, 130. Ruby, Burma, 1, 64; 2, 169; North Carolina, 8, 370.

Salvadorite, Chili, 2, 305. Sapphires, Montana, 4, 417, 421, 424. Scheelite, Nova Scotia, 4, 78. Schizolite, Greenland, 10, 325. Senaite, 7, 75. Silver, No. Carolina, 7, 242. Smithsonite, 6, 123. Sperrylite, North Carolina, 6, 381; Ontario, 1, 110. Sphalerite, crystals, Kansas, 9, 134. Spodiophyllite, Greenland, 10, 324. Stelznerite, Chili, 8, 468. Stibiodomeykite, 10, 445. Stokesite, 8, 469. Stromeyerite, British Columbia, 4, 78. Sulphate, fibrous, Montana, 7, 57. Sulphohalite, 6, 511; chemical composition, 9, 425. Sylvanite, Colorado, 10, 419.

Tainolite, Greenland, 10, 324. Talc, St. Lawrence Co., N. Y., Smyth, 3, 76. Tantalite, crystallized, 6, 128. Tapiolite, crystallized, 6, 121. Tellurides, Colorado, 10, 419. Tetragophosphite, Sweden, 2, 306. Tetradymite, British Columbia, 4, 78. Tetrahedrite, British Columbia, 2, 88. Thalenite, 7, 320. Thaumasite, West Paterson, N. J., 1, 229. Tiffanyite, 1, 72. Tilasite, Sweden, 1, 71. Topaz, Utah, 4, 107; supposed pseudomorphs, 6, 121. Torbernite, etching figures, 6, 41. Tourmaline, chemical constitution, Penfield and Foote, 7, 97; F. W. Clarke, 8, 111; Penfield, 10, 19; etching-figures, 5, 178; secondary enlargement, 5, 187. Triphylite, N. Hampshire, 9, 20. Tripuhyite, Brazil, 5, 316. Turquois, chemical composition, 10, 346. Tysonite, 7, 51.

Uraninite, gases from, 3, 242; radioactive substances from, see radio-active.

Valléite, 7, 75. Von Diestite, Colorado, 8, 469.

Wardite, 2, 154. Weldite, Tasmania, 3, 352. Wellsite, Clay Co., North Carolina, 3, 443; Whitneyite, 10, 446. Wolfsbergite, 4, 27. Wollastonite, Oneida Co., N. Y., 1, 328.

Xenotime, Ontario, 5, 285.

MINERALS—
Zeolites, chemical experiments on, Friedel, **2**, 88; **9**, 117, 345. Zinnwaldite, **7**, 202. Zircon, twinned crystals, **6**, 323 ; No. Carolina, **5**, 127; California, **5**, 426. Zirkelite, Brazil, **1**, 71 ; **5**, 153.

Mines and Mining in the United States, Law of, Barringer and Adams, **6**, 436.

Minks, A., die Protrophie, **3**, 855.

Minnesota, Geology, vol. iii, part 2, **3**, 349 ; vol. iv, **9**, 149 ; 24th Ann. Rep., **9**, 456.

Mississippi river, floods, **5**, 240.
— hydrology of, Greenleaf, **2**, 29.

Missouri Botanical Garden, 7th report, **2**, 89 ; 8th report, **5**, 78 ; 11th report, **9**, 283.
— Devonian fossil in, Broadhead, **2**, 237.
— flora of Lower Coal Measures, D. White, **10**, 166.
— Geol. Survey, 1894, **1**, 149.

Mixter, W. G., electrosynthesis, **4**, 51 ; **6**, 217.
— experiments with endothermic gases, **7**, 323 ; partial non-explosive combination of explosive gases and gaseous mixtures, **7**, 327.
— products of the explosion of acetylene, **9**, 1 ; **10**, 299.

Molecules and Molecular Theory of Matter, Risteen, **1**, 57.

Mollusks, morphology of, Verrill, **2**, 91.

Mont Blanc, geology and petrography, Duparc and Mrazec, **7**, 242.

Montana, Geology of Little Belt Mts., Weed and Pirsson, **10**, 466.
— vertebrate fossils from, E. Douglass, **10**, 428.

Moon, relation to aurora, Clayton, **5**, 81.

Moore, B., physiology, **7**, 473.

Moore, J. E., electrical discharge and the kinetic theory of matter, **6**, 21.

Moraines, see GEOLOGY.

Morgan, T. H., Development of the Frog's Egg, **4**, 161.

Morgan, W. C., determination of tellurium, **2**, 271.

Morley, F. H., iodometric determination of gold, **8**, 261.

Morozewicz, experiments on the formation of minerals, **8**, 80.

Morphology, Experimental, Davenport, **4**, 397 ; **7**, 474.

Morris, J. C., iodometric estimation of arsenic acid, **10**, 151.

Moses, A. J., drawing of crystal forms, **1**, 462.
— Characters of Crystals, **8**, 84.
— Mineralogy, **10**, 405.

Mudge, E. H., pre-glacial drainage in Michigan, **4**, 383, **10**, 158; mouth of Grand River, Michigan, **8**, 31.

Murray, G., Study of Seaweeds, **1**, 74.

Museum, see American, National.
— Cycads in Yale, Ward, **10**, 827.

N

Nagaoka, H., velocity of seismic waves, **10**, 471.

Nagel, W. A., der Lichtsinn augenloser Tiere, **3**, 162.

Narragansett Basin, geology, Shaler, Woodworth and Foerste, **10**, 163; Carboniferous fauna, Packard, **10**, 164.

National Museum, U. S., Director appointed, **3**, 252.
— Report for 1896, **7**, 321 ; for 1897, **9**, 233 ; for 1898, **10**, 470.

Nebraska, Bulletin of the University of, **10**, 472.
— Phytogeography of, Pound and Clements, **5**, 471.

Nehrling, H. N., American Birds, **1**, 404 ; **3**, 358.

Neural terms, Wilder, **3**, 425.

Newberry, J. S., Extinct Floras of North America, **8**, 394.

New Guinea, etc., Zoological studies, Willey, **7**, 79, 322 ; **8**, 398 ; **10**, 89.

New Jersey Geol Survey, 1894, **1**, 60 ; vol. iv, **5**, 468 ; **8**, 394.
— nepheline - syenite, Ransome, **8**, 417.
— trap of Rocky Hill, Phillips, **8**, 267.
— physical geography, Salisbury, **5**, 468.

Newth, G. S., Manual of Chemical Analysis, **7**, 67.

Newton, H. A., obituary notice of, by J. W. Gibbs, **3**, 359.
— on the worship of meteorites, **3**, 1.

New York Academy of Sciences, vol. xiv, **1**, 77.
— State geol. survey, 1893, **1**, 61.

Niagara Falls, geology of, Spencer, **1**, 398 ; **6**, 439.

Nichols, E. L., Elements of Physics, **1**, 319 ; **2**, 454 ; **4**, 73
— Outlines of Physics, **3**, 420.

Nicol prisms, new, Leiss, **4**, 475.

Nicolson, J. T., flow of marble under pressure, **10**, 401.

Nies, Crystallography, **1**, 402.

Night clouds, height of luminous, **2**, 89.

Nile delta, deposits of, Judd, **4**, 74.

Nipher, F. E., measurement of pressure of wind, **5**, 468; gravitation in gaseous nebulæ, **7**, 459; properties of light-struck photographic plates, **10**, 78.

— Electricity and Magnetism, **6**, 482.

North America, Later Extinct Floras of, Newberry, **8**, 394.

North Carolina and its resources, **3**, 252.

Norton, J. T., Jr., iodometric determination of molybdenum, **6**, 168; hydrochloric acid in titrations by sodium thiosulphate, **7**, 287; estimation of iron in the ferric state, **8**, 25; titration of mercury by sodium thiosulphate, **10**, 48.

Norway, Crustacea of, Sars, **7**, 79.

— North Atlantic Expedition, **3**, 494; **10**, 170.

Nova Scotia, geology of southwest, Bailey, **6**, 510.

Noyes, W. A., Organic Chemistry, **5**, 147.

O

OBITUARY—

Argyll, Duke of, **9**, 462.

Bebb, M. S., **1**, 78. Bertrand, J. **7**, 462. Blanchard, Emile, **7**, 895. Brinton, Daniel Garrison, **8**, 318. Bunsen, Robert Wilhelm, **8**, 318.

Castillo, A., **1**, 78. Clark, A. G., **4**, 83. Collier, P. **2**, 246. Cope, E. D., **3**, 427.

Daubrée, A., **2**, 90. Dawson, Sir John William, **8**, 475; **9**, 82. Des Cloizeaux, A., **4**, 164.

Egleston, T., **9**, 160.

Fizeau, H., **2**, 398. Flower, Sir William Henry, **8**, 238. Frankland, Sir Edward, **8**, 318. Fresenius, C. R., **4**, 84.

Geinitz, Hanns Bruno, **9**, 236. Goode. G. B., **2**, 318. Gould, B. A., **3**, 81. Green, A. H., **2**, 246. Grove, Sir W. R., **2**, 314.

Hall, James, **6**, 284, 437; Hauer, Franz Ritter von, **7**, 474. Haughton, S., **5**, 80. Hazen, Henry Allen, **9**, 235. Hicks, Dr. Henry, **9**, 84. Hubbard, Oliver Payson. **9**, 396. Hughes, David E., **9**, 236.

James. J. F., **3**, 428.

Keeler, J. E., **10**, 325. Kekulé, A., **2**, 314. Krueger, **1**, 494.

OBITUARY—

Lawson, G., **1**, 78. Lea, M. C., **3**, 428.

Marcou, J., **5**, 398. Marsh, O. C., **7**, 403. Mayer, A. M., **4**, 161. Meyer, V., **4**, 398. Milne-Edwards, A., **9**, 462. Mivart, St. George, **9**, 395. Mueller, F. von, **2**, 464. Müller, J., **1**, 826.

Newton, H. A., **2**, 245; **3**, 359.

Orton, Edward, **8**, 400.

Palmieri, S., **2**, 398. Peck, L. W., **7**, 248. Preston, Thomas, **9**, 395. Prestwich, J., **2**, 90, 170.

Rogers, W. A., **5**, 322.

Sachs, J., **4**, 164. Schrauf, A., **5**, 160. Sylvester, J. J., **3**, 358. Symons, George James, **9**, 395.

Ulrich, G. H. F., **10**, 250.

Waagen, Wilhelm, **9**, 395. Wachsmuth, C., **1**, 250. Walker, F. A., **3**, 164. Whitney, J. D., **2**, 246, 312. Wiedemann, G., **7**, 402. Winnecke, Prof., **5**, 80.

Observatories, Mountain, Holden, **3**, 358.

Observatory, see Astronomical, and Astrophysical.

Optical instruments of, R. Fuess, Leiss, **7**, 396.

Ore Deposits of the United States and Canada, Kemp, **9**, 303.

Orthoptera, North America, Scudder, **4**, 250.

Ortmann, A. E., climatic zones in Jurassic times, **1**, 257.

— separation and its bearing on geology and zoögeography, **2**, 68.

— Crangopsis vermiformis of Kentucky, **4**, 283; Linuparus atavus of Dakota, **4**, 290; large oysters of Patagonia, **4**, 355.

— new marine Tertiary horizons near Punta Arenas, Chile, **6**, 478.

— fauna of the Magellanian beds of Chile, **8**, 427.

— invertebrate fossils from Patagonia, **10**, 368.

Osborn, H. F., origin of mammals, **7**, 92; Tertiary mammal horizons of Europe and America, **10**, 400.

Oscillations, see Electric.

Oscillatory currents, Seiler, **4**, 71.

— discharge of a large accumulator, Trowbridge, **4**, 194.

Osmotic pressure and electrolytic dissociation, Compton, **5**, 65; Traube, **5**, 463.

Ostwald, W., Lehrbuch der allgemeinen Chemie, **3**, 357; **5**, 222; **8**, 74; **9**, 64, 65.

Ostwald, W., supersaturation, etc., 4, 151.
— Analytische Chemie, 5. 222.
— grundlinien der anorgan. Chemie, 10, 394.
Ostwald's Klassiker der Exacten Wissenschaften, 2, 397; 3, 494; 4, 398; 5, 80; 6, 200; 7, 170; 8, 475; 10, 90, 406.
Oysters of Patagonia, Ortmann, 4, 355.

P

Pacific Ocean, explorations of the "Albatross," Agassiz, 9, 33, 109, 193, 369, 390.
Packard, A. S., Text-book of Entomology, 6, 108; Carboniferous fauna of Narragansett Basin, 10, 164.
Palache, C., crocoite from Tasmania, 1, 389; method of crystal measurements, etc., 2, 279; powellite, 7, 367; epidote and garnet from Idaho, 8, 299; tellurides from Col., 10, 419.
Paleobotany, Zeiller, 10, 88; Potonié, 10, 88.
— See GEOLOGY.
Paleontology, contributions to, Lucas, 6, 399.
— prize for, American, 2, 85.
— Text-book, von Zittel, translation by C. R. Eastman, 2, 394; 9, 388.
Palmer, A. deF., rate of condensation in the steam jet, 2, 247; pressure-coefficient of mercury resistance, 4, 1; apparatus for measuring very high pressures, 6, 451.
Panama, geology, Bertrand, 10, 82.
— and Costa Rica, geology, Hill, 6, 435, 505.
Parker, T. J., Text-book of Zoology, 5, 319; Manual of Zoology, 9, 390.
Parsons, C. L., Mineralogy, 10, 405.
Patagonia, geology of, J. B. Hatcher, 4, 246, 327; 9, 85.
— invertebrate fossils from, Ortmann, 10, 368.
— mollusks from, Pilsbry, 7, 126.
— oysters from, Ortmann, 4, 355.
— sedimentary rocks of southern, Hatcher, 9, 85.
Peach, B. N., Silurian Rocks of Britain, 9, 300.
Peckham, S. F., Trinidad pitch, 1, 193.
Peirce, A. W., iodometric determination of selenious and selenic acid, 1, 31; separation of selenium from tellurium, 1, 181; gravimetric determination of selenium, 1, 416; selenium monoxide, 2, 163.

Peirce, B. O., induction coefficients of hard steel magnets, 2, 347; properties of seasoned magnets, 5, 334.
Penck, Geographische Abhandlungen, 3, 492.
Pendulum, new form, Stevens, 5, 14.
— observations, Putnam, 1, 186; on Adriatic, 1, 76.
Penfield, S. L., chrysolite-fayalite group, 1, 129; thaumasite, West Paterson, N. J., 1, 229.
— pearceite and polybasite, 2, 17.
— Revision of Brush's Determinative Mineralogy, 2, 459.
— rœblingite, Franklin Furnace, N. J., 3, 413.
— identity of chalcostibite, etc., from Bolivia, 4, 27; bixbyite and topaz, 4, 105; composition of ilmenite, 4, 108; chemical composition of hamlinite, 4, 313.
— clinohedrite from Franklin, N. J., 5, 289.
— Revision of Brush's Determinative Mineralogy, 6, 436; (2, 459).
— composition of tourmaline, 7, 97.
— composition of parisite, 8, 218; new minerals from Franklin, N. J., 8, 389.
— graftonite from New Hampshire, 9, 20; siliceous calcites from S. Dakota, 9, 352; chemical composition of sulphobalite, 9, 425.
— interpretation of mineral analyses, and constitution of tourmaline, 10, 19; calcite crystals from Union Springs, N. Y., 10, 237; composition of turquois, 10, 346.
Penniman, T. D., new method of measurement of self-inductance, 6, 97; electrical measurements, 8, 35.
Pennsylvania, Brownstones of, Hopkins, 5, 78.
— Geol. Survey of, 1, 488.
Periodic current curves, Wehnelt and Donath, 9, 148.
Perkins, C. A., Electricity and Magnetism, 3, 246.
Perry, J., Applied Mechanics, 5, 80.
— Calculus for Engineers, 4, 398.
Perry, J. H., physical geography of Worcester, Mass., 6, 485.
Peters, C. A., titration of oxalic acid, 7, 461; tellurous acid in presence of haloid salts, 8, 122; determination of mercury as mercurous oxalate, 9, 401; volumetric estimation of copper, etc., 10, 359.
Peters, E. D., Jr., Copper Smelting, 1, 54.

Petrography, new term proposed (anhedron), Pirsson, 1, 150.
— methods of, Cohen, 1, 400.
— See ROCKS.
Petroleum in Burma, Noetling, 6, 102.
Petrology, International Journal of, proposed, 8, 470.
— for Students, Harker, 5, 317.
Petterd, W. F., Minerals of Tasmania, 3, 852.
Pfeffer, W., Pflanzenphysiologie, 5, 317.
Phase Rule, Bancroft, 4, 67.
Phelps, I. K., iodometric method for determination of carbon dioxide, 2, 70; combustion of organic substances in the wet way, 4, 372.
Phillips, A. H., structure and composition of the trap rock of Rocky Hill, N. J., 8, 267.
Phosphorescence of inorganic chemical preparations, Goldstein, 10, 459; produced by X-rays, Burbank, 5, 54; by electrification, Trowbridge and Burbank, 5, 55; at low temperatures, Lumière, 7, 472.
Photochromie, Zenker, 10, 162.
Photoelectric relations of fluorspar and selenium, Schmidt, 4, 474; properties of colored salts, Elster and Geitel, 6, 95.
Photographic plates, properties of light-struck, Nipher, 10, 78.
Photometer, flicker, Rood, 8, 194,258.
Photometry, Manual, Stone, 10, 320.
Physical Chemistry, Journal, 2, 90, 392.
— Experiments, Gage, 5, 222; History of, Traumüller, 8, 162.
— Geography, Tarr, 1, 76; Davis, 7, 248.
— Society, American, 8, 398; address before, Rowland, 8, 401.
— of Germany, Transactions, 8, 75.
Physics, Brief Course in General, Hoadley, 10, 465.
— Deductive, Rogers, 5, 148.
— Elementary, Aldous, 6, 100.
— Elements, Crew, 9, 146; Nichols and Franklin, 1, 319; 2, 454; 4, 73.
— Experimental, Loudon and McLennan, 1, 141; Stone, 5, 222.
— Experiments in General, Stratton and Millikan, 5, 389.
— General, Hastings and Beach, 7, 314.
— History of, Cajori, 7, 394; of Experiments, Traumüller, 8, 162.
— Manual, Cooley, 4, 390.
— Manual of Experiments in, Ames and Bliss, 5, 302.

Physics, Modern, Glazebrook, 1, 404.
— Outlines of, Nichols, 3, 420.
— Practical, Kohlrausch, 10, 320.
— School, Avery, 1, 57.
— Text-book, Watson, 9, 296.
— Theoretical, Christiansen, 3, 419.
— Theory of, Ames, 3, 420.
Physikalisch-chemische Propädeutik, Griesbach, 2, 450; 5, 321; 10, 461.
Physikalische Zeitschrift, 8, 386.
Physiology, Moore, 7, 478; Huxley, 10, 90.
— American Journal of, 4, 481.
Pierce, G. W., radio-micrometer applied to the measurement of short electric waves, 9, 252.
Pilsbry, H. A., "Florencia formation," 5, 232.
— mollusks from Patagonia, 7, 126.
Pinchot, G., Primer of Forestry, 8, 399.
Pirsson, L. V., new petrographical term (anhedron), 1, 150; Bearpaw Mts., Montana, 1, 283, 351,
— Bearpaw Mts., Montana, 2, 136, 188; Missourite, Highwood Mts., Montana, 2, 315.
— geology of Castle Mt. district, Montana, 3, 250.
— corundum-bearing rock from Montana, 4, 421.
— geology of Judith Mts., Montana, 6, 508.
— phenocrysts of intrusive igneous rocks, 7, 271.
— ægirite-granite, Miask, Ural Mts., 9, 199.
— determination of minerals by maximum birefringence, 10, 260.
— Little Belt Mts., Montana, 10, 466.
Plants, Evolution of, Campbell, 9, 79.
— See BOTANY.
Polarization capacity, Gordon, 4, 71.
— of light, rotatory magneto-optic, Wright and Kreider, 6, 416; method of detecting, Kreider, 8, 133; produced by torsion, Ewell, 8, 89.
Potonié, Pflanzenpalæontologie, 10, 88.
Potts, L. M., Rowland's method of electric measurements with alternating currents, 10, 91.
Pratt, J. H., thaumasite, West Paterson, N. J., 1, 229.
— northupite, pirssonite, etc., 2, 123.
— wellsite, a new mineral, 3, 443.
— crystallography of Montana sapphires, 4, 424.

Pratt, J. H., mineralogical notes from North Carolina, 5, 126, 429; rhodolite, 5, 294.

— origin of corundum in North Carolina, 6, 49; twinned crystals of zircon, 6, 323; associated minerals of rhodolite, 6, 463.

— chromite, origin, etc., of, 7, 281.

— separation of alumina from molten magmas, 8, 227; crystallography of the rubies of North Carolina, 8, 379.

— two new occurrences of corundum in North Carolina, 10, 295.

Precious Stones, production in 1895, Kunz, 3, 352.

Predazzo, eruptive rocks, Brögger, 1, 399.

Pressures, apparatus for measuring very high, Palmer, 6, 451.

Preston, E. D., latitude determinations, etc., in Hawaiian Is., 1, 75.

Preston, H. L., San Angelo meteorite, 5, 269; iron meteorites, structure and origin, 5, 62; Illinois gulch, Montana, meteorite, 9, 201; two new American meteorites, 9, 283; new meteorite from Oakley, Kansas, 9, 410.

Pribiloff Islands, plants of, Macoun, 9, 232.

Prinz, W., Esquisses Sélénogiques II., 4, 396; geological experiments, 5, 392.

Prisms, longitudinal aberration of, Abbot and Fowle, 2, 255.

Pseudomorphs from New York, Smyth, 4, 309.

Psychrometer, Leavitt, 5, 440.

Putnam, G. R., pendulum observations, 1, 186.

Pyrenees, granite of, Lacroix, 6, 511.

Q

Quebec, geology and auriferous deposits, Chalmers, 8, 394.

Quincke's rotations in an electrical field, Graetz, 9, 382.

R

Rabot, C., variations in length of Arctic glaciers, 4, 395.

Races of Europe. W Z. Ripley, 8, 474.

Radiant heat, transmission by gases, Brush, 5, 222.

Radiation of a dark body, law of, Wien and Lummer, 1, 56.

— phenomena, irreversible, Planck, 9, 219.

Radiation, See Röntgen rays.

Radio-active substances (polonium and radium), Mme. Curie, 8, 159; Debierne, 8, 463, 9, 143; (actinium), 9, 444; M. and Mme. Curie, 9, 143, 144, 145; Becquerel, 9, 147, 443; Giesel, 9, 147, 463; Haën, 8, 386; Rutherford, 9, 220; (barium) Lengyel, 10, 74; M. and Mme. Curie, 10, 392; Debierne, 10, 393; (uranium), Crookes, 10, 318; barium and polonium, 10, 460.

— See Becquerel rays and Uranium radiation.

Radiometer, registering solar, Isham, 6, 160.

Radio-micrometer applied to the measurement of short electric waves, Pierce, 9, 252.

Radium, Curie, 8, 159, 463; Becquerel, 9, 443; spectrum, 9, 143; Runge, 10, 396.

Rafinesque, Icthyologia Ohiensis, 7, 473.

Rammelsberg, C. F., Mineralchemie, 1, 151.

Ramsay, argon, helium. 2, 300, 3, 241, 7, 310; metargon, neon, krypton, 9, 62.

Ransome, F. L., lava flows of the Sierra Nevada, 5, 355; nepheline-syenite in New Jersey, 8, 417; carnotite, etc., of Colorado, 10, 120.

Rarified gases, behavior of, Ebert and Wiedemann, 4, 391.

Rayleigh, limits of audition, 4, 69; nature of the X-rays, 5, 467; possibilities of flight, 10, 77; viscosity of gases as affected by temperature, 9, 375, 10, 461; weight of hydrogen desiccated by liquid air, 10, 459.

Rays, see Becquerel, Cathode, Röntgen.

Reflection, irregular, Hutchins, 6, 373.

Refraction, relation to density, Traube, 3, 479; of air, oxygen, etc., Ramsay and Travers, 5, 227.

Regnault's calorie, Starkweather, 7, 13.

Resistance of mercury, pressure-coefficient, Palmer, 4, 1.

— of thin films, Longden, 9, 407.

— standards, 5, 391.

— See Electric.

Reynolds, O., the dryness of saturated steam and the condition of steam gas, 2, 450.

Rhode Island Brachiopod fauna, Walcott, 6, 327.

Rice, C. B., reading deflections of galvanometers, **2**, 276.

Rice, H. L., theory and practice of interpolation, **9**, 394.

Rice, M. E., broadening of sodium lines, etc., **3**, 472.

Rice, W. N., Dana's Text-book of Geology, revised, **5**, 393.

— use of terms anticlinorium and synclinorium, **2**, 168.

Richards, H. M., increase of respiration after injury, **2**, 464.

Richards, T. W., spectra of argon, **3**, 15; multiple spectra of gases, **3**, 117; temperature and ohmic resistance of gases, **3**, 327; conductivity of electrolytes, **3**, 391; transition temperature of sodic sulphate, **6**, 201; electro-chemical equivalents of copper and silver, **9**, 218.

Richter, E., Seestudien, **6**, 108.

Richter, M. M., Lexikon der Kohlenstoff-Verbindungen, **9**, 445.

Riggs, E. S., skull of Amphictis, **5**, 257.

Ripley, W. Z., Races of Europe, **8**, 474.

Risteen, A. D., Molecules and Molecular Theory of Matter, **1**, 57.

Rivers of North America, Russell, **7**, 72.

Robb, W. L., solarization effects on Rontgen ray photographs, **4**, 243.

Robinson, F. C., Crookes tubes, **1**, 463.

Robinson, H. H., determination of minerals by maximum birefringence, **10**, 260.

Rock specimens distributed by the U. S. Geol. Survey, Diller, **7**, 74.

Rockall Island, petrology, Judd, **7**, 241.

Rocks, Handbook of, Kemp, **3**, 76.

— Treatise on, Merrill, **3**, 423.

ROCKS—

Aegirite-granite, Miask, Ural Mts., Pirsson, **9**, 199.

Alnoite, Manheim, N. Y., Smyth, **2**, 290.

Amphibole-pyroxene rocks, California, Turner, **5**, 423.

Analyses by Hillebrand and Stokes, collated by Clarke, **10**, 250; statement of, Washington, **10**, 59.

Andesite, Tuscany, Washington, **9**, 51.

Andesites from Maine, Gregory, **8**, 359.

Argillaceous, with quartz veins, in Brazil, Derby, **7**, 343.

ROCKS—

Augite-andesite, Smyrna, Washington, **3**, 41.

Augite-syenite, Montana, Weed and Pirsson, **2**, 136.

Bacteria, supposed action on rocks, Branner, **3**, 488.

Basalt in Virginia, Darton and Keith, **6**, 305.

Biotite-dacite, Pergamon, Washington, **3**, 47.

Biotite-tinguaite, Essex Co., Mass., Eakle, **6**, 489.

Ciminite, Viterbo, Italy, Washington, **9**, 44.

Classification of, Brögger, **9**, 456.

Clay slates, phyllites, etc., contact metamorphism of, Clements, **7**, 81.

Corundum-bearing rock from Montana, Pirsson, **4**, 421.

Decay of pre-glacial rocks of E. Canada, Chalmers, **5**, 272.

Diffusion of rocks, Becker, **3**, 21, 280.

Diorite, California, Turner, **5**, 422; and gabbro, California, Lindgren, **3**, 312.

Dunite, West Massachusetts, G. C. Martin, **6**, 244.

Euctolite, Rosenbusch, **7**, 399.

Felsophyre in Virginia, Darton and Keith, **6**, 305.

Flow and fracture as related to structure, Hoskins, **2**, 213; of marble under pressure, **10**, 401.

Fractional crystallization, Becker, **4**, 257.

Gabbro in St. Lawrence Co., N. Y., metamorphism, Smyth, **1**, 273.

Gabbro du Pallet, Lacroix, **8**, 81.

Gneisses, etc., of S. W. Minnesota, Hall, **10**, 168.

Granite of the Pyrenees, Lacroix, **6**, 511.

Granitic breccias of the Cripple Creek region, Stone, **5**, 21; of Grizzly Peak, Colorado, Stone, **7**, 184.

— rocks, California, Lindgren, **3**, 301.

Granodiorite, California, Lindgren, **3**, 308, **9**, 269.

Hatherlite, Leo Henderson, **7**, 318.

Igneous, composition of, Walker, **6**, 410.

— of Christiania, Brögger, **6**, 273.

— containing Paleotrochis, Diller, **7**, 337.

— Pre-Cambrian of Wisconsin, Weidman, **7**, 398.

ROCKS—

Igneous, of Tasmania, Twelvetrees and Petterd, **6**, 511.

— of Temiscouata Lake, Gregory, **10**, 14.

— of Wyoming, Cross, **4**, 115.

— of the Yellowstone, Hague, **1**, 445.

Ijolite, Kuusamo, Finland, **10**, 249.

Itacolumite, accessory elements of, Derby, **5**, 187.

Keratophyre dike near New Haven, Ct., Hovey, **3**, 287.

Kyshtymite and corundum-syenite, of the Urals, **8**, 81.

Latite, Sierra Nevada, Ransome, **5**, 355.

Laurdalite, Brögger, **6**, 273.

Lava beds at Meriden, Ct., Davis, **1**, 1.

— flows of the Sierra Nevada, Ransome, **6**, 509.

Leucite rocks in Wyoming, **4**, 115.

Leucitite, Alban Hills, Italy, Washington, **9**, 53; Montana, Weed and Pirsson, **2**, 143.

Madupite, W. Cross, **4**, 129.

Marble, flow of, under pressure, Adams and Nicolson, **10**, 401.

Metamorphism, Van Hise, **6**, 75; Clements, **7**, 81.

Mica-peridotites, Bengal, Holland, **1**, 400.

Mica-trachyte, Tuscany, Washington, **9**, 46.

Missourite, Highwood Mts., Missouri, Weed and Pirsson, **2**, 315.

Mont Blanc petrography, Duparc and Mrazec, **7**, 242.

Monzonite, Bearpaw Mts., Weed and Pirsson, **1**, 355; of Predazzo, Brögger, **1**, 399.

Nephelite-basalt, Montana, Weed and Pirsson, **2**, 140.

Nephelite - syenite, New Jersey, Ransome, **6**, 417.

Norites, Gabbros, etc., of Transvaal, Leo Henderson, **7**, 317.

Olivine-melilite-leucite rock, Sabatini, **7**, 399.

Orendite, Wyoming, W. Cross, **4**, 123.

Peridotite, occurrence of corundum with, Pratt, **6**, 49.

Petrography of Boston Basin, White, **5**, 470.

Phenocrysts of intrusive igneous rocks, Pirsson, **7**, 271.

Pilandite, Leo Henderson, **7**, 318.

ROCKS—

Pseudo - leucite - sodalite - tinguaite, Montana, Weed and Pirsson, **2**, 194.

Quartz-alunite rock, California, Turner, **5**, 424.

Quartz-amphibole-diorite, California, Turner, **5**, 421.

Quartz - muscovite rock, Belmont, Nevada, Spurr, **10**, 351.

Quartz-syenite, Bearpaw Mts., Weed and Pirsson, **1**, 295.

Quartz - tinguaite - porphyry, Montana, Weed and Turner, **2**, 194.

Rhyolitic lavas of South Mt., Pennsylvania, Bascom, **3**, 160.

Rockallite, Judd, **7**, 241.

Scapolite rocks, Alaska, Spurr, **10**, 310.

Schists of gold and diamond regions, Brazil, Derby, **10**, 207.

Selagite, Tuscany, Washington, **9**, 46.

Shonkinite, Bearpaw Mts., Weed and Pirsson, **1**, 360.

Sölvsbergite, Essex Co., Mass., Washington, **6**, 176.

Syenite, Bearpaw Mts., Weed and Pirsson, **1**, 352.

Theralite, Costa Rica, Wolff, **1**, 271.

Tinguaite, Essex Co., Mass., Washington, **6**, 176; Eakle, **6**, 489.

Tinguaite porphyry, Montana, Weed and Pirsson, **2**, 189.

Trachyte, Bearpaw Mts., Weed and Pirsson, **1**, 291; **2**, 137.

Trachytes, Ischian, Washington, **1**, 375; Italian, Washington, **8**, 286.

Trap of Rocky Hill, N. J., Phillips, **8**, 267.

Vanadium and molybdenum in rocks, Hillebrand, **6**, 209.

Venanzite, Sabatini, **7**, 399.

Wyomingite, W. Cross, **4**, 120.

Yogoite, Bearpaw Mts., Weed and Pirsson, **1**, 355.

Rogers, A. F., sphalerite crystals from Kansas, **9**, 184; mineralogical notes, **9**, 364.

Rogers, F. J., Deductive Physics, **5**, 148.

Rollins, W., regenerating vacuum tubes, **7**, 159; cathode stream, and X-light, **10**, 382.

Romanes, G. J., Essays by, **3**, 358.

Röntgen-rays, absorption by air, **7**, 396; by chemical compounds, **1**, 483.

— charge of electricity in ions caused by, Thomson, **7**, 158.

— chemical action, Villard, **9**, 146.

Röntgen-rays, and color blindness, Dorn, **7**, 159.
— Crookes tubes for, Hutchins and Robinson, **1**, 463.
— experiments on, Doelter, **1**, 319; Goldhammer, **1**, 485; Hutchins and Robinson, **1**, 463; Mayer, **1**, 467; Rowland, Carmichael, and Briggs, **1**, 247; Thomson, **1**, 318; J. Trowbridge, **1**, 245; A. W. Wright, **1**, 235; see also, **1**, 394, 486; **2**, 452; **3**, 71, 152.
— heat produced by, Dorn, **5**, 148.
— impulse theory, Thomson, **5**, 301.
— influence of combination of ions, Rutherford, **5**, 386.
— literature, Bibliography, Phillips, **5**, 71.
— in magnetic field, Strutt, **9**, 376.
— and mineral phosphorescence, Burbank, **5**, 53.
— nature of, Thomson, **2**, 381; Stokes, **5**, 301; Rollins, **10**, 382.
— and ordinary light, Rayleigh, **5**, 467.
— original papers on, Röntgen, **5**, 223.
— penetrative values, Swinton, **3**, 484.
— and Phenomena of the Anode and Cathode, E. P. Thompson, **2**, 392.
— photographs, solarization effects, Robb, **4**, 243.
— physiological effects, Sorel, **3**, 484.
— produced by battery current, J. Trowbridge, **9**, 439.
— refraction of, Haga and Wind, **8**, 385.
— specular reflection of, Rood, **2**, 178.
— solar, search for, on Pike's Peak, Cajori, **2**, 289; not present in sunlight, Lea, **1**, 363.
— source of, Trowbridge and Burbank, **5**, 129.
— in surgery, **4**, 72.
— theory of, Michelson, **1**, 312.
— vacuum tubes for, Rollins, **7**, 159.
Rood, O. N., specular reflection of the Rontgen rays, **2**, 178; flicker photometer, **8**, 194, 258; experiments on high electrical resistance, **10**, 285.
Rosell, C. R., heat of solution of resorcinol in ethyl alcohol, **10**, 449.
Rosenbusch, H., Gesteinslehre, **7**, 73; Mikroskopische Physiographie, **1**, 63, **2**, 460; Euctolite, **7**, 399.
Rotation, optical, in crystalline and liquid states, Traube, **3**, 148; of circular-polarizing crystals, Landolt, **3**, 416; new substance for increasing, Walden, **5**, 463; thermal phenomena attending change of, Brown and Pickering, **4**, 470.

Rotatory polarization, structural and magneto-optic, Wright and Kreider, **6**, 416.
— detection of, Kreider, **8**, 133.
— produced by torsion, Ewell, **8**, 89.
Roth, S., publications of Fl. Ameghino, **9**, 261.
Rothpletz, Glarner Alps, **9**, 303.
Rothwell, R. P., Mineral Industry, vol. iv, **2**, 396.
Rowland, H. A., Röntgen rays, **1**, 247.
— electrical measurements by alternating currents, **4**, 429.
— methods for the measurement of self-inductance, etc., **6**, 97.
— electrical measurements, **8**, 35; address before the American Physical Society, **8**, 401.
— electric measurements with alternating currents, **10**, 91.
Royal Society catalogue of Scientific papers, **1**, 327.
Rubens, H., absorption of infra-red rays by rock salt and sylvine, **5**, 33.
Ruby, Burma, **1**, 64, **2**, 169; N. Carolina, **8**, 370.
Russell, I. C., geology of southwestern Washington, **3**, 246; "plasticity" of glacial ice, **3**, 344.
— Glaciers of North America, **3**, 428.
— Rivers of N. America, **7**, 72.
— Volcanoes of N. America, **5**, 74.
Rutley, F., Mineralogy, **1**, 401.

S

Safford, J. M., phosphates in Tennessee, **2**, 462; Camden chert of Tennessee, **7**, 429.
— Geology of Tennessee, **10**, 399.
Salisbury, R. D., Physical Geography of New Jersey, **5**, 468.
— surface geology of Wisconsin, **10**, 248.
San Clemente Island, geological sketch, Smith, **7**, 315.
Sapphires in Montana, **4**, 417, 421, 424.
Sardeson, F. W., What is the Loess? **7**, 58; Lichenaria typa, W. and S., **8**, 101.
Scale, division errors of a straight, Jacoby, **1**, 333.
Schuchert, C., Lower Silurian fauna of Baffinland, **10**, 81.
Science Abstracts, **5**, 398.
— Introduction to, Hill, **10**, 406.
Scientia, **8**, 86; **10**, 406.
Scientific Periodicals, Catalogue, 1665–1895, Bolton, **6**, 513.

Schneider, P. F., fault at Jamesville, N. Y., 3, 458.

Schott, C. A., magnetic declination in Alaska, 1895, 1, 75.

Schweinitz, E. A. de, meteorite from Forsyth Co., N. C., 1, 208.

Scott, W. B., Introduction to Geology, 3, 422.

Scripture, E. W., Yale Psychological Laboratory Studies, 6, 512.

Scudder, S. H., North American Orthoptera, 4, 250.

Seals and Seal Islands of N. Pacific, 9, 390.

Secondary undulations registered on tide gauges, Denison, 4, 82.

See, T. J. J., Researches of the Evolution of Stellar Systems, 3, 491.

Seestudien, Richter, 6, 108.

Seiches on Bay of Fundy, Duff, 3, 406.

Self inductance, measurement of, Guthe, 5, 141.

— new methods of measurement. Rowland, 4, 429 ; Rowland and Penniman, 6, 97.

Setchell, W. A., Phycotheca Boreali-Americana, 1, 73, 493 ; 3, 78, 354.

— Laboratory Practice for Beginners in Botany, 3, 490.

Seward, A. C., Fossil Plants, 5, 472; Jurassic Plants, 10, 323, Maidenhair tree (Gingko), 10, 323 Wealden flora of Bernissart, 10, 322.

Shaler, N. S., geology of Narragansett Basin, 10, 163.

Shellheaps of Block Island, Eaton, 6, 187.

Shenstone, W. A., life of Liebig, 1, 76.

— Inorganic Chemistry, 10, 395.

Schimizu, S., string alternator, 10, 64.

Shunt box, Stine, 5, 124.

Sierra Nevada, lava flows of, Ransome, 6, 509.

Siren and organ pipe, combination tones of, Barus, 5, 88.

Smithsonian Institution, History, Goode, 5, 158.

— Report, Langley, 5, 239 ; 7, 80, 246, 321, 402 ; 9, 233 ; 10, 469, 470.

— Astrophysical Observatory, 9, 214 ; 10, 470.

— Physical Tables, Gray, 3, 252.

Smyth, C. H., Jr., metamorphism of a gabbro in St. Lawrence Co., N. Y., 1, 273.

— dikes of alnoite at Manheim, N. Y., 2, 290 ; pseudomorphs from New York, 4, 309.

Smyth, H. L., iron-bearing district, Crystal Falls, Michigan, 9, 451.

Soils, alkali in, Hilgard, 2, 100.

— analysis of Hawaiian, Lyons, 2, 421.

— method of analysis, Means, 7, 264.

Solar corona, nature of, Pringsheim, 10, 77 ; recognition independent of total eclipse, Deslandres, 10, 463.

— eclipse of May 28th, 1900, 9, 391 ; 10, 89.

— spectrum, photometry of the ultra-violet portion, Simon, 2, 380.

— X-rays on Pike's Peak, search for, Cajori, 2, 289 ; see also Lea, 1, 363.

— See Sun.

Somali-land, geology of, 2, 393.

Sound, diminution of the intensity of, with the distance, 1, 487.

Sound-waves, stationary, Davis, 10, 281.

South Africa Geological Society, vol. i, Draper, 2, 169 ; vol. ii, 4, 78.

South Dakota Geological Survey, 7, 316.

— huge Cretaceous turtles, Wieland, 9, 237.

Spark discharges, Warburg, 4, 474.

Species, origin of, in relation to separation, Ortmann, 2, 63.

Specific heat, determination by the method of mixtures, F. L. O. Wadsworth, 4, 265.

Spectra of argon, 3, 15 ; multiple, of gases, 3, 117.

— of certain stars, Vogel and Wilsing, 4, 475 ; of hydrogen, Trowbridge, 10, 222 ; of metals in an atmosphere of hydrogen, Crew, 10, 463.

— prismatic and diffraction, Ames, 7, 69.

Spectroscope, new form (echelon), Michelson, 5, 215 ; see also 9, 380.

Spectroscopic Tables, Engelmann, 5, 72.

Spectroscopie, Handbuch der, Kayser, 10, 464.

Spectrum, of aqueous vapor, Trowbridge, 10, 222.

— dispersion of electric, Marx, 7, 68.

— of a gas, influence of small impurities on, Lewis, 9, 65.

— of hydrogen, Trowbridge, 10, 222.

— of lightning, Toepler, 7, 68.

— normal lines of iron, Kayser, 10, 463.

— waves, new formula, Balmers, 3, 245.

— infra-red, Rubens and Aschkinass, 5, 391 ; absorption by rock salt, etc., Rubens and Trowbridge, 5, 83.

m, See also **Zeeman** effect.
', **A. C.**, Devonian strata in
do, **9**, 125.
, **J. W.**, duration of Niagara
etc., **1**, 398.
ges of level in Jamaica, 6,
high plateaus and submarine
ean Valleys, **6**, 272; episode
history of Niagara Falls, 6,

', **L. J.**, diaphorite from Wash-
and Mexico, 6, 316.
, **C. L.**, Physical Chemistry,
; boiling point curves, **9**, 341.
ion of resorcinol in ethyl al-
10, 449.
of Burma, Thorell, **1**, 398.
origin of, Beecher, **6**, 1, 125,
39.
, thermo-mineral, De Launay,

. **E.**, Economic Geology of
ercer Mining District, Utah,
; Geology of Aspen Mining
t, **8**, 465.
lite rocks from Alaska, **10**,
quartz-muscovite rocks, Bel-
Nevada, **10**, 351.
, **T. W.**, Cretaceous section
Paso, Texas, **1**, 21; Cretace-
aleontology of Pacific Coast,

eather, **G. P.**, Regnault's cal-
nd specific volumes of steam,
; thermo-dynamic relations
am, **7**, 129.
catalogue of, Porter, **8**, 87;
ngton catalogue of, **7**, 321.
ra of certain, Vogel and Wil-
t, 475.
telescopes, Todd, **8**, 87.
jet, rate of condensation in,
r, **2**, 247.
ated, dryness of, and the con-
of steam gas, Reynolds, **2**, 450.
fic volumes of, Starkweather,
; thermodynamic relations
arkweather, **7**, 129.
agnets, induction coefficients
d, Peirce, **2**, 347; properties
384.
G., experiments with pecto-
tc., **8**, 245; action of ammo-
chloride on analcite and leu-
, 117; on natrolite, etc., **9**,

Systems, Researches on the
ion of, See **3**, 491.
, **J. S.**, new form of pendu-
i, 14; method of measuring
e tension, **10**, 245.

Stevenson, J. J., Cerrillos coal field,
1, 148.
Stine, W. M., simple compensated
shunt box, **5**, 124.
— Manual of Photometry, **10**, 320.
Stokes, H. N., analyses of biotites
amphiboles, **7**, 294.
Stone, G. H., granitic breccias of the
Cripple Creek region, **5**, 21; of
Grizzly Peak, Colo.. **7**, 184; gla-
ciation of Central Idaho, **9**, 9;
glacial gravels of Maine, **10**, 247.
Stone, W. A., Physics, **5**, 222.
Storage Battery, Treadwell, **6**, 101.
— oscillating discharge from, J.
Trowbridge, **4**, 194; X-rays from
current of, J. Trowbridge, **9**, 439.
Stratton, S. W., new harmonic
analyzer, **5**, 1; Experiments in
Physics, **5**, 389.
Strutt, R. J., behavior of Becquerel
and Röntgen rays in a magnetic
field, **9**, 376.
Submarine peaks, form of, Little-
hales, **1**, 15.
Suess, E., La face de la terre, **5**,
152; **10**, 167.
Sun, Eclipses of, May 28, 1900, **9**,
391.
— Place in Nature, Lockyer, **5**,
300.
— presence of carbon and oxygen in,
Trowbridge, **1**, 329; **10**, 222.
— temperature of the, Scheiner, **9**, 65;
Wilson and Gray, **3**, 152.
— Total Eclipses of, M. L. Todd, **9**,
393.
— See **Solar**.
Sunshine recorder, Isham, **6**, 160.
Surface tension of liquids, Linebar-
ger, **2**, 108, 226; Mayer, **3**, 253;
Stevens, **10**, 245.
— and density of aqueous solutions,
etc., Mac Gregor, **7**, 313.
Switzerland, glacial deposits, **2**, 301.
Sylvester medal, **5**, 240.

T

Taff, J. A., an albertite-like asphalt,
Indian Territory, **8**, 219.
Tapirs, recent and fossil, Hatcher, **1**,
161.
Tarleton, F. A., Mathematical Theory of Attraction, **8**, 88.
Tarr, R. S., Physical Geography, **1**,
76.
— Arctic Sea ice as a geological agent,
3, 223; climate of Davis' and Baffin's Bay, **3**, 315; Elementary Geology, **3**, 351.

Tasmania, igneous rocks. Twelve-trees and Petterd, 6, 511; minerals, Petterd, 3. 352.

Taylor, F. B., scoured bowlders of the Mattawa Valley, 3, 208.

Telegraphy, wireless, Vallot and Lecarme, 10, 396.

Telephone, excursions of diaphragm, Barus, 3, 219.

Telescope objective, new type, Hastings, 7, 267; Harkness, 9, 287.

Temiscouata Lake, igneous rocks of, Gregory, 10, 14.

Températures élevées, mesure des, Chatelier et Boudouard, 9, 395.

Temperatures, experiments in low, Hempel, 7, 392; effect on seeds, 9, 74.

— measurement of high, Holborn and Day, 8, 165, 303; 10, 171.

— underground, Hallock, 4, 76; in the Dakotas, Darton, 5, 161; in Michigan, Lane, 9, 434.

Tennessee, Camden chert of, Safford and Schuchert, 7, 429.

— geology, Safford and Killebrew, 10, 399.

— phosphates in, Safford, 2, 462.

— University Record, 7, 170; 9, 462.

Terrestrial Magnetism, 1, 141.

Texas, geology of, Stanton and Vaughan, 1, 21; Hill and Vaughan, 7, 70. 315.

Thermal expansion, residual viscosity, Day, 2, 342.

Thermodynamic relations of hydrated glass, Barus, 7, 1; 9, 161.

— relations for steam, Starkweather, 7, 13, 129.

Thermodynamics, Duhem, 7, 68.

— of swelling of starch, Rodewall, 5, 297.

Thermoelectricity in certain metals, Holborn and Day, 8, 303.

Thermometer, gas, experiments with, Cady, 2, 341; Holborn and Day, 8, 165; 10, 171.

Thermostat, electrical, Duane and Lory, 9, 179.

Thompson, E. P., Rontgen Rays, 2, 392.

Thompson, S. P., Light, Visible and Invisible, 5, 71.

Thorp, F. H., Inorganic Chemical Preparations, 3, 357; 5, 222.

— Industrial chemistry, 7, 157.

Thorpe, T. E., Humphrey Davy, Poet and Philosopher, 2, 449.

Tide gauge observations, 4, 82.

Tierreich, Schulze, 1, 491; Hartert, 4, 250; Schulze, 8, 397; 10, 89.

Tilden, W. A., Progress of Scientific Chemistry, 8, 385.

Tin deposits, Temescal, So. California, 4, 39.

Todd, D. P., Stars and Telescopes, 8, 87.

Todd, J. E., revision of the moraines of Minnesota, 6, 469; moraines of So. Dakota, 10, 249.

Todd, M. L., Total Eclipses of the Sun, 9, 393.

Torsion, producing rotatory polarization, Ewell, 8, 89.

Transformer, Principles of the, Bedell, 2, 453.

Transvaal, auriferous conglomerate of, Becker, 5, 193; geological survey, Hatch, 5, 393.

— norites, gabbros, and pyroxenites of, Leo Henderson, 7, 317.

Treadwell, A., Storage Battery, 6, 101.

Trinidad pitch, Peckham and Linton, 1, 193.

Trowbridge, A., absorption of infra-red rays in rock salt and sylvine, 5, 33.

— investigation of the coherer, 8, 199.

Trowbridge, J., triangulation by cathode photography, 1, 245; carbon and oxygen in the sun, 1, 329.

— spectra of argon, 3, 15; multiple spectra of gases, 3, 117; temperature and ohmic resistance of gases, 3, 327; does a vacuum conduct electricity? 3, 343; electrical conductivity of the ether, 3, 387; conductivity of electrolytes, 3, 391.

— electrical discharges in air, 4, 190; oscillatory discharge of a large accumulator, 4, 194.

— phosphorescence produced by electrification, 5, 55; electromotive force, 5, 57; source of X-rays, 5, 129.

— explosive effect of electrical discharges, 8, 239.

— production of X-rays by battery current, 9, 439.

— spectra of hydrogen and of aqueous vapor, 10, 222.

True, R. H., toxic action of acids on Lupinus albus, 9, 183.

Turner, H. W., rocks and minerals from California, 5, 421; rock-forming biotites and amphiboles, 7, 294; roscoelite, 7, 455.

Turpin, G. S., Inorganic Chemistry, 1, 317.

Tutorial Chemistry, Bailey, 3, 357; 5, 390.

Tutorial Statics, Briggs and Bryan, 3, 426.

Type specimens, value of, Marsh, 6, 401.

— of American Museum, New York, 9, 69.

Tyrrell, J. B., is land around Hudson Bay at present rising? 2, 200; the Cretaceous of Athabasca river, 5, 469.

U

Ultra-red rays, Rubens and Trowbridge, 3, 484.

Ultra-violet light, effect on gases, 9, 381; 10, 464.

Undulations, secondary, in Bay of Fundy, Duff, 3, 406.

Ungava, Trenton rocks at, Whiteaves, 7, 433.

United States. See GEOL. REPORTS; also Coast Survey and National Museum.

Uranium radiation, Rutherford, 7, 238; Becquerel, 7, 471; source of, Crookes, 7, 472.

— radio-activity, Crookes, 10, 318.

— See Becquerel rays and radio-active.

V

Vacùum tubes for Róntgen rays, regenerating, Rollins, 7, 159.

Valentine, W., analysis of biotites and amphiboles, 7, 294.

Van Hise, C. R., North American pre-Cambrian geology, 2, 205.

— earth movement, 5, 230; metamorphism of rocks and rock flowage, 6, 75.

Van Name, R. G., sulphocyanides of copper and silver in gravimetric analysis, 10, 451.

Van't Hoff, Chimie Physique, 7, 157; 10, 461; Doppelsalzen, etc., 4, 68; Arrangement of Atoms in space, 5, 388.

Van Tieghem, new system of classification of phænogamia, 4, 79.

Vapors, fluorescence of, Wiedemann and Schmidt, 1, 393.

Vaughan, T. W., Cretaceous section at El Paso, Texas, 1, 21; outlying areas of the Comanche series, 4, 43; Lower Cretaceous Gryphæas of Texas, 7, 70.

Velocity, means of producing a constant angular, Webster, 3, 379.

— of seismic waves, 10, 471.

— of electric waves, 8, 1.

Verrill, A. E., the Opisthoteuthidæ, 2, 74; molluscan archetype, 2, 91.

— Ledidæ and Nuculidæ of N. Atlantic coast, 3, 51; protective coloration in mammals, birds, etc., 3, 132; changes in the colors of certain fishes, 3, 185; supposed giant cephalopod on the Florida coast, 3, 79, 162, 355.

— new American Actinians, 6, 493; 7, 41, 148, 205, 375.

— geology of the Bermudas, 9, 318.

Vibration of high notes, time of, 5, 302; 7, 471.

Viscosity of gases as affected by temperature, Rayleigh, 9, 375; 10, 461.

— of mixtures of liquids, Linebarger, 2, 331; Thorpe and Rodger, 4, 65.

— of rubber, Day, 2, 342.

Volcanoes of North America, Russell, 5, 74.

Voltameter, silver, Kahle, 7, 239.

W

Waddell, J., School Chemistry, 10, 461.

Wadsworth, F. L. O., cathetometer, 1, 41; determination of specific heat by the method of mixtures, 4, 265.

Wadsworth, M. E., zirkelite, 5, 153.

Wakker, J. H., propagation of sugar-cane, 1, 324.

Walcott, C. D., genus Lingulepis, 3, 404.

— brachiopod fauna of Rhode Island, 6, 327; fossil Medusæ, 6, 509.

— Pre-Cambrian fossiliferous formations, 8, 78.

— Lower Cambrian in Atlantic Province, 9, 302.

— Reports of U. S. Geological Survey, see GEOL. REPORTS (United States).

Waldo, F., Elementary Meterology, 3, 80.

Walker, C. F., iodic acid in the analysis of iodides, 3, 293; titration of sodium thiosulphate with iodic acid, 4, 235; iodine in the analysis of alkalies, etc., 6, 455.

Walker, T. L., sperrylite, 1, 110; percussion figures on cleavage plates of mica, 2, 5; etching figures on triclinic minerals, 5, 176; crystalline symmetry of torbernite, 6, 41; composition of igneous rocks, 6, 410; crystal symmetry of the micas, 7, 199.

Ward, H. A., Australian meteorites, 5, 135.

Ward, H. L., new Kansas meteorite, 7, 233 ; new iron meteorite, North Carolina, 8, 235 ; new meteorite at Allegan, Michigan, 8, 412.

Ward, L. F., Cretaceous formations of the Black Hills, 9, 70 ; Cycads from the Wyoming Jurassic, 9, 384 ; Mesozoic lavas of the U. S., 10, 320; fossil cycads in the Yale Museum, 10, 327.

Warming, E., Ökologische Pflanzengeographie, 2, 89.

Warren, C. H., mineralogical notes, 6, 116 ; composition of parisite, 8, 21 ; new minerals from Franklin, N. J., 8, 339.

Washburn Observatory, Univ. of Wisconsin, publications, 2, 90.

Washington, H. S., Ischian trachytes, 1, 375.

— igneous rocks from Smyrna and Pergamon, 3, 41 ; Yale collection of meteorites, 3, 83.

— meteorite from Jerome, Gove Co., Kansas, 5, 447.

— sölvsbergite and tinguaite, Essex Co., Mass., 6, 176.

— analyses of Italian volcanic rocks, 8, 286.

— analyses of Italian volcanic rocks, 9, 44 ; notice of Brögger's lectures at Baltimore, 9, 456.

— statement of rock analyses, 10, 59.

Water, absence of coloration in, Spring, 7, 313.

— dielectric constant, Calvert, 9, 382.

— movements of ground, King, 9, 157.

Watson, W., Text-book of Physics, 9, 296.

Wave-length, effect of pressure on, Humphreys, 4, 392.

Way, A. F., separation of iron from chromium, etc., 8, 217.

Weather, March, in the United States, Fassig, 8, 319.

— seven-day period, Clayton, 2, 7.

Webster, A. G., means of producing constant angular velocity, 3, 379 ; rapid break for large currents, 3, 383.

— Theory of Electricity and Magnetism, 4, 72.

Weed, W. H., geology of Bearpaw Mts., Mont., 1, 283, 351 ; 2, 136, 188.

— Missourite, Highwood Mts., Montana, 2, 315.

— geology of Castle Mt. district, Mont., 3, 250.

— geology of Judith Mts., Mont., 6, 508.

Weed, W. H., Enrichment of mineral veins, 10, 82.

— Little Belt Mountains, Mont., 10, 466.

Weeks, F. B., Bibliography of N. A. geology, etc., 2, 303 ; 6, 510 ; 8, 393 ; 9, 448.

Weisbach, Determinative tables, 5th Ed., 10, 84 ; translated and edited by Fraser, 3, 162.

— Synopsis Mineralogica, 5, 78.

Wells, temperatures of deep, Hallock, 4, 76 ; Darton, 5, 161 ; Lane, 9, 484.

Wells, H. L., double fluorides of cæsium and zirconium, 1, 18.

— Double halogen salts of cæsium and rubidium, 3, 461 ; double fluorides of zirconium with lithium, etc., 3, 466.

— translation of Fresenius's Qualitative Analysis, 4, 473.

— Qualitative Chemical Analysis, 6, 269.

Wells, J. S. C., Inorganic Qualitative Analysis, 6, 269.

West Indies, flora, Urban, 7, 244.

— submarine plateaus and valleys, Spencer, 6, 272.

— See Jamaica.

West Virginia, geol. survey, 7, 398, 399.

White, D., age of the lower coals of Missouri, 3, 158; flora of the Lower Coal Measures of Missouri, 10, 166.

White, T. G., original Trenton rocks, 2, 430 ; petrography of the Boston Basin, 5, 470.

Whiteaves, J. F., Trenton rocks at Ungava, 7, 433.

Whiteley, R. L., Organic Chemistry, Fatty Compounds, 1, 53.

Whitfield, R. P., mollusca and crustacea of N. Jersey Miocene, 1, 61.

Wieland, G. R., Archelon ischyros from South Dakota, 2, 399.

— Currituck Sound, Virginia and North Carolina, 4, 76 ; depth of peat in the Dismal Swamp, 4, 76 ; eopaleozoic hot springs and siliceous oölite, 4, 262.

— the protostegan plastron, 5, 15.

— American fossil cycads, 7, 219, 305, 383.

— terminology of vertebral centra, 8, 163 ; cycadean monœcism, 8, 164 ; note on Cycadofilices, 8, 309.

— huge Cretaceous turtles of South Dakota, 9, 237 ; evolution of the Testudinate humerus, 9, 418.

Wilcox, E. M., winter conditions of reserve food substances of certain deciduous trees, 6, 69.

Williams, E. H., Manual of Lithology, 1, 150.

Williams, George H., Memorial Lectures on Geology, Brögger, 9, 456.

Williams, H. S., Geological Biology, 1, 63.

— Southern Devonian formations, 3, 393.

— Paleotrochis in Mexico, 7, 385.

— Devonian interval in northern Arkansas, 8, 139.

— Silurian - Devonian boundary in North America, 9, 203.

Williams, R. P., Chemical Experiments, 1, 317 ; Inorganic Chemistry, 3, 357.

Willis, J. C, Flowering Plants and Ferns, 3, 343.

Wilson, E. B., the Cell in Development and Inheritance, 3, 161.

Wind pressure measured, Nipher, 5, 468.

Wisconsin, building stones of, Buckley, 9, 69.

— Copper-bearing rocks, U. S. Grant, 10, 249.

Wolff, J. E., theralite in Costa Rica, 1, 271.

Woodward, A. S., catalogue of Fossil Fishes, 1, 396.

Wright, A. W., cathode rays and their effects. 1, 235 ; structural and magneto-optic rotation, 6, 416.

Wright, L., Induction Coil in practical Work, 4, 324.

X

X-rays, see Röntgen Rays.

Y

Yale Museum, cycads in, Ward, 10, 327; collection of meteorites, 3, 83.

— Collections presented by O. C. Marsh, 5, 156.

Yale Psychological Laboratory, studies from, Scripture, 6, 512.

Yellowstone National Park, Chittenden, 1, 327; geology of, Hague, Iddings, et al., 9, 297.

— age of igneous rocks of, Hague, 1, 445.

— Tertiary floras, Knowlton, 2, 51.

Yerkes Observatory, publications, 9, 311.

Yucatan, analysis of specimens from, Howe and Campbell, 2, 418.

Z

Zeeman effect, Zeeman, 3, 486; Dunstan, Rice and Kraus, 3, 472; Reese, 6, 99.

— investigated with echelon spectroscope, Blythswood and Marchant, 9, 880.

Zeiller, Paleobotany, 10, 88.

Zenker, Lehrbuch der Photochromie, 10, 162.

Zirkel, F., Naumann's Mineralogy, 13th Ed., 3, 424.

Zittel, K. A. von, Text-Book of Paleontology, translation by Eastman, 2, 394 ; 9, 388.

ZOOLOGY.

Actinians, new American. Verrill, 6, 493; 7, 41, 143, 205, 875.

Asterias pallida, metamorphosis, Goto, 7, 78,

Bibliotheca Zoologica II., Taschenberg, 1, 77; 6, 103.

Birds, see Birds.

Catalogus mammalium, 3, 351; 7, 79; 8, 397.

Cephalopod of Florida, supposed, Verrill, 3, 79, 162, 355.

Coloration, protective, in mammals, birds, etc., Verrill, 3, 132.

Congress of Zoology, 1898, 5, 288; 8, 398.

Crustacea of Norway, Sars, 7, 79.

Embryology of Invertebrates, Korschelt and Heider, 8, 471.

Fauna, History of European, Scharff, 8, 395.

Fishes of North and Middle America, Jordan and Evermann, 7, 79, 169.

— changes in the color of certain, Verrill, 3, 135.

Glow beetle, light of, Muraoka, 3, 151.

Ichthyologica Ohiensis by Rafinesque, Call, 7, 473.

Insects, Structure and Life, Carpenter, 8, 473.

Ledidæ and Nuculidæ of N. Atlantic coast, Verrill and Bush, 3, 51.

Lepidoptera Phalænæ in the British Museum, Hampson, 7, 246.

Lichtsinn augenloser Tiere, Nagel, 3, 162.

Mammals, catalogue of, Trouessart, 3, 351; 7, 79; 8, 397.

Manual of Zoology, Parker and Haswell. 9, 390.

Mollusca of the Chicago area, Baker, 7, 79.

Molluscan archetype, Verrill, 2, 91.

ZOOLOGY—
Mollusks from Patagonia, Pilsbry, 7, 126.
Odonata of Ohio, Kellicott, 8, 88.
Opisthoteuthidæ, Verrill, 2, 74.
Oysters, Patagonia, Ortmann, 4, 355.
Spiders of Burma, Thorell, 1, 398.
Spines, study of, Beecher, 6, 1, 125, 249, 329.
Text-book of Zoology, Parker and Haswell, 5, 319.

ZOOLOGY—
Tierreich, 1, 491; 4, 250: 8, 397; 10, 89.
Vertebrate Zoology, Kingsley, 8, 472.
Zoological Bulletin, 4, 83.
Zoological studies in New Guinea, etc., Willey, 7, 79, 322; 8, 398; 10, 89.
Zoologisches Addressbuch, 1, 77.
See also GEOLOGY.